FOOD BYPRODUCTS MANAGEMENT AND THEIR UTILIZATION

FOOD BYPRODUCTS MANAGEMENT AND THEIR UTILIZATION

Edited by

Ricardo Gómez-García, PhD

Ana A. Vilas-Boas

Débora A. Campos, PhD

Maria Manuela Pintado, PhD

Cristóbal Noé Aguilar, PhD

AAP | APPLE ACADEMIC PRESS

First edition published 2024

Apple Academic Press Inc.
1265 Goldenrod Circle, NE,
Palm Bay, FL 32905 USA

760 Laurentian Drive, Unit 19,
Burlington, ON L7N 0A4, CANADA

CRC Press
2385 NW Executive Center Drive,
Suite 320, Boca Raton FL 33431

4 Park Square, Milton Park,
Abingdon, Oxon, OX14 4RN UK

Library and Archives Canada Cataloguing in Publication

Title: Food byproducts management and their utilization / edited by Ricardo Gómez-García, PhD, Ana A. Vilas-Boas, Débora A. Campos, PhD, Maria Manuela Pintado, PhD, Cristóbal Noé Aguilar, PhD.
Names: Gómez-García, Ricardo, editor. | Vilas-Boas, Ana A., editor. | Campos, Débora A. editor. | Pintado, Maria Manuela, editor. | Aguilar, Cristóbal Noé editor.
Description: First edition. | Includes bibliographical references and index.
Identifiers: Canadiana (print) 20230452019 | Canadiana (ebook) 20230452043 | ISBN 9781774912959 (hardcover) | ISBN 9781774912966 (softcover) | ISBN 9781003377801 (ebook)
Subjects: LCSH: Food industry and trade—By-products. | LCSH: Food industry and trade—Waste minimization. | LCSH: Bioactive compounds. | LCSH: Food—Biotechnology.
Classification: LCC TP373.8 .F66 2024 | DDC 664/.08—dc23

Library of Congress Cataloging-in-Publication Data

Names: Gómez-García, Ricardo, editor.

Title: Food byproducts management and their utilization / edited by Ricardo Gómez-García, PhD, Ana A. Vilas-Boas, Débora A. Campos, PhD, Maria Manuela Pintado, PhD, Cristóbal Noé Aguilar, PhD.

Description: First edition. | Palm Bay, FL : Apple Academic Press, 2024. | Includes bibliographical references and index. | Summary: "Food byproducts derived from industrial processing is a serious worldwide problem because it generates environmental pollution and results in significant food and economic losses from food waste. This new volume shows how food byproducts can be value-added renewable sources with the application of novel biotechnologies that avoid hazardous chemicals. The volume discusses the importance of valorizing food wastes and illustrates their value-added properties for industry. It explains the significant progress in bioresources processing for compound extraction and production as well as the increasing interest of food ingredients development, in which health care, environment, and economics play an essential part in biotechnological research. It considers the waste byproducts of various crops, such as tomato, melon, maize, berries, soybean, coffee, and their uses in the generation of health-benefiting bioactive compounds. The volume goes on to explore the various biotechnological strategies to extract, produce, and recover bioactive compounds along with the cost-effectiveness of these methods. This volume, Food Byproducts Management and Their Utilization, presents strategies that are of interest in food engineering, green chemistry, biotechnology, and some other areas, while paying special attention to biorefinery approaches and new challenges that industries are dealing with in the era of sustainable development. It aims to encourage not only researchers but also governmental and enterprise sectors to recognize the value and applications of food byproducts and waste"-- Provided by publisher.

Identifiers: LCCN 2023023230 (print) | LCCN 2023023231 (ebook) | ISBN 9781774912959 (hardback) | ISBN 9781774912966 (paperback) | ISBN 9781003377801 (ebook)

Subjects: LCSH: Food industry and trade--By-products. | Agricultural wastes. | Food processing by-products industry.

Classification: LCC TP373.8 .F626 2024 (print) | LCC TP373.8 (ebook) | DDC 664/.08--dc23/eng/20230712

LC record available at https://lccn.loc.gov/2023023230

LC ebook record available at https://lccn.loc.gov/2023023231

ISBN: 978-1-77491-295-9 (hbk)
ISBN: 978-1-77491-296-6 (pbk)
ISBN: 978-1-00337-780-1 (ebk)

About the Editors

Ricardo Gómez-García, PhD
Universidade Católica Portuguesa, CBQF-Centro de Biotecnologia e Química Fina–Laboratório Associado, Escola Superior de Biotecnologia, Porto, Portugal

Ricardo Gómez-García, PhD, completed his doctorate program in Biotechnology at the Catholic University of Portugal (2017–2022). He graduated as a chemical engineer (2016) from the School of Chemistry, Autonomous University of Coahuila (UA de C), Mexico. He started his scientific career at the UA de C Food Research Department, where he gained great expertise in food engineering and technology, food chemistry, and biotechnology. For his PhD, he worked to develop innovative and sustainable technologies within the context of a zero-waste and circular economy approach for the extraction and identification of functional bioactive compounds with industrial interest from fruit byproducts for their valorization and reincorporation into the industrial value chains. Dr. Gómez-García is the author and co-author of 16 articles in indexed journals, seven chapters in books of international circulation with prestigious publishers, and one European provisional and one national patent applications. He has participated in more than 15 international congress communications in areas related to biotechnology, food science and technology, and bioengineering. He has also been participating in several national and international entrepreneurship programs in Portugal and across Europe through the European Institute of Innovation and Technology (EIT Food).

Ana A. Vilas-Boas
Universidade Católica Portuguesa, CBQF-Centro de Biotecnologia e Química Fina–Laboratório Associado, Escola Superior de Biotecnologia, Porto, Portugal

Ana A. Vilas-Boas holds a BSc degree in Biotechnology, an MSc degree in Food Engineering, and is a PhD student in Biotechnology at the Catholic University of Portugal. She has great expertise as a researcher within the framework of diverse national and European projects such as ESSENCE (2017), FODIAC (2018), Nature Bioactive Food (2020), and MEDISMART (present) project at the Centre for Biotechnology and Fine Chemistry (CBQF) of Catholic University of Portugal. She has been working directly with some of the Portuguese industries in the context of food byproducts and noncompliant food management from the food industrial chains to characterize and identify value-added bioactive compounds. Currently, her PhD work intends to develop innovative, integrative, and sustainable process to valorize orange byproducts within the circular economy context for the extraction of hesperidin and pectin to develop a novel nutraceutical to enhance the gut microbiota health. She has published 12 articles in indexed journals, five chapters in books of international circulation with prestigious publishers, and 10 international congress communications. In addition, she participated in different entrepreneur programs with innovative ideas and was awarded with the first place at Ecotrophelia Portugal (2018), Innovation Track (2018), and Born from Knowledge (BfK) (2019).

Débora A. Campos, PhD
Universidade Católica Portuguesa, CBQF-Centro de Biotecnologia e Química Fina–Laboratório Associado, Escola Superior de Biotecnologia, Porto, Portugal AgroGrIN Tech, Porto, Portugal

Débora A. Campos, PhD, is a postdoctoral Researcher since 2020 at Universidade Católica Portuguesa-Escola Superior de Biotecnologia (UCP-ESB). She has developed scientific work in waste/byproducts valorization from agroindustrial origin, where she developed innovative and sustainable green technologies for the extraction/separation of compounds of industrial interest. This work led her to the submission and acceptance of an European patent for a new industrial technology application. She carried out scientific research works in different areas, green nanotechnology, and microbiology. She has been working in different European projects and programs within the framework of entrepreneurism approach such as LAB2BUSINESS (Porto, 2017), Global Food Venture–EIT Food (2018), and EIT Food Innovator Fellowship (2019). Her expertise and experience in the subject field include a BSc degree in Bioscience (2010), a MSc studies in Applied Microbiology (2013), and a PhD program in Biotechnology (2019) which was achieved through the Catholic University of Portugal. Currently, she is working for her postdoctoral degree at the same institution within the MIRACLE–H2020 EU project (ending in 2022) which was awarded in a national Scientific Career Financial Support to support her research and developments. Dr. Campos has published 34 scientific articles in international journals, six book chapters, coadvisor of PhD thesis, one granted European patent, one national patent and participation in eight scientific European projects, as well as over 25 participations in international congresses in the area of foods and functional ingredients.

Maria Manuela Pintado, PhD
Universidade Católica Portuguesa, CBQF-Centro de Biotecnologia e Química Fina–Laboratório Associado, Escola Superior de Biotecnologia, Porto, Portugal

Maria Manuela Pintado, PhD, is the Director of the CBQF Research Center for Biotechnology and Fine Chemistry at the School of Biotechnology at the Catholic University of Portugal. As an Associate Professor, she works in the interface of microbiology with health. Prof. Pintado has developed numerous projects of I & D+I (with companies and scientific institutions, both national and international) in the areas of valuing of byproducts and residues, production and characterization of bioactive compounds, functional foods, and natural antimicrobials. Additionally, she has been the advisor of 18 PhD theses, 16 postdoctoral fellowships, 36 MSc theses, and 54 BSc theses. Her expertise and experience in the subject field include pharmaceutics (1991) by the Universidade do Porto (Portugal), her MSc (1994) program in Food Science and Engineering, and her PhD (1999) in Biotechnology, both of them at Escola Superior de Biotecnologia da Universidade Católica Portuguesa (Portugal). She worked for her postdoctoral research project (2003) in the same institution. Dr. Pintado has published more than 320 papers in indexed journals, 35 book chapters, 34 proceedings, more than 150 contributions in scientific meetings, and 20 patents.

Cistóbal N. Aguilar, PhD
Director of Research and Postgraduate Programs at Universidad Autonoma de Coahuila, Mexico

Cristóbal Noé Aguilar, PhD, is a Director of Research and Postgraduate Programs at the Universidad Autonoma de Coahuila, Mexico. Dr. Aguilar has published more than 360 papers in indexed journals, and 450 contributions in scientific meetings. He has also edited more than 30 books. He has been awarded several prizes and awards, the most important of which are the National Prize of Research (2010) from the Mexican Academy of Sciences; the Prize CCC-2008 from the Mexican Society of Biotechnology and

Bioengineering; National Prize AgroBio-2005; and the Mexican Prize in Food Science and Technology. In 2019, he received the Outstanding Research Award by the International Association of Bioprocessing (IBA). Dr. Aguilar is a member of the Mexican Academy of Science, the Central Council of the International Bioprocessing Association, the Mexican Academy of Sciences, the Mexican Society for Biotechnology and Bioengineering, and the Mexican Association for Food Science and Biotechnology. He has developed more than 41 research projects, including 25 international exchange projects.

Contents

Contributors

Cristóbal Noé Aguilar
Bioprocesses and Bioproducts Research Group, Food Research Department, School of Chemistry, Autonomous University of Coahuila, Saltillo, Mexico

Mayra Aguilar-Zárate
Facultad de Ciencias Químicas-CIEP, Universidad Autónoma de San Luis Potosí, San Luis Potosí, Mexico

Pedro Aguilar-Zárate
Engineering Department, Tecnológico Nacional de México/Instituto Tecnológico de Ciudad Valles, San Luis Potosí, México

Emmanuel Aispuro-Hernández
Centro de Investigación en Alimentación y Desarrollo, A.C., Coordinación de Tecnología de Alimentos de Origen Vegetal, C.P. Hermosillo, Sonora, México

José de Jesús Espinoza Arellano
Graduate Studies Department, School of Accounting and Administration, Autonomous University of Coahuila, Torreon, Coahuila, Mexico

R. Baeza-Jiménez
Laboratorio de Biotecnología y Bioingeniería, Centro de Investigación en Alimentación y Desarrollo, Delicias, Chihuahua, México

Lillian Barros
Centro de Investigação de Montanha (CIMO), Instituto Politécnico de Bragança, Campus de Santa Apolónia, Bragança, Portugal

Isabel Bascón-Villegas
Food Science and Technology Department, Universidad de Córdoba, Córdoba, Spain

Manuel J. Bernal-Millán
CIAD, Centro de Investigación en Alimentación y Desarrollo, AC, Functional Foods and Nutraceuticals, Sinaloa, México

J. J. Buenrostro-Figueroa
Laboratorio de Biotecnología y Bioingeniería, Centro de Investigación en Alimentación y Desarrollo, Delicias, Chihuahua, México

L. Ángel Cabanillas-Bojórquez
Centro de Investigación en Alimentación y Desarrollo A.C., Col. Campo El Diez, Culiacán, Sinaloa, México

Paola Camiscia
Instituto de Procesos Biotecnológicos y Químicos. Consejo Nacional de Investigaciones Científicas y Técnicas de Argentina, Suipacha, Rosario, Argentina

Márcio Carocho
Centro de Investigação de Montanha (CIMO), Instituto Politécnico de Bragança, Campus de Santa Apolónia, Bragança, Portugal

Elena Carrasco
Food Science and Technology Department, Universidad de Córdoba, Córdoba, Spain

Cristian Salvador Cossio-Carrillo
Department of Chemical Engineering, Faculty of Chemical Sciences, Autonomous University of Coahuila, Boulevard Venustiano Carranza SN, Saltillo, Coahuila, México

Elizabeth Ximena Castrillón-Duque
Department of Chemical Engineering, Faculty of Chemical Sciences, Autonomous University of Coahuila, Saltillo, Coahuila, México

Mónica L. Chávez-González
Bioprocesses and Bioproducts Research Group, Food Research Department, School of Chemistry, Universidad Autónoma de Coahuila, Saltillo, Mexico;
Nanobioscience Research Group, Food Research Department, School of Chemistry, Autonomous University of Coahuila, Mexico

Juan Carlos Contreras-Esquivel
Food Research Department, Faculty of Chemical Sciences, Autonomous University of Coahuila, Saltillo, Coahuila, México

Natalia Montellano Duran
Biotechnology, Universidad Católica Boliviana San Pablo, Santa Cruz, Bolivia

Ana Ferreira
Polytechinc Institute of Viseu, School of Agriculture, Quinta da Alagoa–Estrada de Nelas, Ranhados, Viseu, Portugal

Isabel C.F.R. Ferreira
Centro de Investigação de Montanha (CIMO), Instituto Politécnico de Bragança, Campus de Santa Apolónia, Bragança, Portugal

Élia Fogeiro
Polytechinc Institute of Viseu, School of Agriculture, Quinta da Alagoa–Estrada de Nelas, Ranhados, Viseu, Portugal

Hugo S. García
UNIDA, Tecnológico Nacional de México/Instituto Tecnológico de Veracruz, México

Aarón F. González-Córdova
Centro de Investigación en Alimentación y Desarrollo, Coordinación de Tecnología de Alimentos de Origen Animal, Laboratorio de Calidad, Autenticidad y Trazabilidad de los Alimentos/Química y Biotecnología de Productos Lácteos, Hermosillo, Sonora

Mayela Govea-Salas
Nanobioscience Research Group, Food Research Department, School of Chemistry, Autonomous University of Coahuila, Mexico

Erick P. Gutiérrez-Grijalva
Catedras CONACyT, Col. Campo El Diez, Culiacán, Sinaloa, México

Yasunori Hamauzu
Shinshu University, Department of Agricultural and Life Sciences, Faculty of Agriculture, Nagano, Japan

Sandrina A. Heleno
Centro de Investigação de Montanha (CIMO), Instituto Politécnico de Bragança, Campus de Santa Apolónia, Bragança, Portugal

Adrián Hernández-Mendoza
Centro de Investigación en Alimentación y Desarrollo, Coordinación de Tecnología de Alimentos de Origen Animal, Laboratorio de Calidad, Autenticidad y Trazabilidad de los Alimentos/Química y Biotecnología de Productos Lácteos, Hermosillo, Sonora

J. Basilio Heredia
CIAD, Centro de Investigación en Alimentación y Desarrollo, AC, Functional Foods and Nutraceuticals, Campo el Diez, Culiacán, Sinaloa, México

Adriana Montserrat Fabela Hernández
School of Accounting and Administration, Autonomus University of Coahuila, Torreon, Coahuila, Mexico

María Luisa Carrillo Inungara
Food Research Laboratory, Multidisciplinary Academic Unit Huasteca Zone, Autonomous University of San Luis Potosí, Cd. Valles, S.L.P., México

L. Alfonso Jiménez-Ortega
CIAD, Centro de Investigación en Alimentación y Desarrollo, AC, Functional Foods and Nutraceuticals, Culiacán, Sinaloa, México

Leticia X. López-Martínez
CONACYT-CIAD (Centro de Investigación en Alimentación y Desarrollo A.C), Coordinación de Alimentos de Origen Vegetal, Col. La Victoria, Hermosillo, Sonora, México

Karla A. Luna
Bioprocesses and Bioproducts Research Group, Food Research Department, School of Chemistry, Universidad Autónoma de Coahuila, Saltillo, Mexico;
Nanobioscience Research Group, Food Research Department, School of Chemistry, Universidad Autónoma de Coahuila, Saltillo, Mexico

Sara Marçal
Centro de Biotecnologia e Química Fina, Laboratório Associado, Escola Superior de Biotecnologia, Universidade Católica Portuguesa, Porto, Portugal

Miguel Ángel Martínez-Téllez
Centro de Investigación en Alimentación y Desarrollo, A.C., Coordinación de Tecnología de Alimentos de Origen Vegetal, Hermosillo, Sonora, México

Mariela R. Michel
Engineering Department, Tecnológico Nacional de México/Instituto Tecnológico de Ciudad Valles, San Luis Potosí, México

Ali Moayedi
Department of Food Science and Technology, Gorgan University of Agricultural Sciences and Natural Resources, Gorgan, Iran

Julio Montañez
Department of Chemical Engineering, Faculty of Chemical Sciences, Autonomous University of Coahuila, Saltillo, Coahuila, México

Víctor Manuel Molina Morejón
Graduate Studies Department, School of Accounting and Administration, Autonomus University of Coahuila, Torreon, Coah., Mexico

Lourdes Morales-Oyervides
Department of Chemical Engineering, Faculty of Chemical Sciences, Autonomous University of Coahuila, Saltillo, Coahuila, México

Diana B. Muñiz-Márquez
Facultad de Estudios Profesionales Zona Huasteca, Universidad Autónoma de San Luis Potosí, Cd. Valles, S.L.P., México

Cláudia M. B. Neves
Polytechinc Institute of Viseu, School of Agriculture, Quinta da Alagoa–Estrada de Nelas, Ranhados, Viseu, Portugal

J. A. Noriega-Rodríguez
Laboratorio de Biotecnología, Departamento de Ingeniería Química y Metalurgia, Universidad de Sonora, Hermosillo, Sonora, México

Majid Nooshkam
Department of Food Science and Technology, Faculty of Agriculture, Ferdowsi University of Mashhad (FUM), Mashhad, Iran

Mariana C. Pedrosa
Centro de Investigação de Montanha (CIMO), Instituto Politécnico de Bragança, Campus de Santa Apolónia, Bragança, Portugal

Filipa Pereira
Polytechinc Institute of Viseu, School of Agriculture, Quinta da Alagoa–Estrada de Nelas, Ranhados, Viseu, Portugal

Guillermo Picó
Instituto de Procesos Biotecnológicos y Químicos, Consejo Nacional de Investigaciones Científicas y Técnicas de Argentina, Rosario, Argentina

Manuela Pintado
Centro de Biotecnologia e Química Fina, Laboratório Associado, Escola Superior de Biotecnologia, Universidade Católica Portuguesa, CBQF, Porto, Portugal

Eber A. Quintana-Obregón
CONACYT-CIAD (Centro de Investigación en Alimentación y Desarrollo A.C), Coordinación de Alimentos de Origen Vegetal, Hermosillo, Sonora, México

Liliana Angélica Guerrero Ramos
Graduate Studies Department, School of Accounting and Administration, Autonomus University of Coahuila, Torreon, Coahuila, Mexico

Rodolfo Ramos-González
CONACYT–Universidad Autónoma de Coahuila, Saltillo, Mexico

Filipa S. Reis
Centro de Investigação de Montanha (CIMO), Instituto Politécnico de Bragança, Campus de Santa Apolónia, Bragança, Portugal

Tânia Bragança Ribeiro
Centro de Biotecnologia e Química Fina, Laboratório Associado, Escola Superior de Biotecnologia, Universidade Católica Portuguesa, CBQF, Porto, Portugal

M. Magdalena Rivera-Salas
CIAD, Centro de Investigación en Alimentación y Desarrollo, AC, Functional Foods and Nutraceuticals. Culiacán, Sinaloa, México

Martina de la Rosa-Hernandez
Food Research Laboratory, Multidisciplinary Academic Unit Huasteca Zone, Autonomous University of San Luis Potosí, Cd. Valles, S.L.P., México

Iván Salmerón
School of Chemical Science, Autonomous University of Chihuahua, Circuit 1, New University Campus, Chihuahua, Chihuahua, México

Mónica Sánchez-Gutiérrez
Food Science and Technology Department, Universidad de Córdoba, Córdoba Darwin Building (Spain)

Diva Santos
Centro de Biotecnologia e Química Fina, Laboratório Associado, Escola Superior de Biotecnologia, Universidade Católica Portuguesa, Porto, Portugal

Rita Sardão
Centro de Biotecnologia e Química Fina, Laboratório Associado, Escola Superior de Biotecnologia, Universidade Católica Portuguesa, Porto, Portugal

Jazel Doménica Sosa-Martínez
Department of Chemical Engineering, Faculty of Chemical Sciences, Autonomous University of Coahuila, Saltillo, Coahuila, México

Jonata M. Ueda
Centro de Investigação de Montanha (CIMO), Instituto Politécnico de Bragança, Campus de Santa Apolónia, Bragança, Portugal

Octavio Valdez-Baro
Centro de Investigación en Alimentación y Desarrollo, AC. Functional Foods and Nutraceuticals Department, Culiacán, Sinaloa, Mexico

Belinda Vallejo-Cordoba
Centro de Investigación en Alimentación y Desarrollo, Coordinación de Tecnología de Alimentos de Origen Animal, Laboratorio de Calidad, Autenticidad y Trazabilidad de los Alimentos/Química y Biotecnología de Productos Lácteos, Hermosillo, Sonora

Nadia Woitovich Valetti
Instituto de Procesos Biotecnológicos y Químicos, Consejo Nacional de Investigaciones Científicas y Técnicas de Argentina, Rosario, Argentina

Manuel A. Vargas-Ortiz
Centro de Investigación en Alimentación y Desarrollo, A.C., Coordinación de Tecnología de Alimentos de Origen Animal, Laboratorio de Calidad, Autenticidad y Trazabilidad de los Alimentos/Química y Biotecnología de Productos Lácteos, Hermosillo, Sonora, México

María de los Ángeles Vázquez-Núñez
Engineering Department, Tecnológico Nacional de México/Instituto Tecnológico de Ciudad Valles, San Luis Potosí, México

Fabiola Veana
Engineering Department, Tecnológico Nacional de México/Instituto Tecnológico de Ciudad Valles, San Luis Potosí, México

Miguel Ángel Villegas-Méndez
Department of Chemical Engineering, Faculty of Chemical Sciences, Autonomous University of Coahuila, Boulevard Venustiano Carranza SN, Saltillo, Coahuila, México

Dulcineia F. Wessel
Polytechinc Institute of Viseu, School of Agriculture, Quinta da Alagoa–Estrada de Nelas, Ranhados, Viseu, Portugal; CITAB, University of Trás-os-Montes e Alto Douro, Vila Real, Portugal; LAQV-REQUIMTE, Department of Chemistry, University of Aveiro, Aveiro, Portugal

Jorge E. Wong-Paz
Facultad de Estudios Profesionales Zona Huasteca, Universidad Autónoma de San Luis Potosí, México

Zahra Zareie
Department of Food Science and Technology, Gorgan University of Agricultural Sciences and Natural Resources, Gorgan, Iran

Abbreviations

ABTS	2, 2'-azino-bis-3-ethylbenzothiazoline-6-sulfonic acid (biochemical reagent)
ACE	angiotensin-converting enzyme
ADF	antioxidant dietary fiber
APCR	anaerobic-packed column reactor
AX	arabinoxylans
BBI	Bowman–Birk inhibitor
BOD	biological oxygen demand
BSG	brewery spent grains
Ca	calcium
CAT	catalase
CBD	cannabidiol
CH	coffee pulp
CNS	central nervous system
CO2	carbon dioxide
COD	chemical oxygen demand
COX	cyclooxygenase
CS	coffee husk
CSE	conventional solvent extraction
CVD	cardiovascular diseases
DF	dietary fiber
DPPH	2,2-diphenyl-1-picrylhydrazyl (biochemical reagent)
DW	dry weight
EU	European Union
FAO	Food and Agriculture Organization
FDA	Food and Drug Administration
Fe	iron
FEL	liquid-state fermentation
FES	solid-state fermentation
FLW	food losses and waste
FRAP	ferric reducing antioxidant power
FVE	flash vacuum expansion
GAE	gallic acid equivalent
GRAS	generally recognized as safe

GSH-Px	glutathione peroxidase
HAT	hydrogen atom transfer
HDL	high-density lipoprotein
HP	high pressure
HPH	high-pressure homogenization
IDF	insoluble dietary fiber
IL-10	interleukin 10
IL-1β	interleukin-1 beta
IL-8	interleukin-8
K	potassium
KTI	Kunitz soybean trypsin inhibitor
LC-MS	liquid chromatography with tandem mass spectrometry
LDL	low-density lipoprotein
MAE	microwave-assisted extraction
MALDI-TOF MS	matrix-assisted laser desorption ionization-time-of-flight mass spectrometry
MBC	minimum bactericidal concentration
MDA	malondialdehyde
MHA	Mueller–Hinton agar
MIC	minimum inhibitory concentration
Mn	manganese
MW	molecular weight
NF-Kb	nuclear factor kappa light chain enhancer of activated B cells
OHAE	ohmic heating-assisted extraction
OHC	oil holding capacity
ORAC	oxygen radical antioxidant capacity
P	phosphorus
PLE	pressurized liquid extraction
PR	pathogenesis related
PS	particle size
PUFA	polyunsaturated fatty acids
ROS	reactive oxygen species
SBH	soybean hull
SCFAs	short-chain fatty acids
SCG	spend coffee grounds
SDF	soluble dietary fiber
SDS-PAGE	sodium dodecyl sulfate-polyacrylamide gel electrophoresis
SOD	superoxide dismutase

SFE	supercritical fluid extraction
TEAC	trolox equivalent antioxidant capacity
TNF-α	tumor necrosis factor alpha
TOC	total organic carbon
UAE	ultrasonic-assisted extraction
UASB	up-flow anaerobic sludge blanket
USA	United States of America
VLDL	very low-density lipoprotein
W	watts
WHC	water-holding capacity
WRC	water-retention capacity
WSC	water swelling capacity
ZN	zinc

Preface

Over the last decade, some pressing issues have been affecting the balance between human society and the environment worldwide. Food byproducts and waste that are overproduced by the food processing industries because of their poor sustainable management have turned into a big global concern causing environmental pollution and economic losses. In this context, biotechnological researchers have been redirecting their objectives to focus on the valorization of these renewable resources attributed to the richness of bioactive compounds, which have shown attractive potential applications in different industries. Furthermore, several methodologies are being developed to extract and recover these value-added products, mainly employing green chemistry approach due to its considerably lower implications of harming the environment. Moreover, biotechnology coupled with green chemistry have shown some interesting results in the development of novel techniques that lead to the creation of commercial products, namely functional ingredients, food additives, and supplements.

Value-added compounds obtained from these renewable resources through green extractive methodologies such as protein precipitation, flash vacuum, microwave and supercritical methods, enzymatic hydrolysis, or fermentation process are considered as attractive technological tools that could be implemented in the industries, particularly in pharmaceutics, nutraceuticals, and food areas and consequently improve the traditional industrial processes allowing a circular economy implementation. Furthermore, due to the green chemistry through the zero-waste context approach, it will be possible to decrease or eliminate the constant overproduction of food waste thus avoiding soil, air, and water pollution. As a result, the application of all these involved concepts (biotechnology, green chemistry, zero-waste, and circular economy) together may need to grow as a multidisciplinary option with regards to biomass utilization and transformation into useful value-added products in an attempt to generate new food supplies with beneficial properties and their application for industrial economic revenue. Therefore, this book provides an insight and understanding of consolidated process systems for waste management by using tools of biotechnology, microbiology, food, and chemical engineering, among other branches of science.

Functional foods are described as food stuff or food ingredients that could improve health by helping in the prevention or risk reduction of certain diseases such as cancers, cardiovascular and inflammatory disorders and diabetes. These benefits are well-attributed to the content of bioactive compounds, which comprise a wide array of natural molecules with bioactivities such as antioxidant, antimicrobial, anticancer, antihypertensive, prebiotic, among many others. Enzymes are biomolecules with natural biological properties and are essential to many industrial processes such as high added-value compounds release, enzymatic synthesis of useful compound, degradation of residual materials, along with many other processes. Production and extraction of these biomolecules are significant factors to improving the industrial processes by increasing the production of high-quality products, boosting human life stability and world sustainability.

All the topics that are described in this book aim to create a fundamental professional awareness due to the significant evidences and facts on food waste management and valorization which could increase scientific, social, and industrial interests to recover value-added compounds with health beneficial bioactivities through developing sustainable technologies. This in turn can contribute to decreasing the negative effects on the environment, reducing cost, energy and time consumption, improving profitability and efficiency for industries. Thus, this book provides significantly accepted and interesting knowledge for readers interested in the fields of nutrition, chemistry, food processing and engineering, and biotechnology.

EDITORS' CRITICAL OPINION

The history of the human behavior related with food consumption and all the involved steps for its production has been changing from the middle of 20th century to the beginning of the 21st century. Such changes have arisen by the conscious issues that society is facing and nowadays are having an evident and significant negative impact on the three main pillars of sustainability: society, economy, and environment. Since human beings learned how to treat soil and some cultivation techniques (agriculture) for production of their own food, they has been repeating the same linear behavior, which involves the traditional linear economy "take-make-dispose", in other words, "production-manufacturing-discard" of food. Through the application of this linear behavior in the food industry, some issues related with food loss and wastage were detected, which consequently are linked to huge economic expenses and environmental pollution. Based on these issues, the linear economy approach is no longer acceptable within the modern food industry. Therefore, a couple of years ago a new approach was introduced—"circular economy" with the aim to reduce these issues and create a more sustainable food supply chain, avoiding food byproducts and waste production. Moreover, the circular economy approach has gained interest not only between researchers of biotechnology, environment and food science, and technology areas, but also for industrial and governmental sectors, looking/working for the same objective, decreasing the current issues by promoting a suitable and correct management of food byproducts. In this regard, such byproducts must be recycled and valorized due to their richness in biofunctional compounds (polyphenols, carotenoids, oils, protein, fibers, among others) with the purpose to (1) minimize the exhaustion of natural resources and (2) satisfy the current demand of nutritional functional food ingredients with health benefits, while the environmental pollution is mitigated, and a bioeconomy is created within industries.

Even though all the chapters presented in this book possess different and unique information that can be used as references to develop new innovative research on the topics of food waste and byproducts production along the food value chain as well as on research on their bioactive constituents extracted/recovered by novel and consolidated technologies, there is still one thing in common among the chapters: their usefulness in large-scale commercial and industrial applications in real food waste management instances in order to exploit the great hidden value that these renewable sources can offer as raw materials of biofunctional ingredients. This will, hopefully, lead to important work that must be carried out during the next years, beyond only research, to highly encourage the generation of new business, revenue streams, and jobs for human well-being and to contribute to sustainable development.

PART I

Food By-Products Generation and Their Valuable Bioactive Compounds

CHAPTER 1

Tomato Agro-Industrial Wastes as a Rich Source of Bioactive Compounds

ZAHRA ZAREIE[1], MAJID NOOSHKAM[2], and ALI MOAYEDI[1]

[1]*Department of Food Science and Technology, Gorgan University of Agricultural Sciences and Natural Resources, Gorgan, Iran*

[2]*Department of Food Science and Technology, Faculty of Agriculture, Ferdowsi University of Mashhad (FUM), Mashhad, Iran*

ABSTRACT

Tomato (*Lycopersicon esculentum*) is a widely cultivated vegetable crop and it is mainly processed to tomato paste, juice, ketchup, purée, and canned tomatoes. However, tomato processing at the industrial scale leads to large quantities of wastes, such as pulp, seeds, peels, and fibrous part residues. The tomato waste management is therefore a worldwide problem in terms of environmental and economic aspects. It is generally agreed that reuse or recycle of these byproducts could decrease processing costs and the corresponding wastes could be considered rich sources of nutrients and bioactive compounds.

This chapter will therefore focus on the various bioactive compounds extracted from tomato wastes, such as carotenoids, proteins, peptides, oil, and dietary fibers. The extraction methods for the release and extraction of these bioactive compounds are also highlighted in this book chapter.

1.1 INTRODUCTION

Tomato (*Lycopersicon esculentum*) is likely originated in Peru and then introduced into Italy as an ornamental plant at the beginning of the 16th

Food Byproducts Management and their Utilization. Ricardo Gómez-García, Ana A. Vilas-Boas, Débora A. Campos, Maria Manuela Pintado, & Cristóbal Noé Aguilar (Eds.)

century; however, until the middle of that century, it did not begin to be grown for food applications. The cultivation of tomato has become widespread over the next centuries and is now considered one of the world's most commonly commercially produced vegetables.[1] United States, Spain, Italy, and Turkey are the leading tomato-growing countries and the world production of tomato reached 124,111,781 metric ton.[2]

Tomato contains remarkable levels of various traditional nutrients. Its folate content is similar to the levels found in carrots and potatoes, and about 10% of the concentrations observed in spinach. It is also rich in vitamin C and potassium.[3] Tomato is consumed at high levels and is therefore a significant source of vitamin C, vitamin A, carotenoids (e.g., β-carotene and lycopene), and flavonoids (e.g., quercetin and kaempferol) in the Western diet. The β-carotene content in fresh tomato is reported to be 0.39 mg 100 g^{-1} and its content in canned tomato sauce is found to be 0.41 mg 100 g^{-1}, while its lycopene content is higher than β-carotene (15.91 mg and 3.02 mg 100 g^{-1} in canned tomato sauce and fresh tomato, respectively).[4] Tomato is mainly consumed as a fresh fruit and after processing into different products including tomato sauce, paste, juice, ketchup, and puree (Figure 1.1).[5] In addition, its dried products are commonly applied in pizza and various vegetable and spicy dishes.[2]

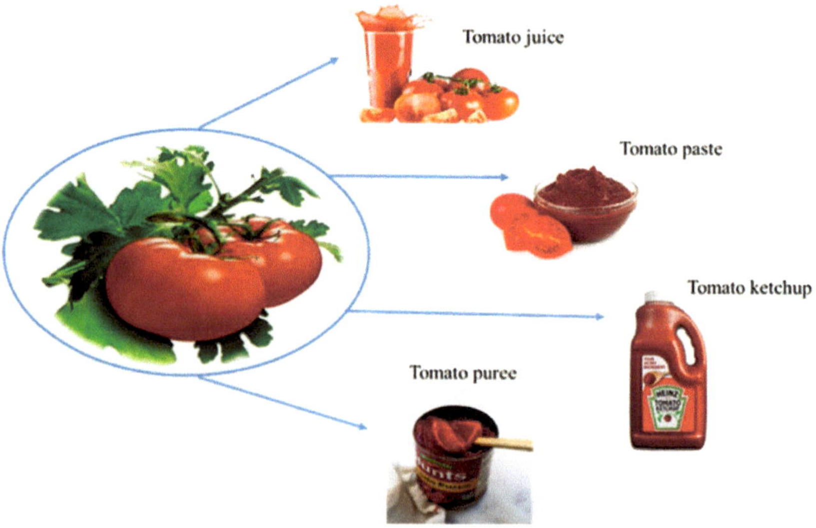

FIGURE 1.1 Tomato products.

Tomato pomace is a byproduct generated during tomato processing, representing approximately 4% of the fruit weight.[6,7] Tomato pomace consists mainly of tomato peels and seeds, and is one of the most promising candidate materials.[8] It contains mainly fiber (25.4–50.0%), protein (15.4–23.7%), fat (5.4–20.5%), and mineral (4.4–6.8%).[6] Therefore, tomato wastes consist of good quality nutrients with the potential to be widely used as food, feed, or fertilizer. Indeed, economic and technological limitations lead to nonutilization of the tomato waste, thereby contributing to environmental pollution.[9] In this context, the use of byproducts of tomato processing industry could provide extra income and decrease the waste disposal problem.[6] This chapter, therefore, focuses on the value-added compounds isolated from the tomato pomace. The extraction procedures of these bioactive molecules are also highlighted in this book chapter.

1.2 TOMATO WASTE RECOVERY

Tomato waste usually represents an environmental problem for the industry. It is potential reutilization for the recovery of various bioactive compounds, such as lycopene, protein, pectin, oil, and dietary fiber could be of benefit (Figure 1.2).[10]

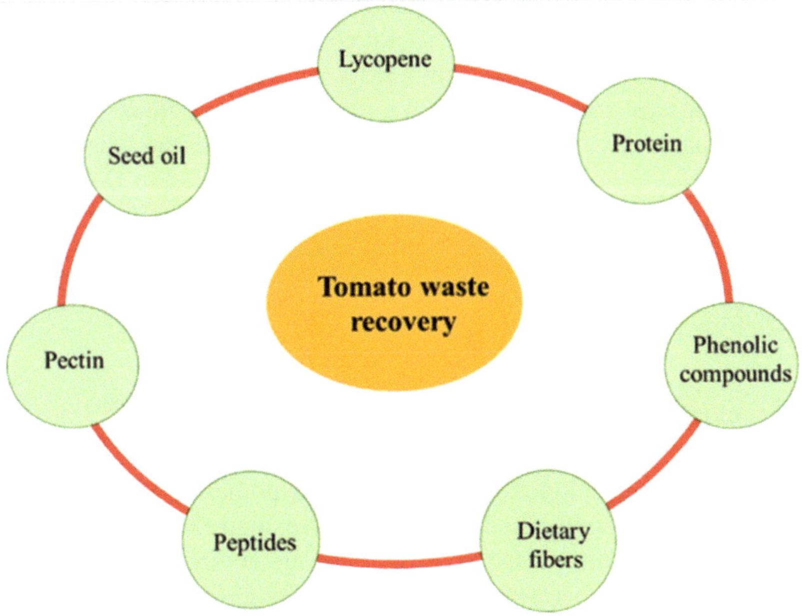

FIGURE 1.2 Bioactive compounds of tomato waste.

1.2.1 LYCOPENE

Tomato contains versatile antioxidative molecules like vitamin E, vitamin C, phenolics, flavonoids, and carotenoids. Among them, lycopene has recently attracted remarkable attention for its possible health-promoting effects. It is the main pigment, with a deep red color, of ripe tomatoes,[8] and the processing, heating, and homogenization treatments lead to lycopene release from cell wall carotenoid–protein complexes. Therefore, processed products have a better lycopene bioavailability compared with fresh tomatoes.[11] Lycopene is classified chemically as a hydrocarbon. It is an oligomer of isoprenoid type consisting of 8 isoprene units joined together into a symmetrical chain containing 13 double bonds (11 conjugated and 2 nonconjugated double bonds).[12] A high degree of conjugated double bonds confers strong antioxidative activities to lycopene molecule, making it one of the most powerful antioxidant compounds.[7]

Lycopene, due to its antioxidant activity, has a high potential to provide protection against cancer and other degenerative diseases and to reduce the incidence of ischemic heart diseases.[13] It has been reported that lycopene is mainly accumulated in the tomato peel and the water-insoluble fractions (72–90%) than in the flesh. The valorization of tomato-processing byproducts is therefore of high interest of researchers and manufacturers.[14,15]

It is also worthwhile of noting that lycopene is mainly found in the plant chromoplast (and further in chromoplasts, during ripening-induced chloroplast-chromoplast transformation). Lycopene is also a highly hydrophobic compound and, therefore, arranged exclusively within the inner part of the lipid bilayer of the vesicles. Moreover, lycopene is present in a complexed form with proteins within the vegetable cells. Thus, more intensive thermal or mechanical treatments and organic solvents are required for lycopene recovery from tomato peels, leading to environmental problems and lycopene thermal degradation.[14]

1.2.2 SEED OIL

Seed is the major component of the tomato pomace (~60%, 1.3 million tons), which is usually used as a low-value material for livestock feed or is otherwise dumped in landfills, causing environmental problems.[16] Therefore, tomato processors are pressingly searching for alternate uses of tomato seeds to protect the environment and increase the economic value of tomato seeds.[17] Tomato seeds contain approximately 20% oil and fatty acid composition of the oil is similar to the low-linolenic soybean oil; accordingly, tomato seeds would be considered a good source of salad oil.[18] Palmitic acid (C16:0; 7–24%) is the main saturated fatty acid of the tomato seed oil. The oil also contains stearic acid (C18:0; 4–13%), oleic acid (C18:1; 18–30%), myristic acid (C14:0; 0.1–2.3%), linolenic acid (C18:3; 1–6%), margaric acid (C17:0; 0.1–0.3%), palmitoleic acid (C16:1; 0.3–7%), arachidic acid (C20:0; 0.2–3%), and behenic, lignoceric, and gondoic acids in very small amounts.[19] Tomato seed oil could be therefore extracted and used as an edible oil with high nutritional quality.[20]

1.2.3 PECTIN

Pectin is abundant in most vegetables and fruits and considered the major part of the primary cell wall of nonwoody plants. From a functional viewpoint, it is a hydrocolloid with the ability to trap water molecules and form gels at low concentrations. Thanks to its excellent safety/health profile and water solubility, pectin is extensively used to add desirable textures to various food products.[21] Tomato waste is a good source of pectin and it has been therefore subjected to different extraction methods for the extraction of high-quality pectin.[22,23] It has been suggested that pectin extracted from tomato wastes has a remarkable quality and could be used as a valuable additive in food industry[22]

1.2.4 PHENOLIC COMPOUNDS

The phenolic compounds are dietary phytochemicals present in plant tissues and an essential part of the human diet, which are of considerable interest because of their antioxidative features. It is well known that tomato byproducts have higher concentrations of phenolic compounds compared with pulp and the removal of peel/seed during tomato processing could lead to a loss of the antioxidant property of these bioactive compounds.[10] The main phenolic compounds of tomato are quercetin, naringenin, rutin, and chlorogenic acid.[24] These phenolic compounds (e.g., flavonoids and hydroxycinnamic acid) show multiple biological properties including metal chelation, free radical-scavenging, modulation of signal transduction pathways and enzymatic activity, and inhibition of cellular proliferation.[25] Phytochemicals with remarkable antioxidant property extracted from byproducts of tomato processing industry could be successfully used as functional ingredient for the development of novel functional foods with health-enhancing effects.[26]

1.2.5 PROTEINS AND PEPTIDES

Tomato seed protein contains globulin, prolamine, gluteline, and albumin, with a high nutritional quality comparable with soybean proteins.[27] It is a rich source of lysine (8–10 g/16 g N) which makes tomato seed protein an ideal candidate for supplementing proteins in cereal products. Moreover, tomato seed proteins could be used as a bioactive ingredient in functional food systems and used for human consumption.[28] Additionally, the lack of antinutritional factor or harmful constituent in the tomato seed makes it a better protein source over other nonconventional sources.[27] It could be also noteworthy that tomato seeds contain almost all essential amino acids, especially leucine and lysine,[29] and protein hydrolysates and peptides of the tomato seeds show antioxidant and ACE inhibitory activities.[30]

1.2.6 DIETARY FIBERS

Dietary fibers have various technological (e.g., gel formation, water/lipid binding, texture modification, and increasing sensory attributes) and biological (e.g., anticancer, antidiabetic, antiobesity, hypocholesterolemia effect, and blood glucose attenuation) functions.[31] Dietary fibers are generally classified as water-insoluble dietary fiber (IDF) and water-soluble dietary fiber (SDF). Compared with IDF, SDF (such as oligosaccharides, pectin, and β-glucans) indicate higher viscosity and have the potential to reduce blood cholesterol and glucose, and improve the commercial values of related functional foods.[32] Tomato peels could be employed as a cheap source of SDF,[33] and it has been reported that they contain 48.5% IDF and 8.9% SDF.[32]

1.3 EXTRACTION METHODS

Extraction, especially conventional solvent extraction (CSE), is generally applied to provide the bioactive molecules from fruit and vegetables. Nonetheless, CSE needs long processing time and high solvent volumes, thereby causing environmental pollution. Thus, new, modern, faster, and more effective extraction methods have been recently developed to increase the extraction efficiency and lower the volume of organic solvents (Table 1.1).[34]

TABLE 1.1 Novel Extraction Methods used for Bioactive Extraction from Tomato and Its By-Products.

Extraction method	Bioactive components	Part of tomato	Conditions	Reference
Microwave-assisted extraction	Lycopene	Peel	Solvent ratio solid liquid ratios, microwave power, delivered energy equivalents	35
	Polyphenols	Peel	Solvents, temperatures (25, 55, and 90°C), times (5 and 10 min)	36
	Phenolic compounds	Tomato	Ethanol concentration (20–80%), temperatures (60–100°C), times (0–4 min)	37
Supercritical fluid	*Trans*-lycopene	Tomato	Carbon dioxide at 40°C without modifier	38
	lycopene	Dried tomato skin	Temperature (40 and 70°C), pressure (25 and 45 MPa), modifier concentration (5 and 15%)	39
	lipids, lycopene, and, β-carotene	Tomato-processing waste (skins and seeds)	Pressures (250 and 300 bar), temperatures (60 and 80°C), particle sizes (80 and 345 µm, solvent flow rates (0.792 and 1.35 kg/h)	12
Ultrasonic extraction	Lycopene	Tomato-processing waste	Ternary solvent system composed of 31% acetone, 38% ethyl acetate, and 31% hexane	40
	Carotenoids	Dry tomato pomace	Hexane proportion in the solvent (25, 50, 75%), 25–45°C, 6 min	41
Ohmic extraction	Oil	Tomato seed	Electric field strength (10–14 V/cm), end point temperature (40–60°C), holding time (5–15 min)	42
	Phenolic compounds	Tomato wastes	70°C, 15 min, 70% ethanol	43
High-pressure/High-pressure homogenization	Flavonoids Lycopene	Tomato pulp	450 MPa, 60% hexane concentration	44
	Proteins Polyphenols	Peel	20,000 RPM, 5 min, 10 passes	14

1.3.1 MICROWAVE-ASSISTED EXTRACTION (MAE)

The MAE is considered an advanced extraction procedure that uses microwave energy to heat solvents effectively. The analytes could be subsequently partitioned from the sample matrix into the extraction medium (i.e., solvent). The MAE is able to significantly reduce both solvent consumption and extraction time.[37] Ho et al.[35] determined the optimal MAE conditions for lycopene extraction from tomato peels. Optimal conditions were: 400 W at 0:10 solvent ratio with an all-trans-lycopene yield of 13.592 mg/100 g. Moreover, ethyl acetate was found to be a better MAE solvent for the recovery of lycopene than hexane. In addition, significant structural disruption was observed in MAE-treated samples, likely leading to the improved lycopene extraction. In another study, the MAE process was optimized to maximize flavonoids/phenolic acids recovery from tomato. The tomato extract obtained in optimum processing conditions showed a high potential to be used as nutraceuticals or active ingredients in the development of functional foods.[45] The application of MAE as an innovative technique has been also studied for the isolation of polyphenols from tomato peel waste. The results showed that the extraction time (5 and 10 min) had no significant effect on total phenols, total flavonoids, and phenolic compounds recovery. On the other hand, the temperature and the solvent had significant effects on the polyphenols yield, and the tomato peel waste was therefore suggested as a sustainable resource of polyphenols for MAE extraction purposes.[36]

1.3.2 SUPERCRITICAL FLUID EXTRACTION (SFE)

The SFE uses supercritical fluids to separate the extractant from the matrix. CO_2 is considered the main extraction solvents for botanicals. CO_2 supercritical extraction is performed above the critical pressure of 74 bar and the critical temperature of 31°C. Supercritical fluids are highly compressible gases with attractively combined gas and liquid properties. These fluids have a potential to initiate reactions, which are not usually achievable in conventional solvents. The SFE is a fast process (10–60 min) and the supercritical fluid is separated from the analyte by simply releasing pressure; a pure residue with almost no trace is therefore yielded.[46] Rozzi et al.[47] used SFE for lycopene extraction from tomato, ant the extraction yield of 61% was obtained under optimum conditions of 34.47 MPa, 86°C, and 500 mL of CO_2 (2.5 mL/min flow rate). In another study, the highest carotenoid concentration (with 90.1% lycopene) from industrial tomato waste was achieved by SFE at 80°C and 460 bar, while tocopherols- and phytosterols-rich products were obtained above 300 bar and 40°C.[48]

1.3.3 ULTRASONIC-ASSISTED EXTRACTION (UAE)

Ultrasound received the most attention among agitation techniques mainly due to its ability to either disrupt cell wall of biological cells or provide a better solvent diffusion into cellular materials and enhance the mass transport. This is due to the cavitation phenomenon, which formed tiny bubbles in the liquid bulk collapse (nonsymmetric) and exploded during a compression cycle. Therefore, the ultrasound-assisted extraction (UAE) is considered a proficient extraction technique capable of reducing working times and increasing extraction yields.[49] Tomato waste was treated with ammonium oxalate/oxalic acid by CSE and UAE at 37 kHz and temperatures of 60°C and 80°C for the pectin extraction purposes. The CSE at 60°C resulted in the highest yield, but UAE during a sonication time of 15 min produced the pectin of better quality. The pectin yields obtained after the extraction at 80°C were similar at times of 24 h for CSE and 15 min for UAE. Thus, it could be deduced that the main advantages of UAE are the lower extraction time and environmentally friendly status.[22] Moreover, the extraction yield of lycopene from tomato paste-processing wastes by UAE and CSE showed that UAE requires lower solvent, lower temperature, and less time than CSE.[50] The UAE has been also used for the efficient extraction of carotenoids from dry tomato pomace.[41]

1.3.4 OHMIC EXTRACTION

Ohmic heating, also called Joule heating, is an alternative fast heating method, which occurs when an alternating electrical current is passed through food and the electrical resistance of food leads to food heating. It is a rapid, uniform, and more environmentally friendly heating method with a high potential to yield a high-quality product with minimal changes in nutritional, structural, or organoleptic properties.[51,52] Coelho et al.[53] optimized the extraction of phenolic compounds from tomato byproducts using ohmic heating technology. The best extraction conditions of polyphenols were 70°C for 15 min, using 70% ethanol as a solvent, and the recovery of rutin was 77%

higher than control samples. Ohmic heating led to recover up to 4.93 μg/g lycopene from tomato byproducts without resorting to organic solvents. Thus, ohmic heating could be a fast, economical, and environmental-friendly process to recover polyphenols from industrial tomato byproducts. Similar observations were outlined in the literature.[43]

1.3.5 HIGH-PRESSURE AND HIGH-PRESSURE HOMOGENIZATION

High-pressure (HP) processing can effectively reduce the extraction time and increase process efficiency. The HP is able to cause some structural changes, such as cell membrane damage or cell deformation that increase the permeability of cell and the diffusion of secondary metabolite and in turn mass transfer rates. Interestingly, it has minimal effects on low-molecular-weight components such as flavors, vitamins, and pigments. Therefore, HP has been recently applied to extract biologically active compounds from natural sources.[55]

The high-pressure homogenization (HPH) technology pressurizes a fluid to flow through a narrow gap valve. This leads to a marked increase in its velocity and in turn a pressure drop with high shear stress and cavitation. The suspended particles and macromolecules in the fluid undergo twisting and deformation under high mechanical stresses applied.[57] HPH is an effective and fast method for the micronization of plant tissue in suspension and to release active biomolecules from cells, resulting in high extraction yields. Thanks to its ease of operation, reproducibility, scalability, and high throughput, HPH is considered a method particularly suitable for industrial applications.[14] HP and HPH technologies have been successfully applied to extract bioactive compounds from tomato wastes. Briones-Labarca et al.[44] studied the effect of operating HP and solvent polarity (solvent mixture) on extraction yield, flavonoid, and lycopene content from tomato pulp. The HP and solvent mixture had significant effects on the responses. The optimum HPH extraction condition was acquired at 60% hexane and 450 MPa, provided the maximum extraction yield of 8.71%, flavonoid content of 21.52 ± 0.09 mg QE/g fresh weight and lycopene content of 2.01 ± 0.09 mg QE/100 g fresh weight. In another study, it has been reported that the tomato peels treatment by the HPH method could increase the release of intracellular compound, such as proteins (+70.5%) and polyphenols (+32.2%), and improve the antioxidant activity (+23.3%) and reduce the oil–water interfacial tension (−15.0%).[14] Thus, HPH extraction could be a powerful tool to increase the extraction and release of bioactive compounds from tomato wastes.

1.4 TOMATO PRODUCTS

1.4.1 TOMATO PASTE

The tomato paste-manufacturing industry is known as one of the most important sections in the food industry.[58] The production pipeline of the tomato paste involves multiple uses of heat (Figure 1.3). In brief, the intact fruit is sorted, washed, and then crushed. The crushed tomato goes through a heat exchanger and is subjected to steam heating (88–99°C) to inactivate pectinase enzymes. This process is known as the hot break. Then, the juice of crushed tomatoes is separated from skins and seeds by an extractor. The juice goes through a multieffect evaporation step to lower its moisture content and form the paste. The paste is subjected to sterilization process *via* steam heating and is then packaged aseptically.[59]

1.4.2 KETCHUP

Ketchup, as one of the most popular products of tomato in the global market, needs lower equipment and simple processing.[5] Tomato paste (23–35° Brix) is generally used for ketchup production.[60] Ketchup is a heterogeneous spiced product and its consistency is an important feature from the consumer and engineering views.[61] Thickening agents are also added to the formulation to increase the consistency and inhibit serum separation from ketchup.[5] The microbial stability of tomato ketchup is based on final pH ($pH < 4.0$), on pasteurization or the addition of preservatives. Ketchup subjects to several relatively important heat treatments and the product can be therefore stored at room temperature for one or more years.[62]

1.4.3 TOMATO JUICE

Tomato juice consists of suspended particles (>150 μm) which include intact or broken cells, water-insoluble pectic, long-chain polymers of cellulose, hemicellulose, and lignin material.[63] Tomato juice is a well-known healthy beverage containing 93.1% moisture, 4.89% carbohydrate, minerals, and vitamins, and is low in fat and protein.[64] The color of fruit juice is a primary factor in evaluating juice quality and sensory acceptance by consumers. It could be also noteworthy that the nutritional quality of tomato juice is mainly attributed to the presence of vitamin C and other biologically active compounds, such as lycopene, which is responsible for the color of tomato juice. And epidemiological studies show that tomatoes and the related products, such as tomato juice, could decrease the probability of cardiovascular diseases and developing prostate cancers.[65]

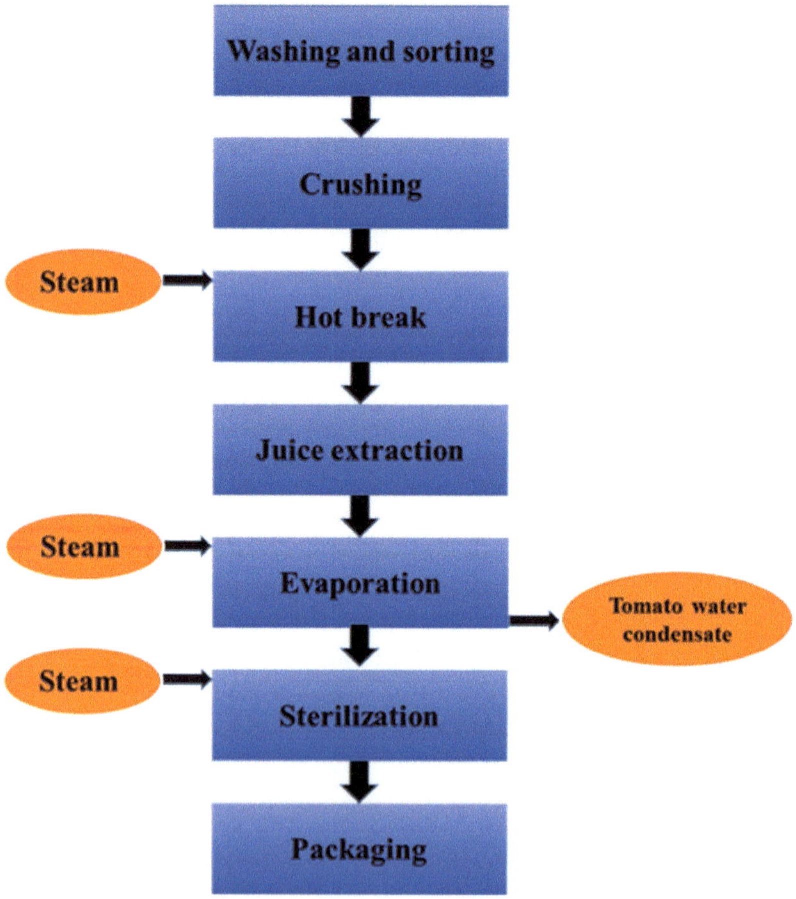

FIGURE 1.3 Flow diagram of processes in tomato paste production under conventional processing.[59]

Source: Adapted from Ref. [59].

1.4.4 TOMATO SAUCE

Tomatoes are generally consumed in the form of industrially processed products, especially tomato sauces. Dietary antioxidants (such as hydroxytyrosol, virgin olive oil), phenolic compounds from vegetables (such as garlic, peppers, carrots, and onions), spices, and herbs could be added to tomato sauces to improve its health benefits and sensory properties.[66] It is worthwhile noting that tomato sauce represents 40.8% of all sauce consumption in Spain.[67]

1.4.5 TOMATO PUREE

Cold-break and hot-break processes are conventionally applied for tomato puree processing. In this context, cold-break process (<66°C) leads to partial inactivation of polygalacturonase, and the corresponding paste has a good color and taste quality; however, the viscosity instability during storage and the relative low consistency of the product are the main problems related to this technique. Indeed, a tomato product with low consistency could not retain its solid fraction in suspension, and the product undergoes syneresis during the storage period. The hot-break process is generally applied to overcome syneresis problems, and tomatoes are processed at about 75–100°C for 30 min. The enzymes polygalacturonase and pectin-methyl esterase are therefore completely inactivated and a paste with firm consistency is achieved after concentration.[68]

1.5 CONCLUSIONS

The ever-growing level of agricultural waste has motivated the manufacturers and researchers to utilize these waste materials to develop green-processing technologies and extract bioactive compounds with health-promoting effects from this waste. This book chapter summarized the main bioactive components of tomato wastes and the advanced technologies to extract them. It was shown that different valuable compounds could be extracted by various technologies from tomato wastes and used as functional ingredients to develop novel functional foods. However, further research and optimization studies on the extraction techniques of tomato bioactive compounds might be necessary to increase their relevance for industrial applications.

KEYWORDS

- **bioactive compounds**
- **extraction**
- **lycopene**
- **polyphenols**
- **tomato waste**

REFERENCES

1. Willcox, J. K.; Catignani, G. L.; Lazarus, S. Tomatoes and Cardiovascular Health. *Crit. Rev. Food Sci. 43*, 1–18.
2. Doymaz, I. Air-drying Characteristics of Tomatoes. *J. Food Eng.* **2007,** *78*, 1291–1297.
3. Beecher, G. R. Nutrient Content of Tomatoes and Tomato Products. *Proc. Soc. Exp. Biol. Med.* **1998,** *218*, 98–100.
4. Koh, E.; Charoenprasert, S.; Mitchell, A. E. Effects of Industrial Tomato Paste Processing on Ascorbic Acid, Flavonoids and Carotenoids and their Stability Over One-year Storage. *J. Sci. Food Agr.* **2012,** *92*, 23–28.
5. Torbica, A.; Belović, M.; Mastilović, J.; Kevrešan, Ž.; Pestorić, M.; Škrobot, D.; Hadnađev, T. D. Nutritional, Rheological, and Sensory Evaluation of Tomato Ketchup with Increased Content of Natural Fibres Made from Fresh Tomato Pomace. *Food Bioprod. Process.* **2016,** *98*, 299–309.
6. Del Valle, M.; Cámara, M.; Torija, M. E. Chemical Characterization of Tomato Pomace. *J. Sci. Food Agr.* **2006,** *86*, 1232–1236.
7. Papaioannou, E. H.; Karabelas, A. J. Lycopene Recovery from Tomato Peel Under Mild Conditions Assisted by Enzymatic Pre-treatment and Non-ionic Surfactants. *Acta Biochim. Pol.* **2012,** *59*, 71–74.
8. Zuorro, A.; Fidaleo, M.; Lavecchia, R. Enzyme-assisted Extraction of Lycopene from Tomato Processing Waste. *Enzyme Microb. Technol.* **2011,** *49*, 567–573.
9. Sogi, D. S.; Garg, S. K.; Bawa, A. S. Functional Properties of Seed Meals and Protein Concentrates from Tomato-processing Waste. *J. Food Sci.* **2002,** 67, 2997–3001.
10. Grassino, A. N.; Djaković, S.; Bosiljkov, T.; Halambek, J.; Zorić, Z.; Dragović-Uzelac, V.; ...; Brnčić, S. R. Valorisation of Tomato Peel Waste as a Sustainable Source for Pectin, Polyphenols and Fatty Acids Recovery Using Sequential Extraction. *Waste Biomass Valori.* **2020,** *11*, 4593–4611.
11. Silaste, M. L.; Alfthan, G.; Aro, A.; Kesäniemi, Y. A.; Hörkkö, S. Tomato Juice Decreases LDL Cholesterol Levels and Increases LDL Resistance to Oxidation. *Brit. J. Nutr.* **2007,** *98*, 1251–1258.
12. Sabio, E.; Lozano, M.; Montero de Espinosa, V.; Mendes, R. L.; Pereira, A. P.; Palavra, A. F.; Coelho, J. A. Lycopene and *β*-carotene Extraction from Tomato Processing Waste Using Supercritical CO_2. *Ind. Eng. Chem. Res.* **2003,** *42* (25), 6641–6646.

13. Ranveer, R. C.; Patil, S. N.; Sahoo, A. K. Effect of Different Parameters on Enzyme-assisted Extraction of Lycopene from Tomato Processing Waste. *Food Bioprod. Process* **2013,** *91*, 370–375.
14. Jurić, S., Ferrari, G., Velikov, K. P.; Donsì, F. (2019). High-pressure Homogenization Treatment to Recover Bioactive Compounds from Tomato Peels. *J. Food Eng. 262*, 170–180.
15. Pataro, G.; Carullo, D.; Falcone, M.; Ferrari, G. Recovery of Lycopene from Industrially Derived Tomato Processing byproducts by Pulsed Electric Fields-assisted Extraction. *Innov. Food Sci. Emerg.* **2020,** *63*, 102369.
16. Lazos, E. S.; Tsaknis, J.; Lalas, S. Characteristics and Composition of Tomato Seed Oil. *Grasas y Aceites.* **1998,** *49*, 440–445
17. Shao, D.; Atungulu, G. G.; Pan, Z.; Yue, T.; Zhang, A.; Li, X. Study of Optimal Extraction Conditions for Achieving High Yield and Antioxidant Activity of Tomato Seed Oil. *J. Food Sci.* **2012,** *77*, E202–E208.
18. Eller, F. J.; Moser, J. K.; Kenar, J. A.; Taylor, S. L. Extraction and Analysis of Tomato Seed Oil. *J. Am. Oil Chem. Soc.* **2010,** *87*, 755–762.
19. Botineştean, C.; Gruia, A. T.; Jianu, I. Utilization of Seeds from Tomato Processing Wastes as Raw Material for Oil Production. *J. Mater. Cycles Waste Manag.* **2015,** *17*, 118–124.
20. Shao, D.; Venkitasamy, C.; Li, X.; Pan, Z.; Shi, J.; Wang, B.; ...; McHugh, T. H. Thermal and Storage Characteristics of Tomato Seed Oil. *LWT-Food Sci. Technol.* **2015,** *63*, 191–197.
21. Ciriminna, R.; Chavarría-Hernández, N.; Inés Rodríguez Hernández, A.; Pagliaro, M. Pectin: A New Perspective from the BIORefinery Standpoint. *Biofuel Bioprod. Biorefin.* **2015,** *9*, 368–377.
22. Grassino, A. N.; Halambek, J.; Djaković, S.; Brnčić, S. R.; Dent, M.; Grabarić, Z. Utilization of Tomato Peel Waste from Canning Factory as a Potential Source for Pectin Production and Application as Tin Corrosion Inhibitor. *Food Hydrocolloid.* **2016,** *52*, 265–274.
23. Sengar, A. S.; Rawson, A.; Muthiah, M.; Kalakandan, S. K. Comparison of Different Ultrasound Assisted Extraction Techniques for Pectin from Tomato Processing Waste. *Ultrason. Sonochem.* **2020,** *61*, 104812.
24. Martínez-Valverde, I.; Periago, M. J.; Provan, G.; Chesson, A. Phenolic Compounds, Lycopene and Antioxidant Activity in Commercial Varieties of Tomato (Lycopersicum Esculentum). *J. Sci. Food Agr.* **2002,** *82*, 323–330.
25. Tulipani, S.; Huelamo, M. M.; Ribalta, M. R.; Estruch, R.; Ferrer, E. E.; Andres-Lacueva, C.; ...; Lamuela-Raventós, R. M. Oil Matrix Effects on Plasma Exposure and Urinary Excretion of Phenolic Compounds from Tomato Sauces: Evidence from a Human Pilot Study. *Food Chem.* **2012,** *130*, 581–590.
26. Kalogeropoulos, N.; Chiou, A.; Pyriochou, V.; Peristeraki, A.; Karathanos, V. T. Bioactive Phytochemicals in Industrial Tomatoes and their Processing Byproducts. *LWT-Food Sci. Technol.* **2012,** *49*, 213–216.
27. Mechmeche, M.; Kachouri, F.; Ksontini, H.; Hamdi, M. Production of Bioactive Peptides from Tomato Seed Isolate by Lactobacillus Plantarum Fermentation and Enhancement of Antioxidant Activity. *Food Biotechnol.* **2017,** *31*, 94–113.
28. Liadakis, G. N.; Tzia, C.; Oreopoulou, V.; Thomopoulos, C. D. Protein Isolation from Tomato Seed Meal, Extraction Optimization. *J. Food Sci.* **1995,** *60*, 477–482.

29. Kartal, C.; Türköz, B. K.; Otles, S. Prediction, Identification and Evaluation of Bioactive Peptides from Tomato Seed Proteins using in Silico Approach. *J. Food Meas. Charact.* **2020,** *14*, 1865–1883.
30. Moayedi, A.; Mora, L.; Aristoy, M. C.; Hashemi, M.; Safari, M.; Toldrá, F. ACE-inhibitory and Antioxidant Activities of Peptide Fragments Obtained from Tomato Processing Byproducts Fermented using Bacillus Subtilis: EFFEct of Amino Acid Composition and Peptides Molecular Mass Distribution. *Appl. Biochem. Biotechnol.* **2017,** *181*, 48–64.
31. Cepeda, E.; Collado, I. Rheology of Tomato and Wheat Dietary Fibers in Water and in Suspensions of Pimento Purée. *J. Food Eng.* **2014,** *134*, 67–73.
32. Li, N.; Feng, Z.; Niu, Y.; Yu, L. Structural, Rheological and Functional Properties of Modified Soluble Dietary Fiber from Tomato Peels. *Food Hydrocolloid.* **2018,** *77*, 557–565.
33. Niu, Y.; Li, N.; Xia, Q.; Hou, Y.; Xu, G. Comparisons of Three Modifications on Structural, Rheological and Functional Properties of Soluble Dietary Fibers from Tomato Peels. *LWT.* **2018,** 88, 56–63.
34. Baltacıoğlu, H.; Baltacıoğlu, C.; Okur, I.; Tanrıvermiş, A.; Yalıç, M. Optimization of Microwave-assisted Extraction of Phenolic Compounds from Tomato: Characterization by FTIR and HPLC and Comparison with Conventional Solvent Extraction. *Vib. Spectrosc.* **2021,** 113, 103204.
35. Ho, K. K.; Ferruzzi, M. G.; Liceaga, A. M.; San Martín-González, M. F. Microwave-assisted Extraction of Lycopene in Tomato Peels: Effect of Extraction Conditions on All-trans and Cis-isomer Yields. *LWT-Food Sci. Technol.* **2015,** *62*, 160–168.
36. Bakić, M. T.; Pedisić, S.; Zorić, Z.; Dragović-Uzelac, V.; Grassino, A. N. Effect of Microwave-assisted Extraction on Polyphenols Recovery from Tomato Peel Waste. *Acta Chim. Slov.* **2019,** *66*, 367–377.
37. Li, H.; Deng, Z.; Wu, T.; Liu, R.; Loewen, S.; Tsao, R. Microwave-assisted Extraction of Phenolics with Maximal Antioxidant Activities in Tomatoes. *Food Chem.* **2012,** *130*, 928–936.
38. Gómez-Prieto, M. S.; Caja, M. M.; Herraiz, M.; Santa-María, G. Supercritical Fluid Extraction of All-trans-lycopene from Tomato. *J. Agric. Food Chem.* **2003,** *51*, 3–7.
39. Kassama, L. S.; Shi, J.; Mittal, G. S. Optimization of Supercritical Fluid Extraction of Lycopene from Tomato Skin with Central Composite Rotatable Design Model. *Sep. Purif. Technol.* **2008,** *60*, 278–284.
40. Anarjan, N. Evaluation the Effects of Ultrasonic Parameters on Simultaneously Extraction and Size Reduction of Lycopene from Tomato Processing Waste. *Waste Biomass Valori.* **2020,** *11*, 1929–1940.
41. Luengo, E.; Condón-Abanto, S.; Condón, S.; Álvarez, I.; Raso, J. Improving the Extraction of Carotenoids from Tomato Waste by Application of Ultrasound Under Pressure. *Sep. Purif. Technol.* **2014,** *136*, 130–136.
42. Karunanithi, S.; Pare, A.; Sunil, C. K.; Loganathan, M. Optimization of Process Parameters of Ohmic Heating for Improving Yield and Quality of Tomato Seed Oil. *Int. J. Pure. App. Biosci.* **2019,** *7*, 104–114.
43. Coelho, M. I.; Pereira, R. N. C.; Teixeira, J. A.; Pintado, M. E. Valorization of Tomato Wastes: Influence of Ohmic Heating Process on Polyphenols Extraction Time. *ISANH* **2017,** *5*, 37–40.
44. Briones-Labarca, V.; Giovagnoli-Vicuña, C.; Cañas-Sarazúa, R. Optimization of Extraction Yield, Flavonoids and Lycopene from Tomato Pulp by High Hydrostatic Pressure-assisted Extraction. *Food Chem.* **2019,** *278*, 751–759.

45. Pinela, J.; Prieto, M. A.; Carvalho, A. M.; Barreiro, M. F.; Oliveira, M. B. P.; Barros, L.; Ferreira, I. C. Microwave-assisted Extraction of Phenolic Acids and Flavonoids and Production of Antioxidant Ingredients from Tomato: A Nutraceutical-oriented Optimization Study. *Sep. Purif. Technol.* **2016,** *164*, 114–124.
46. Sapkale, G. N.; Patil, S. M.; Surwase, U. S.; Bhatbhage, P. K. Supercritical Fluid Extraction. *Int. J. Chem. Sci.* **2010,** *8*(2), 729–743.
47. Rozzi, N. L.; Singh, R. K.; Vierling, R. A.; Watkins, B. A. Supercritical Fluid Extraction of Lycopene from Tomato Processing Byproducts. *J. Agric. Food Chem.* **2002,** *50*, 2638–2643.
48. Vági, E.; Simándi, B.; Vásárhelyiné, K. P.; Daood, H.; Kéry, Á.; Doleschall, F.; Nagy, B. Supercritical Carbon Dioxide Extraction of Carotenoids, Tocopherols and Sitosterols from Industrial Tomato Byproducts. *J. Supercrit. Fluids* **2007,** *40*, 218–226.
49. Arabani, A. A.; Hosseini, F.; Abbaspour, F.; Anarjan, N. The Effects of Ultrasound Pretreatment Processes on Oil Extraction from Tomato Wastes. *Int. J. Biosci.* **2015,** *4*, 8–15.
50. Kumcuoglu, S.; Yilmaz, T.; Tavman, S. Ultrasound Assisted Extraction of Lycopene from Tomato Processing Wastes. *J. Food Sci. Technol.* **2014,** *51*, 4102–4107.
51. Saberian, H.; Hamidi-Esfahani, Z.; Gavlighi, H. A.; Barzegar, M. Optimization of Pectin Extraction from Orange Juice Waste Assisted by Ohmic Heating. *Chem. Eng. Process: Process Intensif.* **2017,** *117*, 154–161.
52. Pereira, R. N.; Rodrigues, R. M.; Genisheva, Z.; Oliveira, H.; de Freitas, V.; Teixeira, J. A.; Vicente, A. A. Effects of Ohmic Heating on Extraction of Food-grade Phytochemicals from Colored Potato. *LWT* **2016,** *74*, 493–503.
53. Coelho, M.; Pereira, R.; Rodrigues, A. S.; Teixeira, J. A.; Pintado, M. E. Extraction of Tomato Byproducts' Bioactive Compounds using Ohmic Technology. *Food Bioprod. Process* **2019,** *117*, 329–339.
54. Darvishi, H.; Saba, M. K.; Behroozi-Khazaei, N.; Nourbakhsh, H. Improving Quality and Quantity Attributes of Grape Juice Concentrate (Molasses) using Ohmic Heating. *J. Food Sci. Technol.* **2020,** *57*, 1362–1370.
55. Strati, I. F.; Gogou, E.; Oreopoulou, V. Enzyme and High Pressure Assisted Extraction of Carotenoids from Tomato Waste. *Food Bioprod. Process* **2015,** *94*, 668–674.
56. Naviglio, D.; Pizzolongo, F.; Ferrara, L.; Aragon, A.; Santini, A. Extraction of Pure Lycopene from Industrial Tomato Byproducts in Water using a New High-pressure Process. *J. Sci. Food Agr.* **2008,** 88, 2414–2420.
57. Augusto, P. E.; Ibarz, A.; Cristianini, M. Effect of High Pressure Homogenization (HPH) on the Rheological Properties of Tomato Juice: Creep and Recovery Behaviours. *Food Res. Int.* **2013,** *54*, 169–176.
58. Bacenetti, J.; Duca, D.; Negri, M.; Fusi, A.; Fiala, M. Mitigation Strategies in the Agro-food Sector: The Anaerobic Digestion of Tomato Puree Byproducts. An Italian Case Study. *Sci. Total Environ.* **2015,** *526*, 88–97.
59. Amón, R.; Maulhardt, M.; Wong, T.; Kazama, D.; Simmons, C. W. Waste Heat and Water Recovery Opportunities in California Tomato Paste Processing. *Appl. Therm. Eng.* **2015,** *78*, 525–532.
60. McCarthy, K. L.; McCarthy, M. J. Relationship Between In-line Viscosity and Bostwick Measurement during Ketchup Production. *J. Food Sci.* **2009,** *74*, E291–E297.
61. Gujral, H. S.; Sharma, A.; Singh, N. Effect of Hydrocolloids, Storage Temperature, and Duration on the Consistency of Tomato Ketchup. *Int. J. Food Prop.* **2002,** *5*, 179–191.

62. Rajchl, A.; Voldřich, M.; Čížková, H.; Hronová, M.; Ševčík, R., Dobiáš, J.; Pivoňka, J. Stability of Nutritionally Important Compounds and Shelf Life Prediction of Tomato Ketchup. *J. Food Eng.* **2010,** *99*, 465–470.
63. Wu, J.; Gamage, T. V.; Vilkhu, K. S.; Simons, L. K.; Mawson, R. Effect of Thermosonication on Quality Improvement of Tomato Juice. *Innov. Food Sci. Emerg.* **2008,** *9*, 186–195.
64. Yoon, K. Y.; Woodams, E. E.; Hang, Y. D. Probiotication of Tomato Juice by Lactic Acid Bacteria. *J. Microbiol.* **2004,** *42*, 315–318.
65. Adekunte, A. O.; Tiwari, B. K.; Cullen, P. J.; Scannell, A. G. M.; O'Donnell, C. P. Effect of Sonication on Colour, Ascorbic Acid and Yeast Inactivation in Tomato Juice. *Food Chem.* **2010,** *122*, 500–507.
66. Motilva, M. J.; Macià, A.; Romero, M. P.; Labrador, A.; Domínguez, A.; Peiró, L. Optimisation and Validation of Analytical Methods for the Simultaneous Extraction of Antioxidants: Application to the Analysis of Tomato Sauces. *Food Chem.* **2014,** *163*, 234–243.
67. Martínez-Huélamo, M.; Tulipani, S.; Estruch, R.; Escribano, E.; Illán, M.; Corella, D.; Lamuela-Raventós, R. M. The Tomato Sauce Making Process Affects the Bioaccessibility and Bioavailability of Tomato Phenolics: A Pharmacokinetic Study. *Food Chem.* **2015,** *173*, 864–872.
68. Krebbers, B.; Matser, A. M.; Hoogerwerf, S. W.; Moezelaar, R.; Tomassen, M. M.; van den Berg, R. W. Combined High-pressure and Thermal Treatments for Processing of Tomato Puree: Evaluation of Microbial Inactivation and Quality Parameters. *Innov. Food Sci. Emerg.* **2003,** *4*, 377–385.

CHAPTER 2

Quantification and Use of Food Losses: The Case of the Cantaloupe Melon in a North-Central Region of Mexico

JOSÉ DE JESÚS ESPINOZA ARELLANO,
ADRIANA MONTSERRAT FABELA HERNÁNDEZ,
VÍCTOR MANUEL MOLINA MOREJÓN, and
LILIANA ANGÉLICA GUERRERO RAMOS

Graduate Studies Department, School of Accounting and Administration, Autonomus University of Coahuila, Torreon, Coahuila, Mexico

ABSTRACT

One-thirds of the world's food production is lost or wasted. These wastes reduce the availability of food for the population, increase its prices, and generate environmental problems due to the generation of greenhouse gases. A common problem in developing countries is the paucity of data on the amount, causes, and uses of the food losses. This work was carried out an estimate of the losses of cantaloupe melon, its reasons, and uses in the production, packaging, and transportation links in a region of North-Central Mexico known as the Comarca Lagunera. This region is the largest producer of cantaloupe melon in Mexico, with an area of more than 4565 hectares and a production of 165,663 tons. In the Comarca Lagunera there are three planting stages of cantaloupe melon: Early (January–February), intermediate (March–April), and late (June–August). A structured questionnaire was applied to a statistical sample of 47 producers and 8 packers. It was found

Food Byproducts Management and their Utilization. Ricardo Gómez-García, Ana A. Vilas-Boas, Débora A. Campos, Maria Manuela Pintado, & Cristóbal Noé Aguilar (Eds.)

that the losses in production vary according to the planting season, with more significant losses in late sowings, 17%, followed by early ones with 11% and intermediate ones, 10%. The main causes of losses are inadequate fruit size, deformed, excessive maturity spotted, and cracked. The product that is wasted in the farm is destined to feed domestic animals or as fertilizer to the land. At the same time, the one that is discarded in the packaging is resold to small local intermediaries for its commercialization. More specialized trainings to producers on production technologies are necessary to obtain higher production yields and better quality.

2.1 INTRODUCTION

In 2000, the FAO adopted 17 Millennium Development Goals, which revolved around reducing poverty worldwide. Fabi and English[1] indicates that Goal 12 leads to goal 12.3 which proposes by 2030 to halve losses and waste in food supply chains. Solutions must be found to feed the population, taking into account that we live in a world with limited natural resources. One way is by reducing food losses.[2]

Around the world, 1.3 billion tons of food are lost and wasted each year, representing about a third of world production.[2] In high- and middle-income countries like USA, Germany, France, Canada, and others, food is wasted significantly at the consumption stage, while in low-income countries losses occur in the production, storage, and distribution stages.[2] Losses occur mainly during production, post-harvest, storage, and transportation. In contrast, waste occurs during distribution and consumption, in direct relation to the behavior of wholesalers and retailers, food service providers, and consumers who decide to discard food that still has value.[3]

In Latin America, it is estimated that the lost and wasted food reaches 28% of production.[4] Other more specific studies estimate that losses and waste by food groups in Latin America concerning meat, dairy products, and oilseeds reach 20%; cereals are wasted by 25%; fish and shellfish by 33%; and fruit and vegetables by 55%.[5]

Reducing food losses and waste is a fundamental means of ensuring food availability to supply, in a sustainably way, the world's growing population.[6] Food losses should be reduced to a minimum level in any country, regardless of its level of economic development and the maturity of its systems; however, the problem of food loss and waste is likely to increase in the short term unless countries adopt strict policies.[7]

In Mexico, several secretariats of the federal government, the private sector, and civil society have joined intending to reduce food losses and waste, which has allowed the issue of waste to be positioned at the highest level and formulated a comprehensive national strategy with a vision of socioeconomic, environmental, and food security.[5] In Mexico, there have been few studies on food losses and waste but none on cantaloupe melons.

A study was carried out in the state of Jalisco, Mexico, to quantify food losses, the causes, and their uses. The state's main crops, such as sugar cane, agave, corn, tomato, banana, and avocado, were studied. The average loss in agricultural products was 10.75%; however, in the case of tomato the losses were 24.64% and in avocado 15.15%. The greatest losses were in small producers since their processes are not very technical and they have little transport and storage infrastructure. The main causes of the losses were pests, variations in climate, poor infrastructure, and lack of training.[8]

Del Angel et al.[9] conducted a study to estimate food losses of three fruit trees in the state of Veracruz, Mexico. They analyzed the cases of chayote (*Figure*), manila mango (*Figure*), and manzano pepper (*Figure*). The loss values found in this study were 13.7% in chayote, 10% in mango, and 30% in manzano pepper.

Inegi[10] conducted a nationwide study on the causes of food losses and waste (FLW) in the primary sector. It was found that 44.2% of the production units have suffered losses due to biological reasons (pests and diseases), while 74.7% of the units have suffered losses due to climatic factors such as droughts, strong winds, hailstorms, frosts, and floods, among other factors.

In this work, the case of the cantaloupe melon in the Lagunera region, located in the North Central of Mexico, was analyzed. The cantaloupe melon is a variety of the *Figure* species of the *Figure* family. It is an annual creeping plant with rough pubescent stems with tendrils that can reach 3 m in length.[11] The weight of its fruits varies from 0.5 to 4 kg. The melon fruit is generally consumed raw in its natural state when it has reached perfect maturity. The melon is not very nutritious, but it has an abundance of sugary and mucilaginous matters, has refreshing properties, and facilitates secretions. Besides being consumed fresh, jams, juices and smoothies, sweets, and jams are also made with it; however, it is estimated that more than 95% is consumed fresh.[12]

Cantaloupe melon production in the world in 2019 was 27.5 million tons. China is the largest producer with 49% of the total, followed by Turkey with 6.47% and India with 4.62%.[13] In Mexico, the cantaloupe melon cultivated area in 2020 was 19,104 ha with a production of 591,574 tons.[14] The study region, the Comarca Lagunera, contributed in year 2020 with an area of 4565 hectares and a production of 165,663 tons, with 28% of the national production and is recognized as the main cantaloupe melon-producing region of the country.[15]

According to FAO[16] there are large data gaps about food loss and waste in the world. For this reason, further research must be carried out in this field. The objective of this work was to estimate the losses of cantaloupe melon, its causes and uses in the production, packaging, and transportation links. The work was carried out in a region of North-Central Mexico known as the Comarca Lagunera. Based on a quantification of melon losses and their causes, specific policies can be implemented for their management and reduction.

2.2 METHODOLOGY

The study region is located in north-central Mexico and is known as the Lagunera Region. The region is made up of fifteen municipalities: Ten of them from the state of Durango and five from the state of Coahuila. The cities of Matamoros and Torreón in the state of Coahuila, and Gómez Palacio and Lerdo, in the state of Durango form the Laguna Metropolitan Area with about 1.5 million inhabitants. The region is located between meridians 102° 22′ and 104° 47′ west longitude and parallels 24° 22′ and 26° 23′ north latitude, and an average altitude of 1150 masl.[17] The region's climate is dry desert and is characterized by an average annual temperature of 20.9°C, hot summers, with average maximum temperatures of 30.2°C (±1.9), minimum temperatures of 10.5°C (±3.0), and average annual precipitation of 287 (±98) mm.[18]

There are various methods reported to quantify food losses and waste.[19] Some of them are: The weighing, counting, volume evaluation, file, journal, modeling, survey, among others.

For this work, the survey method was used, which has some advantages such as not requiring technical specialization in the food area, the investment is less than other methods and no equipment, laboratories, or qualified personnel are required to quantify the losses. Surveys were applied at the producer level and on packaging in which information was not only obtained on waste at their facilities but also on transportation to consumption centers.

To measure the losses in the production stage, questionnaires were applied to 47 cantaloupe melon producers from the municipalities of Matamoros and Viesca, of Coahuila state and to 8 packers established in the region. The study population was determined based on the register of melon producers of SADER Región Lagunera[15] from which the statistical sample was calculated.

The questionnaires applied to the producers were made up of 18 questions, with which information was obtained on planted area, sowing and harvesting period, percentage of losses, causes of losses, uses of the rejected product, transportation methods and type of packaging, among others. The questionnaires applied covered early sowing (planting period January–February), intermediate (planting period March–April), and late (planting period June–August) given that in each stage the climatic conditions, the availability of water, and other factors such as pests and diseases are different, so the different conditions of the agricultural year were covered.

The sample calculation was performed by using simple random sampling[20] according to the following formula:

n = *Figure*

where

Figure = sample size, *FIGURE* = population size = 671 producers, *FIGURE*^{2} = 1.96 (value of *FIGURE* for a 95% confidence level), *Figure* = 1 (maximum admissible error), *Figure*^{2} = 6.62 (variance previously obtained based on the farmer price variable). The sample (*Figure*) obtained was of 47 melon producers to whom the questionnaires corresponding to the production stage were applied.

As for the packers, a sample of eight of them was interviewed during visits to their operation centers. To determine the population of packers, the Municipal Presidency of the city of Matamoros was contacted to request a registry since in that municipality a harvest mobility tax is collected (transport guide). According to the information obtained, there is a registry of 13 formal packers, which is why 67% of the population was covered. The questionnaire was made up of 16 questions, with which information was obtained on the period of the year in which the packaging operates, quantities of melon packed, losses in packaging and transportation, main destinations of the melon, uses and causes of losses, among others.

2.3 RESULTS

2.3.1 *CHARACTERISTICS OF THE PRODUCERS AND OWNERS OF PACKAGING*

Cantaloupe melon producers in the Comarca Lagunera are small-scale with surfaces of between 1 and 2 hectares, elderly and with little education. Packaging owners, by contrast, are slightly younger and more educated.

Producers generally do not have their own packaging. The intermediaries have packaging and are in charge of distributing the melon to the country's large consumption centers such as those in the cities of Monterrey, Mexico city, Guadalajara, León, San Luis Potosí, Querétaro and Puebla, among others.

The age of the producers is an important factor because it is related to the adoption of technologies and their ability to negotiate in the sale of their products. Nineteen percent of melon producers in the region are ≤40 years old, 51% are between 41 and 60 years old, and 30% are in the 61–80 years range. Regarding schooling, 19% have incomplete primary, 26% complete primary, and 32% with secondary education; this implies that 77% of the producers have at most a high-school education (nine years of schooling).

The age of the packaging owners indicates that 24% are between 21 and 40 years old, 63% between 41 and 60 years old, and 13% between 61 and 80 years old. The educational level of the melon packaging owners in the study region is as follows: 25% have completed elementary school, 50% finished high school, 0% high school, and 25% finished a bachelor's degree. Melon packaging operates in the region from the beginning of April when the harvest begins, until the early November, the month in which the cantaloupe melon harvest in the region ends. Depending on the size of the package, melon marketers send an average of 40 to 120 tons to consumption centers daily.

2.3.2 QUANTIFICATION OF LOSSES, CAUSES, AND USES OF CANTALOUPE MELON AT THE FARM LEVEL

Figure 2.1 shows the melon losses in the farm for the early, intermediate, and late planting stages. By losses at this stage, we understand that the producer himself discards in the farm is aware that it will be rejected and no longer takes it to the packaging. A higher percentage of waste (17%) is seen in late plantings, a period in which production conditions are unfavorable for the melon plant due to high temperatures and rains during August and September, which generates a high incidence. of pests and diseases that reduce the quality of production. In the early and intermediate sowings the losses are lower with values of 10–11%.

These data are lower than the losses found in the state of Jalisco, Mexico by[8] in tomato (24.64%) and similar to the avocado crop (15.15%). Without considering the tomato garden management, perhaps the highest percentages of losses are due to its greater degree of perishability compared to the cantaloupe melon. In the same way, the results in melon in the study region are similar to those found in the chayote (13.7%) and mango (10%) crops in the state of Veracruz, Mexico, but lower than the manzano pepper (30%).[9] The melon waste data found in this study are lower than the 55% found in Latin America for fruits and vegetables.[5] However, the results are not comparable because the FAO study covers distribution and consumption, while in the case of melon, only the production, packing, and transportation stages were studied.

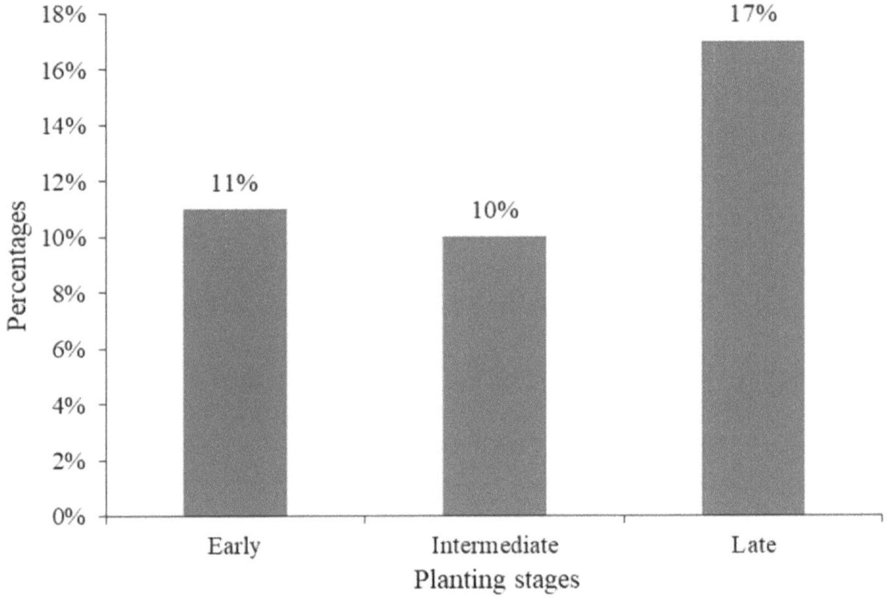

FIGURE 2.1 Average percentage of cantaloupe melon losses in the farm by planting stage.

The characteristics for which the producer himself discards the melon are diverse (Table 2.1), but the inadequate size of the fruit (very small or very large) (26%), fruit with cracks (21%), spots (17%), and fruit deformities (16%) stand out. The most popular fruits are sizes between 23 and 36 (1.8–1.2 kg per melon), so larger or smaller fruits have little demand. The melon discarded in the farm is used as organic fertilizer and/or as food for domestic animals, mainly pigs, cattle, and goats. The results found by Sipra et al.[8] in tomato and avocado crops and Del Angel et al.[9] in chayote, mango and chile also refer to physical defects in the fruit caused by climatic factors, pests, and diseases as the causative factors of the losses of the fruits.

TABLE 2.1 Cantaloupe Melon Characteristics Why Producers Discard It

Characteristics	Percentage
Misshapen fruit	16
Stains on the fruits due to contact with moisture	17
Fruit with cracks	21
Stains on fruits due to pest and disease damage	8
Inappropriate size	26
Excess ripening	9
Smooth (Lack of mesh)	3

2.3.3 QUANTIFICATION OF LOSSES, CAUSES, AND USES OF CANTALOUPE MELON AT THE PACKAGING LEVEL

Figure 2.2 shows the percentages of melon rejection in the packages by planting stage. As in the case of losses in the farm, in the case of packaging, it is in late plantings where the greatest rejections are recorded due to the lower quality of the fruit, reaching 19%. These rejection percentages in the different sowing stages are averages; however, when the melon goes to retail chain stores, the percentages reach up to 30%.[21] In the early sowing stage, the rejection percentage is lower (8%) since production is low in relation to demand.

Regarding the characteristics for which the melon is rejected in the packaging (Table 2.2), the inadequate size of the fruit (22%), melon with pitting and cracks (21%), and excess ripening (17%) stand out. The causes of rejection in the packaging are very similar to those that occur in the farm, that is, they are physical characteristics of the fruit. Neither in the packaging nor in the garden tests are carried out, for example, to evaluate the amount of sugar. The melon rejected in the packaging is sold to other small intermediaries called "pachangueros" at very low prices and is sold in plastic bags in the most important cities of the region such as Torreón, Gómez Palacio, Lerdo, Francisco I. Madero, Matamoros, and San Pedro.

2.3.4 QUANTIFICATION OF LOSSES, CAUSES, AND USES OF MELON DURING TRANSPORT

The transport of melon to the destination markets is carried out mainly in two types of transport: (a) truck with a thermoking-type refrigerator with a capacity of 20–25 tons where the interior temperature is around 7°C and (b) thorton-type truck does not refrigerate with a capacity of 12–15 tons. In the latter case, the melon is regularly transported in bulk, that is, unpacked, and is covered at the top with a layer of crushed ice approximately 20 cm thick. In the thorton-type truck, the temperature is irregular since the melons at the top are colder compared to the bottom ones due to their proximity to the ice sheet. The journey to the destination markets is an average of 650 kilometers with distances ranging from 400 km as is the city of Monterrey, to 1200 km as is the case of the city of Puebla. Figure 2.3 shows the percentage of melon losses in the two types of transport: In the truck with a refrigerator the losses are an average of 1.75% of the load, while in the truck with ice it is 2.2%. The melon that arrives in poor condition at the wholesale market is collected by trucks, along with other fruits and vegetables, to be used for animal consumption, as well as for the production of compost for agricultural uses.

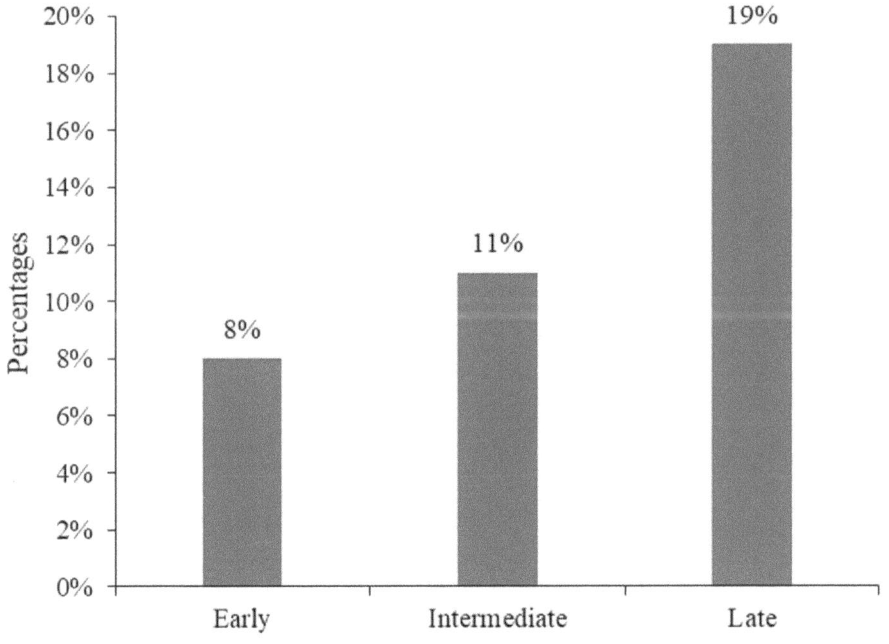

FIGURE 2.2 Percentage of cantaloupe melon by planting stages that packaging rejects to farmers.

TABLE 2.2 Reasons and Characteristics of Why Packers Rejects Melons

Characteristics	Percentage
Misshapen fruit	9
Stains on the fruits due to contact with moisture	9
Chopped and cracked fruit	21
Stains on fruits due to pest and disease damage	13
Inappropriate size	22
Excess ripening	17
Soft	9
Total	100

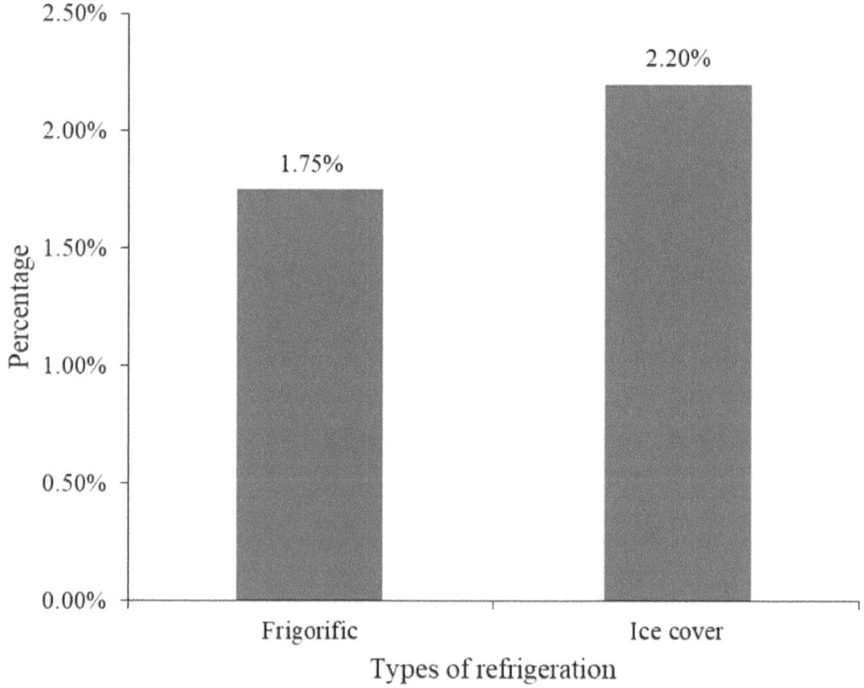

FIGURE 2.3 Average cantaloupe melon losses during transport to destination market.

2.4 CONCLUSIONS

The main causes of cantaloupe melon losses in the Lagunera region are the physical aspects of the fruit such as range from spots and cracks to inappropriate sizes and excess ripening. Some of the reasons for these damages are related to climatic aspects. However, better crop management, from seed selection, adequate control of pests and diseases, irrigation, fertilization, and timely harvest can substantially reduce losses. The losses or rejections that the buyer makes in the producers' melon packaging are also greater in late plantings due to the lower quality of the product obtained on those dates. This lower quality comes from the orchard, so farmers must work to obtain a better quality product. In the case of transportation, the losses are low, on average 2%, so the substantial losses originate in the low quality of the product obtained from the orchard and, in particular, in the late plantings. However, in the latter case, the greater losses of the product are offset by a higher sale price since in late harvests, the supply is lower, so it is sold at higher prices. To obtain a greater quantity and quality of the product, greater producer training is recommended in the technological aspects of cantaloupe melon production. This training should include seed selection, irrigation, fertilization, pest and disease control, and timely harvest. Since small producers predominate, training must be subsidized by government entities. In addition, the cold chain infrastructure must be improved, particularly warehouses in the production and refrigerated transport areas to extend the shelf life of melon fruits.

KEYWORDS

- **cantaloupe melon losses**
- **melon waste**
- **early sowings**
- **late sowings**

REFERENCES

1. Fabi, C.; English, A. SDG 12.3.1: Global Food Loss Index. November, 2018, 54. http://www.fao.org/3/CA2640EN/ca2640en.pdf
2. FAO. Pérdidas y Desperdicio de Alimentos en el Mundo–Alcance, Causas y Prevención. In *Figure*. 2012. https://doi.org/10.3738/1982.2278.562
3. Parfitt, J.; Barthel, M.; MacNaughton, S. Food Waste Within Food Supply Chains: Quantification and Potential for Change to 2050. *Figure* **2010,** *Figure* (1554), 3065–3081. https://doi.org/10.1098/rstb.2010.0126

4. FAO. *Figure*. Recuperado de: http://www.fao.org/americas/noticias/ver/en/c/239392/
5. FAO. *Figure*. Boletín 4, 2017. Organización de Las Naciones Unidas Para La Alimentación y La Agricultura (FAO), 44 p. Retrieved from: http://www.fao.org/3/a-i5504s.pdf
6. Chen, C. R.; Chen, R. J. C. Using Two Government Food Waste Recognition Programs to Understand Current Reducing Food Loss and Waste Activities in the U.S. *Figure* **2018,** *Figure* (8), 1–23. https://doi.org/10.3390/su10082760
7. European Commission. *Figure*, 1–11. Retrieved from http://ec.europa.eu/environment/archives/eussd/pdf/IA_Summary.PDF
8. SIPRA. *Figure*. 2019, Retrieved from: https://transparencia.info.jalisco.gob.mx/sites/default/files/3. DiagnosticoPerdidadDesperdiciosAlimentosJalisco.pdf
9. Del Angel, O. A.; Morales, E.; Castillo, R. I.; Luna, G.; Ramírez, A. Análisis de Pérdidas y y Desperdicios de Alimentos Producidos en el Estado de Veracruz. In Aguilar, G. (Coord.) *Figure* Instituto Politécnico Nacional-CONACYT-Miguel Ángel Porrúa Ed., México, 2019; p 225. ISBN 978-607-524-297-2
10. Instituto Nacional de Estadística y Geografía-INEGI. *Figure* 2017. [Diapositivas de PowerPoint]. Retrieved from: https://www.inegi.org.mx/contenidos/programas/ena/2017/doc/ena2017_pres.pdf.
11. Tiscornia, R. Julio. *Figure*. Buenos Aires: Editorial Albatros, 1979; p 146.
12. Tamaro, D. *Figure*. Madrid, España: Editorial Gustavo Gili, S.L., 1977; p 510. ISBN 10: 8425200598.
13. FAO. *Figure*. Rome; 2020. https://doi.org/10.4060/cb1329en.
14. SIAP, Sistema de Información Agroalimentaria y pesquera. *Figure*. Retrieved from: https://www.gob.mx/siap/. Fecha de consulta: 18 de abril de 2021.
15. SADER Región Lagunera, *Figure Figure*. Disponible en: https://www.gob.mx/sagarpa/regionlagunera. Fecha de consulta: 26 de marzo de 2021.
16. FAO. Pérdidas y Desperdicio de Alimentos En El Mundo–Alcance, Causas y Prevención. In *Figure* 2012; http://www.fao.org/3/i2697s/i2697s.pdf
17. Orona Castillo, I.; Espinoza Arellano, J. J.; González Cervantes, G.; Murillo Amador, B.; García Hernández, J.; Santamaría César, J. Aspectos Técnicos y Socioeconómicos de la Producción de Nuez (Carya illinoensis Koch.) En La Comarca Lagunera, México. *Figure* **2006,** *Figure* (3), 295–301.
18. Figueroa, V. U.; Núñez, G.; Sánchez, J. I.; López, H. E. Regional Nitrogen Balance in the Milk-forage Production System in the Comarca Lagunera, Mexico. *Figure* **2015,** 6 (4), 377–392. https://doi.org/10.22319/rmcp.v6i4.4099
19. Hanson, C.; Lipinski, B.; Robertson, K.; Dias, D.; Gavilan, I.; Gréverath, P.; Ritter, S.; Fonseca, J.; van Otterdijk; R., Dawe; A., Berger; V., Reddy; M., Tran, B., Leach, B. *Figure*. Food and Agriculture Organization of the United Nations, 2016; p 86.
20. Lind, D. A.; William G. M.; Mason, R. *Figure*. Alfaomega Ed. 11a edición, México, D.F. 2004; p 830.
21. Ávila, R. Gerente Técnico Del Empaque de Melón de la "Unión de Productores de Hortalizas del Municipio de Viesca, Coah.", personnel comunication del 12 de Julio de 2017.

CHAPTER 3

Biotechnological Potential of Nejayote: A Residue of the Tortilla Industry

L. X. LÓPEZ-MARTÍNEZ[1], J. A. NORIEGA-RODRÍGUEZ[2], HUGO S. GARCÍA[3], J. J. BUENROSTRO-FIGUEROA[4], and R. BAEZA-JIMÉNEZ[4]

[1]Laboratorio de Antioxidantes y Alimentos Funcionales, CONACYT-Centro de Investigación en Alimentación y Desarrollo, Sonora, México

[2]Laboratorio de Biotecnología, Departamento de Ingeniería Química y Metalurgia, Universidad de Sonora, Blvd. Luis Encinas y Rosales s/n. Col. Centro, Sonora, México

[3]UNIDA, Tecnológico Nacional de México/Instituto Tecnológico de Veracruz, Veracruz, México

[4]Laboratorio de Biotecnología y Bioingeniería, Centro de Investigación en Alimentación y Desarrollo, Delicias, Chihuahua, México

ABSTRACT

Talking about maize in Mexico means history, tradition, legacy, economic, political, and social organization, and above all, identity. Mexico is recognized as the center of origin and diversification of maize. In addition to its economic importance and varied industrial applications, maize constitutes the basis of the diet of the indigenous communities of Latin America, being tortillas as its main form of consumption. To obtain tortillas, maize goes through a nixtamalization process in which at least 3 m^3 of wastes are generated for each processed ton of corn. Usually, these wastes are disposed of through the municipal drainage system without

Food Byproducts Management and their Utilization. Ricardo Gómez-García, Ana A. Vilas-Boas, Débora A. Campos, Maria Manuela Pintado, & Cristóbal Noé Aguilar (Eds.)

any treatment or management. For this reason, it is necessary to develop bioprocesses that allow handling and revaluation of this residue, nejayote, because it contains grain-specific residues such as polyphenols, sugars, gums, and calcium components, which could have important applications for the food and pharmaceutical industries.

3.1 INTRODUCTION

Corn (*Zea mays*) is one of the most important cereals around the world. It constitutes a nutritional source for humans (white corn), animals (yellow corn), and is a versatile raw material in food industry (yellow corn). In Mexico, corn importance is related not only to economic aspects, but also to social and cultural aspects, as well as tradition and identity.

Corn belongs to the Gramineae family. It is a domesticated and highly productive crop, and is completely dependent on human care. Its origin occurred in the central region of Mexico through a domestication process. It is estimated that corn was cultivated approximately 10,000 years BC, with the oldest evidence of 6250 years ago in the Mexican state of Oaxaca. Native Mesoamerican civilizations named corn as "*tlaolli*" which means "*our sustenance*." Nowadays, its relevance and importance in Mexico can be noted in the 7.4 million hectares cultivated, with an annual yield of 27,000 tons.

The genetic diversity of the corn cultivated in Mexico can be appreciated in both the different shapes and sizes of the ears, as well as in the variety of textures and colors of its grains. Among the 64 varieties existing in Mexico, their range of colors include red, black, and blue and the highest production corresponds to white and yellow colors.[5,52,54]

Corn has been used to prepare tortillas since our pre-Hispanic civilizations. Today, tortilla is the main product obtained; however, other derived traditional foods can also be prepared, namely, atole, chimole, corundas, panuchos, chilaquiles, esquites, peneques, garnachas, molotes, pinole, quesadillas, sopes, tacos, tamales, tlayudas; and the festive delight, pozole, in the national holidays. Additionally, it is worth noting the applications of corn as an essential ingredient in food industry for the production of syrup, oil, flour or starch, protein, alcoholic beverages, and food sweeteners. Recently, corn has been employed in the biofuels area to obtain bioethanol.[22]

Corn is submitted to a nixtamalization process to prepare tortillas, where two products are obtained: Nixtamal and nejayote. The latter is the resulting cooking water which still contains some nutritional and bioactive compounds present in corn. However, the nixtamalization process occurs under certain conditions that confer physicochemical properties that turn nejayote into a polluting potential residue due to the lack of regulations for its disposal and treatment and it is dumped in rivers, lakes, soils, and public sewer. In the present chapter, the biotechnological potential of nejayote is described in terms of its composition, the bioprocesses developed for its handling, and revaluation as well as the possible applications for the food and pharmaceutical industries.

3.2 IMPORTANCE OF THE TORTILLA

Corn is the most popular cereal crop globally, with 1.14 billion tons produced in 2018.[20] Tortilla is one of the leading products of corn representatives of Mexico and perhaps the most important food due to its caloric contribution at the diet.[53] The average daily tortilla consumption in the urban and rural areas of Mexico is 5 and 7 tortillas, which represents a caloric intake of 340 and 477 kcal, respectively, and translates into an annual tortilla consumption of 56.7 kg in urban areas and 80 kg in rural areas. To satisfy the production and consumption of tortillas for approximately 120 million people, more than 4 million tons of corn grains are used.[24]

According to the scenario described, the generation of residues is evident. In this sense, bioprocesses are needed to revalorize those residues or byproducts in order to avoid their disposal to the environment and then, this new way to obtain foods will be lined with the circular economy which claims novel processes for producing and consuming and on the same hand, it allows us to respond to environmental and social challenges and on the other hand, generates opportunities for economic development, employment and contributes to food security.

3.2.1 ELABORATION OF TORTILLA

Clean and mature maize kernels are lime-cooked in the water previously mixed with or calcium hydroxide or food-grade lime. Pigmented varieties such as red, blue are lime-cooked until the solution reaches 90°C for 5 min, whereas yellow or white corns 100°C, 20 min and soaking for 12 h. Then, the mixture is washed with water to remove the pericarp and excess calcium hydroxide or food-grade lime (nejayote). The mixture that remains after washing is the nixtamal. The fresh nixtamal is ground using a stone mill into a fine dough (masa), which formed into discs and is baked on a comal (a smooth and flat griddle used in traditional cuisine, which is made with clay or iron and it is of round or elongated oval shapes) which is heated to about 220°C for 1.5 min on its surface and then cooled at room temperature to obtain the tortillas.[29]

3.2.2 NIXTAMALIZATION

Nixtamalization is the primary step during the production of some maize products such as tortillas, tostadas, tortilla chips[35] and consists basically of two main steps: Cooking and steeping.

In traditional nixtamalization, the concentration of calcium hydroxide [$Ca(OH)_2$] varies based on kernel's weight; the most common concentration used is 1% (w/w). The ratio grain:water is between 1:0.9 and 1:4 (w/w). Cooking is done at temperatures higher than 70°C during 5–100 min. The grains are usually allowed to steep in the cooking solution for 1–12 h. The product, known as "nixtamal," is rinsed to eliminate the excess of lime or $Ca(OH)_2$ and pericarp producing "nejayote" which is rich in some bioactive compounds, including vitamins, phenolics compounds, carotenoids sugars, gums, and calcium components, that are leached out during cooking.[7] The nixtamal is milled to form masa (dough), which is used to prepare tortillas, tostadas, tortilla chips, taco shells, tamales, and other dishes.

3.2.2.1 CHANGES/ALTERNATIVES

The traditional nixtamalization process and the effects on tortillas' nutritional, bioactive, and chemical characteristics have been extensively studied. But new food production technologies and trends provide more importance to nutritional and bioactive compounds and environmental problems. Besides traditional nixtamalization, several alternative processes to produce tortillas have been reported.

More ecological calcium sources have been recommended for use in nixtamalization, such as calcium carbonate ($CaCO_3$), calcium chloride ($CaCl_2$), calcium lactate ($C_6H_{10}CaO_6$), or calcium sulfate ($CaSO_4$), as well as weak acids, including propionic and acetic acids.[41] Besides, some alternative processes have been studied, such as extrusion,[15] ultrasound-assisted nixtamalization,[46] infrared heating,[61] as well as the other processes listed in Table 3.1.

It is important to mention that even when these processes present the clear advantage of being ecofriendly, they also present some disadvantages to consider; therefore, their applications are still limited (see Table 3.1).

3.3 NEJAYOTE

Nejayote is a byproduct that remains after a nixtamalization process, considered as wastewater highly polluting due to its high chemical content and biological oxygen demand (BOD).[10] According to Gutiérrez-Uribe et al.,[26] around 1.2. million m^3 of nejayote are generated monthly in Mexico. This effluent has a $pH > 10$ and contains a remarkable amount of organic soluble and insoluble solids such as polysaccharides, starch, fiber, lipids, proteins, as well as some bioactive compounds, including polyphenols and carotenoids, namely, lutein and zeaxanthin.[14,58]

3.3.1 COMPOSITION

3.3.1.1 PHYSICOCHEMICAL CHARACTERIZATION

Some physicochemical, nutritional, and wastewater properties of nejayote are shown in Table 3.2. These properties result from the natural components present in the maize, which are dragged into the water during the nixtamalization process. These properties are directly related to the parameters of the process, such as type of corn, endosperm hardness, grain quality, type and concentration of lime, cooking and soaking time and temperature, handling of corn during washing and transportation, equipment, and process practices.[49,50] Nejayote is dumped into water (rivers or lakes), soils, and public sewer system,[17] constituting an important source of water pollution that impacts in several aspects, namely, economic, environmental, nutritional, and health.[41] The treatment of nejayote helps to decrease pollution and also allows to obtain several compounds with high-added value and potential uses in technological applications (see section 3.3): Enzymes, bioactive compounds, biofuels, gums, among others.[8,19,23,47,60] The presence of arabinose, xylose, glucose, galactose, and *D*-glucuronic acid has been reported, as well as an important fraction of dietary fiber, which can be used as a prebiotic ingredient.[43] According to García-Depraect et al.,[23] the BOD/COD (chemical oxygen demand) ratio of nejayote is 0.6, which suggests that the organic matter of this wastewater can be biodegraded.

TABLE 3.1 Characteristics of Different Nixtamalization Alternative Processes, Advantages, and Disadvantages to Produce Masa and Tortillas.

Type of nixtamalization	Processing specifications	Advantages	Disadvantages	References
Extrusion	Extrusor (single or twin screw), calcium hydroxide 0.3–1.5% (w/w). Corn grifts or flour are conditioned with water and calcium salts, passed through different heating zones and crew speeds to obtain flour.	Reduction of use of water until 90–95%.	The texture of the tortilla is modified by the effect on the starch granules.	15
Ultrasound-assisted or bathing heat	Stone grinder, calcium hydroxide 1.2% (w/v) ultrasonic unit or bath, A stainless-steel transducer tank is set at 80°C, and the nixtamal is cooked with an ultrasound frequency of 25–42 kHz, varying the power and intensity.	Reduction of processing time.	NR	46
Microwave	Microwave oven, calcium hydroxide (0–2%). Whole maize is mixed with water (70–100%), cooking times from 10 to 20 min. 800 W and 2.450 MHz of power. The obtained flours are rehydrated to generate masa.	Cooking time is reduced by 50%.	NR	30
Ohmic heating	Ohmic cooker, calcium hydroxide 0.3%. Milled corn is mixed with calcium salts and water (50–60%), screw speed (10–20 rpm), 75–85°C and the power of 1 W/g.	Decrease in dough-making time. Nejayote production decreases.	High energy consumption. Expensive equipment.	33
Infrared heating	Convection infrared oven, stone grinder, and temperature of 85°C.	Reduction of aflatoxin contents.	No reduction of water use.	61
Enzymatic	Stone grinder, 0.1% of alkaline protease. Calcium hydroxide. Kernels are cooked in water at time and temperature adjusted depending on the kernel hardness. The grain is digested in a 0.1% solution of alkaline protease (60°C), pH is adjusted with calcium salts. After 4 h, corn kernels are washed and ground. The material is dehydrated and re-ground to obtain flour.	Allows the recycling of water.	The enzymes could be expensive.	27

TABLE 3.2 Physicochemical, Nutritional, and Wastewater Properties in Nejayote

Parameter	Range values
Physical and chemical	
pH	9–13.48
Total suspended solids (mg/L)	1810–25,000
°Brix	1.53
Moisture (%)	97.72
Volatile solids (%)	0.24–0.84
Density (Kg/m^3)	1030.85–1047.32
Turbidity (NTU)	538–1072
Conductivity (μS/cm)	2424–4510
Hardness (mg/L)	5768.67
Nutritional	
Ash (%)	0.767
Carbohydrate (%)	0.862
Crude protein (%)	0.113–7.42
Fat (%)	0.008–1.48
Fibre (%)	0.581
Total polyphenols (mg/L)	1190.74
Calcium (mg/L)	1526.21
Ammonia nitrogen (mg/L)	2–4.65
Phosphate (mg/L)	7.6–58.75
Sulfate (mg/L)	13–22.50
Waste water properties	
Total organic carbon (TOC, mg/L)	2984–9836
Total nitrogen (mg/L)	95.33–428
Biochemical oxygen demand (BOD, mg O_2/L)	190–7875
Chemical oxygen demand (COD, mg O_2/L)	3430–40,058

Adapted from: Castro-Muñoz et al.;[10] Castro-Muñoz et al.;[12] Castro-Muñoz; Yáñez-Fernández;[14] Díaz-Montes et al.;[17] España-Gamboa et al.;[19] López-Maldonado et al.;[28] Meraz et al.;[34] Ramírez-Araujo et al.;[41] Valderrama-Bravo et al.;[58] Valero et al.[60]

3.3.1.2 BIOACTIVE CHARACTERIZATION

During the nixtamalization process, the pericarp removal releases valuable compounds, such as arabinoxylans (AX) and phenolic acids, which possess antioxidant activity.[42] The phytochemical profile of nejayote depends on the source of extraction and the maize genotypes. The phytochemical composition of nejayote solids obtained from six different types of kernels (white, yellow, white quality protein, red, high-carotenoid, and blue) was reported by Gutiérrez-Uribe et al.,[26] with free and bound ferulic acid as the main component. The authors also reported that 91% of the phenolic associated with the nejayote were bound, with range values of 84.95–155.19 and 1349–3539 mg ferulic acid/100 g of dry matter for free and bound ferulic acid, respectively.

Concerning the kernel variety, the content of bound ferulic acid was: White quality protein > blue > yellow > red > high carotenoid > white. Compared to kernels and masa, the content of phenolic acids in nejayote was 2-folds higher. As a consequence of nixtamalization, the content of ferulic acid increases up to 125- and 10-fold higher free ferulic acid and 15- and 53-fold higher bound ferulic acid than those values obtained in kernel and masa, respectively. These values are directly correlated with the antioxidant capacity values, reporting 40- and 8-fold higher free, and 191- and 61-fold higher bound antioxidant capacities than those obtained in kernels and masa, respectively. In other work, Ayala-Soto et al.[6] evaluated the total phenolic content, where the hydroxycinnamic acids profile and the antioxidant capacities of AX were extracted from nejayote. The total phenolic content of 195.51 mg gallic acid equivalent (GAE)/100 g dry AX extract was found, where *p*-coumaric acid (0.108 μg/mg dry AX extract), ferulic acid (0.75 μg/mg dry AX extract), dihydrodiferulic acid (0.15 μg equivalents of ferulic acid/mg dry AX extract), and dehydrotriferulic acid (0.04 μg equivalents of ferulic acid/mg dry AX extract) were the main hydroxycinnamic acids observed. Extract of AX showed an antioxidant activity of 6.58 mM TE/100 g dry extract. The antioxidant capacity of nejayote AX was greatly affected by the concentration of dehydrotriferulic and *p*-coumaric acids. These studies concluded that nejayote is a promising source of valuable compounds that have been proven to exert health benefits (for instance, against oxidative stress), and on the same time, this use can be helpful for reducing its BOD level.

3.3.2 TREATMENT

Nejayote is rarely used as an animal supplement and discharged to the drain without any treatment, containing many organic compounds and wasted nutrients.[12] Several processes have been evaluated to minimize the effluents to revalorize this residue through biological (both liquid and solid fermentations) and physicochemical (adsorption, gelification, and precipitation) treatments. Regarding the treatment, steps such as decantation, centrifugation, and filtration are necessary to recover the maximal number of value-added compounds of industrial interest.

3.3.2.1 FERMENTATION

Fermentation processes have been conducted to evaluate the biodegradation of the organic matter present in nejayote using bioreactors, and its effect on removal efficiency. The use of a single-stream multi-stage bioreactor under aerobic conditions allowed reducing 86.4 and 80.9%, the levels of COD and dissolved organic carbon, respectively.[51]

In anaerobic conditions, nejayote has been used for bioenergy generation. Rapid two-stage digestion of nejayote was performed, with values of COD removed of 62.7%.[60] Higher removal values of COD (96%) and biogas generation with 84% methane were obtained using two-phase anaerobic digestion in a coupled system (Anaerobic-packed column reactor with an up-flow anaerobic sludge blanket) to optimize the acidogenic phase and increase the methanogenic process.[19]

Fermentation processes can be employed to produce several metabolites from nejayote as a substrate. Ramírez et al.[44] evaluated the use of *Bacillus claussi* on the antioxidant and antimicrobial activities of nejayote. After a culture of 4.8 days at 36°C, values of total phenolic content of 524 mg GAE/100 mL and antioxidant activity of 2078 μmol GA/100 mL were determined. This bioprocess allowed to increase 150% of the total phenolic content, 30% of the antioxidant activity, as well as to increase the content of several compounds, such as ferulic acid and its derivates (8-O-4-hydrodiferulic, 8-O-4-hydrotriferulic), and synthesized caffeic acid when compared to the nonfermented material (control). Fermented nejayote was successfully used to inhibit up to 3.5- and 2.72-fold the growth of *Salmonella typhimurium* and *Enterococcus faecalis*, compared to nonfermented nejayote. Solid-state fermentation (SSF) can be considered a bioeconomic process suitable for the application and transformation of nejayote, aiming to add value to this abundant byproduct, converting it into a renewable and low-cost source of bioactive compounds. It has been reported the use of SSF of nejayote solids with four different strains of edible fungi, including *Aspergillus oryzae*, *Pleurotus ostreatus* (Pearl and Blue), and *Hericium erinaceus*.[2] The authors reported an increase of free phenolic compounds (31–325%), antioxidant activity (2-fold), and soluble dietary fiber (45.49%) after three days of culture. Results indicated that fermented nejayote could be used as an antioxidant and antimicrobial agent incorporated into different food products.

Xylanase, arabinofuranosidase, and acetyl esterase enzymes have been detected in fermented nejayote by *Streptococcus infantarius* isolated from pozol.[16] García-Zamora et al.[25] applied laccase enzyme in the presence of chitosan to reduce COD and ferulic acid content of nejayote. Under optimal conditions (pH 6, 35°C, 24 h, and 0.3025 nmol laccase), removal levels of 70 and 78% in phenolics and COD were obtained, respectively. Another study reported the use of nejayote combined with swine wastewater as a cost-effective culture medium for microalgae *Athrospira maxima* and *Chlorella vulgaris* growth, which induced the removal of polluting agents.[25]

3.3.2.2 PHYSICOCHEMICAL TREATMENTS

Filtration is the process where a solution passes through a porous membrane assisted or not assisted by an external force. The membrane retains solid particles, according to its porous size, allowing the pass of fluid. Variables, including density, viscosity, pH, solids content, and particle size, affect filtration yield. The filtration of nejayote has been evaluated by designing an intermittent filtration cell at constant pressure, using diatomaceous earth (diatomite) as a filter aid. Filtration of nejayote produces changes in the physicochemical composition of nejayote, reducing the contents of both organic and inorganic components.[58]

Microfiltration and ultrafiltration processes have been also applied to nejayote, with high retention values in turbidity (72.76 and 95.3%, respectively), total solid content (23.74 and 37%), carbohydrates (14.69 and 46.6%), and calcium (11.36 and 14.7%). On the same hand, important reductions in electrical conductivity (21.75%) and total suspended solids (15.14%) were observed after microfiltration.[12,13] Based on these studies, Castro-Muñoz and Yáñez-Fernández[14] developed an integrated membrane process to fractionate and revalorize nejayote, consisting in sequential microfiltration followed by ultrafiltration. Three valuable fractions were obtained: Enriched in carbohydrates, high content of calcium components and enriched in phenolic compounds. Fractions enriched in carbohydrates and phenolic compounds could be useful in food and pharmaceutical industries.

Other treatments employed for nejayote include the use of biodegradable biopolymers that facilitates the coagulation, gelification, precipitation, and flocculation steps to remove the organic matter. These eco-friendly processes are further submitted to centrifugation or decantation. The use of chitosan (3 g/L, pH 5.5) allowed an efficiency of 80% in turbidity removal, and a reduction of 46% in total solids content.[34] López-Maldonado et al.[28] attained removal values of 80% for total suspended solids and turbidity, using an optimal dose of 1250 mg chitosan/L at pH 5. The use of alginate and chitosan to reduce nejayote pollutants was reported by Vacio-Muro et al.[57] Alginate involves adsorption and gelification to trap suspended and dissolved compounds, and the authors achieved a COD removal of 61.43%. Chitosan works in adsorption–precipitation phenomena and allowed to obtain a COD removal of 59.74%. Sequential use improved the COD removal, where the higher value (70.19%) was obtained with alginate-chitosan, compared to the chitosan-alginate (67.21%), indicating that the resultant effluent after this treatment can be discharged to sewerage systems or used for agricultural irrigation, according to the maximum permissible values of total solid content defined in the Official Mexican Standard (NOM-002-SEMARNAT-1996 and NOM-001-SEMARNAT-1996). These treatment processes represent an excellent alternative to reduce COD levels in nejayote, purifying it before discharge, and benefits to the environment.

3.3.3 APPLICATIONS

A maize processing plant can produce between 2.5 and 3.5 m^3 of nejayote for each metric tonne of produced nixtamal per day. As has been mentioned, nejayote is considered an alkaline pollutant with high pH and organic load; however, for its bioactive content, which includes polyphenols, carbohydrates, sugars, gums, and calcium components, it can be suitable for food, pharmaceutical, and biotechnological applications (see Figure 3.1).

Figure 3.1 shows that corn is not only crucial for traditional Mexican food. Its nutritional and bioactive contents are relevant for the development of bioprocess to revalorize this residue and analytical protocols for the complete characterization and quantification of the components present in nejayote. This is important for the techno-economic feasibility of those bioprocesses at industrial level.

Some important applications of nejayote as a rich source of valued compounds are described below.

3.3.3.1 PROTEINS AND CALCIUM

Proteins and calcium provided by the solids from nejayote play a significant role in cattle nutrition, growth, and development. Evidence shows that the solids can be used as fodder in the feeding of livestock, including benefits not only to performance due to the weight gain of the animals during the rearing period besides to the environment.[62]

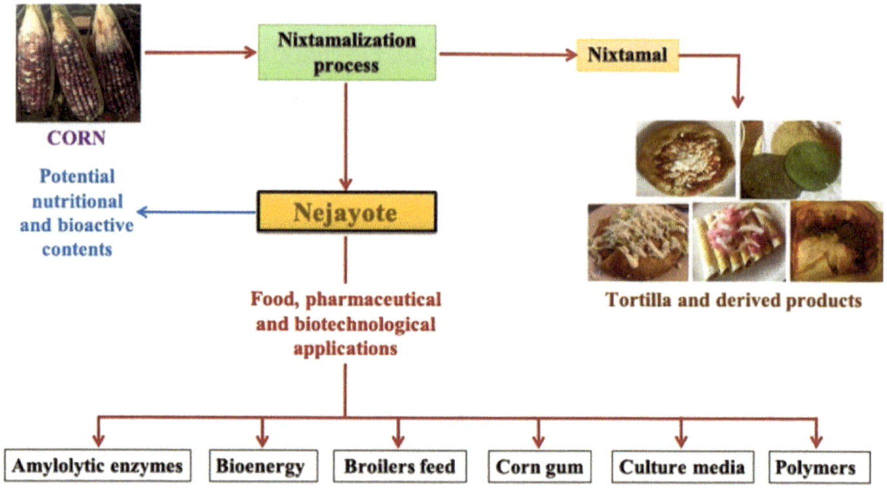

FIGURE 3.1 Food, pharmaceutical, and biotechnological applications of nejayote.

3.3.3.2 POLYMERS

Nejayote could be used as an agglutinant of different materials with applications in the wiring or insulator industries, an interesting application in the polymer industry. Carvajal-Millán et al.[9] patented the obtaining of corn gum from the nixtamalization residue, which includes the filtration and centrifugation of the extract, followed by the acidification of the supernatant. After that, precipitation occurs and then the gum was filtrated, washed, and dried. Based on that methodology, Niño-Medina et al.[37] characterized a gum with 75% of purity in AX (w/w), constituting of 47% xylose and 28% arabinose. In other studies, Niño-Medina et al.[36] and Paz-Samaniego et al.[40] extracted the nejayote-containing solids to determine the AX and polyphenol contents. They found that the yields are related to the quality of the raw materials (corn and alkali), as well as to the nixtamalization process conditions.

3.3.3.3 NUTRACEUTICALS

Diverse biological properties of nejayote are attributed to the presence of bioactive compounds. For instance, AX are polymers of xylose replaced by an arabinofuranosyl, usually present in grasses and cereals without cellulose.[32] These components are important for food industry as stabilizers, because they can confer viscosity, texture, and structure in bakery products.[31]

Other studies have cited that polyphenols content, as well as other compounds such as carotenoids or proteins, can vary depending on the variety employed and nixtamalization process conditions.[26,48]

With respect to solids, they can be incorporated into bread, as well as in other food products, as a source of dietary fiber, calcium, and antioxidants. Solids can also be useful in medicine to reduce cardiovascular diseases and to prevent some types of cancer due to the presence of ferulic acid.[1]

3.3.3.4 PHARMACEUTIC FORMULATIONS

Blanco-Gámez et al.[7] used nejayote as a culture medium, and according to their findings, the microorganisms isolated and characterized were *Bacilli* gram (+) with catalytic activity. At the same time, the results showed that the bacteria were very similar to *Bacillus megaterium*, which is of pharmaceutical importance as a penicillin G acylase producer.[38]

Buitimea Cantúa et al.[8] recovered from nejayote by ultrafiltration, enriched fractions in protein, phenolic acids, and feruloyl putrescine with anti-inflammatory activity due to the presence of proteins with molecular weight lower than 13.4 kDa, indicating the presence of small protein-phenolic aggregates or peptides that showed the capacity of inhibiting the action of nitric oxide. These aggregates remained solubilized in the aqueous effluent, suggesting the potential of nejayote in diverse applications such as pharmaceutical, cosmetical, and food applications to enhance their protein content phytochemicals and the anti-inflammatory activity.

3.3.3.5 BIOFUELS

Durán-de Bazúa et al.[18] used an aerobic system of low capacity (0.5 to 50 m^3/day) and an anaerobic–aerobic system for mills with larger capacity (2500 m^3/day) to treat the byproduct of nixtamal industries in order to obtain reusable water and to produce energy from the methane production. However, the authors cited that the treatment was barely profitable in small mills because high energy expenditure. In the higher capacity mill, methane production was from 9.6 to 16.8 m^3 per ton of maize, which was used to generate energy, 23 kg of solid residue, and 10.6 kg of biomass.

España-Gamboa et al.[19] enhanced the methanogenic process by means of a two-phase anaerobic digestion system, consisted of one anaerobic-packed column reactor (APCR) for the acidogenic phase optimisation and other up-flow anaerobic sludge blanket (UASB) reactor. This allowed a 96% COD removal after 19 days, yielding a biogas containing 84% methane. The yield reached for methane was 282 L/kg of COD removed.

3.3.3.6 CULTURE MEDIA

Salmerón-Alcocer et al.[51] treated nejayote with microorganisms (*Paenibacillus amylolyticus*, *Pseudomonas putida*, and *Acinetobacter* spp.) isolated from soil contaminated by the same effluent. A triple-cascade bioreactor system was employed and the resulting residue was used as culture medium. From their measurements, COD decreased from 28,000 to 3612 ppm after treatment, whereas the COD produced by providing concentrations of $(NH_4)_2SO_4$ (1 g/L) and KH_2PO_4 (0.3 g/L), to enrich the medium, decreased 87.3 and 87.6%, respectively. Furthermore, due to the nature of the isolated bacteria, they could be of interest to the oil industry since they are used to degrade aromatic hydrocarbons and their derivatives, namely, benzene, phenol, toluene, and xylene.[45]

Nejayote has also been employed as culture medium for lactic acid bacteria (*L. rhamnosus, L. casei, L. helveticus*) to produce bacteriocins and assayed against *E. coli* and *Listeria innocua.*[43] According to their results, *L. helveticus* exhibited the highest inhibitions at 13 and 10% against *L. innocua* and *E. coli*, respectively.

3.3.3.7 BIOFERTILIZER

Tellez et al.[55] assessed raw and treated nejayote (raw nejayote submitted to a coagulation-flocculation process) as foliar and edaphic fertilizers for native blue maize. The authors evaluated plant height, number of leaves, stem diameter, and grain yield. The response variables affected at the early stage of plant development and their effects were observed in the stem diameter and the number of leaves for foliar treatment with treated nejayote.

3.4 CONCLUDING REMARKS

The importance of corn goes beyond the nutritional and gastronomic sphere, and is inserted into the tradition, culture, and identity of Mexico. However, a residue is obtained during corn processing, which is a suitable resource for food, pharmaceutical, and biotechnological applications.

Indeed, there is still research work to do to develop feasible and low-cost technological platforms to process and manage nejayote for the recovery of its nutrient (fiber fractions, polysaccharides, starch, protein, lipids) and bioactive compounds (polyphenols, carotenoids, cellulose, hemicellulose, and germ). This is mandatory to avoid the loss of value-added components demanded by the population, and also to avoid the disposal of a residue that impacts the economic, environmental, health, and nutritional aspects.

KEYWORDS

- **corn**
- **nejayote**
- **treatment**
- **applications**

REFERENCES

1. Abbas, S. R.; Sabir, S. M.; Ahmad, S. D.; Boligon, A. A.; Athayde, M. L. Phenolic Profile, Antioxidant Potential and DNA Damage Protecting Activity of Sugarcane (*Saccharum Officinarum*). *Food Chem.* **2014,** *147*, 10–16.
2. Acosta-Estrada, B. A.; Villela-Castrejón, J.; Perez-Carrillo, E.; Gómez-Sánchez, C. E.; Gutiérrez-Uribe, J. A. Effects of Solid-state Fungi Fermentation on Phenolic Content, Antioxidant Properties and Fiber Composition of Lime Cooked Maize Byproduct (Nejayote). *J. Cereal Sci.* **2019,** *90*, 102837.
3. Acosta-Estrada, B. A.; Lazo-Vélez, M. A.; Nava-Valdez, Y.; Gutiérrez-Uribe, J. A.; Serna-Saldívar, S. O. Improvement of Dietary Fiber, Ferulic Acid and Calcium Contents in Pan Bread Enriched with Nejayote Food Additive from with Maize (*Zea Mays*). *J. Cereal Sci.* **2014,** *60*, 264–269.
4. Argun, M. S.; Argun, M. E. Treatment and Alternative Usage Possibilities of a Special Wastewater: Nejayote. *J. Food Process Eng.* **2018,** *41*, e12609.
5. ASERCA. 2018. *Agencia de Servicios a la Comercialización y Desarrollo de Mercados Agropecuarios*. https://www.gob.mx/aserca/articulos/conoces-el-origen-del-maiz?idiom=es (accessed May 1, 2021).
6. Ayala-Soto, F. E.; Serna-Saldívar, S. O.; García-Lara, S.; Pérez-Carrillo, E. Hydroxycinnamic Acids, Sugar Composition and Antioxidant Capacity of Arabinoxylans Extracted from Different Maize Fiber Sources. *Food Hydrocoll.* **2014,** *35*, 471–475.

7. Blanco-Gámez, E. A.; Sánchez-González, E. A.; Valladares, M. N.; Olvera, C. Identificación de Microorganismos Aislados Del Nejayote. *Revista Salud Pública y Nutrición* **2008,** *13*, 1–5.
8. Buitimea-Cantúa N.; Antunes-Ricardo M.; Gutiérrez-Uribe J. A.; Rocha-Pizaña M. R.; De la Rosa-Millán J.; Torres-Chávez P. Protein-phenolic Aggregates with Anti-inflammatory Activity Recovered from Maize Nixtamalization Wastewaters (Nejayote). *LWT-Food Sci. Technol.* **2020,** *134*, 1–10.
9. Carvajal-Millán, E.; Rascón-Chu, A.; Márquez-Escalante, J. A. Método Para la Obtención De Goma De Maíz a Partir Del Líquido Residual De La Nixtamalización Del Grano De Maíz. Mexican Patent PA/a/2005/008124
10. Castro-Muñoz, R.; Fíla, V.; Durán-Páramo, E. A Review of the Primary Byproduct (Nejayote) of the Nixtamalization During Maize Processing: Potential Reuses. *Waste Biomass Valori.* **2019,** *10*, 13–22.
11. Castro-Muñoz, R.; Barragán-Huerta, B. E.; Yáñez- Fernández, J. The Use of Nixtamalization Waste Waters Clarified by Ultrafiltration for Production of a Fraction Rich in Phenolic Compounds. *Waste Biomass Valori.* **2016,** *7*, 1167–1176.
12. Castro-Muñoz, R.; Orozco, C.; Ceron, G.; Yañez-Fernández, J. Characterization of the Microfiltration Process for the Treatment of Nixtamalization Wastewaters. *Ingeniería Agrícola y Biosistemas* **2015b,** *7*(1), 23–34.
13. Castro-Muñoz, R.; Cerón-Montes, G. I.; Barragán-Huerta, B. E.; Yáñez-Fernández, J. Recovery of Carbohydrates from Nixtamalization Wastewaters (Nejayote) by Ultrafiltration. *RMIQ* **2015a,** *14*, 735–744.
14. Castro-Muñoz, R.; Yáñez-Fernández, J. Valorization of Nixtamalization Wastewaters (Nejayote) by Integrated Membrane Process. *Food Bioprod. Process* **2015,** *95*, 7–18.
15. Contreras-Jiménez, B.; Gaytán-Martínez, M.; Morales-Sánchez, E.; Figueroa-Cardenas, J. D. D.; Pless, R.; González-Jasso, E.; …; Velazquez, G. Effects of Tempering Time, $Ca(OH)_2$ Concentration, and Particle Size on the Rheological Properties of Extruded Corn Flour. *Cereal Chem.* **2017,** *94*, 230–236.
16. Cooper-Bribiesca, B.; Navarro-Ocaña, A.; Díaz-Ruiz, G.; Aguilar-Osorio, G.; Rodríguez-Sanoja, R.; Wacher, C. Lactic Acid Fermentation of Arabinoxylan From Nejayote by *Streptococcus Infantarius* ssp. *Infantarius* 25124 Isolated From Pozol. *Front Microbiol.* **2018,** *9*, 3061–3061.
17. Díaz-Montes, E.; Castro-Muñoz, R.; Yáñez-Fernández, J. An Overview of Nejayote, A Nixtamalization Byproduct. *Ingeniería Agrícola y Biosistemas* **2016,** *8*, 41–60.
18. Durán-de Bazúa, C.; Sánchez-Tovar, S. A.; Hernández-Morales, M. R.; Bernal-González, M. (2007). Use of Anaerobic-Aerobic Treatment Systems for Maize Processing Installations: Applied Microbiology in Action. *Communicating Current Research and Educational Topics and Trends in Applied Microbiology*, pp 1–10. http://www.formatex.org/microbio/pdf/Pages3-12.pdf
19. España-Gamboa, E.; Domínguez-Maldonado, J. A.; Tapia-Tussell, R.; Chale-Canul, J. S.; Alzate-Gaviria, L. Corn Industrial Wastewater (Nejayote): A Promising Substrate in Mexico for Methane Production in a Coupled System (APCR-UASB). *Environ. Sci. Pollut. Res.* **2018,** *25*, 712–722.
20. FAO. El Futuro De La Alimentación y La Agricultura: Vías Alternativas Hacia El 2050. Versión Resumida. (accessed January 2021) 2018. Disponible en: http://www.fao.org/3/CA1553ES/ca1553es.pdf

21. Ferreira-Rolón, A.; Ramírez-Romero, G.; Ramírez-Vives, F. Aumento De La Actividad Metanogénica En Los Lodos Granulares, Precipitando Calcio En El Nejayote Mediante El Burbujeo de CO_2. *RMIQ* **2014,** *13*, 517–525.
22. FIRCO. *Fideicomiso de Riesgo Compartido*. 2017. https://www.gob.mx/firco/articulos/el-maiz-base-de-la-dieta-mexicana-desde-epoca-prehispanica (accessed May 1, 2021).
23. García-Depraect, O.; Gómez-Romero, J.; León-Becerril, E.; López-López, A. A Novel Biohydrogen Production Process: Co-digestion of Vinasse and Nejayote as Complex Raw Substrates Using a Robust Inoculum. *Int. J. Hydrog. Energy* **2017,** *42*, 5820–5831.
24. García-Lara, S.; Chuck-Hernandez, C.; Serna-Saldivar, S. O. Development and Structure of the Corn Kernel. *Corn* **2019,** 147–163.
25. García-Zamora, J. L.; Sánchez-González, M.; Lozano, J. A.; Jáuregui, J.; Zayas, T.; Santacruz, V.; Hernández, F.; Torres, E. Enzymatic Treatment of Wastewater from the Corn Tortilla Industry Using Chitosan As An Adsorbent Reduces the Chemical Oxygen Demand and Ferulic Acid Content. *Process Biochem.* **2015,** *50*, 125–133.
26. Gutiérrez-Uribe, J. A.; Rojas-García, C.; García-Lara, S.; Serna-Saldivar, S. O. Phytochemical Analysis of Wastewater (Nejayote) Obtained After Lime-cooking of Different Types of Maize Kernels Processed into Masa for Tortillas. *J. Cereal Sci.* **2010,** *52*, 410–416.
27. Jackson, D. S.; Sahai, D. *Enzymatic Process for Nixtamalization of Cereal Grains*. 2002.
28. López-Maldonado, E. A.; Oropeza-Guzmán, M. T.; Suárez-Meraz, K. A. Integral Use of Nejayote: Characterization, New Strategies for Physicochemical Treatment and Recovery of Valuable Byproducts. In *Physico-Chemical Wastewater Treatment and Resource Recovery*, Farooq, R.; Ahmad, Z. Eds.; IntechOpen; 2017.
29. Lopez-Martinez, L. X.; Parkin, K. L.; Garcia, H. S. Phase II-inducing, Polyphenols Content and Antioxidant Capacity of Corn (*Zea mays* L.) from Phenotypes of White, Blue, Red and Purple Colors Processed into Masa and Tortillas. *Plant Foods Hum. Nutr.* **2011,** *66*, 41–47.
30. Martínez-Bustos, F.; García, M. N.; Chang, Y. K.; Sánchez-Sinencio, F.; Figueroa, C. J. D. Characteristics of Nixtamalised Maize Flours Produced with the Use of Microwave Heating During Alkaline Cooking. *J. Sci. Food Agric.* **2000,** *80*, 651–656.
31. Matos-Chamorro, A.; Chambilla-Mamani, E. Importancia De La Fibra Dietética, Sus Propiedades Funcionales En La Alimentación Humana y En La Industria Alimentaria. *Revista de Investigación en Ciencia y Tecnología de Alimentos* **2010,** *1*(1).
32. Mendis, M.; Simsek, S. Arabinoxylans and Human Health. *Food Hydrocoll.* **2014,** *42*, 239–243.
33. Ménera-López, I.; Gaytán-Martínez, M.; Reyes-Vega, M. L.; Morales-Sanchez, E.; Figueroa, J. D. C. Physico-chemical Properties and Quality Assessment of Corn Flour Processed by a Continuous Ohmic Heating System and Traditional Nixtamalization. *CyTA-J. Food* **2013,** *11*, 8–14.
34. Meraz, K. A. S.; Vargas, S. M. P.; Maldonado, J. T. L.; Bravo, J. M. C.; Guzman, M. T. O.; Maldonado, E. A. L. Eco-friendly Innovation for Nejayote Coagulation–flocculation Process Using Chitosan: Evaluation Through Zeta Potential Measurements. *Chem. Eng. J.* **2016,** *284*, 536–542.
35. Milán-Carrillo, J.; Gutiérrez-Dorado, R.; Cuevas-Rodríguez, E. O.; Garzón-Tiznado, J. A.; Reyes-Moreno, C. Nixtamalized Flour from Quality Protein Maize (*Zea Mays* L). Optimization of Alkaline Processing. *Plant Foods Hum. Nutr.* **2004,** *59*, 35–44.

36. Niño-Medina, G.; Carvajal-Millán, E.; Lizardi, J.; Rascon-Chu, A.; Márquez-Escalante, J. A.; Gardea, A.; Martínez-López, A. L.; Guerrero, V. Maize Processing Waste Water Arabinoxylans: Gelling Capability and Cross-linking Content. *Food Chem.* **2009,** *115*, 1286–1290.
37. Niño-Medina, G.; Carvajal-Millán, E.; Gardea-Bejar, A.; Rascón-Chu, A.; Márquez-Escalante, J. A. Caracterización Composicional, Fisicoquímica y Funcional De Una Goma De Maíz Recuperada Del Nejayote. *XVII Congreso Nacional de Biotecnología y Bioingeniería*, 2007.
38. Nucci, E. R.; Silva, R. G.; Gomes, T. C.; Giordano, R. C.; Cruz, J. G. A Fuzzy Logic Algorithm for Identification of the Harvesting Threshold During PGA Production by *Bacillus Megaterium. Braz. J. Chem. Eng.* **2005,** *22*, 521–527.
39. Pappa, M. R.; de Palacios, P. P.; Bressani, R. Effect of Lime and Wood Ash on the Nixtamalization of Maize and Tortilla Chemical and Nutritional Characteristics. *Plant Foods Hum. Nutr.* **2010,** *65*, 130–135.
40. Paz-Samaniego, R.; Carvajal-Millan, E.; Brown-Bojorquez, F.; Rascón-Chu, A.; López-Franco, Y. L.; Sotelo-Cruz, N.; Lizardi-Mendoza, J. Gelation of Arabinoxylans from Maize Wastewater— Effect of Alkaline Hydrolysis Conditions on the Gel Rheology and Microstructure. In *Wastewater Treatment Engineering*, Samer, M., Ed.; IntechOpen, 2015, pp 101–114.
41. Ramírez-Araujo, H.; Gaytán-Martínez, M.; Reyes-Vega, M. L. Alternative Technologies to the Traditional Nixtamalization Process: Review. *Trends Food Sci. Technol.* **2019,** *85*, 34–43.
42. Ramírez-Jiménez, A. K.; Rangel-Hernández, J.; Morales-Sánchez, E.; Loarca-Piña, G.; Gaytán-Martínez, M. Changes on the Phytochemicals Profile of Instant Corn Flours Obtained by Traditional Nixtamalization and Ohmic Heating Process. *Food Chem.* **2019,** *276*, 57–62.
43. Ramírez-Romero, G.; Reyes-Velazquez, M.; Cruz-Guerrero, A. Estudio Del Nejayote Como Medio De Crecimiento De Probióticos y Producción De Bacteriocinas. *RMIQ* **2013,** *12*, 463–471.
44. Ramírez, K.; Quintero-Soto, M. F.; Rochín-Medina, J. J. Enhancement of the Antioxidant and Antimicrobial Activities of Maize Wastewater by an Eco-friendly Process. *J. Food Meas. Charact.* **2020,** *14*, 1682–1689.
45. Riser-Roberts, E. *Remediation of Petroleum Contaminated Soils: Biological, Physical and Chemical Processes*. Lewis Publishers: Boca Raton, Florida, 1998.
46. Robles-Ozuna, L. E.; Ochoa-Martínez, L. A.; Morales-Castro, J.; Gallegos-Infante, J. A.; Quintero-Ramos, A.; Madera-Santana, T. J. Effect of Nixtamalization Conditions Ultrasound Assisted on Some Physicochemical, Structural and Quality Characteristics in Maize Used for Pozole. *CyTA-J. Food.* **2016,** *14*, 324–332.
47. Rocha-Pizaña, M. d. R.; Chen, W. N.; Lee, J. J. L.; Buitimea-Cantúa, N. E.; González-Nimi, E.; Gutierrez-Uribe, J. A. Production of a Potential Collagenolytic Protease by Nejayote Fermentation with *Aspergillus Oryzae. Int. J. Food Sci. Technol.* **2020,** *55*, 3289–3296.
48. Rojas-García, C.; García-Lara, S.; Serna-Saldívar, S. O.; Gutiérrez-Uribe, J. A. Chemopreventive Effects of Free and Bound Phenolics Associated to Steep Waters (Nejayote) Obtained after Nixtamalization of Different Maize Types. *Plant Foods Hum. Nutr.* **2012,** *67*, 94–99.

49. Roque-Maciel, L.; Arámbula-Villa, G.; López-Espíndola, M.; Ortiz-Laurel, H.; Carballo-Carballo, A.; Herrera-Corredor, J. A. Nixtamalización De Cinco Variedades De Maíz Con Diferente Dureza De Grano: Impacto En Consumo De Combustible y Cambios Fisicoquímicos. *Agrociencia*. **2016,** *50*, 727–745.
50. Rosentrater, K. A. A Review of Corn Masa Processing Residues: Generation, Properties, and Potential Utilization. *Waste Manage*. **2006,** *26*, 284–292.
51. Salmerón-Alcocer, A.; Rodríguez-Mendoza, N.; Pineda-Santiago, V.; Cristiani-Urbina, E.; Juárez-Ramírez, C.; Ruiz-Ordaz, N.; Galíndez-Mayer, J. Aerobic Treatment of Maize-processing Wastewater (Nejayote) in a Single-stream Multi-stage Bioreactor. *J. Environ. Eng. Sci*. **2003,** *2*, 401–406.
52. SENASICA. *Servicio Nacional de Sanidad, Inocuidad y Calidad Agroalimentaria*. 2020. https://www.gob.mx/senasica/articulos/maiz-mexicano-sustento-ancestral (accessed May 1, 2021).
53. Serna-Saldivar, S. O. History of Corn and Wheat Tortillas. In *Tortillas*; AACC International Press. 2015, pp 1–28.
54. SIAP. 2018. Servicio de Información Agroalimentaria y Pesquera. https://www.gob.mx/siap/articulos/con-los-colores-del-maiz-mexico-se-pinta-solo. (accessed May 1, 2021).
55. Téllez, V.; López, J. F.; Aragón, A.; Zayas, T. Application of Nejayote as a Foliar and Edaphic Fertiliser to Native Blue Maize (*Zea mays* L.) Crops. *Am. J. Plant Sci*. **2016,** *7*, 2221–2238.
56. Tortora-Gerard, J.; Funke-Berdell, R.; Case-Christine, L. *Introducción a la Microbiología*. Editorial Médica Panamericana: Buenos Aires, Argentina, 2007.
57. Vacio-Muro, K. J.; Lozano-Álvarez, J. A.; Sánchez-González, M. N.; Chávez Vela, N. A.; Torres-Ramírez, E.; Jáuregui-Rincón, J. Remoción de Contaminantes Del Nejayote Con Alginato y Quitosano. *Rev. Int. Contam. Ambient*. **2020,** *36*, 497–515.
58. Valderrama-Bravo, C.; Gutiérrez-Cortez, E.; Contreras-Padilla, M.; Oaxaca-Luna, A.; Del Real López, A.; Espinosa-Arbelaez, D. G.; Rodríguez-García, M. E. Physico-mechanic Treatment of Nixtamalization Byproduct (Nejayote). *CyTA–J. Food* **2013,** *11*(Supp. 1), 75–83.
59. Valderrama-Bravo, C.; Gutiérrez-Cortez, E.; Contreras-Padilla, M.; Rojas-Molina, I.; Mosquera, J. C.; Rojas-Molina, A.; Beristain, F.; Rodríguez-García, M. E. Constant Pressure Filtration of Lime Water (Nejayote) Used to Cook Kernels in Maize Processing. *J. Food Eng*. **2012,** *110*(3), 478–486.
60. Valero, D., Rico, C.; Tapia-Tussell, R.; Alzate-Gaviria, L. Rapid Two Stage Anaerobic Digestion of Nejayote through Microaeration and Direct Interspecies Electron Transfer. *Processes* **2020,** *8*(12), 1614.
61. Vázquez-Durán, G. A. V. A.; Moreno-Martínez. Impact of Different Alkaline-heating Processes on Technological and Nutritional Properties of Maize Tortillas. *J. Food Nutr. Res*. **2014,** *53*, 60–70.
62. Velasco-Martínez, M.; Angulo, O.; Vázquez-Couturier, D. L.; Arroyo-Lara, A.; Monroy-Rivera, J. A. Effect of Dried Solids of Nejayote on Broiler Growth. *Poult. Sci*. **1997,** *76*, 1531–1534.
63. Voet, D.; Voet, J. G. *Bioquímica*. Editorial Médica Panamericana: Buenos Aires, Argentina, 2006.

CHAPTER 4

Byproducts of the Berry Juice Industry: Phytochemical Recovery and Valuable Applications

CLÁUDIA M. B. NEVES[1], ÉLIA FOGEIRO[1], ANA FERREIRA[1], FILIPA PEREIRA[1], and DULCINEIA F. WESSEL[1,2,3]

[1]*Polytechinc Institute of Viseu, School of Agriculture, Quinta da Alagoa–Estrada de Nelas, Ranhados, Viseu, Portugal*

[2]*CITAB, University of Trás-os-Montes e Alto Douro, Vila Real, Portugal*

[3]*LAQV-REQUIMTE, Department of Chemistry, University of Aveiro, Aveiro, Portugal*

ABSTRACT

The fruit juice industry is considered one of the largest markets in the agro-industrial sector; however, due to its massive fruit processing it generates a large quantity of byproducts, leading to environmental and economic issues. These food byproducts are currently used for thermal energy recovery, compost, and animal feed. Nevertheless, these byproducts contain bioactive compounds that can be used to obtain high added-value products.

Berry pomace, the byproduct resulting from berry juice production, encompasses the skins, pulp, seeds, and stems of the fruit. Each constituent is rich in distinct chemical compounds, for example, seeds are rich in lipophilic compounds such as polyunsaturated fatty acids, tocopherols, tocotrienols, sterols, and carotenoids, while the other parts are rich in cell wall dietary fiber, anthocyanins, and other polyphenols. These compounds present anti-oxidant, anti-inflammatory, and antimicrobial properties, and therefore can

Food Byproducts Management and their Utilization. Ricardo Gómez-García, Ana A. Vilas-Boas, Débora A. Campos, Maria Manuela Pintado, & Cristóbal Noé Aguilar (Eds.)

be used as food additives and as valuable ingredients for functional foods, nutraceuticals, pharmaceuticals, and other products.

In this chapter, we intend to compile scientific information about berry juice byproducts, their management practices, and current valorization procedures.

4.1 INTRODUCTION

The human health benefits of consuming fruits and vegetables are well established. There are several evidences of the relationship between a diet rich in this type of products and a low incidence of chronic diseases, such as cancer and cardiovascular diseases.[1,2] Among the different eaten fruits, berries have attracted the interest of consumers in recent years as they are considered one of the highest dietary sources of antioxidants and bioactive compounds.[3] Their high antioxidant power is intrinsically linked with the presence of phenol-type phytochemicals, such as flavonoids (particularly anthocyanins, flavonols, and flavanols), tannins, and phenolic acids.[4] In addition, they are a good source of nutritive components such as vitamins, sugars, unsaturated fats, dietary fibers, essential minerals, and organic acids.[5]

Berries are often consumed in fresh, however due to their short shelf life, and an increasingly busy society make their consumption in the form of juice very popular among consumers who are looking for readily available and healthy food products. With the success of functional products over the last decade, it is expected that the production of juices with health-beneficial properties, such as those based on berries, will continue to increase.[6]

Industrial berry juice production inevitably results in a huge amount of pomace that needs to be processed. This byproduct is often treated as waste and is simply discarded or used as fertilizer; however, direct disposal of pomace to soil brings environmental issues, because of its susceptibility to microbial growth.[7] Other practice commonly adopted for pomace mitigation includes their use as animal feed; however, its low protein content makes it unappealing for this purpose.[7,8] After extracting the juice, a part of the phytochemicals initially present in the berries remains in the pomace.[9] Thus, an efficient management of this byproduct may involve the recovery of these compounds that, due to their bioactivities, may result in products with greater economic value overcoming the currently applied solutions.[9,10]

This chapter will focus on the recovery of berry pomace as a source of high value-added components. It aims to provide comprehensive information on the chemical composition, bioactives, and the potential applications of this biowaste.

4.2 BERRY JUICE PROCESSING AND BYPRODUCT RECOVERY STRATEGIES

Fruit juice is defined as the extractable fluid content of cells and tissues of fruits. The edible part of fruits has cells with thin cell walls and vacuoles that store the cell sap, the main component of fruit juice.[11] Berry juices are rich in phenolic compounds, which are responsible for their intense color, and other sensorial qualities.[12] In addition, these compounds are associated with health-promoting properties, such as antioxidant, anticarcinogenic, anti-inflammatory, neuroprotective, antiobesity, antidiabetic, and cardiovascular disease prevention,[13] thus making berry juices popular among consumers.

Juices can legally have several designations depending on how they are prepared.[14] In the case of berry juices, depending on its processing, they can fit in one of the following categories: "fruit juice" if obtained directly from the edible part of the fruit; "concentrated fruit juice" if a concentration step takes place to remove a specific proportion of the water (at least 50%); "fruit juice from concentrate" when the concentrated fruit juice is diluted with water.[11,14]

The main stages of industrial berry juice production are shown schematically in Figure 4.1. Processing begins with the harvesting of the fruit, which can be done manually or mechanically. Then, the berries are sorted and washed to be used in juice production, which is initiated by the mashing of the fruit allowing the water and nutritional compounds to exit out of the cell.[12] The mash can be submitted to enzymatic treatments. Afterward, the mash is pressed to separate the liquid fraction (juice) from the solid fraction. To eliminate the fine haze that can be formed, some berry juices are subjected to clarification and/or filtration steps. When the aim is to obtain a concentrated fruit juice for easy conservation and storage, a concentration step occurs normally achieved by thermal treatment under reduced pressure.[12] Finally, to ensure microbial safety and acceptable shelf-life, juice can be pasteurized. During the juice production the various treatment steps influence the concentration and type of bioactive compounds extracted into the juice and also impact those that still remain in the byproducts.

Several processes can be applied before mashing, such as freezing and subsequent thawing or heating of the berries to facilitate the juice recovery from the berries.[9,12,15] When the berries are frozen, the water inside the cells expands and causes cell rupture that is necessary to the juice extraction. The cell degradation caused by freezing is also responsible for the liberation of valuable compounds, such as anthocyanins, allowing an increase of phenolic compounds in the berry juice. The conditions used during freezing and thawing influence the anthocyanin content of the juice. Long storage (10 months) at freezing temperatures (−18°C) caused losses in the anthocyanin content (13% for blueberries, 19% for blackberries, and 16% for raspberries).[12]

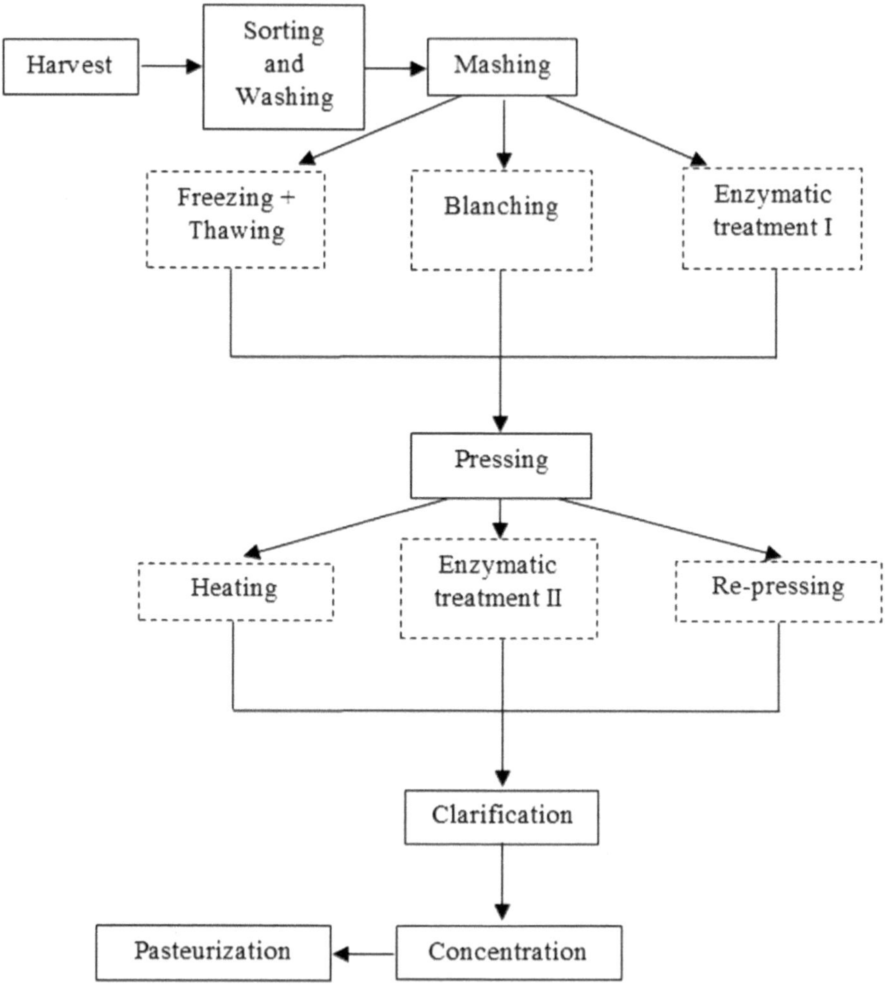

FIGURE 4.1 Berry juice processing flowchart.

The aim of heating the whole berries prior to mashing is to preserve and increase the content of valuable compounds through the disruption of cells and the inhibition of native enzymes, such as polyphenol oxidase, peroxidases, and glycosidases that are responsible for the phenolic compounds' degradation, especially anthocyanins.[16–18] By inactivating these enzymes, the valuable compounds are maintained and can be extracted, increasing its concentration in the juice. For blueberries, an increase of 11% in total anthocyanin content was reported due to the blanching of the fruit before the juice extraction.[12,19] Heating the berries or the mash can be beneficial to the extraction of compounds but can also have deleterious effects on the anthocyanin content due to the different thermal stability of different anthocyanins in various matrixes.[12]

The enzymatic treatments applied during berry juice processing aim to depolymerize pectin and other cell wall polysaccharides, facilitating the release of the juice. Cell wall-bound anthocyanins are also released, increasing their concentration in the juice extracted. The commercial enzymes applied often are polygalacturonase, pectin lyase, cellulose, and hemicellulose. Even though the application of these enzymes helps to improve the juice yield and to increase the phenolic compounds content in the juice, their use may change the anthocyanin profile due to enzymatic side activities.[20–22] The use of high dosage of commercial cell wall degrading enzymes (1000 nkat/g based on endopolygalacturonase activity) caused a total or pronounced decrease of anthocyanidin galactosides content in bilberry juice as a result of their high galactosidase activities.[21] Heffels et al.[22] compared the effect of a dosage level of 0.5 nkat/g commonly used in industrial processing with a considered excessive dosage (10 nkat/g) on juice yield and anthocyanin content and profile during bilberry juice processing. The common industrial dosage caused a small increase in juice yield and higher anthocyanin content, whereas the excessive dosage significantly improved both juice yield and total anthocyanin content. However, the changes in the anthocyanin profile were more pronounced when the excessive dosage was applied.

Since some steps in berry juice processing include temperature, nonthermal innovative processes are being studied as an alternative. These processes include high-pressure processing, pulsed electric fields, ultrasound, and flash vacuum expansion. These technologies generally use low temperatures or no temperature at all, leading to berry juices richer in anthocyanins and other valuable compounds due to their preservation and null degradation, while assuring the inactivation of damaging enzymes and guarantying microbiological and sensorial quality of these juices.[11,12]

The processing of berry fruits into juice leads to an unintentional production of berry byproducts, which should be valorized as raw materials rich in bioactive compounds with high potential industrial interest. When the juice is separated from the solid fraction during pressing step, a press cake or pomace is formed, consisting of skins, pulp, seeds, and stems. This accounts for around 20–30% of the berries used and constitutes a manufacturing waste that the agri-food sector must manage sustainably.[9] Since the berries are rich in bioactive compounds present in the skins, pulp, seeds, and stems of the fruit, a part of these valuable compounds is still retained in the berry pomace. The discard of this byproduct is costly and leads to the loss of many important and valuable compounds with health-promoting abilities. For instance, around 25–50% of procyanidins are trapped in blueberry pomace.[23] The same happens in the production of chokeberries juice, since around 30% of procyanidins remain in the pomace.[24] Also, from the anthocyanins present in blueberries and cranberries, 18%[25] and 13–17%[26] are not extracted and stay in the pomace, respectively.

A strategy to use berry pomace is its inclusion in animal feed supplementation to reduce antibiotics in animal production, while maintaining meat quality.[27] The use of cranberry pomace at 1% and 2% in broiler chicken feed improved the birds' immunity compared with a control feed by reducing the expression of inflammation-related genes in the liver and upregulating the expression of anti-inflammatory genes in the bursa.[27] Another study evaluated the use of sea buckthorn pomace as a supplement in lamb feed. It was concluded that the use of this pomace helped lamb meat production due to the increase in muscle mass and the improvement of meat tenderness, water retention, and antioxidant capacity.[28] However, due to the low digestible energy, the utilization of fruit pomace as animal feed is limited. Composting is also frequently applied to use berry juice byproducts. Still, the significant quantity of phytochemicals makes composting difficult due to their antimicrobial activity that inhibits beneficial microorganisms' growth.[29]

Other options that are gaining interest for berry pomace valorization are the conversion in berry pomace powders, which can be used directly as an ingredient in the production of new food products enriched in phenolic compounds and the extraction of bioactive compounds that can later be used in the food, nutraceutical, and cosmetic industries.

For the recovery of these valuable compounds the conventional solid-liquid extraction by maceration with shaking and the Soxhlet extraction, using different conditions and solvents, are the most frequently applied processes. These extraction methods are simple; however, the exhaustive recovery of phenolic compounds requires large amounts of solvents, which must be evaporated at the end and usually long times of extraction are required. The extraction time is dependent on the temperature used since the compounds are thermally sensitive. For temperatures above 60°C, extraction time should not surpass 1 h, but for extraction at room temperature this time can go up to 24 h.[9] These aspects limit the technological and economic viability of extraction upscaling for industrial pomace valorization. In addition, the solvent toxicity, the security, and the environmental impact of the process are other issues to be considered. Since the recovery of the compounds is often intended for food, nutraceutical, and personal care applications, water and food grade solvents are preferable to the common hazardous organic solvents. Other methods, such as CO_2 supercritical fluid extraction,[30,31] enhanced solvent extraction,[32,33] ultrasound-assisted extraction, and microwave-assisted extraction,[34–37] have gained importance for the recovery of added-value compounds from berry byproducts. Moreover, these technologies either are applied at room temperature or use mild conditions, allowing a better preservation of the phenolic compounds during their extraction from the berry pomace.[9]

For an effective utilization of the pomace an initial processing of the pomace must occur. Pomace contains high moisture percentage and water activity and therefore is prone to rapid microbial spoilage. Consequently, utilizing pomace requires, in most cases, the removal of the excess water until around 6% (w/w) of moisture, to allow a significant increase in its shelf life.[38] Several methods can be used to dry the pomace, such as hot-air convectional drying, low-temperature vacuum drying, freeze-drying, spray drying, infrared drying, or microwave drying.[9] The drying conditions should be studied for each pomace because each one has different compounds with different susceptibilities to temperature or drying conditions. For example, in the production of aronia pomace powder, conventional drying resulted in a powder with darker and more intense color than the ones obtained by freeze-drying or spray-drying.[39] Generally, temperatures of around 60°C should be used for drying, since it does not significantly affect the phenolic compounds present in the pomace.[40] A milling step should also be added to ensure that the particles are all about the same size, allowing a better consistency of the obtained berry pomace powder. This milling can also allow the breakdown of seeds, liberating valuable compounds and making them available.[9] Fractioning can also occur, mainly by sieving, to separate particles with different sizes. This can also be important to separate the berry seeds from the remaining powder that can then be used to extract oil.[9]

4.3 BERRY JUICE BYPRODUCTS AS A SOURCE OF NUTRIENTS AND BIOACTIVE COMPOUNDS

4.3.1 NUTRITIONAL COMPOSITION

Berry juice byproducts contain high amounts of dietary fiber, bioactive compounds, and nutritionally valuable substances.[29] These components make berry pomace a product liable to be valued as beneficial in terms of human nutrition.[29] Hence, knowledge of berry pomace chemical composition and functional properties facilitates its utilization in new value-added applications, namely as food additives, ingredients for functional foods, and nutraceuticals while increasing the overall profit from fruit processing and allowing the development of clean production processes.[41] The distribution of the different fractions of berry byproducts relies considerably on berry species and varieties. For example, in Sea buckthorn (*Hippophae rhamnoides* L.) whole berry, skin, and seeds percentage by weight account for 31.9% and 10.7%, respectively,[42] while in blueberries (*Vaccinium corymbosum*, cv. Rubel) the respective mass fractions are 19.0% and 1.5%.[43] The heterogeneity observed in berry pomace constituents greatly influences their chemical composition, once skin and pulp are considered a source of dietary fiber and bioactive compounds, while seeds influence mainly the fat and protein content.[44,45]

The chemical composition of macrocomponents of pomace from the berries is listed in Table 4.1. In general, the main component in berry pomace is dietary fiber, which accounts for 26.2–72.6% of total dry mass (DM), followed by lipids (1.2–20.0% DM) and protein (2.2–15.7% DM).

In berry pomace, proteins are mostly concentrated in the seeds, although they are also part of the complex structures that compose plant cell walls.[41,46] This macronutrient accounts for up to 15.7% of the total dry mass of berry pomace, being generally higher than the values found in pomace obtained from other fruits, namely apple.[47] The total protein content and amino acid composition of berries pomace may vary significantly depending on fruit species and variety. Pieszka et al.[48] studied the amino acid content in chokeberry and blackcurrant pomace. The sum of essential amino acids is higher in blackcurrant pomace (40.16 g/kg DM) than in chokeberry pomace (28.90 g/kg DM). Both pomaces are rich in aspartic acid and glutamic acid, being blackcurrant pomace particularly abundant in proline, lysine, and cysteine; and chokeberry pomace in glycine, alanine, and histidine.[48] Mokhtar et al.[49] performed a similar study focusing on goldenberry pomace,

TABLE 4.1 Proximate Chemical Composition of Berry Pomaces in % (w/w) Dry Mass (DM).

Berry	Moisture	Lipids	Protein	Dietary fiber		Carbohydrates	Ash	References
				IDF	SDF			
Blueberry	13.0	4.0	6.6	26.2		60.9[b]	2.1	45
Cranberry (*Vaccinium oxycoccos* L.)	4.5	12.0	2.2	65.5	5.7	8.4[b]	1.1	46
Cranberry (*Vaccinium macrocarpon* L.)	4.0	1.2	8.2	-	-	85.8[c]	0.8	132
Raspberry (*Rubus idaeus* L.)	96.9[a]	14.3	8.7	72.6		-	1.5	133
Blackcurrant (*Ribes nigrum* L.)	3.4	20.0	15.7	55.2	4.0	2.2[b]	2.7	134
Redcurrant (*Ribes rubrum* L.)	5.2	14.2	11.8	51.1	7.0	12.6[b]	3.0	134
Gooseberry (*Ribes uva-crispa* L.)	5.0	10.9	12.4	49.6	7.0	16.3[b]	3.4	134
Chokeberry (*Aronia melanocarpa*)	2.7	3.6	6.0	52.5	7.0	28.9[b]	1.9	134

[a]% (w/w) fresh mass. [b]Carbohydrate content was calculated by 100 – (Protein + Fat + Fiber + Ash). [c]Carbohydrate content was calculated by 100 –(Protein + Fat + Ash) IDF, insoluble dietary fiber; SDF, soluble dietary fiber.

being this byproduct rich in essential amino acids (30.79 g/100 g of protein), particularly leucine and valine; and nonessential amino-acids, namely glutamic acid and arginine.[49] The richness of fruit pomace in amino acids, such as lysine and histidine, as well as its adequate balance between essential and nonessential amino acids, makes it a suitable product to be incorporated into food and feed formulations.

Furthermore, berries are known to be a good source of many vitamins and minerals. Nour et al.[50] determined the content of mineral elements in dried sea buckthorn pomace, being Ca (724 mg/kg), Fe (62.9 mg/kg), Mn (12.6 mg/kg), and Zn (22.3 mg/kg) the most prevalent elements, with higher values than the ones reported in fresh berries.[51] Likewise, chokeberry, black-currant, and goldenberry pomaces have shown to be good sources of mineral elements like Fe, Mn, Zn, Ca, P, and K.[48,49] This reveals the potential of berry pomaces as a source of minerals to be used, for example, as a supplement in food products containing low levels of minerals.

As previously described, numerous reports outline the substantial amounts of important nutrients and bioactive compounds retained in berry pomace and could potentially be recovered to be used as novel ingredients, particularly for functional foods, feed formulations, and nutraceuticals. Therefore, dietary fiber, fatty acids, and phenolic compounds from berry pomace will be discussed with more detail in the following sections.

4.3.2 DIETARY FIBER

In 2009, the CODEX Alimentarius Commission defined dietary fiber as carbohydrate polymers that are not hydrolyzed by the endogenous enzymes in the small intestine and that fit in one of the subsequent groups: a) exist naturally in the food as consumed; b) are obtained from food raw material through physical, enzymatic, or chemical processes; or c) have synthetic origin and demonstrate a beneficial effect on health.[52] It can be classified according to its solubility in water into soluble fiber (e.g., pectin and some hemicelluloses) or insoluble fiber (e.g., cellulose and lignin).[53] The soluble DF forms viscous solutions and is resistant to digestion in the small intestine, but is easily fermented by the microbiota of the large intestine, while insoluble DF is linked to laxative properties. Dietary fiber intake is associated with several positive health effects, namely reducing the glycemic response and plasma cholesterol, prolonged gastric emptying, and the decrease of transit time through the small intestine.[54]

In berries pomace, there is a demonstrated dominance of insoluble DF over the soluble one, although some berries (e.g., blackcurrant) are particularly rich in soluble fiber.[53,54] The DFs from berries are predominantly lignin and carbohydrate polymers (e.g., pectins, hemicelluloses, and cellulose) located within the plant cell wall. Nawirska et al.[55] determined the structural constituents of DF in chokeberry pomace and blackcurrant, being the content in cellulose, hemicellulose, and lignin (g/100 g DM) as follows: 34.56, 30.24, 22.68 and 21.01, 15.87, 26.66, respectively. However, in berry pomace, the content of the soluble and insoluble portions, as well as fractions of cellulose, hemicellulose, and lignin, may highly depend on the berry variety, juice pretreatment, and extraction methodologies.[41,55] For instance, the content in DF fractions in pomace from different berry species may vary between 12.0–34.6% for cellulose, 6.0–25.3% for hemicellulose, 24.1–59.3% for lignin, and 2.7–7.9% for pectin.[53,56–58]

Additionally, in berry pomace some phenolic compounds, such as flavonoids, can be tightly bound to the dietary fiber complex. Tagliani et al.[45] evaluated the antioxidant activity and phenolic compounds content of the extractable and hydrolysable fractions of blueberry pomace. The hydrolysate fraction, related to antioxidant fiber compounds, accounted for 64% of total polyphenol content and for 69% of the $ABTS^{\bullet+}$ scavenging activity.[45] This explains why different types of berry pomaces show higher levels of phenolic compounds and higher antioxidant capacity than the corresponding fresh fruit, once polyphenols remain associated with wall compounds present in skins and seeds after juice pressing.[9,24,46]

Because of its well-balanced proportion between soluble and insoluble DF and the presence of a significant amount of polyphenols, such as flavonoids and anthocyanins, berries pomace is a promising ingredient to be used in innovative functional products, as will be explained further.

4.3.3 PHENOLIC COMPOUNDS

Phenolic compounds, often referred to as polyphenols, are considered one of the major classes of secondary metabolites found in plants. From a chemical point of view, they possess at least one benzene ring with one or more hydroxyl substituents, and may occur as functionalized derivatives such as esters, methyl-esters, or glycosides.[59] Nevertheless, polyphenols are a complex group of compounds with diverse structures ranging from simple phenolic molecules to highly polymerized compounds.[60] Research interest in phenolic compounds present in food and in agro-industrial byproducts has increased greatly, specially due to their antioxidant capacity and health benefits.[60,61] Berry juice byproducts are abundant in phenolic compounds, mainly flavonoids (principally anthocyanins and flavonols), phenolic acids, and tannins (Fig. 4.2). Thus, they are a valuable source of natural antioxidants with potential applications in different areas.

4.3.3.1 FLAVONOIDS

Flavonoids are a large group of phenolic compounds highly abundant in plant kingdom. Their chemical structure is based on a diphenylpropane skeleton comprising two benzene rings (A and B) connected by a pyrane ring (C). They are grouped into various classes, which differ in the level of oxidation and pattern of substitution of the C ring.[62] In this chapter, only the flavonoid classes that are predominant in berry juice byproducts will be addressed.

4.3.3.2 ANTHOCYANINS

Anthocyanins are a dominating group of flavonoids present in berry pomaces. These polyphenolic compounds comprise a group of naturally occurring pigments responsible for the great diversity of colors in flowers and fruits. Anthocyanins occur as glycosides and their aglycones are called anthocyanidins. From the naturally occurring anthocyanidins that have already been identified, cyanidin, delphinidin, petunidin, malvidin, pelargonidin, and peonidin are the most frequently found in plants.[63] Anthocyanins in fruits are usually glycosylated in the C3 position (3-*O*-monoglycosides) and, in less extension, in both positions C3 and C5 (3,5-*O*-diglycosides).[63] The sugars commonly bonded to anthocyanidins are monosaccharides (glucose, galactose, rhamnose, and arabinose), disaccharides (mainly rutinose, sambubiose, and sophorose), and trisaccharides.[63]

Besides their coloring properties, which make them attractive to be used as natural dyes and food colorants,[16] anthocyanins present high antioxidant activity and have been linked to a wide range of health-promoting properties, including cytoprotective, antimicrobial, anticancer, neuroprotective, antiobesity, and antidiabetic properties.[64–67]

The most common anthocyanins in berries and their byproducts are cyanidin derivatives; however, in some berries high amount of malvidin-, delphinidin-, and peonidin-based anthocyanins can be found.

Jara-Palacios et al.[68] studied the anthocyanin profile of pomaces from blueberries, red raspberries, red currants, and blackberries. The highest amounts of anthocyanins were found in blueberries (11.9 mg/g); however, red currant pomaces extracts exhibited the highest total phenolic content (34.5 mg GAE/g DM) and $ABTS^{\bullet+}$ scavenging activity (61 mmol TE/100 g DM).[68] In blueberry pomaces the most abundant anthocyanins were malvidin-3-*O*-galactoside and malvidin-3-*O*-arabinoside, with delphinidin-,

Flavonoids

Anthocyanins

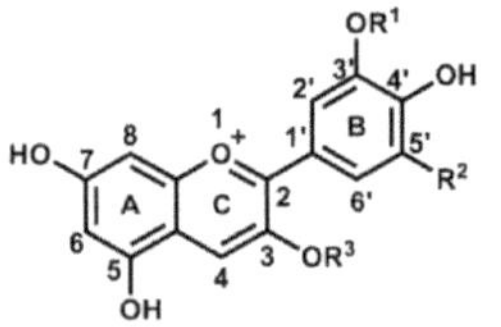

Compound	R¹	R²	R³	Berry
Cyanidin-3-*O*-glucoside	H	H	glucose	Elderberry, Blackberry
Cyanidin-3-*O*-sambubioside	H	H	sambubiose	Elderberry, Red currant
Cyanidin-3-*O*-rutinoside	H	H	rutinose	Blackberry, Blackcurrant
Cyanidin-3-*O*-arabinoside	H	H	arabinose	Cranberry
Cyanidin-3-*O*-sophoroside	H	H	sophorose	Raspberry
Delphidin-3-*O*-galactoside	H	OH	galactose	Blueberry
Delphidin-3-*O*-rutinoside	H	OH	rutinose	Blackcurrant
Petunidin-3-*O*-glucoside	CH_3	OH	glucose	Blueberry

Flavonols

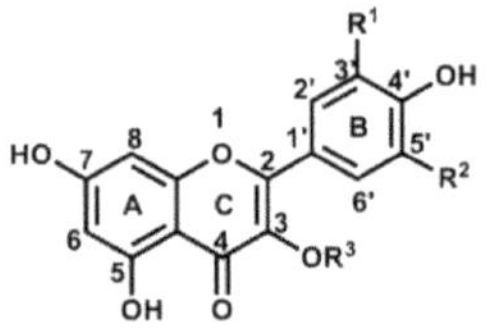

Compound	R¹	R²	R³
Quercentin	H	OH	H
Kaempferol	H	H	H
Myricetin	OH	OH	H
Isorhamnetin	H	OCH_3	H
Quercentin-3-*O*-glucoside	H	OH	glucose
Quercentin-3-*O*-rutinoside (**Rutin**)	H	OH	rutinose

Phenolic acids

Derivatives of benzoic acid

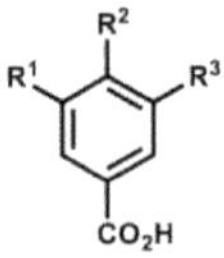

Compound	R¹	R²	R³
Benzoic acid	H	H	H
p-Hydroxybenzoic acid	H	OH	H
Gallic acid	OH	OH	OH
Protocatechuic acid	OH	OH	H
Syringic acid	OCH_3	OH	OCH_3

Derivatives of cinnamic acid

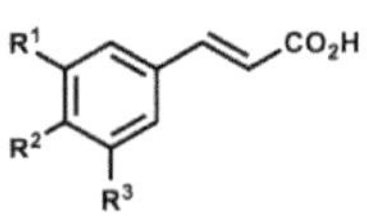

Compound	R¹	R²	R³
Cinnamic acid	H	H	H
p-Coumaric acid	H	OH	H
Caffeic acid	OH	OH	H
Ferulic acid	OCH_3	OH	H
Sinapic acid	OCH_3	OH	OCH_3

Esters of caffeic acid

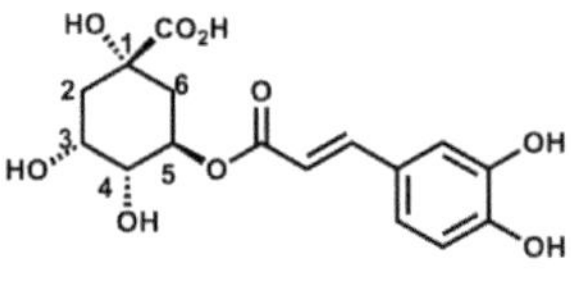

5-*O*-caffeoylquinic acid
(Chlorogenic acid)

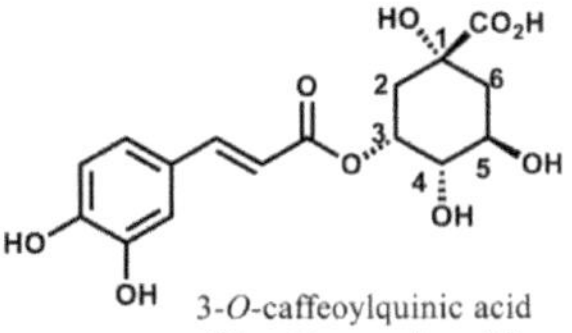

3-*O*-caffeoylquinic acid
(Neochlorogenic acid)

FIGURE 4.2 *(Continued)*

Tannins

Hydrolyzable tannins

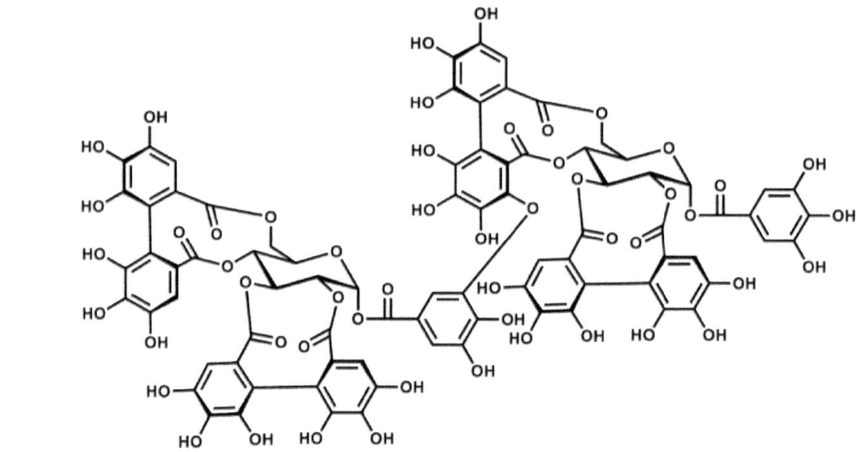

Sanguiin H-6 (ellagitannin)

Condensed tannins or proanthocyanidins

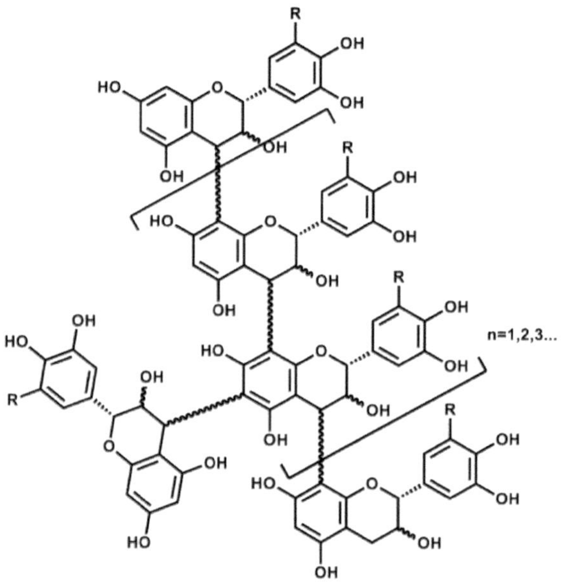

(Polymerized proanthocyanidin)

Flavan-3-ol monomeric units

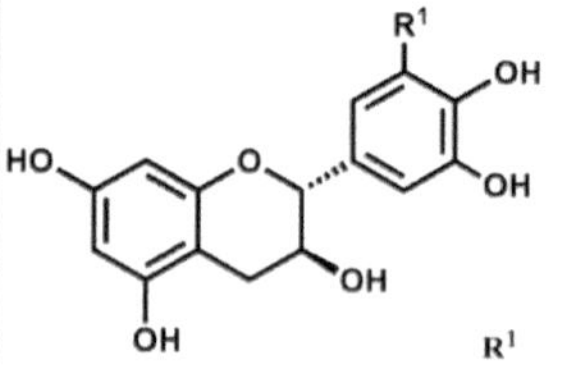

	R^1
(+)-Catechin	H
(+)-Gallocatechin	OH

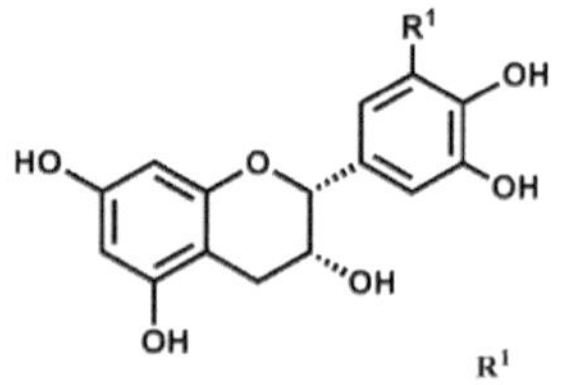

	R^1
(-)-Epicatechin	H
(-)-Galloepicatechin	OH

FIGURE 4.2 Chemical structures of the main representative phenolic compounds present in berry juice byproducts.

cyanidin-, petunidin-, peonidin, and other malvidin-based anthocyanins also identified in smaller amounts, while in the other pomaces only cyanidin derivates were present. The main anthocyanins in the red raspberry pomaces were cyanidin-3-*O*-sophoroside and cyanidin-3-*O*-glucoside.[68,69] Cyanidin-3-*O*-(2″-*O*-glucosyl)-rutinoside and cyanidin-3-*O*-rutinoside can also be detected in substantial amounts in pomaces of some varieties of raspberry.[68,69] Cyanidin-3-*O*-rutinoside was the major anthocyanin found in black raspberry seeds.[70] Cyanidin-3-*O*-sambubioside was predominant in red currant pomace extracts, corresponding to about 82% of the anthocyanin content.[68] Cyanidin-3-*O*-sambubioside and cyanidin-3-*O*-glucoside represented around 90% of the total monomeric anthocyanin contents in elderberry pomace.[32,33] In blackberry pomaces, cyanidin-3-*O*-glucoside was the main anthocyanin (1.7 mg/g DM), which represented 86% of anthocyanin content. Delphinidin-3-*O*-rutinoside and cyanidin-3-*O*-rutinoside were identified as the most abundant anthocyanins in blackcurrant pomaces, followed by smaller amounts of delphinidin-3-*O*-glucoside and cyanidin-3-*O*-glucoside.[71,72] Anthocyanin extraction from blackcurrant pomace by microwave-assisted extraction had a relative distribution of 39.8–44.3% for delphinidin-3-*O*-rutinoside, 23.3–26.9% for cyanidin-3-*O*-rutinoside, 20.2–28.1% for delphinidin-3-*O*-glucoside, while cyanidin-3-*O*-glucoside and malvidin-3-*O*-rutinoside were present in lower concentrations (7.1–7.3% and 0.9–1.4%, respectively).[73] The main representative anthocyanins in chokeberry pomace were cyanidin-3-*O*-galactoside and cyanidin-3-*O*-arabinoside.[24,74,75] In cranberry pomace, the anthocyanins identified were glycosides of cyanidin and peonidin.[46] Cyanidin-3-*O*-arabinoside was the most prominent, followed by peonidin-3-*O*-arabinoside, peonidin-3-*O*-galacoside, and cyanidin-3-*O*-galactoside.[46]

4.3.3.3 FLAVONOLS

Flavonols belong to the group of 4-oxo-flavonoids which carry a carbonyl group on C4 of C ring. They are also characterized by the presence of a hydroxyl group at position C3, and approximately 90% of plant flavonols are hydroxylated in positions C5 and C7. The most common flavonols found in plants and foods are quercetin, kaempferol, myricetin, and isorhamnetin (Figure 4.2), and usually they occur as O-glycosides.[76]

Thirteen flavonols were identified in dried cranberry pomace in about 3.6 mg/g DM.[46] Most of the flavonols were glycosides of quercetin and the aglycones myricetin and quercetin were also found in high amounts, of about 0.6 mg/g and 1.5 mg/g of DM, respectively.[46] Sójka et al.[77] obtained three fractions with different phenolic compositions from blackcurrant pomace. The fraction more concentrated in polyphenols > 50% (w/w) was rich in anthocyanins, followed by a fraction composed by flavonol aglycones > 20% (w/w), and finally a flavanol glycosides-containing fraction > 9% (w/w). Interestingly, this last fraction, which contained myricetin, quercetin, kaempferol, and isorhamnetin glycosides, had the highest antioxidant capacity, despite having the lowest polyphenol content.[77]

4.3.3.4 PHENOLIC ACIDS

Phenolic acids contain at least one carboxylic acid group attached to a phenol ring and, according to the carbon chain, can be divided into two categories: Hydroxybenzoic acids (C6–C1), which were derived from benzoic acid such as *p*-hydroxybenzoic acid, gallic acid, protocatechuic acid, and syringic acid, or hydroxycinnamic acids (C6–C3), which were derived from cinnamic acid such as *p*-coumaric acid, caffeic acid, ferulic acid, and sinapic acid. Natural hydroxycinnamic acids usually appear as esters formed by linkage with quinic acid, for example chlorogenic acid and neochlorogenic, which are abundant in some berries. After anthocyanins, phenolic acids were the second most abundant class of phenolic compounds identified in the blueberry pomace extract.[78] The main phenolic acid identified was chlorogenic acid followed by *p*-hydroxycinnamic acid and caffeic acid.[78] In bilberry and blackberry pomaces the predominant phenolic acids were protocatechuic acid (3.36–35.18 mg/g) and gallic acid (9.57–31.98 mg/g).[80] For dried blackcurrant and chokeberry pomaces the phenolic acids identified were chlorogenic, caffeic, and ellagic acids.[48] The sum of the three phenolic acids resulted in a content of 0.13 mg/g in blackcurrant and 6.30 mg/g in chokeberry.[48] *p*-Coumaric acid was the main phenolic acid in the cold-pressed oils of seeds from cranberry and blueberry with contents of 7.73 and 3.25 mg/g, respectively.[80]

4.3.3.5 TANNINS

Tannins are classified into two major groups based on their structure as the hydrolysable and the condensed tannins or proanthocyanidins.

Hydrolysable tannins include the gallotannins and the ellagitannins. They are polymeric compounds containing a polyol, usually glucose, esterified with gallic acid—gallotannins, or with hexahydroxydiphenic acid—ellagitannins.[81] In the presence of acids or bases, gallotannins are hydrolyzed to gallic acid, and ellagitannins to hexahydroxydiphenic acid, which spontaneously lactonizes to ellagic acid. Hydrolysable tannins have been described to possess several biological properties, including antioxidant, anticancer, estrogenic/antiestrogenic, antimicrobial, antiviral, and anti-inflammatory effects.[82,83]

The content of ellagitannins in raspberry pomace extracts obtained with 80% methanol and 80% acetone from different plant cultivars, depending on the extraction solvent and raspberry cultivar, has been reported to vary from 31.57 to 62.42 mg/g of dry extract.[84] Moreover, ellagitannins contents were higher by 63–80% in the raspberry pomace extracts than in the pulp extracts of the corresponding cultivars.[84] Fotschki et al. were able to obtain a total content of ellagitannins of 477 mg/g from raspberry pomace after the extraction with 60% acetone and subsequent purification.[85]

Condensed tannins are oligomeric and polymeric proanthocyanidins consisting of coupled flavan-3-ol units that are present in a great diversity of structures in nature. The main representative flavan-3-ol monomeric units are (+)-catechin, (+)-gallocatechin, (-)-epicatechin, and (-)-galloepicatechin (Fig. 4.2). Proanthocyanidins can be divided into different groups depending on the variation in hydroxylation pattern and stereochemistry of the monomeric units, the location and type of interflavan linkage, and the degree of polymerization.[86] Like other polyphenols, proanthocyanidins show antioxidant properties among several health benefits.[87]

Oszmiański and Wojdylo reported that proanthocyanidins were the main polyphenolic compounds in chokeberry pomace with a content up to 81.92 mg/g DM. Similar results were obtained by Mayer-Miebach et al. (48–258 mg/g DM)[24] and by Sójka et al. (73–129 mg/g DM).[75] Proanthocyanidins in chokeberry pomace are composed predominantly of (−)-epicatechin units, which are the main constituents in the extension units and chain-terminating units.[74,75] In raspberry pomace, Fotschki et al. obtained 107.13 mg/g of proanthocyanidins composed by 40% of (+)-catechin in the extension units and 54% of (−)-epicatechin in the terminal units.[85]

4.3.4 LIPOPHILIC COMPOUNDS

In some berries, the seeds may represent a significant part of the pomace. For example, seeds have been reported to represent about 55% of dried blackcurrant pomace,[44] and about 90% and 95%, in raspberry and blackberry pomace, respectively.[88]

The seeds of berries are a valuable source of oil rich in unsaturated fatty acids, such as linoleic (C18:2*n*-6 or ω-6), α-linolenic (C18:3*n*-3 or ω-3), and oleic (C18:1*n*-9) acids. Solvent extraction and cold pressing are the traditional processes used to extract the oil from seeds. However, the required use of toxic organic solvents in solvent extraction and the usual low yields obtained by cold pressing led to the search for alternative extraction technologies, such as supercritical fluid extraction.[89] The oil content of berry seeds ranged from 7.6–22.2% in bilberry to 29.4% in mulberry (*Morus alba*). C18:2*n*-6 is the most abundant fatty acid with contents ranging from 32.7% of total fatty acids in red currant to 81.4% in mulberry (Table 4.2). High levels of C18:3*n*-3 are usually present in the seed oil of berries, except for mulberry seeds that have the lowest content (0.6%) from the berries presented.[90] The sum of C18:2*n*-6 and C18:3*n*-3 corresponds to the highest portion of the total fatty acids from 56–58% in gooseberry to 87.1% in black raspberry. Bilberry, cranberry, and elderberry (*S. nigra*) seed oils have almost the same proportion of C18:2*n*-6 and C18:3*n*-3 and the sum of the two fatty acids contributes to about 70–75% of the total fatty acids.[89,91–93] C18:1*n*-9 is the third most abundant fatty acid in berry seed oils. The lowest content of C18:1*n*-9 was found in mulberry (*Morus alba*) with 5.8% of the total fatty acids. Substantial amounts were found in bilberry, blueberry, elderberry (*S. ebulus*), and cranberry (20.3–23.6%) and up to 26.5% in blackberry. Currant seed oils, especially those from blackcurrants, also contain a considerable amount of γ-linolenic acid (C18:3*n*-6).[93–98] This fatty acid is an intermediate in the biosynthesis of compounds such as prostaglandins, prostacyclins, and thromboxanes and is believed to possess antitumor and anti-inflammatory activities.[99–101] Šavikin et al.[95] studied the fatty-acid composition in 28 currant varieties (*Ribes sp.*) and blackcurrant showed higher contents of C18:3*n*-6 acid (10.7–18.5%) than red and white currant varieties (3.3–7.6% and 5.1–6.9%, respectively). Remarkable C18:3*n*-6 contents (22.0–24.6%) were found in the seeds from selected blackcurrant genotypes.[97] Low ratio of omega-6/omega-3 (*n*-6/*n*-3) fatty acids was observed, generally in the range 1–2:1 (Table 4.2). Humans' diet in industrialized countries is usually deficient in *n*-3 fatty acids with *n*-6/*n*-3 ratios usually as high as 15:1, which is associated with a higher incidence of cancer and many other diseases, such as cardiovascular, inflammatory, and autoimmune diseases.[102] Therefore, berry seed oils can be a balanced source of *n*-3 fatty acids, contributing to a lower *n*-6/*n*-3 ratio and a healthy human diet.

Sea buckthorn juice industry is an interesting case of lipophilic compounds recovery from the berry press-cake residue.[103] This berry, unlike other species, accumulates fat in all morphological parts of the fruit; therefore after removing

TABLE 4.2 Average Fatty Acid Composition of Some Berry Seed Oils.

Berry	Oil content % (w/w) dry mass	Fatty acid (%)[a]							n6/n3	References
		C16:0	C18:0	C18:1n-9	C18:2n-6	C18:3n-3	C18:3n-6	C18:4n-3		
Blackcurrant (*R. nigrum*)	13.7–27.7	5.6–8.1	1.1–2.0	6.6–13.5	39.6–53.3	10.0–19.2	10.7–24.6	2.4–4.4	2.8–5.0	93–97
Red currant (*R. rubrum* L.)	18.2–24.2	3.9–6.2	1.2–2.4	12.9–19.8	32.7–44.8	21.0–32.0	3.3–9.5	2.2–4.7	1.1–2.2	93–95, 109
White currant (*R. rubrum*)	19.8–22.8	4.4–6.1	1.4–2.0	16.7–19.9	35.9–39.0	22.1–28.3	5.1–6.9	3.4–4.0	1.3–1.8	95
Gooseberry (*R. uva-crispa* L.)	15.6–17.9	7.5–8.1	1.5–1.6	14.3–15.8	33.9–38.0	20.5–21.8	8.5–9.8	4.3–5.0	1.8	94, 109
Raspberry (*R. idaeus* L.)	10.7	2.6–2.7	0.9–1.0	11.3–12.0	55.4–54.5	28.8–29.1	—	—	1.9	93, 135
Red raspberry	—	2.4	0.9	10.9	53.7	31.7	—	—	1.7	106
Black raspberry	—	2.0	0.6	9.7	53.4	33.7	—	—	1.6	107
Mulberry (*Morus alba*)	29.4	8.4	3.5	5.8	81.4	0.6	—	—	—	90
Mulberry (*Cudrania tricuspidate*)	16.7	10.2	3.0	11.4	74.4	0.6	—	—	—	90
Bilberry (*Vaccinium myrtillus* L.)	7.6–22.2	4.5–6.1	1.1–1.6	21.8–23.4	33.7–35.9	36.1–36.3	nd-0.07	nd-0.04	0.9–1.0	89, 93
Blackberry (*Rubus ulmifolius* S.)	10.5	3.3–7.8	1.6–3.0	14.4–26.5	48.8–63.4	13.4–17.6	—	—	3.5–3.8	91, 106, 107
Blueberry (*Vaccinium corymbosum*)	—	5.2–5.7	1.3–2.8	21.4–22.9	41.9–43.5	25.1–28.3	—	—	1.5–1.7	106, 107, 110

TABLE 4.2 *(Continued)*

Berry	Oil content % (w/w) dry mass	Fatty acid (%)[a]							n6/n3	References
		C16:0	C18:0	C18:1n-9	C18:2n-6	C18:3n-3	C18:3n-6	C18:4n-3		
Cranberry (*Vaccinium oxycoccos* L.)	—	5.2–5.4	1.0–1.3	23.6–25.3	35.3–37.7	34.3–30.1	—	—	1.0–1.3	93, 106
Elderberry (*S. nigra*)	15.9–22.4	7.9–13.1	2.1–2.3	12.8–13.5	34.3–38.4	32.1–40.8	—	—	0.8–1.2	91, 92
Elderberry (*S. ebulus*)	24.9	5.7	2.9	20.3	41.4	27.5	—	—	1.5	92
Sea buckthorn (*H. rhamnoides* L.)	5.5–14.2	6.7–9.6	1.9–4.1	12.9–26.1	36.2–43.6	20.2–36.3	—	—	—	93, 136

[a]*Expressed as weight % of total fatty acids. Palmitic acid (C16:*0); stearic acid (C18:0); oleic acid (C18:1*n*-9); linoleic acid (C18:2*n*-6); α-linolenic acid (C18:3*n*-3); γ-linolenic acid (C18:3*n*-6); stearidonic acid (C18:4n-3).

nd, Not detected.

the juice, the press-cake (composed by the seeds, peels, and the remaining pulp) is very rich in lipids and valuable fat-soluble phytochemicals. Different oils can be obtained from the seeds and the pulp/peel (without the seeds). The seeds are a source of a very unsaturated oil, with up to 73% or more of the total fatty acids corresponding to the sum of C18:2*n*-6 and C18:3*n*-3, whereas the pulp/peel oil is more saturated, containing about 38% of C16:0.[103,104] In addition, palmitoleic acid (C16:1*n*-7), which is practically absent in the seed oil, comprises 14–50% of the fatty acids in pulp/peel oil.[103,104]

In addition to essential fatty acids, seed oils are excellent sources of natural vitamin E. Vitamin E includes two groups of fat-soluble compounds with similar chemical structures, the tocopherols, and tocotrienols, which are also known as tocols. The tocopherols have a saturated side chain, while the tocotrienols have an unsaturated side chain containing three double bonds. Tocopherols and tocotrienols have four homologues, named α, β, γ, δ, depending on the position and number of the methyl substitutions on the aromatic side of the chromanol ring (Figure 4.3).[105] Tocols are strong antioxidants and are reported to have neuroprotective and anticancer properties.

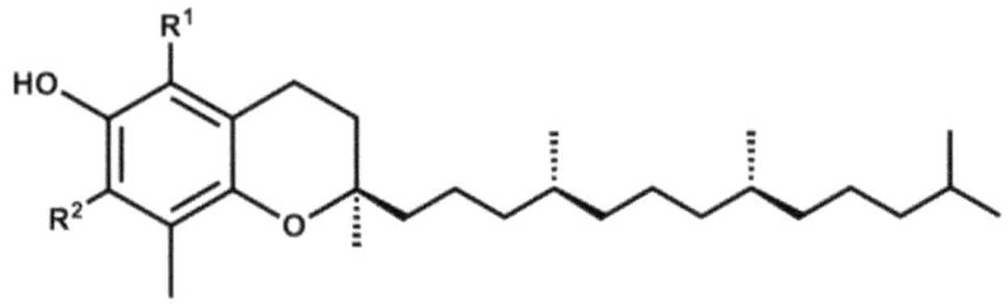

Tocopherol

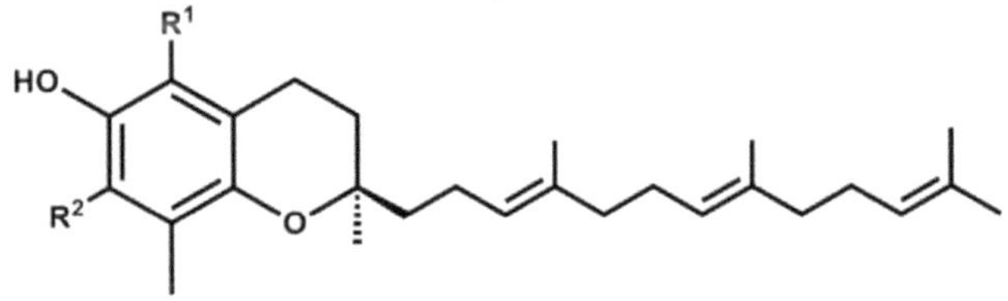

Tocotrienol

	R^1	R^2
α	CH_3	CH_3
β	CH_3	H
γ	H	CH_3
δ	H	H

FIGURE 4.3 Chemical structures of tocopherol and tocotrienol derivatives. Adapted from Ref. 105.

The tocopherol composition of berry seed oils has been widely studied (Table 4.3). The highest concentrations of tocopherols were found in raspberry (*Rubus idaeus* L.) seed oils in the range of 198–400 mg/100 g of oil and high amounts were also observed in blackcurrant, red currant, and white currant seed oils (110.0–245.8, 85.7–248.1, and 131.5–182.4 mg/100

TABLE 4.3 Tocopherol Composition of Some Berry Seed Oils.

Berry	Total mg/100g oil	%[a]				References
		α-T	β-T	γ-T	δ-T	
Blackcurrant (*R. nigrum*)	112.8–245.8[b]	29.2–43.7	nd	53.6–65.1	2.5–8.6	98
	110.0[c]	9.1	—	81.8	9.1	93
Red currant (*R. rubrum* L.)	85.7–248.1[b]	12.7–35.8	0.6–4.3	44.5–67.6	12.0–31.2	98
	90[c]	nd	—	66.7	33.3	93
White currant (*R. rubrum*)	131.5–182.4[b]	9.1–18.1	0–0.8	59.5–68.8	14.0–27.7	98
Red-black gooseberry (*R. grossularia*)	55.9–119.1[b]	16.9–28.2	nd	63.3–79.4	4.5–15.4	98
Yellow gooseberry (*R. grossularia*)	71.5–94.6[b]	17.2–37.6	nd	59.4–73.8	3.0–9.0	98
Rasberry (*Rubus idaeus* L.)	400[c]	17.5	—	80.0	2.5	93
	360[d]	19.7	tr	75.6	4.8	135
	198[e]	23.3	nd	72.7	3.6	135
Red raspberry (*Rubus idaeus*)	88.9[e]	17.0	—	62.9	20.1	110
Black raspberry	72.6[e]	18.8	—	61.4	19.8	107
Mulberry (*Morus nigra* L.)	28[f]	2.6	—	89.3	7.8	137
Mulberry (*Morus alba*)	99.64	6.25	—	18.02	75.72	90
Mulberry (*Cudrania tricuspidata*)	16.07	27.7	—	46.4	25.9	90
Blackberry (*Rubus ulmifolius* S.)	4.37	0.8	nd	99.2	nd	91
	28.6[e]	18.3	—	65.5	16.2	107
Blueberry (*Vaccinium corymbosum*)	11.1[e]	64.8	—	30.3	5.4	110
Elderberry (*S. nigra*)	0.31	15.7	nd	84.3	nd	91
Sea buckthorn (*H. rhamnoides* L.)	170.0	52.9	—	47.1	—	93

[a]*Expressed as % of total tocopherols;* [b]*Soxhlet extraction with petroleum ether.* [c]*Supercritical CO_2 extraction* [d]*Hexane extracted.* [e]*Cold pressed;* [f]*fresh weight.*

tr, traces; nd, not detected.

g of oil, respectively). Lowest contents of tocopherols were obtained in elderberry (*S. nigra*), blackberry, and blueberry seed oils (0.31, 4.37–28.6, and 11.1 mg/100 g of oil, respectively). γ-Tocopherol is usually present in higher amounts and β-tocopherol is not detected or is present in trace amounts (Table 4.3). However, there are some exceptions: α-tocopherol was the main tocopherol present in blueberry and sea buckthorn seed oils, and δ-tocopherol was predominant in mulberry (*Morus alba*) seed oil. The presence of tocotrienols in berry seeds has been less reported than tocopherols; nevertheless, they were observed in higher amounts in the seed oils of some *Vaccinium* species. Yang et al.[93] obtained a total tocols content of 40, 195, and 130 mg/100 g of oil for bilberry, cranberry, and lingonberry, respectively, and γ-tocotrienol was the main component present (≈75% in bilberry and ≈92% in cranberry and lingonberry). A similar result was obtained by Gustinelli et al.[89] for bilberry seed oils, with γ-tocotrienol as the main tocol compound in the range of 77.6–81.1%. In crowberry (*Empetrum nigrum* L.) seed oil, Yang et al.[93] obtained 10 mg/100 g oil of γ-tocopherol, 30 mg/100 g oil of γ-tocotrienol, and 10 mg/100 g oil of δ-tocotrienol. In a study conducted by Van Hoed et al.[106] the main tocol found in cold-pressed blueberry seed oil was γ-tocotrienol (33 mg/100 g oil, ≈88% of tocols). In another study, Li et al.[107] reported δ-tocotrienol as the main tocol in blueberry seed oil (≈95%) with content of 124.5 mg/100 g.

Sterols are other minor biologically active lipophilic constituents that can be found in berry seed oils. Plant sterols are known to effectively reduce serum low-density lipoprotein (LDL) cholesterol and atherosclerotic risk.[108] Therefore, the recovery of these compounds from berry juice byproducts is particularly attractive for functional food and pharmaceutical sectors. β-Sitosterol is usually the predominant sterol in berry seed oils. Kim et al.[90] have found sterol contents in the range 451.9–507.6 mg/100 g of oil in mulberry seeds. In addition to β-sitosterol, which contributed to about 91% of the total quantified sterols, campesterol and stigmasterol were also identified with contents of 24.5–36.6 and 10.3–16.5 mg/100 g, respectively. Górnaś et al.[109] reported phytosterol contents of 314.0 and 586.0 mg/100 g in red currant and gooseberry seed oils, respectively. The major phytosterol present in red currant seed oil was β-sitosterol (187.0 mg/100 g of oil), followed by Δ7-stigmasterol (36.0 mg/100 g), sitostanol (22.0 mg/100 g), cycloartenol (19.0 mg/100 g), and Δ7-avenasterol (19.0 mg/100 g). In gooseberry seed oil, besides β-sitosterol (240.0 mg/100 g), 24-methylene-cycloartanol was also present in a high amount (213.0 mg/100 g), followed by Δ7-stigmasterol (44.0 mg/100 g), avenasterol (37.0 mg/100 g), campesterol (30.0 mg/100 g), and stigmasterol (18.0 mg/100 g).[109]

Carotenoids are also present in berry seed oils in minor amounts; however, they are important components due to their functional properties. They give strong yellow color to edible oils, may contribute to the antioxidant activity of oils, and some carotenoids, namely β-carotene, α-carotene, and β-cryptoxanthin, have provitamin A activity. Zeaxanthin was the major carotenoid found in blueberry, red raspberry, marionberry, and boysenberry seed oils with contents of 7800–13600 μg/kg, along with β-carotene (82.2–2405.2 μg/kg), cryptoxanthin (717.6–1812.9 μg/kg), and lutein (53.3–97.7 μg/kg).[110]

4.4 VALUE-ADDED APPLICATION FOR BERRY JUICE BYPRODUCTS

The previous sections show that berry pomace is an interesting source of valuable compounds, namely proteins, minerals, fiber, phenolics, and lipophilic compounds. This great diversity highlights berry byproducts as promising raw materials to be valued and applied for diverse industrial purposes, whether due to its nutritional value or to its bioactive compounds content.

The most studied field regarding the valorization of berries pomace is as food fortification to increase the nutritive value and/or balance food composition. Cosmetic and nutraceutical areas are also promising fields for the effective application of the health-beneficial berry pomace constituents. Dried and ground pomace powder is the simplest way to apply this product. However, when increased concentrations of the target groups of compounds are required, extraction and/or fractionation are the preferred methods.[41]

4.4.1 FOOD APPLICATIONS

In recent years, there is growing consumer interest for health-enhancing food products such as nutrition modified and functional foods. Dried and grounded is the predominant way to apply berry pomace in food products. In this industry, berry powder is essentially used to enhance the nutritional value of the food and/or as food colorant.

Several authors have reported the application of pomace from different berries (raspberry, blackcurrant, elderberry, and blueberry pomace) in cookies, aiming to increase the dietary fiber fraction and boost their antioxidant potential.[45,47,111–113] Tańska et al.[111] studied the effect of replacing 20%

of the flour weight with berry pomace, namely rowanberry, blackcurrant, and elderberry, in the organoleptic characteristics of shortbread cookies. The addition of berry pomace originated cookies with significantly higher fiber content and antioxidant capacity when compared with the control product (without pomace). Also, the addition of berry pomace had a significant positive effect upon the color acceptance of the cookies by a sensory evaluation panel. Although the panelists considered the total organoleptic characteristics of the cookies with the addition of pomace acceptable, the variation in cookies' hardness and the sour taste, also observed in other studies, can limit the use of this byproduct as a substitute for flour in cookies.[111] Other cereal-based products, for example, bread and extruded snacks, have also been successfully incorporated with 10–30% of berries pomace powder as a substitute for flour, with similar levels of consumer acceptability as the control products.[114–117]

Synthetic pigments from petrochemicals, notwithstanding the known adverse effects on human health, have been extensively used in food products. Therefore, their applications have been interrupted and preferably avoided, promoting an increasing priority to search for safer, natural, and eco-friendly pigments between the industrial and scientific communities.[118] The anthocyanin-rich fractions of berry byproducts have great potential to be exploited as natural colorants for various food applications.[119–121] The main challenges in this area are the development of an efficient, eco-friendly, and cost-effective method for pigment extraction that can be used in an industrial scale and the stabilization of these natural pigments in a wider variety of food products, since they are usually limited to acidic media and are susceptible to degradation.

4.4.2 PHARMACEUTICAL AND COSMETIC APPLICATIONS

The pharmaceuticals and cosmetics industries are also wide and promising fields for the effective application of health-beneficial berry pomace constituents. The potential of this agro-industrial byproduct to be used in these areas is attributed mainly to its lipophilic compounds and phenolic content.

Although the fatty acid composition in seed oils may vary with berry species and variety, this byproduct is generally rich in essential fatty acids such as linoleic ω-3 and linolenic acids ω-6. These compounds are important in maintaining tissue integrity and are essential for the synthesis of the cell membrane and tissue regeneration, having skin revitalization properties.[109] As an example, oils from Sea buckthorn (*Hippophae rhamnoides* L.) seeds are already used in traditional medicine to treat eczema, psoriasis, lupus erythematosus, and chronic wounds.[122] Other studies performed by Linnamaa et al. demonstrated the effect of dietary supplementation with blackcurrant seed oil on the inhibition of the development of atopic dermatitis and atopic sensitization during the fetal period and infancy.[123,124] The studies showed that supplementation with blackcurrant seed oil at an early stage of life or during pregnancy and breastfeeding reduced the prevalence of atopic dermatitis in both mothers and infants.[123,124] Other constituents of the lipophilic fraction of berry pomace, such as squalene and phytosterols, may also be valuable to the industry due to their emollient and skin protection properties.[109]

Concerning the cosmetics industry, Plainfossé et al.[125] studied the potential of different blackcurrant (*Ribes nigrum* L.) extracts to be used as cosmetic ingredients. The results were promising concerning the extracts anti-hyaluronidase and anti-collagenase activities, as well as anti-inflammatory and antioxidant properties, evidencing the potential of berry pomace to be used as a complete antiaging agent in the cosmetic industry.[125]

4.4.3 OTHER APPLICATIONS AND FUTURE TRENDS

In sustainable agriculture, new environmental-friendly micronutrient fertilizers based on berry seeds, produced via biosorption, arise as an alternative to conventional options, leading to a better assimilation of micronutrients by plants while being nontoxic, biodegradable, and cheap.[126,127] Also, the use of fruit pomaces in animal production is gaining popularity. Berry pomace have been extensively studied as a supplement in animal nutrition, due to its phenolics and insoluble fiber content, aiming to increase meat antioxidant stability, increase feed nutritive value, and reduce the use of antibiotics in animal production.[128,129] More recently, the antimicrobial and antioxidant potential of the byproducts of the berry juice industry allowed exploring their use in active packaging solutions[130] as well as in the reinforcement of packaging boards, leading to a significant improvement in water resistance and mechanical properties of the final product when up to 75% of fruit pomace was used as a substitute to newspaper in pulp boards production.[130,131]

It is evident the numerous valuable applications that berry pomace byproducts exhibit; however, their industrial application to obtain ingredients and new products still present some difficulties that need to be further evaluated, namely the technological upscaling and some economic issues.

4.5 CONCLUSIONS

Berry juice processing byproducts contain appreciable amounts of nutritionally important constituents and valuable bioactive components that can be recovered to be used in the production of value-added ingredients and products, providing benefits to the industry. Several studies have shown the potential of using the whole pomace through its addition in the preparation of functional and healthy foods. The extraction of functional compounds from berry byproducts is also seen as an option for exploiting these byproducts. Conventional extraction is still one of the most studied methods, favoring the use of green extraction solutions, such as water or food grade solvents. In recent years, milder and sustainable technologies have been applied such as supercritical fluid extraction, enhanced solvent extraction, ultrasound-assisted, and microwave-assisted extraction. Regarding the chemical composition of berry pomace, the main component is dietary fiber, reaching up to 73% of total dry mass in some berries. Berry pomace is also rich in phenolic compounds, which are components with special interest due to their health-promoting effects. In particular, anthocyanins remain in pomace to a large extent. Flavonols, phenolic acids, and tannins are other phenolic compounds found in berry pomace and, given their biological properties, their recovery can bring several opportunities. Berry seed oils are a valuable source of unsaturated fatty acids and natural vitamin E and contain other minor bioactive components such as sterols and carotenoids. As berry pomace contains several biologically active compounds, current uses are underestimated and its potential opens doors to new and valuable industrial applications.

ACKNOWLEDGMENTS

Thanks are due to the Polytechnic Institute of Viseu and to FCT/MEC for the financial support to CERNAS-IPV, LAQV-REQUIMTE (UIDB/50006/2020), and CITAB (FCT UIDB/04033/2020) research units, through national funds, and the cofunding by the FEDER, within the PT2020 Partnership Agreement and Compete 2020. The authors thank the financial support of the Project BagaConValor—Criação de valor no processo tecnológico de produção de sumo concentrado de baga de sabugueiro (POCI-01-0247-FEDER-033558).

Funding: This work was financially supported by POCI-01-0247-FEDER-033558.

KEYWORDS

- **fruit byproducts**
- **berry pomace**
- **phytochemicals**
- **valorization approaches**

REFERENCES

1. Steinmetz, K. A.; Potter, J. D. Vegetables, Fruit, and Cancer Prevention. *J. Am. Diet. Assoc.* **1996,** *96* (10), 1027–1039. DOI: 10.1016/S0002-8223(96)00273-8.
2. van't Veer, P.; Jansen, M. C.; Klerk, M.; Kok, F. J. Fruits and Vegetables in the Prevention of Cancer and Cardiovascular Disease. *Public Health Nutr.* **2000,** *3* (1), 103–107. DOI: 10.1017/S1368980000000136.
3. Skrovankova, S.; Sumczynski, D.; Mlcek, J.; Jurikova, T.; Sochor, J. Bioactive Compounds and Antioxidant Activity in Different Types of Berries. *Int. J. Mol. Sci.* **2015,** *16* (10), 24673–24706. DOI: 10.3390/ijms161024673.
4. Seeram, N. P. Berry Fruits: Compositional Elements, Biochemical Activities, and the Impact of Their Intake on Human Health, Performance, and Disease. *J. Agric. Food Chem.* **2008,** *56* (3), 627–629. DOI: 10.1021/jf071988k.
5. Foito, A.; McDougall, G. J.; Stewart, D. Evidence for Health Benefits of Berries. In *Annual Plant Reviews Online*; Wiley, 2018; pp 105–148. DOI: 10.1002/9781119312994.apr0600.
6. Sun-Waterhouse, D. The Development of Fruit-based Functional Foods Targeting the health and Wellness market: A Review. *Int. J. Food Sci. Technol.* **2011,** *46* (5), 899–920. DOI: 10.1111/j.1365-2621.2010.02499.x.
7. Arvanitoyannis, I. S.; Varzakas, T. H. Fruit/Fruit Juice Waste Management: Treatment Methods and Potential Uses of Treated Waste. In *Waste Management for the Food Industries*; Elsevier, 2008; pp 569–628. DOI: 10.1016/B978-012373654-3.50012-2.
8. van Dyk, J. S.; Gama, R.; Morrison, D.; Swart, S.; Pletschke, B. I. Food Processing Waste: Problems, Current Management and Prospects for Utilisation of the Lignocellulose Component Through Enzyme Synergistic Degradation. *Renew. Sustain. Energy Rev.* **2013,** *26*, 521–531. DOI: 10.1016/j.rser.2013.06.016.
9. Struck, S.; Plaza, M.; Turner, C.; Rohm, H. Berry Pomace - A Review of Processing and Chemical Analysis of Its Polyphenols. *Int. J. Food Sci. Technol.* **2016,** *51* (6), 1305–1318. DOI: 10.1111/ijfs.13112.
10. Chockchaisawasdee, S.; Stathopoulos, C. E. Extraction, Isolation and Utilisation of Bioactive Compounds from Fruit Juice Industry Waste. In *Utilisation of Bioactive Compounds from Agricultural and Food Waste*, 2nd Edition; Q. v. Vuong, Ed.; CRC Press: Boca Raton, 2017; pp 272–313.

11. Mihalev, K.; Dinkova, R.; Shikov, V.; Mollov, P. Classification of Fruit Juices. In *Fruit Juices*; Rajauria, G.; Tiwari, B. K., Eds., Academic Press: San Diego, 2018; pp 33–44. DOI: 10.1016/B978-0-12-802230-6.00003-5.
12. Weber, F.; Larsen, L. R. Influence of Fruit Juice Processing on Anthocyanin Stability. *Food Res. Int.* **2017,** *100*, 354–365. DOI: 10.1016/j.foodres.2017.06.033.
13. He, J.; Giusti, M. M. Anthocyanins: Natural Colorants with Health-Promoting Properties. *Annu. Rev. Food Sci. Technol.* **2010,** *1* (1), 163–187. DOI: 10.1146/annurev.food.080708.100754.
14. Directive 2012/12/EU of the European Parliament and of the Council of 19 April 2012 amending Council Directive 2001/112/EC relating to fruit juices and certain similar products intended for human consumption. 2012. Accessed: Jul. 16, 2021. [Online]. Available: https://eur-lex.europa.eu/LexUriServ/LexUriServ.do?uri=OJ:L:2012:115:0001:0011:EN:PDFhttps://eur-lex.europa.eu/LexUriServ/LexUriServ.do?uri=OJ:L:2012:115:0001:0011:EN:PDF
15. Khanal, R. C.; Howard, L. R.; Prior, R. L. Effect of Heating on the Stability of Grape and Blueberry Pomace Procyanidins and Total Anthocyanins. *Food Res. Int.* **2010,** *43* (5), 1464–1469. DOI: 10.1016/j.foodres.2010.04.018.
16. Francis, F. J.; Markakis, P. C. Food colorants: Anthocyanins. *Crit. Rev. Food Sci. Nutr.* **1989,** *28* (4), 273–314. DOI: 10.1080/10408398909527503.
17. Kader, F.; Rovel, B.; Girardin, M.; Metche, M. Mechanism of Browning in Fresh Highbush Blueberry Fruit (Vaccinium corymbosum L). Role of Blueberry Polyphenol Oxidase, Chlorogenic Acid and Anthocyanins. *J. Sci. Food. Agric.* **1997,** *74* (1), 31–34. DOI: 10.1002/(SICI)1097-0010(199705)74:1<31::AID-JSFA764>3.0.CO;2-9.
18. Kader, F.; Irmouli, M.; Nicolas, J. P.; Metche, M. Involvement of Blueberry Peroxidase in the Mechanisms of Anthocyanin Degradation in Blueberry Juice. *J. Food Sci.* **2002,** *67* (3), 910–915. DOI: 10.1111/j.1365-2621.2002.tb09427.x.
19. Rossi, M.; Giussani, E.; Morelli, R.; R. lo Scalzo, Nani, R. C.; Torreggiani, D. Effect of Fruit Blanching on Phenolics and Radical Scavenging Activity of Highbush Blueberry Juice. *Food Res. Int.* **2003,** *36* (9–10), 999–1005. DOI: 10.1016/j.foodres.2003.07.002.
20. Landbo, A.-K.; Meyer, A. S. Effects of Different Enzymatic Maceration Treatments on Enhancement of Anthocyanins and Other Phenolics in Black Currant Juice. *Int. J. Food Sci. Technol.* **2004,** *5* (4), 503–513. DOI: 10.1016/j.ifset.2004.08.003.
21. Buchert, J. et al. Effect of enzyme-aided pressing on anthocyanin yield and profiles in bilberry and blackcurrant juices. *J. Sci. Food. Agric.* **2005,** *85* (15), 2548–2556. DOI: 10.1002/jsfa.2284.
22. Heffels, P.; F. Bührle, Schieber, A.; Weber, F. Influence of Common and Excessive Enzymatic Treatment on Juice Yield and Anthocyanin Content and Profile During Bilberry (Vaccinium Myrtillus L.) Juice Production. *Eur. Food Res. Technol.* **2017,** *243* (1), 59–68. DOI: 10.1007/s00217-016-2722-0.
23. Khanal, R. C.; Howard, L. R.; Brownmiller, C. R.; Prior, R. L. Influence of Extrusion Processing on Procyanidin Composition and Total Anthocyanin Contents of Blueberry Pomace. *J. Food Sci.* **2009,** *74* (2), H52–H58. DOI: 10.1111/j.1750-3841.2009.01063.x.
24. Mayer-Miebach, E.; Adamiuk, M.; Behsnilian, D. Stability of Chokeberry Bioactive Polyphenols during Juice Processing and Stabilization of a Polyphenol-Rich Material from the Byproduct. *Agriculture* **2012,** *2* (3), 244–258. DOI: 10.3390/agriculture2030244.

25. Skrede, G.; Wrolstad, R. E.; Durst, R. W. Changes in Anthocyanins and Polyphenolics During Juice Processing of Highbush Blueberries (Vaccinium corymbosum L.). *J. Food Sci.* **2000,** *65* (2), 357–364. DOI: 10.1111/j.1365-2621.2000.tb16007.x.
26. White, B. L.; Howard, L. R.; Prior, R. L. Impact of Different Stages of Juice Processing on the Anthocyanin, Flavonol, and Procyanidin Contents of Cranberries. *J. Agric. Food Chem.* **2011,** *59* (9), 4692–4698. DOI: 10.1021/jf200149a.
27. Das, Q. et al. Organic Cranberry Pomace and Its Ethanolic Extractives as Feed Supplement in Broiler: Impacts on Serum Ig Titers, Liver and Bursal Immunity. *Poult. Sci.* **2021,** *100* (2), 517–526. DOI: 10.1016/j.psj.2020.09.044.
28. Qin, X.; Zhang, T.; Cao, Y.; Deng, B.; Zhang, J.; Zhao, J. Effects of Dietary Sea Buckthorn Pomace Supplementation on Skeletal Muscle Mass and Meat Quality in Lambs. *Meat Sci.* **2020,** *166*, 108141. DOI: 10.1016/j.meatsci.2020.108141.
29. Rohm, H. et al. Adding Value to Fruit Processing Waste: Innovative Ways to Incorporate Fibers from Berry Pomace in Baked and Extruded Cereal-based Foods—A SUSFOOD Project. *Foods* **2015,** *4* (4), 690–697. DOI: 10.3390/foods4040690.
30. Paes, J.; Dotta, R.; Barbero, G. F.; Martínez, J. Extraction of Phenolic Compounds and Anthocyanins from Blueberry (Vaccinium Myrtillus L.) Residues Using Supercritical CO2 and Pressurized Liquids. *J. Supercrit. Fluids* **2014,** *95*, 8–16. DOI: 10.1016/j.supflu.2014.07.025.
31. Eliasson, L.; Labrosse, L.; Ahrné, L. Effect of Drying Technique and Particle Size of Bilberry Press Cake on the Extraction Efficiency of Anthocyanins by Pressurized Carbon Dioxide Extraction. *LWT - Food Sci. Technol* **2017,** *85*, 510–516. DOI: 10.1016/j.lwt.2017.03.030.
32. Seabra, I. J.; Braga, M. E. M.; Batista, M. T.; de Sousa, H. C. Effect of Solvent (CO2/ethanol/H2O) on the Fractionated Enhanced Solvent Extraction of Anthocyanins from Elderberry Pomace. *J. Supercrit. Fluids* **2010,** *54* (2), 145–152. DOI: 10.1016/j.supflu.2010.05.001.
33. Seabra, I. J.; Braga, M. E. M.; Batista, M. T. P.; de Sousa, H. C. Fractioned High Pressure Extraction of Anthocyanins from Elderberry (Sambucus Nigra L.) Pomace. *Food Bioprocess Technol.* **2010,** *3* (5), 674–683. DOI: 10.1007/s11947-008-0134-2.
34. Pap, N. et al. Microwave-Assisted Extraction of Anthocyanins from Black Currant Marc. *Food Bioprocess Technol.* **2013,** *6* (10), 2666–2674. DOI: 10.1007/s11947-012-0964-9.
35. Zhang, H.; Tchabo, W.; Ma, Y. Quality of Extracts from Blueberry Pomace by High Hydrostatic Pressure, Ultrasonic, Microwave and Heating Extraction: A Comparison Study. *Emir. J. Food Agric.* **2017,** *29* (10), 815. DOI: 10.9755/ejfa.2017.v29.i10.1259.
36. Bamba, B.; Shi, J.; Tranchant, C.; Xue, S.; Forney, C.; Lim, L.-T. Influence of Extraction Conditions on Ultrasound-Assisted Recovery of Bioactive Phenolics from Blueberry Pomace and Their Antioxidant Activity. *Molecules* **2018,** *23* (7), 1685. DOI: 10.3390/molecules23071685.
37. Lončarić, A. et al. Green Extraction Methods for Extraction of Polyphenolic Compounds from Blueberry Pomace. *Foods* **2020,** *9* (11), 1521. DOI: 10.3390/foods9111521.
38. Reque, P. M. et al. Characterization of Blueberry Fruits (Vaccinium spp.) and Derived Products. *Food Sci. Technol. (Campinas)* **2014,** *34* (4), 773–779. DOI: 10.1590/1678-457X.6470.
39. Horszwald, A.; Julien, H.; Andlauer, W. Characterisation of Aronia Powders Obtained by Different Drying Processes. *Food Chem.* **2013,** *141* (3), 2858–2863. DOI: 10.1016/j.foodchem.2013.05.103.

40. Su, M.-S.; Silva, J. L. Antioxidant Activity, Anthocyanins, and Phenolics of Rabbiteye Blueberry (Vaccinium Ashei) Byproducts as Affected by Fermentation. *Food Chem.* **2006,** *97* (3), 447–451. DOI: 10.1016/j.foodchem.2005.05.023.
41. Venskutonis, P. R. Berries. In *Valorization of Fruit Processing Byproducts*, 1st Edition, Galanakis, C., Ed.; Academic Press: San Diego, 2020; pp 95–125. DOI: 10.1016/B978-0-12-817106-6.00005-8.
42. Stobdan, T.; Korekar, G.; Srivastava, R. B. Nutritional Attributes and Health Application of Seabuckthorn (Hippophae Rhamnoides L.) – A Review. *Curr. Nutr. Food Sci.* **2013,** *9* (2), 2013, DOI: 10.2174/1573401311309020008.
43. Lee, J.; Wrolstad, R. E. Extraction of Anthocyanins and Polyphenolics from Blueberry Processing Waste. *J. Food Sci.* **2006,** *69* (7), 564–573. DOI: 10.1111/j.1365-2621.2004.tb13651.x.
44. Sójka, M.; Król, B. Composition of Industrial Seedless Black Currant Pomace. *Eur. Food Res. Technol.* **2009,** *228* (4), 597–605. DOI: 10.1007/s00217-008-0968-x.
45. Tagliani, C.; Perez, C.; Curutchet, A.; Arcia, P.; Cozzano, S. Blueberry Pomace, Valorization of an Industry Byproduct Source of Fibre with Antioxidant Capacity. *Food Sci. Technol.* **2019,** *39* (3), 644–651. DOI: 10.1590/fst.00318.
46. White, B. L.; Howard, L. R.; Prior, R. L. Proximate and Polyphenolic Characterization of Cranberry Pomace. *J. Agric. Food Chem.* **2010,** *58* (7), 4030–4036. DOI: 10.1021/jf902829g.
47. Bhushan, S.; Kalia, K.; Sharma, M.; Singh, B.; Ahuja, P. S. Processing of Apple Pomace for Bioactive Molecules. *Crit. Rev. Biotechnol.* **2008,** *28* (4), 285–296. DOI: 10.1080/07388550802368895.
48. Pieszka, M.; Gogol, P.; Pietras, M.; Pieszka, M. Valuable Components of Dried Pomaces of Chokeberry, Black Currant, Strawberry, Apple and Carrot as a Source of Natural Antioxidants and Nutraceuticals in the Animal Diet. *Ann. Anim. Sci.* **2015,** *15* (2), 475–491. DOI: 10.2478/aoas-2014-0072.
49. Mokhtar, S. M.; Swailam, H. M.; Embaby, H. E.-S. Physicochemical Properties, Nutritional Value and Techno-functional Properties of Goldenberry (Physalis Peruviana) Waste Powder Concise Title: Composition of Goldenberry Juice Waste. *Food Chem.* **2018,** *248*, 1–7. DOI: 10.1016/j.foodchem.2017.11.117.
50. Nour, V.; Panaite, T. D.; Corbu, A. R.; Ropota, M.; Turcu, R. P. Nutritional and Bioactive Compounds in Dried Sea-Buckthorn Pomace. *Erwerbs-Obstbau* **2021,** *63* (1), 91–98. DOI: 10.1007/s10341-020-00539-1.
51. Arif, S. et al. Determination of Optimum Harvesting Time for Vitamin C, Oil and Mineral Elements in Berries Sea Buckthorn (Hippophae Rhamnoides). *Pak. J. Bot.* **2010,** *42* (5), 3561–3568. Accessed: Jul. 17, 2021. [Online]. Available: http://www.pakbs.org/pjbot/PDFs/42(5)/PJB42(5)3561.pdf
52. Joint FAO/WHO Food Standards Programme, Secretariat of the CODEX Alimentarius Commission: CODEX Alimentarius International Food Standards. *Guidelines on Nutrition Labelling CAC/GL 2-1985 as Last Amended 2010*. Accessed: Dec. 21, 2021. [Online]. Available: https://www.fao.org/fao-who-codexalimentarius/sh-proxy/es/?lnk=1&url=https%253A%252F%252Fworkspace.fao.org%252Fsites%252Fcodex%252FStandards%252FCXG%2B2-1985%252FCXG_002e.pdf
53. Alba, K.; MacNaughtan, W.; Laws, A. P.; Foster, T. J.; Campbell, G. M.; Kontogiorgos, V. Fractionation and Characterisation of Dietary fibre from Blackcurrant Pomace. *Food Hydrocolloids* **2018,** *81*, 398–408. DOI: 10.1016/j.foodhyd.2018.03.023.

54. Alba, K.; Campbell, G. M.; Kontogiorgos, V. Dietary Fibre from Berry-processing Waste and Its Impact on Bread Structure: A Review. *J. Sci. Food. Agric.* **2019,** *99* (9), 4189–4199. DOI: 10.1002/jsfa.9633.
55. Nawirska, A.; Uklańska, C. Waste Products from Fruit and Vegetable Processing as Potential Sources for Food Enrichment in Dietary Fibre. *Acta Sci. Pol., Technol. Aliment* **2008,** *7* (2), 35–42. Accessed: Jul. 17, 2021. [Online]. Available: https://www.food.actapol.net/pub/3_2_2008.pdf
56. Laroze, L.; Soto, C.; Zúñiga, M. E. Phenolic Antioxidants Extraction from Raspberry Wastes Assisted By-enzymes. *Electron. J. Biotechnol.* **2010,** *13* (6), 1–11. DOI: 10.2225/vol13-issue6-fulltext-12.
57. Górecka, D.; Pachołek, B.; Dziedzic, K.; Górecka, M. Raspberry Pomace As A Potential Fiber Source for Cookies Enrichment. *Acta Sci. Pol., Technol. Aliment* **2010,** *9* (4), 451–462. Accessed: Jul. 17, 2021. [Online]. Available: https://www.food.actapol.net/pub/5_4_2010.pdf
58. Nawirska, A.; Kwaśniewska, M. Dietary Fibre Fractions from Fruit and Vegetable Processing Waste. *Food Chem.* **2005,** *91* (2), 221–225. DOI: 10.1016/j.foodchem.2003.10.005.
59. Escarpa, A.; Gonzalez, M. C. An Overview of Analytical Chemistry of Phenolic Compounds in Foods. *Crit. Rev. Anal. Chem.* **2001,** *31* (2), 57–139. DOI: 10.1080/20014091076695.
60. Bravo, L. Polyphenols: Chemistry, Dietary Sources, Metabolism, and Nutritional Significance. *Nutr. Rev.* **1998,** *56* (11), 317–333. DOI: 10.1111/j.1753-4887.1998.tb01670.x.
61. Balasundram, N.; Sundram, K.; Samman, S. Phenolic Compounds in Plants and Agri-Industrial Byproducts: Antioxidant Activity, Occurrence, and Potential Uses. *Food Chem.* **2006,** *99* (1), 191–203. DOI: 10.1016/j.foodchem.2005.07.042.
62. Kumar, S.; Pandey, A. K. Chemistry and Biological Activities of Flavonoids: An Overview. *Sci. World J.* **2013,** *2013*, 1–16. DOI: 10.1155/2013/162750.
63. Mazza, G.; Miniati, E. *Anthocyanins in Fruits, Vegetables, and Grains*; CRC Press, 2018. DOI: 10.1201/9781351069700.
64. Konczak, I.; Zhang, W. Anthocyanins—More Than Nature's Colours. *J. Biomed. Biotechnol.* **2004,** *2004* (5), 239–240. DOI: 10.1155/S1110724304407013.
65. Tsuda, T. Anthocyanins as Functional Food Factors—Chemistry, Nutrition and Health Promotion. *Food Sci. Technol. Research* **2012,** *18* (3), 315–324. DOI: 10.3136/fstr.18.315.
66. Smeriglio, A.; Barreca, D.; Bellocco, E.; Trombetta, D. Chemistry, Pharmacology and Health Benefits of Anthocyanins. *Phytother. Res.* **2016,** *30* (8), 1265–1286. DOI: 10.1002/ptr.5642.
67. Khoo, H. E.; Azlan, A.; Tang, S. T.; Lim, S. M. Anthocyanidins and Anthocyanins: Colored Pigments as Food, Pharmaceutical Ingredients, and the Potential Health Benefits. *Food Nutr. Res.* **2017,** *61* (1), 1361779. DOI: 10.1080/16546628.2017.1361779.
68. Jara-Palacios, M. J.; Santisteban, A.; Gordillo, B.; Hernanz, D.; Heredia, F. J.; Escudero-Gilete, M. L. Comparative Study of Red Berry Pomaces (Blueberry, Red Raspberry, Red Currant and Blackberry) as Source of Antioxidants and Pigments. *Eur. Food Res. Technol.* **2019,** *245* (1), 1–9. DOI: 10.1007/s00217-018-3135-z.
69. Tumbas Šaponjac, V. et al. Anthocyanin Profiles and Biological Properties of Caneberry (Rubus spp.) Press Residues. *J. Sci. Food. Agric.* **2014,** *94* (12), 2393–2400. DOI: 10.1002/jsfa.6564.
70. Choi, M.; Shim, S.; Kim, G. Protective Effect of Black Raspberry Seed Containing Anthocyanins Against Oxidative Damage to DNA, Protein, and Lipid. *J. Food Sci. Technol.* **2016,** *53* (2), 1214–1221. DOI: 10.1007/s13197-015-2094-7.

71. Kapasakalidis, P. G.; Rastall, R. A.; Gordon, M. H. Extraction of Polyphenols from Processed Black Currant (Ribes nigrum L.) Residues. *J. Agric. Food Chem.* **2006,** *54* (11), 4016–4021. DOI: 10.1021/jf052999l.
72. Sandell, M. et al. Orosensory Profiles and Chemical Composition of Black Currant (Ribes Nigrum) Juice and Fractions of Press Residue. *J. Agric. Food Chem.* **2009,** *57* (9), 3718–3728. DOI: 10.1021/jf803884y.
73. Pap, N. et al. Microwave-Assisted Extraction of Anthocyanins from Black Currant Marc. *Food Bioprocess Technol.* **2013,** *6* (10), 2666–2674. DOI: 10.1007/s11947-012-0964-9.
74. Oszmiański, J.; Wojdylo, A. Aronia Melanocarpa Phenolics and Their Antioxidant Activity. *Eur. Food Res. Technol.* **2005,** *221* (6), 809–813. DOI: 10.1007/s00217-005-0002-5.
75. Sójka, M.; Kołodziejczyk, K.; Milala, J. Polyphenolic and Basic Chemical Composition of Black Chokeberry Industrial Byproducts. *Ind. Crops Prod.* **2013,** *51*, 77–86. DOI: 10.1016/j.indcrop.2013.08.051.
76. Hollman, P. C.; Arts, I. C. Flavonols, Flavones and Flavanols - Nature, Occurrence and Dietary Burden. *J. Sci. Food. Agric.* **2000,** *80* (7), 1081–1093. DOI: 10.1002/(SICI)1097-0010(20000515)80:7<1081::AID-JSFA566>3.0.CO;2-G.
77. Sójka, M.; Guyot, S.; Kołodziejczyk, K.; Król, B.; Baron, A. Composition and Properties of Purified Phenolics Preparations Obtained From an Extract of Industrial Blackcurrant (*Ribes Nigrum* L.) Pomace. *J. Hortic. Sci. Biotechnol.* **2009,** *84* (6), 100–106. DOI: 10.1080/14620316.2009.11512604.
78. Lončarić, A. et al. Green Extraction Methods for Extraction of Polyphenolic Compounds from Blueberry Pomace. *Foods* **2020,** *9* (11), 1521. DOI: 10.3390/foods9111521.
79. Čanadanović-Brunet, J. et al. Polyphenolic Composition, Antiradical and Hepatoprotective Activities of Bilberry and Blackberry Pomace Extracts. *J. Berry Res.* **2019,** *9* (2), 349–362. DOI: 10.3233/JBR-180362.
80. van Hoed, V. et al. Influence of Filtering of Cold Pressed Berry Seed Oils on Their Antioxidant Profile and Quality Characteristics. *Food Chem.* **2011,** *127* (4), 1848–1855. DOI: 10.1016/j.foodchem.2011.01.134.
81. Koleckar, V. et al. Condensed and Hydrolysable Tannins as Antioxidants Influencing the Health. *Mini Rev. Med. Chem.* **2008,** *8* (5), 436–447. DOI: 10.2174/138955708784223486.
82. Buzzini, P. et al. Antimicrobial and Antiviral Activity of Hydrolysable Tannins. *Mini Rev. Med. Chem.* **2008,** *8* (12), 1179–1187. DOI: 10.2174/138955708786140990.
83. Landete, J. M. Ellagitannins, Ellagic Acid and Their Derived Metabolites: A Review About Source, Metabolism, Functions and Health. *Food Res. Int.* **2011,** *44* (5), 1150–1160. DOI: 10.1016/j.foodres.2011.04.027.
84. Bobinaitė, R.; Viškelis, P.; Šarkinas, A.; Venskutonis, P. R. Phytochemical Composition, Antioxidant and Antimicrobial Properties of Raspberry Fruit, Pulp, and Marc Extracts. *CYTA J. Food* **2013,** *11* (4), 334–342. DOI: 10.1080/19476337.2013.766265.
85. Fotschki, B.; Juśkiewicz, J.; Sójka, M.; Jurgoński, A.; Zduńczyk, Z. Ellagitannins and Flavan-3-ols from Raspberry Pomace Modulate Caecal Fermentation Processes and Plasma Lipid Parameters in Rats. *Molecules* **2015,** *20* (12), 22848–22862. DOI: 10.3390/molecules201219878.
86. Oliveira, J.; Mateus, N.; de Freitas, V. Flavanols: Catechins and Proanthocyanidins. in *Natural Products*; Springer Berlin Heidelberg: Berlin, Heidelberg, 2013; pp 1753–1801. doi: 10.1007/978-3-642-22144-6_58.
87. Koleckar, V. et al. Condensed and Hydrolysable Tannins as Antioxidants Influencing the Health. *Mini-Rev. Med. Chem.* **2008,** *8* (5), 436–447. DOI: 10.2174/138955708784223486.

88. Dimic, E.; Vujasinovic, V.; Radocaj, O.; Pastor, O. Characteristics of Blackberry and Raspberry Seeds and Oils. *Acta Period. Technol.* **2012,** *43* (43), 1–9. DOI: 10.2298/APT1243001D.
89. Gustinelli, G.; Eliasson, L.; Svelander, C.; Alminger, M.; Ahrné, L. Supercritical CO2 Extraction of Bilberry (Vaccinium Myrtillus L.) Seed Oil: Fatty Acid Composition and Antioxidant Activity. *J. Supercrit. Fluids*, **2018,** *135*, 91–97. DOI: 10.1016/j.supflu.2018.01.002.
90. Kim, E.-O. et al. Comparison of Functional Constituents and Biological Activity of the Seed Extracts from Two Mulberry Fruits. *Prev. Nutr. Food Sci.* **2010,** *15* (2), 98–104. DOI: 10.3746/jfn.2010.15.2.098.
91. Fazio, A.; Plastina, P.; Meijerink, J.; Witkamp, R. F.; Gabriele, B. Comparative Analyses of Seeds of Wild Fruits of Rubus and Sambucus Species from Southern Italy: Fatty Acid Composition of the Oil, Total Phenolic Content, Antioxidant and Anti-inflammatory Properties of The methanolic Extracts. *Food Chem.* **2013,** *140* (4), 817–824. DOI: 10.1016/j.foodchem.2012.11.010.
92. Dulf, F.; Oroian, I.; Vodnar, D.; Socaciu, C.; Pintea, A. Lipid Classes and Fatty Acid Regio-distribution in Triacylglycerols of Seed Oils of Two Sambucus Species (S. Nigra L. and S. Ebulus L.). *Molecules* **2013,** *18* (10), 11768–11782. DOI: 10.3390/molecules181011768.
93. Yang, B.; Ahotupa, M.; Määttä, P.; Kallio, H. Composition and Antioxidative Activities of Supercritical CO2-extracted Oils from Seeds and Soft Parts of Northern Berries. *Food Res. Int.* **2011,** *44* (7), 2009–2017. DOI: 10.1016/j.foodres.2011.02.025.
94. Piskernik, S.; Vidrih, R.; L. Demšar, Koron, D.; Rogelj, M.; Žontar, T. P. Fatty Acid Profiles of Seeds from Different Ribes Species. *LWT* **2018,** *98*, 424–427. DOI: 10.1016/j.lwt.2018.09.011.
95. Šavikin, K. P.; Đorđević, B. S.; Ristić, M. S.; Krivokuća-Đokić, D.; Pljevljakušić, D. S.; Vulić, T. Variation in the Fatty-Acid Content in Seeds of Various Black, Red, and White Currant Varieties. *Chem. Biodivers.* **2013,** *10* (1), 157–165. DOI: 10.1002/cbdv.201200223.
96. Dobson, G.; Shrestha, M.; Hilz, H.; Karjalainen, R.; McDougall, G.; Stewart, D. Lipophilic Components in Black Currant Seed and Pomace Extracts. *Eur. J. Lipid. Sci. Technol.* **2012,** *114* (5), 575–582. DOI: 10.1002/ejlt.201100313.
97. Ruiz del Castillo, M. L.; Dobson, G.; Brennan, R.; Gordon, S. Genotypic Variation in Fatty Acid Content of Blackcurrant Seeds. *J. Agric. Food Chem.* **2002,** *50* (2), 332–335. DOI: 10.1021/jf010899j.
98. Goffman, F. D.; Galletti, S. Gamma-Linolenic Acid and Tocopherol Contents in the Seed Oil of 47 Accessions from Several Ribes Species. *J. Agric. Food Chem.* **2001,** *49* (1), 349–354. DOI: 10.1021/jf0006729.
99. Neoptolemos, J. P. Essential Fatty Acids. *Br. J. Surg.* **2005,** *77* (3), 353–354. DOI: 10.1002/bjs.1800770338.
100. Bakshi, A.; Mukherjee, D.; Bakshi, A.; Banerji, A. K.; Das, U. N. γ-linolenic Acid Therapy of Human Gliomas. *Nutrition* **2003,** *19* (4), 305–309. DOI: 10.1016/S0899-9007 (02)00862-6.
101. Sergeant, S.; Rahbar, E.; Chilton, F. H. Gamma-linolenic Acid, Dihommo-gamma linolenic, Eicosanoids and Inflammatory Processes. *Eur. J. Pharmacol.* **2016,** *785*, 77–86. DOI: 10.1016/j.ejphar.2016.04.020.
102. Simopoulos, A. P. The Importance of the Ratio of Omega-6/Omega-3 Essential Fatty Acids. *Biomed. Pharmacother.* **2002,** *56* (8), 365–379. DOI: 10.1016/S0753-3322(02)00253-6.

103. Beveridge, T.; Li, T. S. C.; Oomah, B. D.; Smith, A. Sea Buckthorn Products: Manufacture and Composition †. *J. Agric. Food Chem.* **1999,** *47* (9), 3480–3488. DOI: 10.1021/jf981331m.
104. Oomah, B. D. Sea Buckthorn Lipids. In *Sea Buckthorn (Hippophae Rhamnoides L.): Production and Utilization*, Li, T. S. C.; Beveridge, T. H. J., Eds.; NRC Research Press: Ottawa, 2003; pp 51–68.
105. Ricciarelli, R.; Zingg, J.-M.; Azzi, A. The 80th Anniversary of Vitamin E: Beyond Its Antioxidant Properties. *Biol. Chem.* **2002,** *383* (3–4), 457–465. DOI:10.1515/BC.2002.048.
106. van Hoed, V. et al. Berry Seeds: A Source of Specialty Oils with High Content of Bioactives and Nutritional Value. *J. Food Lipids* **2009,** *16* (1), 33–49. DOI: 10.1111/j.1745-4522.2009.01130.x.
107. Li, Q.; Wang, J.; Shahidi, F. Chemical Characteristics of Cold-Pressed Blackberry, Black Raspberry, and Blueberry Seed Oils and the Role of the Minor Components in Their Oxidative Stability. *J. Agric. Food Chem.* **2016,** *64* (26), 5410–5416. DOI: 10.1021/acs.jafc.6b01821.
108. A. de Jong, Plat, J.; Mensink, R. P. Metabolic Effects of Plant Sterols and Stanols (Review). *J. Nutr. Biochem.* **2003,** *14* (7), 362–369. DOI: 10.1016/S0955-2863(03)00002-0.
109. Górnaś, P.; Rudzińska, M. Seeds Recovered from Industry Byproducts of Nine Fruit Species with a High Potential Utility as a Source of Unconventional Oil for Biodiesel and Cosmetic and Pharmaceutical Sectors. *Ind. Crops Prod.* **2016,** *83*, 329–338. DOI: 10.1016/j.indcrop.2016.01.021.
110. Parry, J. et al. Fatty Acid Composition and Antioxidant Properties of Cold-Pressed Marionberry, Boysenberry, Red Raspberry, and Blueberry Seed Oils. *J. Agric. Food Chem.* **2005,** *53* (3), 566–573. DOI: 10.1021/jf048615t.
111. Tańska, M.; Roszkowska, B.; Czaplicki, S.; Borowska, E. J.; Bojarska, J.; Dąbrowska, A. Effect of Fruit Pomace Addition on Shortbread Cookies to Improve Their Physical and Nutritional Values. *Plant Foods Hum. Nutr.* **2016,** *71* (3), 307–313. DOI: 10.1007/s11130-016-0561-6.
112. Gagneten, M.; Archaina, D. A.; Salas, M. P.; Leiva, G. E.; Salvatori, D. M.; Schebor, C. Gluten-free Cookies Added with Fibre and Bioactive Compounds from Blackcurrant Residue. *Int. J. Food Sci. Technol.* **2021,** *56* (4), 1734–1740. DOI: 10.1111/ijfs.14798.
113. Schmidt, C.; Geweke, I.; Struck, S.; Zahn, S.; Rohm, H. Blackcurrant Pomace from Juice Processing as Partial Flour Substitute in Savoury Crackers: Dough Characteristics and Product Properties. *Int. J. Food Sci. Technol.* **2018,** *53* (1), 237–245. DOI: 10.1111/ijfs.13639.
114. Quiles, A.; Campbell, G. M.; Struck, S.; Rohm, H.; Hernando, I. Fiber from Fruit Pomace: A Review of Applications in Cereal-based Products. *Food Rev. Int.* **2018,** *34* (2), 162–181. DOI: 10.1080/87559129.2016.1261299.
115. Reißner, A.-M.; Beer, A.; Struck, S.; Rohm, H. Pre-Hydrated Berry Pomace in Wheat Bread: An Approach Considering Requisite Water in Fiber Enrichment. *Foods* **2020,** *9* (11), 1600. DOI: 10.3390/foods9111600.
116. Mäkilä, L. et al. Exploiting Blackcurrant Juice Press Residue in Extruded Snacks. *LWT -Food Sci. Technol.* **2014,** *57* (2), 618–627. DOI: 10.1016/j.lwt.2014.02.005.
117. Quiles, A. et al. Use of Berry Pomace to Replace Flour, Fat or Sugar in Cakes. *Int. J. Food Sci. Technol.* **2018,** *53* (6), 1579–1587. DOI: 10.1111/ijfs.13765.
118. Sharma, M.; Usmani, Z.; Gupta, V. K.; Bhat, R. Valorization of Fruits and Vegetable Wastes and byproducts to Produce Natural Pigments. *Crit. Rev. Biotechnol.* **2021,** *41* (4), 535–563. DOI: 10.1080/07388551.2021.1873240.

119. Nemetz, N. J.; Schieber, A.; Weber, F. Application of Crude Pomace Powder of Chokeberry, Bilberry, and Elderberry as a Coloring Foodstuff. *Molecules*, **2021,** *26* (9), 2689. DOI: 10.3390/molecules26092689.
120. Wathon, M. H.; Beaumont, N.; Benohoud, M.; Blackburn, R. S.; Rayner, C. M. Extraction of Anthocyanins from Aronia Melanocarpa Skin Waste As a Sustainable Source of Natural Colorants. *Color. Technol.* **2019,** *135* (1), 5–16. DOI: 10.1111/cote.12385.
121. Yamashita, C.; Chung, M. M. S.; dos Santos, C.; Mayer, C. R. M.; Moraes, I. C. F.; Branco, I. G. Microencapsulation of an Anthocyanin-rich Blackberry (Rubus spp.) Byproduct Extract by Freeze-drying. *LWT* **2017,** *84*, 256–262. DOI: 10.1016/j.lwt.2017.05.063.
122. Guliyev, V. B.; Gul, M.; Yildirim, A. Hippophae Rhamnoides L.: Chromatographic Methods to Determine Chemical Composition, Use in Traditional Medicine and Pharmacological Effects. *J. Chromatogr. B* **2004,** *812* (1–2), 291–307. DOI:10.1016/j.jchromb.2004.08.047.
123. Linnamaa, P. et al. Blackcurrant Seed Oil for Prevention of Atopic Dermatitis in Newborns: A Randomized, Double-blind, Placebo-controlled Trial. *Clin. Exp. Allergy* **2010,** *40* (8), 1247–1255. DOI: 10.1111/j.1365-2222.2010.03540.x.
124. Linnamaa, P. et al. Black Currant Seed Oil Supplementation of Mothers Enhances IFN-γ and Suppresses IL-4 Production in Breast Milk. *Pediatr. Allergy Immunol.* **2013,** *24* (6), 562–566. DOI: 10.1111/pai.12110.
125. Plainfossé, H.; Trinel, M.; Verger-Dubois, G.; Azoulay, S.; Burger, P.; Fernandez, X. Valorisation of Ribes nigrum L. Pomace, An Agri-Food Byproduct to Design a New Cosmetic Active. *Cosmetics* **2020,** *7* (3), 56. DOI: 10.3390/cosmetics7030056.
126. Samoraj, M.; Tuhy, Ł.; Rusek, P.; Rój, E.; Chojnacka, K. Pilot Plant Conversion of Blackcurrant Seeds into New Micronutrient Fertilizer Biocomponents via Biosorption. *BioResources* **2016,** *11* (1), 400–413.
127. Samoraj, M.; Tuhy, Ł.; Chojnacka, K. Valorization of Biomass into Micronutrient Fertilizers. *Waste Biomass Valor.* **2019,** *10* (4), 925–931. DOI: 10.1007/s12649-017-0108-6.
128. Juskiewicz, J. et al. The Fatty Acid Profile and Oxidative Stability of Meat from Turkeys Fed Diets Enriched with n-3 Polyunsaturated Fatty Acids and Dried Fruit Pomaces as a Source of Polyphenols. *PLOS ONE* **2017,** *12* (1), e0170074. DOI: 10.1371/journal.pone.0170074.
129. Das, Q. et al. Gut Microbiota, Blood Metabolites, and Spleen Immunity in Broiler Chickens Fed Berry Pomaces and Phenolic-Enriched Extractives. *Front. Vet. Sci.* **2020,** *7*. DOI: 10.3389/fvets.2020.00150.
130. Singh, A.; Gu, Y.; Castellarin, S. D.; Kitts, D. D.; Pratap-Singh, A. Development and Characterization of the Edible Packaging Films Incorporated with Blueberry Pomace. *Foods* **2020,** *9* (11), 1599. DOI: 10.3390/foods9111599.
131. Gouw, V. P.; Jung, J.; Simonsen, J.; Zhao, Y. Fruit Pomace as a Source of Alternative Fibers and Cellulose Nanofiber as Reinforcement Agent to Create Molded Pulp Packaging Boards. *Compos. Part A Appl. Sci. Manuf.* **2017,** *99*, 48–57. DOI: 10.1016/j.compositesa.2017.04.007.
132. Park, S.; Zhao, Y. Development and Characterization of Edible Films from Cranberry Pomace Extracts. *J. Food Sci.* **2006,** *71* (2), E95–E101. doi: 10.1111/j.1365-2621.2006.tb08902.x.
133. Saad, N. et al. Enzyme-Assisted Extraction of Bioactive Compounds from Raspberry (Rubus Idaeus L.) Pomace. *J. Food Sci.* **2019,** *84* (6), 1371–1381. DOI: 10.1111/1750-3841.14625.

134. Reißner, A.-M. et al. Composition and Physicochemical Properties of Dried Berry Pomace. *J. Sci. Food. Agric.* **2019,** *99* (3), 1284–1293. DOI: 10.1002/jsfa.9302.
135. Oomah, B. D.; Ladet, S.; Godfrey, D. V.; Liang, J.; Girard, B. Characteristics of Raspberry (Rubus Idaeus L.) Seed Oil. *Food Chem.* **2000,** *69* (2), 187–193. DOI: 10.1016/S0308-8146(99)00260-5.
136. Yang, B.; Kallio, H. P. Fatty Acid Composition of Lipids in Sea Buckthorn (Hippophaë Rhamnoides L.) Berries of Different Origins. *J. Agric. Food Chem.* **2001,** *49* (4), 1939–1947. DOI: 10.1021/jf001059s.
137. Gómez-Mejía, E. et al. Valorisation of Black Mulberry and Grape Seeds: Chemical Characterization and Bioactive Potential. *Food Chem.* **2021,** *337* (2020), 127998. DOI: 10.1016/j.foodchem.2020.127998.

CHAPTER 5

Value-Added Lipids from Plant Foods and Respective Applications

MARIANA C. PEDROSA, JONATA M. UEDA, SANDRINA A. HELENO, MÁRCIO CAROCHO, FILIPA S. REIS, ISABEL C. F. R. FERREIRA, and LILLIAN BARROS

Centro de Investigação de Montanha (CIMO), Instituto Politécnico de Bragança, Campus de Santa Apolónia, Bragança, Portugal

ABSTRACT

Since ancient times, man has used plants as a food source, associating their consumption with nutritional and health benefits. Among the nutrients obtained from plants, which the consumer associates more with beneficial health effects than those obtained from the animal diet, we can highlight the lipids. Lipids play a vital role in plant or animal biology, as they are a source of energy and can also act as hormones or signaling molecules. In plants, the lipid content varies according to the species, but this fraction is mainly constituted by polar and nonpolar lipids and acyl glycerides. The main classes of lipids found in plants include glycerols, phospholipids, glycolipids, sterols, free fatty acids, vitamins, waxes, carotenes, terpenes, hydrocarbons, phenolics, tocols, chlorophyll pigments, and their derivatives. Given the importance of these nutrients and their effects on human health, obtaining and using them as value-added molecules with a potential industrial application at different levels can and should be analyzed. Nowadays, several sectors (e.g. food, cosmetic, or pharmaceutical) invest in the search and characterization of different molecules to be used, such as functional and/or nutraceutical ingredients. Hence, promoting the concept of circular

Food Byproducts Management and their Utilization. Ricardo Gómez-García, Ana A. Vilas-Boas, Débora A. Campos, Maria Manuela Pintado, & Cristóbal Noé Aguilar (Eds.)

economy and sustainability of processes and products, byproducts have been identified as an excellent source of interest compounds.

With these aspects in mind, this chapter presents the main vegetable sources of lipids, emphasizing food byproducts, the beneficial effects of these molecules on human health, ways of obtaining and extraction methodologies, and the potential for application in different sectors.

5.1 INTRODUCTION

5.1.1 THE IMPORTANCE OF LIPIDS AND THE MAIN SOURCES

Vegetable oils represent about 25% of the daily caloric intake, an important fraction of the human diet. Despite the risks associated with the intake of fats, namely obesity and cardiovascular diseases, fatty acids are essential for the proper functioning of the human body by supplying energy and providing beneficial health effects.[1,2]

Lipids are generally defined as "substances insoluble in water and soluble in nonpolar solvents, containing large hydrocarbon chains." However, some short-chain lipids (1–4 carbons) are miscible in water, and some authors exclude C1–C3 fatty acids from the definition of lipids, maintaining only C4 (butyric acid) due to its presence in dairy products. Lipids classification can be carried out according to its physical characteristics at room temperature (oils are presented in the liquid form, and fats are solid), polarity, essential and nonessential fatty acids, or structure, the most used classification.[3]

In plants, lipids have two main functions: Performance and regulation of cells and tissues metabolism and energy storage for seed germination.[4] In addition, unsaturated fatty acids also have a protective function in plants, preventing possible pathogens in the food matrix. Examples are phenolic lipids, which act against insects and abiotic stresses.[5] Lipids are considered essential for plants since the matrix of the photosynthetic membrane is formed by lipids. This membrane is located in the chloroplast and is constituted mainly by glycolipids and one phospholipid, phosphatidylglycerol.[6]

As part of plant lipids, there are also other fat-soluble molecules, namely vitamins A and E, and carotenoids, which are often studied because of their ability to protect cells from oxidative damage. These antioxidant molecules contribute to preventing the oxidation of polyunsaturated fatty acids, for example, by breaking the chain reaction propagation through the donation of hydrogen atoms to the peroxyl radical, preventing the attack on other lipids.[7]

Regarding the main sources of lipids in plants, seeds are generally rich in fatty acids, namely omega-6, such as grapes, corn, and sunflower seeds.[8] Carotenoids are among the most widely distributed pigments in nature, responsible for coloring flowers, fruits, and vegetables. They can be found in tomatoes (lycopene), oranges and tangerines (β-cryptoxanthin), carrots (β-carotene), and watercress (lutein).[9] Phenolic lipids, especially resorcinolic lipids, are found in greater quantities in rice, wheat, rye, and other grains and cereals.[10]

Phylloquinones and menaquinones are compounds called vitamin K (K_1 and K_2, respectively). Only vitamin K_1 is present in plants, found in dark green leaves such as spinach.[11] Finally, as for phytosterols, they are found in all vegetable lipid sources, with oils and margarine being the most significant sources, but also found in certain seeds and vegetables.[12]

5.2 LIPID CHARACTERIZATION

As for their classification, lipids in plants can be subdivided into several groups, but the main ones include fatty acids, carotenoids, phytosterols, fat-soluble vitamins, and phenolic lipids (Figure 5.1). Although one group does not exclude the other, this representation covers a greater amount of lipid compounds in plants.

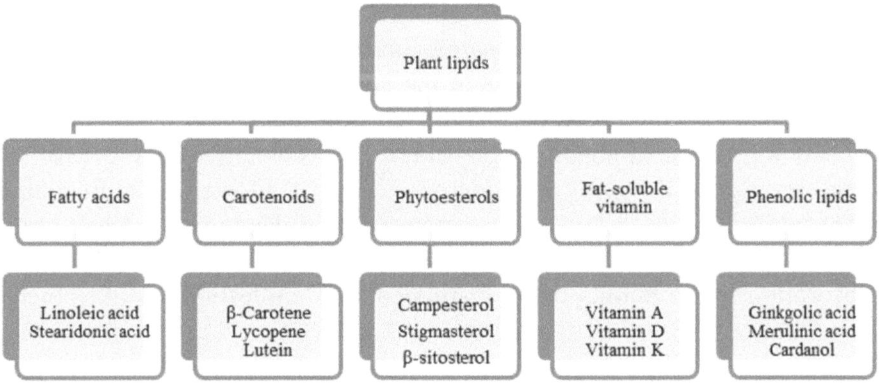

FIGURE 5.1 Classification of plant lipids and examples.

Fatty acids are defined as carboxylic acids composed of several carbon chains and can be classified as short chain (<6 carbon atoms), medium (6–12), long (12–21), or very long (>21 carbon atoms). Plants reserve unsaturated fatty acids since they are generally liquid during the processes of deposition, storage, and consumption, and generate greater fluidity in the membranes and protection against oxidation.[13,14]

Phenolic lipids are characterized by being derivatives of mono and dihydroxyphenols, obtained from the secondary metabolism of plants when subjected to stress conditions, linked to the lipid part, consisting of long chains of aliphatic hydrocarbons. Phenolic lipids have both the hydrophobic and hydrophilic part of the molecule. Their amphiphilic character ensures their greater applicability in the food industry, and a higher absorption by the organism.[10,15]

The human body is not able to synthesize carotenoids, thus being necessary to obtain them from external sources.[16] In plants, they are known and studied for their lipid-soluble pigmentation, found in the membranes of chloroplasts and chromoplasts; they protect plants against photo oxidative processes, help in the control of growth and stability of the cell membrane, while also aiding photosynthesis.[17]

Carotenoids are classified as tetraterpenoids, based on isoprenoid structures with acyclic, monocyclic, or bicyclic forms, and can be divided into carotenes (hydrocarbons) or xanthophylls (containing oxygen in the form of hydroxyl, methoxy, and others).[18,19]

Phytosterols are defined as the unsaponifiable lipid fraction, which can be divided into sterols, stanols (saturated form of sterols), and conjugated forms esterified with phenolic acids and free fatty acids. These molecules belong to the family of triterpenes, and are present in the cell walls of plants, responsible for controlling the permeability and fluidity of membranes, in which esters play a storage role in the cells. In addition, phytosterols are precursors of steroid hormones, responsible for plant defense.[19,20]

Among the most abundant phytosterols, β-sitosterol, campesterol, and stigmasterol are the most common ones, being β-sitosterol the most abundant molecule. Phytosterols are structurally similar to cholesterol, having only an additional methyl or ethyl group in their side chain. Their poor absorption in the human body, compared with cholesterol, is attributed to this feature.[19,21]

The fat-soluble vitamins (A, D, E, and K) are very different in terms of their biosynthesis pathway, bioavailability, physiological function, and other aspects. For instance, vitamins A, D, and E are sensitive to light and oxygen, while vitamin K is the only vitamin stable in the presence of heat and oxygen.[22] Among vitamin D classifications, only vitamin D_2 is produced by plants, namely ergocalciferol, but in low amounts.[23]

5.3 THE BENEFICIAL EFFECTS OF LIPIDS ON HUMAN HEALTH

The effects of lipids on human health are related to the functions of these components in the human body. These molecules are part of the cell membrane structure and act as mediator molecules (signaling and regulation); as transporters of fat-soluble vitamins, besides having the capacity to store energy.[24,25] Although the functions of lipids can be synthesized into these four main pathways, within these groups of functions, these molecules may have several specific roles due to the broad spectrum of lipid structures.[26]

The movement of lipids in the human body is also influenced by diet (quantity ingested and quality), lifestyle, and genetics. These factors interact with each other and define lipid circularity, impact on metabolism and possibility of causing disorders, such as obesity, if the circularity becomes erratic.[27] Despite the benefits of lipids to maintain some body functions, excess intake can cause serious health issues.[25]

Concentration plays an essential role in pathophysiological processes; its variation can affect the functionality of these processes.[24] Therefore, these biomolecules' irregular metabolism is connected with the risk of adverse health effects like chronic diseases.[27] For instance, the low ratio between polyunsaturated (PUFA)/saturated fatty acids (SFA) and between n-3 PUFA/n-6 PUFA are a risk factor to development pathologies.[28] Long-chain polyunsaturated fatty acids provide several health benefits to human metabolisms, such as controlling blood cholesterol levels, mainly omega-3 series and α-linolenic acid (C18:3n3). This family of molecules helps prevent coronary heart disease, inflammation, hypertension, cancer, and also autoimmune disorders. PUFA cannot be synthesized by the organism and must be obtained from the diet, being abundant in nuts, grains, vegetables, and marine foods.[25,29] The composition of lipids in triacylglycerol and phospholipids also influences the nutritional values, stability, and physical properties of these lipids. For example, high quantities of linoleic acid can positively influence human health, such as preventing cancer. In addition, the presence of high levels of tocopherol can defend the body against oxidative stress and protect against coronary heart diseases.[30] Linoleic acid (C18:2n6) is an omega-6 fatty acid that acts in several systems in the body, such as controlling metabolism, the reproductive and nervous system and also acts in the growth of bones, muscles, skin, and hair.[31]

Regarding lipids of cell membranes, phospholipids represent the vast majority.[32] The cell membrane structure is characterized by being a fluid phospholipid bilayer that serves as a matrix for other constituents, mainly proteins and glycoproteins. The amphipathic character provided by the polarity of the lipid and the hydrophobicity of the hydrocarbon chains defines the membrane arrangement.[33–35] Besides protection from cell surroundings, several studies have demonstrated the role of lipids as mediators of membrane signaling and in cellular communication.[36]

Arachidonic (20:4n6), eicosapentaenoic (20:5n3), and docosahexaenoic acids (22:6n3) are unsaturated fatty acids, in which the first is a crucial component of the membrane and metabolized from linoleic acid (18:2n6) and precursor of eicosanoids (controls parts of cell activity). While the last two perform essential functions in the phospholipid membrane, they are metabolized from α-linolenic acid, and eicosapentaenoic acid is also the precursor of some eicosanoids. Linoleic acid can be found in vegetables, while α-linolenic acid can be found in leaves, roots, and stalks. Eicosanoids are substances similar to short-lived hormones and act as communication regulators between cells in multiple human systems, such as cardiovascular, reproductive, respiratory, renal, nervous, immune, and endocrine.[25]

Eicosanoids are obtained from a set of enzymatic reactions having fatty acids as precursors. The metabolism of eicosanoids is part of the body's immune system. They are mediators for the inflammatory response and essential for various immune response processes.[37] Other fatty acids with beneficial effects on health are $\Delta5$-unsaturated polymethylene interrupted (double bond at $\Delta5$ position, such as pinolenic acid, taxoleic acid, sciadonic acid), which can act on the inflammatory response by preventing the production of pro-inflammatory mediators, as well as reducing blood pressure and controlling the concentration of cholesterol in the blood.[38]

Phytosterols and phytostanols are plant sterol and stanols, respectively, that can be derived from vegetable oils, fruits, nuts, and cereals. These substances have the capacity to reduce the level of cholesterol absorption, lowering the low-density lipoprotein (LDL) levels in the blood. However, secondary effects may occur and as previously mentioned, these phytochemicals may act in other biological processes and could trigger unwanted disorders. Therefore, more studies are needed to understand the effects (positives and negatives) of these phytochemicals on the body.[39,40] Vergès and Fumeron[41] studied the relationship between sterols and cardiovascular diseases but could not reach a concrete conclusion about the connection. The available data is ambiguous, showing both positive and adverse effects, and even no correlation between the two factors. Although it was not possible to reach a concrete conclusion, studies have shown a potential risk for the development of cardiovascular diseases associated with high levels of sterols in the blood.

Plant sterols when incorporated in fat-based foods become susceptible to auto-oxidation, resulting in the formation of oxidation products (e.g., 7-hydroxy-(7-OH-), 7-keto-, 5,6-epoxy-(epoxy-), and 3,5,6-triol-(triol) derivatives from plant sterols) that may provide antidiabetic and anticarcinogenic characteristics. However, they can also provide negative outcomes, such as cytotoxicity, oxidative stress, and atherogenicity, according to studies performed in vitro and in vivo. Different oxidation products from plants appear to have distinct biological effects in the body, but information on physiological effects of these products is still limited.[42]

Carotenoids and vitamin E can act as antioxidants in the human organism. Carotenoids can support immune system activity, reduce the risk of cancer and cardiovascular diseases, and are also precursors to vitamin A. Vitamin A has functions in gene regulation and vision, while vitamin K assists in blood coagulation and calcium absorption.[43,44] The antioxidant activity of vegetable oils can be explained by the presence of bioactive compounds, such as phenolics, phytosterols, tocopherols, acting against reactive oxygen species and protecting against chronic and cardiovascular diseases.[30] Previous research indicated that the consumption of oils derived from nut tree might cause better health benefits than the nut itself.[25]

5.4 BIOWASTE RAW MATERIAL AS SOURCES OF VALUABLE LIPIDS

Nowadays, raw materials from industrial waste and processing are gaining more notoriety, mainly due to the search for more sustainable processes and technologies along the supply chain. This sustainable strategy provides waste valuation with the extraction of value-added compounds (bioactive compounds, lipids, proteins, starch, dietary fibers, bio-based chemicals); thus, it can reduce raw material cost for secondary processes and contribute to the improving circular economy.[45–47]

Actions aimed at sustainability are already part of the ideals of the European Union (EU), such as encouraging circular and environmentally neutral products. The *European Green Deal* brings together a set of strategies to achieve the Sustainable Development Goals of the United Nations. In addition to the "New Circular Economy Action Plan," aiming at a more sustainable growth of the EU and less linear economy, the European Commission (EC) has also defined "A Farm to Fork Strategy for a fair, healthy and environmentally-friendly food system." All these efforts are means for the EU to reach the goal of climate neutrality by 2050.[48,49] The strategy specifies the measures that the EC intends to adopt, such as expanding and promoting sustainable production methods with social responsibility and promoting circular business, avoiding major waste and severe impacts on the environment.[49]

Agricultural and agro-industrial activity produces a large amount of waste and byproducts along the production chain, namely bagasse, seeds, peels, stem, bran, pulp, and wastewater. However, these products may contain high content of functional and nutritional compounds and are often discarded in landfills, losing their value and causing negative environmental impacts.[47]

Although some kinds of agricultural waste (e.g., rice bran, spent ground coffee, and flour of oilseed) present considerable amounts of lipids, these quantities are relatively lower than oilseeds and oleaginous biomass. Thus, more efficiency is required, whether in choosing the most appropriate extraction method or/and with the most suitable solvents.[46] Table 5.1 shows some byproducts that are valuable lipid sources. Regarding the classification of lipids, most of the research works found in the literature studied the identification and quantification of fatty acids in byproducts. There are also several studies in which tocopherols, carotenoids, and phenolics were identified, but they are in smaller proportions compared with fatty acids.

Mango is a tropical fruit with an estimated production of 30–50 million metric tons worldwide. It can be used to produce various products, such as juices, jams, flakes, and fruit bars. Consequently, tons of byproducts are generated in the production of all these processed foods. Byproducts of peels, chipped fruits, and kernels can account for around 35–40% of production waste.[54] Cucurbits are plants from the Cucurbitaceae family, including, for instance, watermelon and pumpkin. Their seeds are byproducts usually discarded and present no economically viable destination. However, these seeds have a great nutritional value, since they are sources of vitamins and minerals (e.g., iron, phosphorous, and zinc) and can provide various benefits to human health. These products have the potential to be used in the pharmaceutical and food industry for the production of oils, dietary supplements, and food enrichment.[30]

According to the Olive Council, in 2019–2020, the world production of olive oil was estimated to be around 3.1 million tons. Byproducts from the olive oil industry, such as olive stone and olive skins, are sources of PUFA and triterpene acids, respectively. The olive mill oil wastewater generated from different processing steps can also be a source of carbohydrates, fatty acids, tocopherols, and specially phenolic compounds.[50] Elkacmi et al.[65] confirmed that this wastewater can be a natural source of oleic and linoleic acids by fractional crystallization, being a suitable method of providing value to this raw material. Karrar et al.[30] also reported many benefits from pumpkin seed oil, such as in the digestive and cardiovascular system, in

TABLE 5.1 Lipid Compounds from Industries Byproducts.

Source	Byproduct	Main identified compounds	Reference
Olive oil industry	Pomace	Residual oil, phenolics, tocopherols, fatty acids	50
	Tree leaves	Phenolics compounds	
	Olive stones	PUFA (mainly linoleic acid)	
	Olive skins	Triterpene acid	
	Olive mill oil wastewater	Fatty acids, carotenoids, tocopherols	
	Olive pomace oil	Unsaponifiable matter (sterols, fatty alcohols, tocopherols, triterpene alcohols, and squalene)	51
Broccoli (*Brassica oleracea* L. *var. Italica*)	Leaves	Unsaturated fatty acids (mainly alfa-linolenic (18:3 n-3), linoleic (18:2 n-6), and palmitic palmitic (16:0))	29
Brazil nut *(Bertholletia excelsa* Humb. & Bonpl.*)* beverage	Cake from Brazil nut	Mainly UFA (linoleic and oleic acids (18:1)), SFA (palmitic and stearic acids), and minor lipid compounds (free fatty acids, squalene, tocopherols, and β-sitosterol)	52
Tomato industry	Paste waste	Lycopene and b-carotene	53
	Seeds and skins	Lycopene	
Soybean industry	Waste deodorization distillate	Fatty acids and tocopherols	53
Mango industry	Kernel	UFA (oleic and linoleic) and SFA (Stearic acid).	54
	Seed almonds fat	Fatty acids (mainly oleic, stearic, and palmitic acids) and smaller amounts of linoleic, arachidic, behenic, lignoceric, and linolenic acids	55,56
Pistachio nuts	Low marketable appearance and quality nuts	54.73% oleic, linoleic, and 9.46% palmitic acids	57
Watermelon	Seeds	UFA (61% linoleic acid and oleic acid) and 13% of palmitic acid	31
Corn	Kernels	TAG, phytosterol fatty acyl esters, free phytosterols, ferulate, phytosterol ester, free linoleic acid, and γ-Tocopherol.	58

TABLE 5.1 *(Continued)*

Source	Byproduct	Main identified compounds	Reference
Coffee industry	Spent coffee grounds	Mainly triglycerides, palmitic acid, stearic acid (C18:0), linoleic acid, and unsaponifiable matter	59,60
Tiger nut milk	Tiger nuts byproducts	Oleic acid (62.4–69.0%), palmitic (13.8–17.9%), linoleic (10.9–12.1%), stearic (3.0–4.3%), and tocopherols	61
Cucurbit vegetables	Seeds	MUFA, PUFA, tocopherols, carotenoids, β-carotene, phytosterol, and vitamins.	30
Pine (*Pinus pumila*)	Kernel oil	High content of UFA (pinolenic, linoleic, and oleic acid) and lower content of SFA (palmitic and stearic)	38
Gurum (*Citrulluslanatus var. Colocynthoide*)	Seeds	Linoleic acid, oleic, stearic, palmitic, linolenic and gadoleic acid, tocopherols, sterols, and antioxidant activity	62
Açaí (Euterpe *Oleracea* Mart.)	Pulp	Triglycerides (palmitodiolein and triolein), oleic, palmitic, linoleic, palmitoleic and atearic acids, phospholipids, sterols, sphingolipids, terpenes, and antioxidant potential	63
Rice oil processing	Bran wax	Policosanols (linear saturated fatty alcohols - C_{24}-C_{34})	
	First bran fraction from rice polishing	Policosanol	64
Industrial processing of tomato	Pomace	Carotenoids (mainly in forms of lycopene and ß-carotene)	44

the energetic metabolism, against cancer and urinary infections, as well as antimicrobial, antidiabetic, antihypercholesterolemic, and anti-inflammatory activity. The authors also specified the action of sterols as inducers of the healthy functioning of the bladder and prostate gland.

Besides the other compounds exhibited in Table 5.1, spent coffee grounds possess unsaponifiable matter that presents themselves as important compounds that can be used in the pharmaceutical and cosmetic industries, being rich sources of tocopherols that are essential in providing oxidative stability, and also as functional agents in the food industry.[60]

Roselló-Soto et al.[61] demonstrated the potential use of oils derived from the byproduct of "horchata" (typical tiger nut milk) production in functional or nutraceutical foods, due to the presence of large amounts of monounsaturated fatty acids (MUFA) and PUFA, in smaller quantities, mainly C18:2n6. According to Buratto et al.[63] the presence of large amounts of oleic and linoleic acids (essential fatty acids), and unsaturated fatty acids (UFA) in oils obtained from the açaí pulp (byproduct) makes its use attractive as a nutraceutical substance, mainly because they are fatty acids that are not metabolized by humans. The application of the lipolysis bioprocess can be a necessary tool for the potential use of açaí oils as an omega source because it allows the extraction of oleic and linoleic fatty acids and their possible implementation in products.

While presenting various difficulties and challenges, byproducts can be beneficial for multiple areas (e.g., economic, environmental, and industrial) and applied in the cosmetic, pharmaceutical, and food fields, which reinforces the importance of research related to this theme.[54]

5.5 LIPID BIOSYNTHESIS

Glycerolipids are produced by two different pathways, either from the chloroplast (prokaryotic pathway) or through endoplasmic reticulum (eukaryotic pathway); a plant may have both glycerolipid biosynthesis pathways. In addition, these two pathways also represent the biosynthesis of triacylglycerols, cutins, waxes among other lipidic compounds.[66] Figures 5.2–5.4 show the biosynthetic pathways of fatty acids and carotenoids, phytosterols and fat-soluble vitamins, and phenolic acids, respectively.

The biosynthesis of fatty acids is based on the constant incorporation of two carbon units derived from acetate for initiation and subsequent formation of the acyl group containing 16 or 18 carbons.[66] Initially, acetate $(HCO_3)^-$ produces malonyl-CoA (co-enzyme A), and together with the acyl carrier protein they enter the fatty acid synthase cycle, producing acyl groups composed of 4 carbons. Finally, two carbon units are added, until the formation of 16:0 or 18:0 fatty acids.[67]

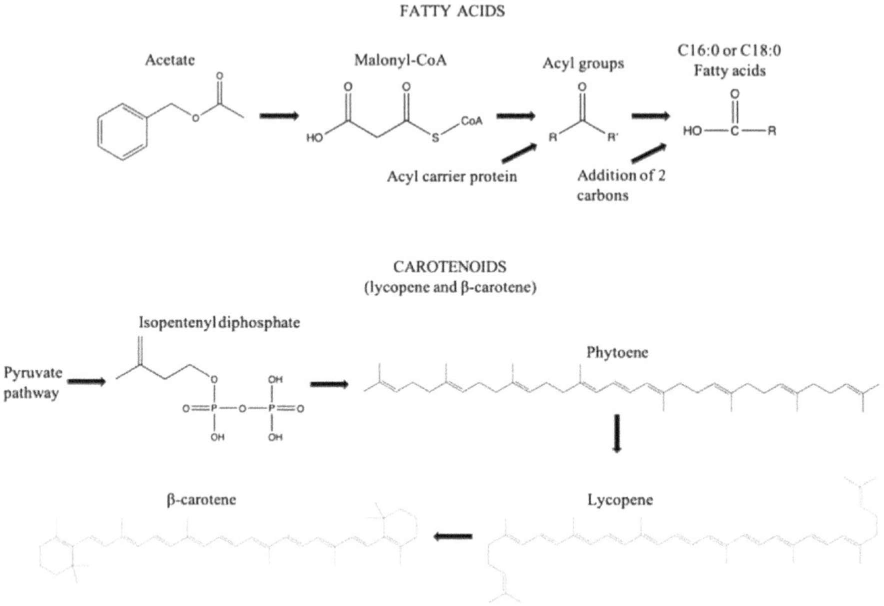

FIGURE 5.2 Fatty acids and carotenoids biosynthesis.

As for carotenoids, they are usually formed by C_{40} terpenoid compounds from the condensation of eight isoprene units, synthetized within the plastids, produced from isopentenyl diphosphate, which is produced from the pyruvate pathway. The carotenoid molecule has a reverse linkage order in the center, making it symmetrical, with double bonds responsible for light absorption. In addition to protecting against photo-oxidation, carotenoids are responsible for providing substrates for the plant's growth regulator (abscisic acid) and some other hormones.[18,68,69]

The precursor responsible for the formation of all phytosterols is a molecule named squalene, formed from the biosynthetic pathway to produce farnesyl diphosphate. Several reactions occur until the formation of 24-methylenelophenol, then forming two compounds: Episterol and avenasterol. While episterol is responsible for the formation of the campesterol, avenasterol sequentially generates fucosterol, *β*-sitosterol, and stigmasterol, respectively.[20]

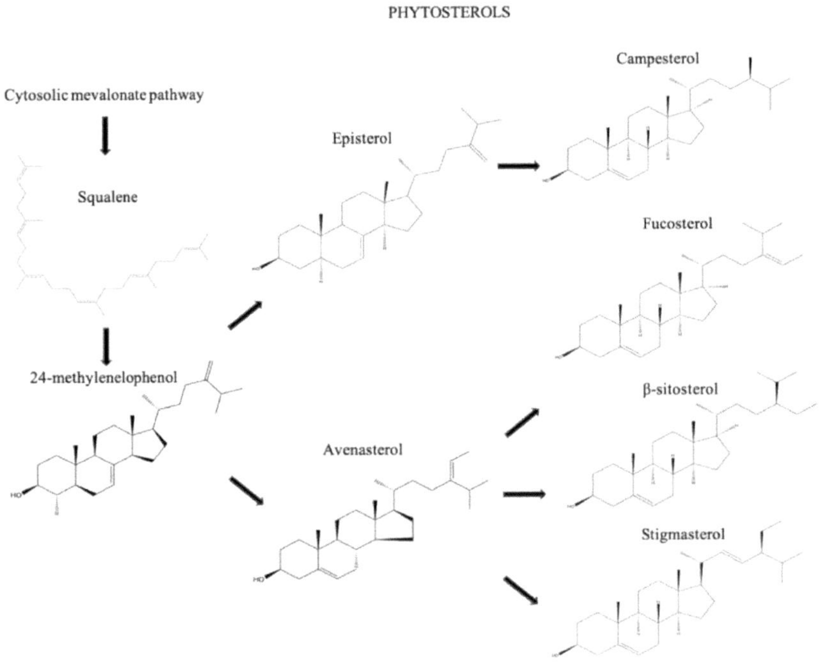

FIGURE 5.3 Phytosterol biosynthesis.

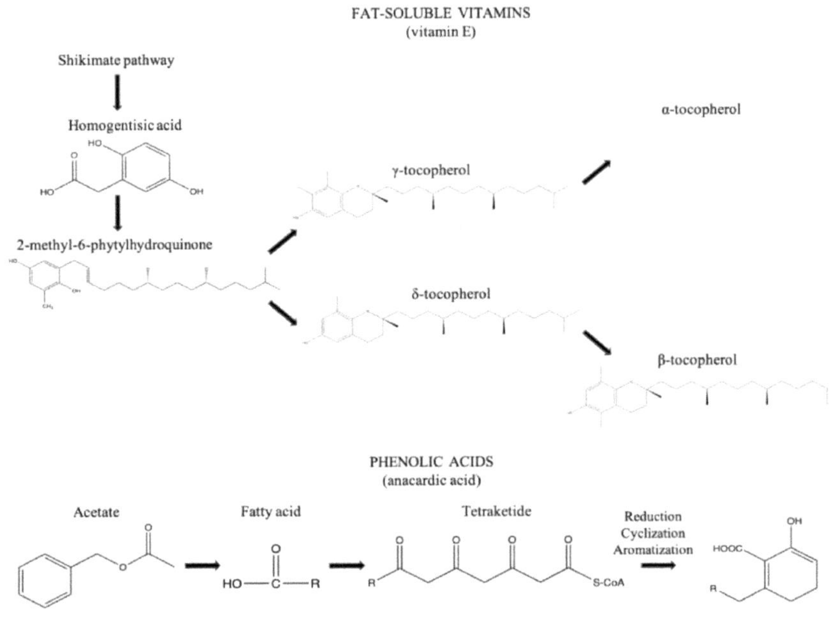

FIGURE 5.4 Fat-soluble vitamins and phenolic acids biosynthesis.

Tocopherols (vitamin E constituents) begin the biosynthesis path through the condensation of homogentisate, until the formation of the first intermediate, 2-methyl-6-phytylplastoquinone, responsible for the formation of all tocopherols. From a common cyclase, γ- and δ-tocoferol are produced, following a methylation reaction to form the last tocopherols, α and β.[70]

As for phenolic lipids, specifically resorcinolic lipids, the formation of these compounds is initiated from the path of acetogenic or polyketide biosynthesis. Initially, acetate is formed from glucose metabolism, being used as a precursor for the formation of long chains of phenolic lipids, incorporating with the phenolic ring of anacardic acid (6-pentadecylsalicylic acid).[71]

5.6 EXTRACTION OF LIPIDS

Selecting an appropriate extraction method and adequate parameters is a crucial stage to obtain lipids with satisfactory yields and purity. These choices will determine the efficiency, quality, and quantity of the final molecules.[72,73] The ideal extraction method should be carried out in short periods, with low energy and solvent consumption, preferably with solvents that are not harmful to human health and do not compromise the stability of the target biomolecules.[73]

An essential step to be performed before the extraction is the previous operational tests and searches to establish the range of parameter values.[74] However, a difficulty related to lipid extractions found in literature is the heterogeneity of the parameters used, such as temperature, pressure, solvent ratio/solid, making the comparison more challenging.[60]

The solvent choice is another important aspect in lipid extraction because it will influence its solubility. Therefore, the polarity (the fraction of polar or nonpolar elements) of lipids guides the choice for the most suitable solvent system for the extraction. These molecules are soluble in organic solvents (e.g., diethyl ether, benzene, methanol, ethanol, hexane, chloroform, petroleum ether, and their mixtures) and poorly soluble or insoluble in water. For instance, phospholipids and glycolipids, which are more polar, are better extracted with polar mixtures, while nonpolar solvents are better extractants of neutral lipids, such as fatty acids, glycerides, and sterols.[32,37,75] Although polarity is an essential parameter, the complex matrix of each plant impacts the extraction of lipids.[58] Moisture content, particle size, time and type of extraction, and solvent mixture are also characteristics that influence the extraction efficiency.[59,72] The solvent's toxicity should also be considered because some food, pharmaceutical, and cosmetic applications require ingestion or exposure. Therefore, the solvents should not be harmful to human health, nontoxic, or either cause negative effects over time.[60]

Natural lipids can be linked in three different ways with other molecules through van der Waals interaction between lipids–proteins and lipid–lipid; electrostatic and hydrogen bonding with proteins; and covalent bonding between lipids, carbohydrates, and proteins. These interactions become obstacles to removing and separating lipids from the cellular matrix, requiring physical and chemical treatments.[32,37]

The temperature can be an ally to increase rates of lipids extraction. Regarding the solvent, the increase in temperature decreases the strength of intermolecular forces which decreases the viscosity and helps the penetration into the plant matrix particles, as well as the decrease in surface tension. Related to the matrix, the temperature breaks the cohesiveness and adhesiveness between oil molecules and oil-plant matrix molecule interactions, consequently improving the solubility of lipids.[59]

Moreau et al.[58] compared the effects of temperature and solvent in the extraction of lipids in corn kernel and ground oats via pressurized liquid extraction. In general, higher amounts of lipids were extracted at 100°C than at 40°C and at 100°C, with no indication of lipid instability. Regarding the polarity of the molecules, methylene chloride and hexane were the solvents that extracted the highest quantities of nonpolar lipids from corn, such as triacylglycerides (TAG), free phytosterols, fatty acyl esters, and ferulate esters phytosterols, except for free linoleic acid that was better extracted with ethanol. For the oats, hexane and methylene chloride were also the best solvents for TAG. However, the study showed a minimal effect of the temperature and solvent. Once more, ethanol was the best solvent to extract free linoleic acid. Concerning polar lipids content, increase in polarity and temperature, in both foods, increased the extraction yields.

Pretreatment is a stage that takes place before extraction, which is usually employed to prepare the material and help to optimize the extraction processing. Cleaning, milling, and thermal treatment are some examples.[76] Furthermore, attention must be paid during the whole process of extraction to prevent changes in the lipids, such as oxidation or hydrolysis caused by enzymes.[32]

Loyao et al.[60] defined two types of lipid extractions from plant-based derivatives, mechanical and solvent extraction. The first one is suggested for samples with more than 20% w/w of lipid contents and provides low yield, but oils of higher quality. The second, with biomass less than 20% w/w of lipids, provides high yields, but with traces of solvent.[31]

Conventional methods of lipids extraction, for example Soxhlet, usually demand long extraction times, large volume of solvents, and a limited number of sample processing. Nevertheless, Soxhlet is one of the main methods applied in studies to determine and compare data. Nonconventional techniques, such as supercritical fluid extraction, microwave-assisted extraction, accelerated solvent extraction, and ultrasound-assisted extraction, are gathering more interest. Among some advantages are the replacement of hazardous solvents, reduction of solvent consumption, time and efficiency increase.[60,75] New demands for "greener" processing for both consumers and the environment are growing worldwide. Industries are increasingly seeking for innovation to meet these demands and remain competitive in the market.[77] Table 5.2 summarizes the most common techniques for lipids extraction.

5.6.1 CONVENTIONAL EXTRACTION TECHNIQUES

The Soxhlet extraction technique has been used as a reference for lipid extraction and applied for solid–liquid extraction in several areas. This procedure, which consists of a continuous percolation (recirculation current) of a selected solvent, according to the desired final characteristics,[81] has been widely used.[29,50,61,72,82] However, it has some disadvantages, such as long extraction times, high amounts of solvents, unsuitability for thermo-labile compounds, and the fact that it requires a step for evaporation and concentration of the final extracts.[77,83] Another conventional method used to extract lipids is pressing, a mechanical method of extraction widely used for seeds with large quantities of oils. Although its efficiency is low, since high amounts of oils can remain on the substrate, it is a straightforward and safe technique. Therefore, and given that it is a methodology used in some industrial processes, researchers have been looking for alternatives both to take advantage of and to extract the components remaining in the byproducts and waste generated in mechanical extraction, and the method itself.[31,84,85]

Nowadays, there is a growing concern about the environmental impacts of using certain solvents, which has led to more sustainable techniques for carrying out extractions. The use of new alternative unconventional, emerging, or green extraction methodologies has thus gained considerable attention in recent years. These new techniques also aim to reduce the consumption of raw materials, solvents, and energy, minimizing extraction times.

5.6.2 SUPERCRITICAL FLUID EXTRACTION (SFE)

The supercritical state of a fluid is achieved when it is brought to a certain temperature and pressure above the critical point. The fluid behaves with

TABLE 5.2 Extraction Methodologies Applied to Industries' Byproducts for Lipid's Recovering.

Byproducts	Estimated lipid content	Extraction technique	Oil yield (w/w)	Extraction parameters	References
Spent coffee grounds	16%	Soxhlet extraction	7–30.4%	480 min, solid/solvent ratio (1:9 w/v);	59
		Accelerated solvent extraction		165°C, 70 mL/min;	
		Microwave-assisted extraction	77.25–82.63%	–	60
		Ultrasound-assisted extraction	83–98%	–	
		Supercritical CO_2 extraction	82.6–98.1%	–	
Pistachio nuts (low marketable appearance and quality)	50%	Pressurized fluid extraction (*n*-hexane)	52.6%	60°C, 30 min, 10 bar. The sample were ground and mixed with crushed glass;	57
		Pressurized fluid extraction (ethanol)	15.0%	50°C, 5 min, 10 bar. The sample were ground and mixed with crushed glass;	
Mango kernel	5–11%	Supercritical CO_2 extraction	3.28% polar lipids	40°C, 180 min, 50 MPa, 30 g/min;	54,56,78
Watermelon seeds	50–51%	Pressurized liquid extraction (ethanol)	25.7–37%	40–80°C, 3–42 min, 1500 psia, sample/solvent ratio (0.30);	79
		Supercritical CO_2 extraction + 10% ethanol	51.83%	60°C, 400 bar, 15 g/min;	31
Olive pomace oil		Ultrasound-assisted extraction (*n*-hexane)	11.03%	60°C, solid/liquid (1:12 g/mL);	51
Açaí pulp (E. oleracea.)		Soxhlet extraction	43.1%	Total 30 h (6 h - hexane, water - 12 h and ethanol - 12 h);	63
Açaí seed			3.5%	–	

TABLE 5.2 *(Continued)*

Byproducts	Estimated lipid content	Extraction technique	Oil yield (w/w)	Extraction parameters	References
Açaí slurry			1.2%	–	
Broccoli (*Brassica oleracea L. var. Italica*)		Supercritical CO_2 extraction + 15% methanol	80.2% UFA and 19.8% SFA	60°C, 90 min, 300 bar, 3 mL/min;	29
Olive cake (pulp, skin and stone)	8%	Microwave-assisted solvent extraction (MASE)	4.95%	1 min, 720 W, solvent/solid ratio (3 cm^3/g);	80
Brazil nut cake (*B. excelsa*)	50%	Supercritical CO_2 extraction	59,20%	60°C, 180 min, 400 bar, 40 g/min;	52
Tiger nuts byproducts		Supercritical CO_2 extraction	7.36%	40°C, 120 min, 40 MPa, 20 g/mL;	61
Tomato paste waste		Supercritical CO_2 extraction + 5% ethanol		55°C, 120 min, 300 bar, 4 kg/h;	53
Tomato seeds and skins		Supercritical CO_2 extraction	61%	86°C, 345 bar, 2.5 mL/min;	
Corn kernels		Accelerated Solvent Extraction (Ethanol)	5.9%	100°C, 10 min + 5 min heating, 1000 psi;	58
Kernel oil from *Pinus pumila* (Pall.) Regel	24%	Homogenate-circulating ultrasound in combination with aqueous enzymatic extraction	31.89%	50°C, 600 W, 25.5 min, aqueous enzyme mixture (cellulase, pectinase and hemicellulase 1:1:1), sample/liquid ratio (7.8 mL/g)	38
Gurum (*Citrulluslanatus var. Colocynthoide*) seeds		Screw-press + Pretreatment with microwave	35.46%	60°C, 25 min, microwave (800 W);	62

TABLE 5.2 *(Continued)*

Byproducts	Estimated lipid content	Extraction technique	Oil yield (w/w)	Extraction parameters	References
Bran wax from rice oil processing		Ultrasound hexane or dilute aqueous NaOH (pH = 12)	2% of policosanol	45°C, 30 min, 300 W;	64
First bran fraction from rice polishing		Sonication - 0.2 M KOH (water/methanol 3:1)	8% of policosanol	60 min (18.2 kHz, 350 W, 60°C)	
Dried tomato pomace		Monosonication-assisted extraction (hexane)		45°C, 6 min, 50 kPa, 94 μm;	44

both gaseous and liquid characteristics; it is compressible, however, with a density similar to a liquid. Between 2001 and 2003, according to the *Food Science and Technology Abstract Database*, 26% of SFE were related to oil and essential oils area.[53] Supercritical carbon dioxide (SC-CO_2) is the most supercritical fluid extraction applied and uses pressurized carbon dioxide as a fluid, and at the end of the process, CO_2 is depressurized and recovered as a gas. The advantages of this method are the absence of organic solvents, CO_2 being cheap, nontoxic, nonflammable, highly diffusible, efficient, and considered a green solvent by European agencies. The processing has the potential to produce high-quality compounds and cause less environmental damage. Commonly, SC-CO_2 is used to obtain nonpolar lipids, providing mostly the recovery of insoluble water molecules. Combining CO_2 and other solvents increases the fluid's polarity, improving the recovery of polar molecules content. On the other hand, it can provide more oxidation stability and nontoxicity of polar lipids when compared with other extractions.[54,86] Among several solvents described in the literature, ethanol is applied because of its price, availability, low toxicity, and low environmental impact.[86] Furthermore, the pressure and temperature adaptability can mold the extraction according to the final objective.[29]

Previous research studied the use of SC-CO_2 to obtain vitamin E from the plant *Silybum marianum* (L). Gaertn., and wheat germ. The results showed better performance and yields compared with conventional methods with *n*-hexane or chloroform/methanol.[53] Arnáiz et al.[29] also studied the application of SC-CO_2 to extract lipids from broccoli leaves and as CO_2 is not suitable for polar lipids, the authors added another extraction using CO_2 + methanol, which resulted in increased fatty acids content. Soxhlet extraction also was performed to compare the contents and SFE has demonstrated higher yields, although the content of the lipids in both techniques was similar. Herrero et al.[53] showed in their work the best efficiency of SFE for the extraction of carotenoids from carrots compared with conventional methods, besides the extraction time being six times less. In general, SFE is a potential alternative method to extract lipids facing the new global demands.

5.6.3 ACCELERATED SOLVENT EXTRACTION (ASE)

ASE or pressurized solvent extraction is a reliable, efficient extraction method that usually requires less time, less quantity of solvents, and it has been used to obtain lipids.[57] The technique principle is to use pressure and high temperature to change solvent characteristics, such as viscosity reduction to facilitate the diffusivity and incorporation of solvent, preventing thermal degradation and improving the efficacy.[79]

Colivet et al.[79] studied the application of ASE in watermelons seed to obtain oils. The authors concluded that higher temperatures provide a greater yield of oils; however, lower temperatures allowed higher yields of phenolic compounds and higher antioxidant activity. The study conducted by Efthymiopoulos et al.[59] pointed out that higher temperatures employed in the ASE significantly improved the amount of oil extracted from spent coffee grounds with less process time compared to Soxhlet extraction. However, besides having improved the oil extraction using polar solvents, regarding nonpolar solvents, this improvement was limited to the optimum temperature of each solvent.

5.6.4 MICROWAVE-ASSISTED EXTRACTION (MAE)

MAE uses microwave energy to heat, penetrate, interact, and disrupt the sample. Its actions can occur through two mechanisms: a) a dipolar polarization that consists of the alternation of the electric field from the microwave energy, influencing the movement of rotating dipoles, generating molecular collisions, and consequently heating the material; and b) ionic conduction that involves the oscillation of the ions due to the electric field generated by the waves, heating the material as a result of the passage of the electric field.[87,88] In both cases, the energy heats the matrix, and the internal pressure generates the rupture of the cell wall of the material. In this way, the components are released.[63]

The use of MAE has proven to be a suitable option for application in food, plants, and animal tissue to obtain fats and oils, organic compounds, phenolics, and aromas, in comparison with conventional techniques. The main advantages are shorter processing times, homogeneous heating, higher extract purity, less energy, and solvent usage.[81] Despite being used and studied at a laboratory scale, MAE also started to be employed on an industrial scale.[87]

According to Mansour et al.,[89] the application of MAE with hexane as a solvent improved the oil yield extraction from byproducts from the olive oil industry, namely olive cake (pulp, skin, and stone), compared with conventional methods. Previous research conducted by Gullón et al.[50] has already demonstrated the beneficial effect on hydroxytyrosol (HT), maslinic acid (MA), and oleanolic acid (OA) extraction from olive pomace, using MAE (600 W, 50°C) in comparison with conventional processes and using ethanol as a solvent in both processes. The MAE allowed a higher yield of the compounds in a shorter time.

Several other studies have already evidenced the effectiveness of the use of MAE as seed pretreatment to obtain high-quality oils, oxidative stability and with strengthened compounds, such as phytosterols, tocopherols, and phenolic compounds.[62,63,84,85,88]

5.6.5 ULTRASOUND-ASSISTED EXTRACTION (UAE)

The UAE is a potential method to reduce time extraction, energy and solvent consumption, improve the efficiency and quality of the extract.[86] Ultrasound can meet new demands and be a "green" technique for extraction. UAE also has great reproducibility and it allows the obtaining of final products with high purity.[77] The ultrasound mechanism is based on the acoustic cavitation phenomenon that consists in generation and collapse of bubbles by ultrasound waves. This collapse causes a rupture on the cell walls, favoring the solvent penetration into the plant matrix and the extraction.[51]

Despite the advantages, UAE can alter some organoleptic and chemical characteristics unwantedly. Pingret and Chemat[82] summarize the principal change that can occur on plant-derived lipids, such as: Oxidation and changes in fatty acids in olive oils, off-flavors, an increase of peroxide molecules and volatile compounds from degradation in sunflower oil; oxidation and off-flavors in kiwi seed oil; oxidation and also changes in fatty acids in soybean oil; and modification in the volatile composition in thyme oil. Overall, the two main mechanisms for lipid degradation in UAE are by hydrolysis or oxidation. Whereas the kind of energy applied, lipids composition, presence of minor compounds, antioxidants, and thermal sensitive compounds are considered possible responsible factors.

Chanioti and Tzia[51] compared the extraction yield, unsaponifiable matter, total phenol content (TPC), and antioxidant activity (AA) of olive pomace oil, using UAE and Soxhlet extraction (SE). The oil yield under UAE was 88.93% lower than SE. The TPC and AA were also lower. Although, when comparing the parameters used, the ultrasound time was eight times shorter than SE, the temperature was reduced from 78°C to 60°C, and the extract presented higher quality. Therefore, considering all these details, the UAE becomes more advantageous, mainly for saving energy and possibly reducing costs.

In a study to extract the oil content of rice byproducts, Cravotto et al.[64] carried out different extraction methods (separator funnel, Soxhlet, UAE), and

UAE (300 W, 30 min, 45°C) was the technique that obtained the highest yield of oils. Previous reports have shown an increase in the yield of carotenoids in dried tomatoes using UAE. Luengo et al.[44] investigated the effects of the addition of pressure in the ultrasound extraction (Manosonication-assisted extraction) and the results indicated an increase in the yield of carotenoids from tomato pulp (byproduct), besides the possibility of decreasing the amount of solvent used with similar yield. Despite the good results obtained by this technique, its industrial scalability still requires further analysis, both scientific and economic.

5.7 MAJOR APPLICATIONS OF LIPIDS

The vast majority of lipid production is intended for human consumption, followed by use in the industrial field.[76] In food products, fats and oils are ingredients that provide flavor and sensory characteristics, such as palatability and creamy mouthfeel. These attributes are part of consumer acceptance and desire of the food. However, there is a global trend toward a healthier lifestyle and more significant consumer concern about the content of what they eat, driving demand for lower fat and functional foods.[25]

Long-chain fatty alcohols as alkyl lactates and carboxylates can be used in the cosmetic and pharmaceutical field.[76] Lecithin is an array of phospholipids that can be traced in vegetable oils, and it has applications as surface-active components in the food, pharmaceutical, and cosmetics industries.

Jahurul et al.[55] studied the addition of mango seed fat (MSF) and palm stearin (PS) as hard cocoa butter replacers, to improve the chocolate production problems related to the butter temperature resistance in warm climate countries. The authors concluded that some MSF/PS blends could be used as replacers, without changing substantially the triglyceride compositions. However, it exhibited a difference in the crystal morphology and solid fat content, whereby the solid fat content turned to 0% at higher temperatures. Therefore, these blends showed a potential use to produce chocolate in tropical countries, improving the temperature resistance of hard butter and the associated drawbacks. Kayathi et al.[54] also pointed out the potential use of mango kernel extracts as a cocoa butter replacement for vegetable oils, due to the greater stability provided by the higher quantity of UFA. The authors studied the economic assessment for industrial-scale plant, and indicated that the payback from the investment would be four years, demonstrating the capacity for economically viable large-scale production.

Tree nut oils and byproducts have already been used in culinary preparations around the world and with the rise of new health trends, their use also grows. Higher levels of MUFA and content of minor antioxidant components provide higher oxidative stability than vegetable oils. Furthermore, the consumption of fats and oils from these nuts has demonstrated positive anti-atherogenic effects.[25]

Nunes et al.[90] pointed out the potential use of oils extracted from olive pomace to manufacture of soaps, excipients in cosmetic formulations, active ingredients, as well as in foods. The authors also indicated the high MUFA content as an advantage for application in the previously mentioned industries, due to greater oxidative stability relative to PUFA-rich oils. According to Roselló-Soto et al.[61] the presence of fatty acids from the omega-6 family (C18:2n6) in byproducts from the production of "Horchata," a tiger nuts (*Cyperus esculentus* L.) beverage, is an indication of the potential use of these oils in functional foods and nutraceuticals. In addition, the application of SC-CO_2 was capable of extracting oil with the presence of bioactive compounds. The size, fast oxidation, and solubility make medium-chain triacylglycerols easily digestible and with potential use in clinical treatment against diseases related to the function of lipids in the organism. Furthermore, they can be used in functional foods as carriers of flavor and vitamins.[25]

The search for more adequate fat nutritional levels in cooked meat products drives the use of vegetable oils, due to their healthier fatty acid and sterol profiles.[28] Heck et al.[91] extracted bioactive compounds from jabuticaba peels (Myrciaria cauliflora (Mart.) O. Berg) and created a blend with linseed and chia oils (hydrogelled emulsions) to replace fat in pork burgers. The replacement of 60% fat with these blends generated sensory (e.g., color, flavor, texture) and oxidation changes in the products. In addition, the analysis made after the hamburgers were cooked showed an increase in lipid oxidation due to the thermal instability of the PUFA-rich oils used. Overall, the addition of 10% of the extract to hydrogelled emulsions was the most effective in controlling lipid oxidation and providing satisfactory sensory characteristics for 60 days.

López-García et al.[92] carried out in vitro tests on the effects of plant sterols in milk-based fruit beverages compared with cholesterol oxidation products. The authors observed the activity against oxidative stress and evidenced the possible application of these beverages to prevent human chronic diseases related to oxidative stress, but human studies are still needed to validate these results. Gupta et al.[93] investigated plant sterols as food supplements (e.g., in fat spreads, milk, mayonnaise, yoghurt, cakes) to decrease blood cholesterol levels. Clinical trials have shown the reducing effect of sterols (intake 2–3 g/day) and stanols (1.5 g/day) on LDL levels, by around 15%. The use of these phytochemicals together with statins has also shown beneficial effects on LDL reduction. However, there are still concerns about the long-term health effects of these biomolecules and the safe amount to intake. Therefore, there is still a need for more studies on the real consequences and effects for the organism.[94]

Byproducts and waste from cork production (bark from the *Quercus suber* L. tree) are promising products that can be applied in the cosmetic and pharmaceutical fields due to their peculiar composition. Carriço et al.[95] pointed out the application in skin products. The waxes generated from the extraction process are great sources of triterpenes and, in smaller quantities, of fatty acids, *n*-alkanes, and *n*-alkanols. The triterpenes can act as antioxidants in these products (sunscreens and as wrinkle control products). The terpenoids can act against the skin ageing, against the degradation of the elastin (a fiber that helps maintain the skin's structure). Suberin (long-chain fatty components, glycerol, hydroxyfatty, and phenolic acids, linked by ester groups), another component of cork, can also be useful in antiageing products, improving the texture and brightness of the skin. The anti-inflammatory characteristics of the group of compounds present in cork, including suberin, triterpenes, phenolics, and polysaccharides, make its byproducts potential candidates for valorization and use in the cosmetic area. However, further and deeper studies on possible effects (e.g., toxic effects) on the skin should be carried out before these substances can be employed in pharmaceutical and cosmetic products.

5.8 LIPID STABILITY

Plant cell membranes often adjust to stress conditions to maintain their stability and optimize photosynthesis like other cellular processes. The composition of the lipid membrane of plants can be changed to achieve such effects, whether by cold, heat, mechanical damage, and others. Like animals, plants also undergo autophagy, that is, the self-degradation of cellular components after suffering some stress condition.[96,97]

Cold exposure increases the content of phospholipids and the degree of unsaturation of fatty acids. The application of heat, on the other hand, increases the degree of saturation of fatty acids and can also alter the genes of plant through a mutation to stabilize the membrane of chloroplasts.[96]

Regarding the incorporation of functional lipids in foods, attention should be paid to some characteristics of the incorporated foodstuff. Some bioactive compounds of lipid matrices may be sensitive to light, oxygen or behave as pro-oxidants. Carotenoids, for instance, are physically unstable and can recrystallize and lose their bioactivity after incorporation.

The method by which lipophilic compounds are digested and absorbed in the human body is also a fundamental step in the development of functional foods. During the whole digestion the food goes through a process of pH variation, temperature shifts, ionic strength, among other factors that reduce the bioavailability of the compounds. As for the factors intrinsic to the lipid matrix that affect its absorption, its availability, droplet size, and interfacial characteristics are mentioned.[98]

Vitamins and other lipophilic compounds have low bioavailability due to their low solubility in small intestine fluids (only 10–30% of fatty acids are released from triacylglycerols from the lipid hydrolysis reaction). Thus, micro and nanoencapsulation are technologies employed to increase bioaccessibility, releasing them throughout the digestive process, most of which is absorbed in the mucus, increasing its absorption.[99]

As the application of heat can degrade phytochemicals, technologies such as spray drying, and freeze drying are considered good alternatives due to the conservation of thermosensitive compounds. The choice of material for encapsulation depends on the food matrix to be applied, since the different materials may have greater or lesser stability in terms of pH, protein composition, humidity, among other characteristics of the food.[16,98]

5.9 TRENDS AND CONCLUSIONS

A different kind of application that is also increasing is microalgae as raw material to produce biodiesel. These organisms possess a high concentration of lipids and a potential replacement for fossil fuels.[100] Lipids from plant byproducts can be recovered and converted into biodiesel. Biodiesel derived directly from vegetable oils can lead to competition with food production from these same vegetables. Therefore, the production of biodiesel from waste and byproducts has great potential of application and there are already several studies on this topic, namely with spent coffee grounds[46,60] and wastewater algae.[89]

Regarding extractive methods, novel approaches are being studied, such as hurdle techniques. Chen et al.[38] studied the combination of homogenate-circulating ultrasound with aqueous enzyme extraction to obtain oils derived from *Pinus pumila* (Pall.) Regel seeds. The process used previous homogenization of the seeds and mixed the enzymes (cellulase, pectinase, and hemicellulase) with the sample and the water solution. Then, the entire solution was inserted into an ultrasound with a circular reflux system. The process was optimized, and the yield obtained was approximately 31.89%, being 93.47% unsaturated fatty acids. Furthermore, the results demonstrated the capability of the process to produce edible oils efficiently, with less impact on the environment and application in pharmaceutical products.

The other example is the Sono-Soxhlet method, which consists of ultrasound inside or outside the Soxhlet chamber to improve the extraction. This technique has already been used for oil and fatty acid extraction of oleaginous seeds. The Sono-Clevenger is another technique used to obtain essential oils. The mechanism is similar to the previous one; however, instead of Soxhlet, it uses Clevenger. The extraction time is reduced and the yields are significantly increased with the application of this method.[77] Despite nonconventional extraction methodologies, there are still challenges for their application on an industrial scale, among them scalability, costs involved in the implementation have to be considered. In this aspect, it becomes important to analyze the economic viability as a whole. Another important detail is the collaborative work between the scientific community and the industry.

In recent years, lipidomics has been gaining relevance in science. This term is characterized by the study of lipids and their interactions with biological systems (species, varieties, biological roles such as pathways and networks),[101] through qualitative and quantitative analyses to understand the effects on metabolism. Technological advances, such as identification techniques (mass spectrometry), computational data analysis, and bioinformatics, are tools that enable the lipidomics studies.[102] The examples of the use of this discipline may include analyses related to pathologies that can be caused by irregular lipid metabolism, food safety and quality monitoring, biomarkers, and drug delivery systems.[103,104]

Lipids have potential applications in various sectors (food, pharmaceuticals, medicinal, cosmetics, energy). However, there are still difficulties and challenges for the application of lipids in innovative products. Moreover, more studies and in vivo validation of the results obtained through in vitro assays are mandatory for the effective use of lipids and to increase their applications in the industry sectors.

ACKNOWLEDGMENTS

FCT, Portugal for financial support through national funds FCT/MCTES to the CIMO (UIDB/00690/2020), and the European Regional Development Fund

(ERDF) through the Regional Operational Program North 2020, within the scope of the projects Project GreenHealth (Norte-01-0145-FEDER-000042) and ValorNatural® (Norte-01-0247-FEDER-024479), whom the author F.S. Reis thanks for her contract. L. Barros thanks the national funding by FCT, P.I., through the institutional scientific employment program-contract for her contract, and M. Carocho (CEECIND/00831/2018), and S.A. Heleno (CEECIND/03040/2017) to the national funding by FCT, P.I., through the individual scientific employment program-contracts.

CONFLICTS OF INTEREST

The authors declare no conflict of interest.

KEYWORDS

- **nutrients**
- **lipids**
- **byproducts**
- **industrial application**
- **extraction methodologies**

REFERENCE

1. Broun, P.; Gettner, S.; Somerville, C. Genetic Engineering of Plant Lipids. *Annu. Rev. Nutr.* **1999,** *19*, 197–216.
2. Lunn, J.; Theobald, H. E. The Health Effects of Dietary Unsaturated Fatty Acids. *Br. Nutr. Found.* **2006,** *31*, 178–224.
3. O'Keefe, S. F.; Sarnoski, P. J. Nomenclature and Classification of Lipids. In *Food Lipids: Chemistry, Nutrition and Biotechnology*; Akoh, C. C., Min, D. B., Eds.; CRC Press, 2017; pp 3–36.
4. Pokorny, J.; Parkanyiová, L. Plant Lipids and Oils. In *Chemical and Functional Properties of Food Lipids*; Sikorski, Z. E., Kolakowska, A., Eds.; CRC Press, 2003.
5. Lim, G.; Singhal, R.; Kachroo, A.; Kachroo, P. Fatty Acid–and Lipid-Mediated Signaling in Plant Defense. *Annu. Rev. Phytopathol.* **2017,** *55*, 505–536.
6. Liu, X.; Ma, D.; Zhang, Z.; Wang, S.; Du, S.; Deng, X. Plant Lipid Remodeling in Response to Abiotic Stresses. *Environ. Exp. Bot.* **2019,** *165*, 174–184, DOI: 10.1016/j.envexpbot.2019.06.005.

7. Livrea, M. A.; Tesoriere, L. Antioxidant Activity of Vitamin A within Lipid Environments. In *Subcellular Biochemistry Volume 30–Fat-Soluble Vitamins*; Quinn, P. J., Kagan, V. E., Eds.; Springer Science+Business Media: New York, 1998; pp 113–144.
8. Simopoulos, A. P. Omega-3 Fatty Acids in Wild Plants, Nuts and Seeds. *Asia Pac. J. Clin. Nutr.* **2002,** *11*, 163–173.
9. Jaswir, I.; Noviendri, D.; Hasrini, R. F.; Octavianti, F. Carotenoids: Sources, Medicinal Properties and Their Application in Food and Nutraceutical Industry. *J. Med. Plants Res.* **2011,** *5*, 7119–7131.
10. Kozubek, A.; Zarnowski, R.; Stasiuk, M.; Gubernator, J. Natural Amphiphilic Phenols as Bioactive Compounds. *Cell. Mol. Biol. Lett.* **2001,** *6*, 351–355.
11. Newman, P.; Shearer, J. Vitamin K Metabolism. In *Subcellular Biochemistry Volume 30–Fat-Soluble Vitamins*; Quinn, P. J., Kagan, V. E., Eds.; Springer Science+Business Media: New York, 1998; pp 455–488.
12. Lagarda, M. J.; Garc, G.; Farr, R. Analysis of Phytosterols in Foods. *J. Pharm. Biomed. Anal.* **2006,** *41*, 1486–1496. DOI: 10.1016/j.jpba.2006.02.052.
13. Heinz, E. Biosynthesis of Polyunsaturated Fatty Acids. In *Biosynthesis of Polyunsaturated Fatty Acids*; Moore, T. S., Ed.; CRC Press, 1993; pp 33–90.
14. Settembre, C.; Ballabio, A. Lysosome: Regulator of Lipid Degradation Pathways. *Trends Cell Biol.* **2014,** *24*, 743–750. DOI: 10.1016/j.tcb.2014.06.006.
15. Stasiuk, M.; Kozubek, A. Biological Activity of Phenolic Lipids. *Cell. Mol. Life Sci.* **2010,** *67*, 841–860. DOI: 10.1007/s00018-009-0193-1.
16. Eun, J.; Maruf, A.; Das, P. R.; Nam, S. A Review of Encapsulation of Carotenoids Using Spray Drying and Freeze Drying. *Crit. Rev. Food Sci. Nutr.* **2020,** *60*, 3547–3572. DOI: 10.1080/10408398.2019.1698511.
17. Tomoaia-Cotisel, M.; Quinn, P. J. Biophysical Properties of Carotenoids. In *Subcellular Biochemistry Volume 30–Fat-Soluble Vitamins*; Quinn, P. J., Kagan, V. E., Eds.; Springer Science+Business Media: New York, 1998; pp 219–242.
18. Bartley, G. E.; Scolnik, P. A. Plant Carotenoids : Pigments for Photoprotection, Visual Attraction, and Human Health. *Plant Cell* **1995,** *7*, 1027–1038.
19. Chen, B.; Mcclements, D. J.; Decker, E. A. Design of Foods with Bioactive Lipids for Improved Health. *Annu. Rev. Food Sci. Technol.* **2013,** *4*, 35–36. DOI: 10.1146/annurev-food-032112-135808.
20. Miras-moreno, B.; Bele, A.; Pedren, M. A.; Almagro, L. Bioactivity of Phytosterols and Their Production in Plant in Vitro Cultures. *J. Agric. Food Chem.* **2016,** *64*, 7049–7058. DOI: 10.1021/acs.jafc.6b02345.
21. Marangoni, F.; Poli, A. Phytosterols and Cardiovascular Health. *Pharmacol. Res.* **2010,** *61*, 193–199. DOI: 10.1016/j.phrs.2010.01.001.
22. Fanali, C.; Orazio, G. D.; Fanali, S.; Gentili, A. Advanced Analytical Techniques for Fat-Soluble Vitamin Analysis. *Trends Anal. Chem.* **2017,** *87*, 82–97. DOI: 10.1016/j.trac.2016.12.001.
23. Jäpelt, R. B.; Jakobsen, J. Vitamin D in Plants: A Review of Occurrence, Analysis, and Biosynthesis. *Front. Plant Sci.* **2013,** *4*, 1–20. DOI: 10.3389/fpls.2013.00136.
24. Alhouayek, M.; Ameraoui, H.; Muccioli, G. G. Bioactive Lipids in Inflammatory Bowel Diseases–From Pathophysiological Alterations to Therapeutic Opportunities. *Biochim. Biophys. Acta - Mol. Cell Biol. Lipids* **2021,** *1866*. DOI: 10.1016/j.bbalip.2020.158854.
25. Shahidi, F. *Nutraceutical and Specialty Lipids and Their Co-Products*; Taylor & Francis, 2006.

26. Fahy, E.; Cotter, D.; Sud, M.; Subramaniam, S. Lipid Classification, Structures and Tools. *Biochim. Biophys. Acta-Mol. Cell Biol. Lipids* **2011,** *1811*, 637–647. DOI: 10.1016/j.bbalip.2011.06.009.
27. De Meester, F.; Watson, R. R. *Handbook of Lipids in Human Function: Fatty Acids*; Academic Press and AOCS Press, 2015.
28. Sánchez-Zapata, E.; Díaz-Vela, J.; Pérez-Chabela, M. L.; Pérez-Alvarez, J. A.; Fernández-López, J. Evaluation of the Effect of Tiger Nut Fibre as a Carrier of Unsaturated Fatty Acids Rich Oil on the Quality of Dry-Cured Sausages. *Food Bioprocess Technol.* **2013,** *6*, 1181–1190. DOI: 10.1007/s11947-011-0733-1.
29. Arnáiz, E.; Bernal, J.; Martín, M. T.; García-Viguera, C.; Bernal, J. L.; Toribio, L. Supercritical Fluid Extraction of Lipids from Broccoli Leaves. *Eur. J. Lipid Sci. Technol.* **2011,** *113*, 479–486. DOI: 10.1002/ejlt.201000407.
30. Karrar, E.; Sheth, S.; Navicha, W. B.; Wei, W.; Hassanin, H.; Abdalla, M.; Wang, X. A Potential New Source: Nutritional and Antioxidant Properties of Edible Oils from Cucurbit Seeds and Their Impact on Human Health. *J. Food Biochem.* **2019,** *43*, 1–10. DOI: 10.1111/jfbc.12733.
31. Rai, A.; Mohanty, B.; Bhargava, R. Modeling and Response Surface Analysis of Supercritical Extraction of Watermelon Seed Oil Using Carbon Dioxide. *Sep. Purif. Technol.* **2015,** *141*, 354–365. DOI: 10.1016/j.seppur.2014.12.016.
32. Wrolstad, R. E.; Acree, T. E.; Decker, E. A.; Penner, M. H.; Reid, D. S.; Schwartz, S. J.; Shoemaker, C. F.; Smith, D. M.; Sporns, P. Handbook of Food Analytical Chemistry: Water, Proteins, Enzymes, Lipids and Carbohydrates. *Handb. Food Anal. Chem.* **2005,** *1–2*, 1–1362. DOI: 10.1002/0471709085.
33. Yeagle, L. Lipid Regulation of Cell Membrane Structure and Function. *Fed. Am. Soc. Exp. Biol. J.* **1989,** *3*, 1833–1842.
34. J. Quinn, P.; Joo, F.; Vigh, L. The Role of Unsaturated Lipids in Membrane Structure and Stability. *Prog. Biophys. Mol. Biol. J.* **1989,** *53*, 71–103.
35. Bretscher, M. S. Membrane Structure: Some General Principles. *Science* **1973,** *181* (80), 622–9.
36. Soteriou, C.; Kalli, A. C.; Connell, S. D.; Tyler, A. I. I.; Thorne, J. L. Advances in Understanding and in Multi-Disciplinary Methodology Used to Assess Lipid Regulation of Signalling Cascades from the Cancer Cell Plasma Membrane. *Prog. Lipid Res.* **2021,** *81.* 101080, DOI: 10.1016/j.plipres.2020.101080.
37. Akoh, C. C.; Min, D. B. *Food Lipids: Chemistry, Nutrition, Biotechnology*; Marcel Dekker, Inc., 2002.
38. Chen, F.; Zhang, Q.; Gu, H.; Yang, L. An Approach for Extraction of Kernel Oil from Pinus Pumila Using Homogenate-Circulating Ultrasound in Combination with an Aqueous Enzymatic Process and Evaluation of Its Antioxidant Activity. *J. Chromatogr. A.* **2016,** *1471*, 68–79.
39. Rodríguez, J. L. F.; Dynarowicz-Latka, P.; Conde, J. M. How Unsaturated Fatty Acids and Plant Stanols Affect Sterols Plasma Level and Cellular Membranes ? Review on Model Studies Involving the Langmuir Monolayer Technique. *Chem. Phys. Lipids.* **2020,** *232*, 104968. DOI: 10.1016/j.chemphyslip.2020.104968.
40. Vanmierlo, T.; Bogie, J. F. J.; Mailleux, J.; Vanmol, J.; Lütjohann, D.; Mulder, M.; Hendriks, J. J. A. Progress in Lipid Research Plant Sterols : Friend or Foe in CNS Disorders? *Prog. Lipid Res.* **2015,** *58*, 26–39. DOI: 10.1016/j.plipres.2015.01.003.
41. Vergès, B.; Fumeron, F. Potential Risks Associated with Increased Plasma Plant-Sterol Levels. *Diabetes Metab.* **2015,** *41*, 76–81. DOI: 10.1016/j.diabet.2014.11.003.

42. Lin, Y.; Knol, D.; Menéndez-carreño, M.; Baris, R.; Janssen, H.; Trautwein, E. A. Oxidation of Sitosterol and Campesterol in Foods upon Cooking with Liquid Margarines without and with Added Plant Sterol Esters. *Food Chem.* **2018,** *241*, 387–396. DOI: 10.1016/j.foodchem.2017.08.118.
43. Damodaran, S.; Parkin, K. L.; Fennema, O. R. *Química de Alimentos de Fennema*; 4 ed.; Artmed: Porto Alegre, 2010.
44. Luengo, E.; Condón-abanto, S.; Condón, S.; Álvarez, I.; Raso, J. Improving the Extraction of Carotenoids from Tomato Waste by Application of Ultrasound under Pressure. *Sep. Purif. Technol.* **2014,** *136*, 130–136.
45. Kakadellis, S.; Harris, Z. M. Don't Scrap the Waste: The Need for Broader System Boundaries in Bioplastic Food Packaging Life-Cycle Assessment–A Critical Review. *J. Clean. Prod.* **2020,** *274*, 122831. DOI: 10.1016/j.jclepro.2020.122831.
46. Go, A. W.; Pham, T. Y. N.; Ju, Y. H.; Agapay, R. C.; Angkawijaya, A. E.; Quijote, K. L. Extraction of Lipids from Post-Hydrolysis Spent Coffee Grounds for Biodiesel Production with Hexane as Solvent: Kinetic and Equilibrium Data. *Biomass and Bioenergy.* **2020,** *140*, 105704. DOI: 10.1016/j.biombioe.2020.105704.
47. Torres-León, C.; Ramírez-Guzman, N.; Londoño-Hernandez, L.; Martinez-Medina, G. A.; Díaz-Herrera, R.; Navarro-Macias, V.; Alvarez-Pérez, O. B.; Picazo, B.; Fillarreal-Vázquez, M.; Ascacio-Valdes, J.; et al. Food Waste and Byproducts: An Opportunity to Minimize Malnutrition and Hunger in Developing Countries. *Front. Sustain. Food Syst.* **2018,** *2*, 1–17. DOI: 10.3389/fsufs.2018.00052.
48. Comission, E. European Green Deal 2019.
49. Comission, E. A Farm to Fork Strategy for a Fair, Healthy and Environmentally-Friendly Food System 2020, 22.
50. Gullón, P.; Gullón, B.; Astray, G.; Carpena, M.; Fraga-Corral, M.; Prieto, M. A.; Simal-Gandara, J. Valorization of Byproducts from Olive Oil Industry and Added-Value Applications for Innovative Functional Foods. *Food Res. Int.* **2020,** *137*. DOI: 10.1016/j.foodres.2020.109683.
51. Chanioti, S.; Tzia, C. Optimization of Ultrasound-Assisted Extraction of Oil from Olive Pomace Using Response Surface Technology: Oil Recovery, Unsaponifiable Matter, Total Phenol Content and Antioxidant Activity. *LWT - Food Sci. Technol.* **2017,** *79*, 178–189. DOI: 10.1016/j.lwt.2017.01.029.
52. Vasquez, W. V.; Hernández, D. M.; del Hierro, J. N.; Martin, D.; Cano, M. P.; Fornari, T. Supercritical Carbon Dioxide Extraction of Oil and Minor Lipid Compounds of Cake Byproduct from Brazil Nut (Bertholletia Excelsa) Beverage Production. *J. Supercrit. Fluids* **2021,** *171*, 105188. DOI: 10.1016/j.supflu.2021.105188.
53. Herrero, M.; Cifuentes, A.; Iban, E. Sub- and Supercritical Fluid Extraction of Functional Ingredients from Different Natural Sources: Plants, Food-byproducts, Algae and Microalgae A Review. *Food Chem.* **2006,** *98*, 136–148. DOI: 10.1016/j.foodchem. 2005.05.058.
54. Kayathi, A.; Chakrabarti, P. P.; Bonfim-Rocha, L.; Cardozo-Filho, L.; Jegatheesan, V. Selective Extraction of Polar Lipids of Mango Kernel Using Supercritical Carbon Dioxide (SC–CO2) Extraction: Process Optimization of Extract Yield/Phosphorous Content and Economic Evaluation. *Chemosphere* **2020,** *260*. DOI: 10.1016/j.chemosphere.2020.127639.
55. Jahurul, M. H. A.; Zaidul, I. S. M.; Nik Norulaini, N. A.; Sahena, F.; Abedin, M. Z.; Mohamed, A.; Mohd Omar, A. K. Hard Cocoa Butter Replacers from Mango Seed Fat and Palm Stearin. *Food Chem.* **2014,** *154*, 323–329. DOI: 10.1016/j.foodchem.2013.11.098.

56. Solís-Fuentes, J. A.; Durán-De-Bazúa, M. C. Mango Seed Uses: Thermal Behaviour of Mango Seed Almond Fat and Its Mixtures with Cocoa Butter. *Bioresour. Technol.* **2004,** *92*, 71–78. DOI: 10.1016/j.biortech.2003.07.003.
57. Sheibani, A.; Ghaziaskar, H. S. Pressurized Fluid Extraction of Pistachio Oil Using a Modified Supercritical Fluid Extractor and Factorial Design for Optimization. *LWT-Food Sci. Technol.* **2008,** *41*, 1472–1477. DOI: 10.1016/j.lwt.2007.09.002.
58. Moreau, R. A.; Powell, M. J.; Singh, V. Pressurized Liquid Extraction of Polar and Nonpolar Lipids in Corn and Oats with Hexane, Methylene Chloride, Isopropanol, and Ethanol. *JAOCS, J. Am. Oil Chem. Soc.* **2003,** *80*, 1063–1067. DOI: 10.1007/s11746-003-0821-y.
59. Efthymiopoulos, I.; Hellier, P.; Ladommatos, N.; Russo-Profili, A.; Eveleigh, A.; Aliev, A.; Kay, A.; Mills-Lamptey, B. Influence of Solvent Selection and Extraction Temperature on Yield and Composition of Lipids Extracted from Spent Coffee Grounds. *Ind. Crops Prod.* **2018,** *119*, 49–56. DOI: 10.1016/j.indcrop.2018.04.008.
60. Loyao, A. S.; Villasica, S. L. G.; Dela Peña, P. L. L.; Go, A. W. Extraction of Lipids from Spent Coffee Grounds with Non-Polar Renewable Solvents as Alternative. *Ind. Crops Prod.* **2018,** *119*, 152–161. DOI: 10.1016/j.indcrop.2018.04.017.
61. Roselló-Soto, E.; Barba, F. J.; Lorenzo, J. M.; Dominguez, R.; Pateiro, M.; Mañes, J.; Moltó, J. C. Evaluating the Impact of Supercritical-CO2 Pressure on the Recovery and Quality of Oil from "Horchata" byproducts: Fatty Acid Profile, α-Tocopherol, Phenolic Compounds, and Lipid Oxidation Parameters. *Food Res. Int.* **2019,** *120*, 888–894. DOI: 10.1016/j.foodres.2018.11.054.
62. Karrar, E.; Sheth, S.; Wei, W.; Wang, X. Effect of Microwave Heating on Lipid Composition, Oxidative Stability, Color Value, Chemical Properties, and Antioxidant Activity of Gurum (Citrulluslanatus Var. Colocynthoide) Seed Oil. *Biocatal. Agric. Biotechnol.* **2020,** *23*. DOI: 10.1016/j.bcab.2020.101504.
63. Buratto, R. T.; Cocero, M. J.; Martín, Á. Characterization of Industrial Açaí Pulp Residues and Valorization by Microwave-Assisted Extraction. *Chem. Eng. Process.-Process Intensif.* **2021,** *160*. DOI: 10.1016/j.cep.2020.108269.
64. Cravotto, G.; Binello, A.; Merizzi, G.; Avogadro, M. Improving Solvent-Free Extraction of Policosanol from Rice Bran by High-Intensity Ultrasound. *Eur. J. Lipid Sci. Technol.* **2004,** *106*, 147–151. DOI: 10.1002/ejlt.200300914.
65. Elkacmi, R.; Kamil, N.; Bennajah, M. Separation and Purification of High Purity Products from Three Different Olive Mill Wastewater Samples. *J. Environ. Chem. Eng.* **2017,** *5*, 829–837.
66. Ohlrogge, J. B.; Jaworski, J. G.; Post-Beittenmiller, D. De Novo Fatty Acid Biosynthesis. In *Lipid Metabolism in Plants*; Moore, T. S., Ed.; CRC Press, 1993; pp 3–32.
67. Thelen, J. J.; Ohlrogge, J. B. Metabolic Engineering of Fatty Acid Biosynthesis in Plants. *Metab. Eng.* **2002,** *4*, 12–21. DOI: 10.1006/mben.2001.0204.
68. Tanaka, Y.; Sasaki, N.; Ohmiya, A. Biosynthesis of Plant Pigments: Anthocyanins, Betalains and Carotenoids. *Plant J.* **2008,** *54*, 733–749. DOI: 10.1111/j.1365-313X.2008.03447.x.
69. Hirschberg, J. Carotenoid Biosynthesis in Flowering Plants. *Curr. Opin. Plant Biol.* **2001,** *4*, 210–218.
70. Hofius, D.; Sonnewald, U. Vitamin E Biosynthesis: Biochemistry Meets Cell Biology. *Trends Plant Sci.* **2003,** *8*, 7–9.
71. Kozubek, A.; Tyman, J. H. P. Resorcinolic Lipids, the Natural Non-Isoprenoid Phenolic Amphiphiles and Their Biological Activity. *Chem. Rev.* **1999,** *99*. DOI: 10.1021/cr970464o.

72. Azmir, J.; Zaidul, I. S. M.; Rahman, M. M.; Sharif, K. M.; Mohamed, A.; Sahena, F.; Jahurul, M. H. A.; Ghafoor, K.; Norulaini, N. A. N.; Omar, A. K. M. Techniques for Extraction of Bioactive Compounds from Plant Materials: A Review. *J. Food Eng.* **2013,** *117*, 426–436. DOI: 10.1016/j.jfoodeng.2013.01.014.
73. Sik, B.; Hanczné, E. L.; Kapcsándi, V.; Ajtony, Z. Conventional and Nonconventional Extraction Techniques for Optimal Extraction Processes of Rosmarinic Acid from Six Lamiaceae Plants as Determined by HPLC-DAD Measurement. *J. Pharm. Biomed. Anal.* **2020,** *184*. DOI: 10.1016/j.jpba.2020.113173.
74. Azmin, S. N. H. M.; Manan, Z. A.; Alwi, S. R. W.; Chua, L. S.; Mustaffa, A. A.; Yunus, N. A. Herbal Processing and Extraction Technologies. *Sep. Purif. Rev.* **2016,** *45*, 305–320. DOI: 10.1080/15422119.2016.1145395.
75. Stikeleather, L. F.; Singleton, J. A. Induction Heated Pressure Vessel System and Method for the Rapid Extraction of Lipids and Trace Polar Micronutrients from Plant Material. *Food Bioprod. Process. Trans. Inst. Chem. Eng. Part C* **2001,** *79*, 169–175. DOI: 10.1205/096030801750425262.
76. Stevens, C. V; Verhé, R. G. *Renewable Bioresources: Scope and Modifications for Non-Food Applications*; Ltd, J.W. & S., Ed.; 1st ed.; 2004;
77. Chemat, F.; Rombaut, N.; Sicaire, A. G.; Meullemiestre, A.; Fabiano-Tixier, A. S.; Abert-Vian, M. Ultrasound Assisted Extraction of Food and Natural Products. Mechanisms, Techniques, Combinations, Protocols and Applications. A Review. *Ultrason. Sonochem.* **2017,** *34*, 540–560. DOI: 10.1016/j.ultsonch.2016.06.035.
78. Ribeiro, S. M. R.; Schieber, A. Bioactive Compounds in Mango (Mangifera Indica L.). *Bioact. Foods Promot. Heal.* **2010,** 507–523. DOI: 10.1016/B978-0-12-374628-3.00034-7.
79. 79. Colivet, J.; Oliveira, A. L.; Carvalho, R. A. Influence of the Bed Height on the Kinetics of Watermelon Seed Oil Extraction with Pressurized Ethanol. *Sep. Purif. Technol.* **2016,** *169*, 187–195. DOI: 10.1016/j.seppur.2016.06.020.
80. Amarni, F.; Kadi, H. Kinetics Study of Microwave-Assisted Solvent Extraction of Oil from Olive Cake Using Hexane. *Innov. Food Sci. Emerg. Technol.* **2010,** *11*, 322–327.
81. Virot, M.; Tomao, V.; Ginies, C.; Visinoni, F.; Chemat, F. Microwave-Integrated Extraction of Total Fats and Oils. *J. Chromatogr. A* **2008,** *1196–1197*, 57–64. DOI: 10.1016/j.chroma.2008.05.023.
82. Pingret, D.; Chemat, F. Degradation during Application of Ultrasound in Food Processing: A Review. *Food Control* **2013,** *31*, 593–606. DOI: 10.1016/j.foodcont.2012.11.039.
83. Virot, M.; Tomao, V.; Colnagui, G.; Visinoni, F.; Chemat, F. New Microwave-Integrated Soxhlet Extraction. An Advantageous Tool for the Extraction of Lipids from Food Products. *J. Chromatogr. A.* **2007,** *1174*, 138–144. DOI: 10.1016/j.chroma.2007.09.067.
84. Yang, M.; Huang, F.; Liu, C.; Zheng, C.; Zhou, Q.; Wang, H. Influence of Microwave Treatment of Rapeseed on Minor Components Content and Oxidative Stability of Oil. *Food Bioprocess Technol.* **2013,** *6*, 3206–3216. DOI: 10.1007/s11947-012-0987-2.
85. Gaber, M. A. F. M.; Tujillo, F. J.; Mansour, M. P.; Juliano, P. Improving Oil Extraction from Canola Seeds by Conventional and Advanced Methods. *Food Eng. Rev.* **2018,** *10*, 198–210. DOI: 10.1007/s12393-018-9182-1.
86. Misra, N. N.; Koubaa, M.; Roohinejad, S.; Juliano, P.; Alpas, H.; Inácio, R. S.; Saraiva, J. A.; Barba, F. J. Landmarks in the Historical Development of Twenty First Century Food Processing Technologies. *Food Res. Int.* **2017,** *97*, 318–339. DOI: 10.1016/j.foodres.2017.05.001.

87. Cravotto, G.; Boffa, L.; Mantegna, S.; Perego, P.; Avogadro, M.; Cintas, P. Improved Extraction of Vegetable Oils under High-Intensity Ultrasound and/or Microwaves. *Ultrason. Sonochem.* **2008,** *15*, 898–902. DOI: 10.1016/j.ultsonch.2007.10.009.
88. Mohamed, M. A.; Knoerzer, K.; Mansour, M. P.; Trujillo, F. J.; Juliano, P.; Shrestha, P. Improved Canola Oil Expeller Extraction Using a Pilot-Scale Continuous Flow Microwave System for Pre-Treatment of Seeds and Flaked Seeds. *J. Food Eng.* **2020,** *284*, 110053. DOI: 10.1016/j.jfoodeng.2020.110053.
89. Mansour, E. A.; Abo El-Enin, S. A.; Hamouda, A. S.; Mahmoud, H. M. Efficacy of Extraction Techniques and Solvent Polarity on Lipid Recovery from Domestic Wastewater Microalgae. *Environ. Nanotechnology, Monit. Manag.* **2019,** *12*, 100271. DOI: 10.1016/j.enmm.2019.100271.
90. Nunes, M. A.; Costa, A. S. G.; Bessada, S.; Santos, J.; Puga, H.; Alves, R. C.; Freitas, V.; Oliveira, M. B. P. P. Science of the Total Environment Olive Pomace as a Valuable Source of Bioactive Compounds: A Study Regarding Its Lipid- and Water-Soluble Components. *Sci. Total Environ.* **2018,** *644*, 229–236. DOI: 10.1016/j.scitotenv.2018.06.350.
91. Heck, R. T.; Ferreira, D. F.; Fagundes, M. B.; Santos, B. A. Dos; Cichoski, A. J.; Saldaña, E.; Lorenzo, J. M.; de Menezes, C. R.; Wagner, R.; Barin, J. S.; et al. Jabuticaba Peel Extract Obtained by Microwave Hydrodiffusion and Gravity Extraction: A Green Strategy to Improve the Oxidative and Sensory Stability of Beef Burgers Produced with Healthier Oils. *Meat Sci.* **2020,** *170*, 108230. DOI: 10.1016/j.meatsci.2020.108230.
92. López-García, G.; Cilla, A.; Barberá, R.; Genovés, S.; Martorell, P.; Alegría, A. Effect of Plant Sterol and Galactooligosaccharides Enriched Beverages on Oxidative Stress and Longevity in Caenorhabditis Elegans. *J. Funct. Foods* **2020,** *65*, 103747. DOI: 10.1016/j.jff.2019.103747.
93. Gupta, A. K.; Savopoulos, C. G.; Ahuja, J.; Hatzitolios, A. I. Role of Phytosterols in Lipid-Lowering: Current Perspectives. *QJM.* **2011,** *104*, 301–308. DOI: 10.1093/qjmed/hcr007.
94. Weingärtner, O.; Baber, R.; Teupser, D. Plant Sterols in Food: No Consensus in Guidelines. *Biochem. Biophys. Res. Commun.* **2014,** *446*, 811–813. DOI: 10.1016/j.bbrc.2014.01.147.
95. Carriço, C.; Ribeiro, H. M.; Marto, J. Converting Cork Byproducts to Ecofriendly Cork Bioactive Ingredients: Novel Pharmaceutical and Cosmetics Applications. *Ind. Crop. Prod.* **2018,** *125*, 72–84. DOI: 10.1016/j.indcrop.2018.08.092.
96. Welti, R.; Shah, J.; Li, W.; Li, M.; Chen, J.; Burke, J. J.; Fauconnier, M.; Chapman, K.; Chye, M.; Wang, X. Plant Lipidomics: Discerning Biological Function by Profiling Plant Complex Lipids Using Mass Spectrometry. *Front. Biosci.* **2007,** *12*, 2494–2506.
97. Tenenboim, H.; Burgos, A.; Willmitzer, L.; Brotman, Y. Using Lipidomics for Expanding the Knowledge on Lipid Metabolism in Plants. *Biochimie* **2016,** *130*, 91–96. DOI: 10.1016/j.biochi.2016.06.004.
98. Mcclements, D. J.; Decker, E. A.; Park, Y.; Weiss, J. Designing Food Structure to Control Stability, Digestion, Release and Absorption of Lipophilic Food Components. *Food Biophys.* **2008,** 219–228. DOI: 10.1007/s11483-008-9070-y.
99. Lv, S.; Gu, J.; Zhang, R.; Zhang, Y.; Tan, H.; Mcclements, D. J. Vitamin E Encapsulation in Plant-Based Nanoemulsions Fabricated Using Dual-Channel Microfluidization: Formation, Stability and Bioaccessibility. *J. Agric. Food Chem.* **2018,** *66*, 10532–10542. DOI: 10.1021/acs.jafc.8b03077.
100. Xu, X.; Zhang, H.; Gao, T.; Teng, J. Impacts of Applied Voltage on Forward Osmosis Process Harvesting Microalgae: Filtration Behaviors and Lipid Extraction Efficiency. *Sci. Total Environ.* **2021,** *773*, 145678. DOI: 10.1016/j.scitotenv.2021.145678.

101. Wei, F.; Wu, B. *Use of Lipidomics for Food Quality Assurance and Authentication*; Elsevier, 2021.
102. Castellanos, D. B.; Martín-Jiménez, C. A.; Rojas-rodríguez, F.; Barreto, G. E.; Gonz Brain Lipidomics as a Rising Field in Neurodegenerative Contexts: Perspectives with Machine Learning Approaches. *Front. Neuroendocrinol.* **2021,** *61*, 100899. DOI: 10.1016/j.yfrne.2021.100899.
103. Wu, B.; Wei, F.; Xu, S.; Xie, Y.; Lv, X.; Chen, H.; Huang, F. Mass Spectrometry-Based Lipidomics as a Powerful Platform in Foodomics Research. *Trends Food Sci. Technol.* **2021,** *107*, 358–376. DOI: 10.1016/j.tifs.2020.10.045.
104. Lam, S. M.; Wang, Z.; Li, B.; Shui, G. High-Coverage Lipidomics for Functional Lipid and Pathway Analyses. *Anal. Chim. Acta.* **2021,** *1147*, 199–210. DOI: 10.1016/j.aca.2020.11.024.

CHAPTER 6

Antioxidant Dietary Fiber from Food Byproducts: Case Studies

TÂNIA BRAGANÇA RIBEIRO, DIVA SANTOS, SARA MARÇAL, RITA SARDÃO, and MANUELA PINTADO

Universidade Católica Portuguesa, CBQF-Centro de Biotecnologia e Química Fina–Laboratório Associado, Escola Superior de Biotecnologia, Rua Diogo Botelho, Porto, Portugal

ABSTRACT

The food industry produces tons of byproducts per year, even before the consumption phase. During the cultivation, postharvest, and processing of fruits, vegetables and cereals were produced large amounts of food byproducts. Food byproducts have been described as a rich source of bioactive compounds associated with several health benefits such as phenolic compounds and dietary fiber. These bioactive compounds are generally studied separately. However, phenolic compounds have been found associated with dietary fiber, as linked components, conferring health-related properties, especially antioxidant capacity. A new dietary fiber concept, the antioxidant dietary fiber, has emerged and is explored in several vegetable byproducts. Antioxidant dietary fiber has become increasingly attractive to food, pharmaceutical, and cosmetic industries due to its ability to confer not only nutritional and technological properties, but also bioactive beneficial health properties to products. In this sense, the purpose of the present chapter is to: (1) Clarify the concept of antioxidant dietary fiber; (2) Identify vegetable byproducts as potential resources of antioxidant dietary fiber; (3) Review the transformation processes of food byproducts into antioxidant dietary fibers as ingredients with potential to develop new functional food

Food Byproducts Management and their Utilization. Ricardo Gómez-García, Ana A. Vilas-Boas, Débora A. Campos, Maria Manuela Pintado, & Cristóbal Noé Aguilar (Eds.)

products; and (4) Illustrate the technological, nutritional, and health-benefit functionalities of antioxidant dietary fiber as a new functional ingredient.

6.1 INTRODUCTION

According to CODEX Alimentarius, dietary fiber can be defined as a group of carbohydrate polymers that are not digested into absorbable compounds by human digestive enzymes.[1] Nowadays, the dietary fiber concept is changing, for example other noncarbohydrates and lignin components as phenolics compounds are considered part of dietary fiber structure.[2] The intake of dietary fiber is essential in a balanced diet and has several health benefits, for instance, it avoids constipation, stimulates the growth of the beneficial intestinal microflora, improves blood glucose control, decreases blood pressure, and contributes to weight loss. These health benefits were not only associated with carbohydrates components, but also with bioactive compounds bound to fiber.[2,3] The adequate intake (AI) of fiber for adults is 25–30 g per day.[4,5] In the United States and Europe, fiber is considered a nutrient of public health concern since the fiber intake of a large part of the population does not reach recommendations.[5,6] In recent years, the health concern of consumers with their fiber intake has been rising. The main sources of fiber in the diet are fruits, vegetables, and cereals. However, the demand for processed food products with high fiber content (e.g., cereal bars, cookies, and bread) has increased significantly. As a consequence, the food industry has been searching for new fiber sources. Recently, byproducts namely peels, bagasse, cores, seeds, husk, from fruits, vegetables, and cereals were highlighted as sustainable dietary fiber sources that could provide the technological, nutritional, and health-beneficial functionalities from all its carbohydrates and noncarbohydrate components (e.g., phenolics).[7,8]

6.2 ANTIOXIDANT DIETARY FIBER CONCEPT

In the last 5–10 years, the dietary fiber definition has been a theme of discussion between several organizations (Codex Alimentarius; European Community; Food and Drug Administration (FDA), and others).[6] In 2009, the Commission on Nutrition and Foods for Special Dietary Uses of the Codex Alimentarius Commission established a definition that has a general global agreement. The definition declares that *dietary fiber means carbohydrate polymers with ten or more monomeric units, which are not hydrolyzed by endogenous enzymes in the small intestine of human beings and belong to the following categories:*

- *Edible carbohydrate polymers naturally occurring in food as consumed.*
- *Carbohydrate polymers, which have been obtained from raw material in food by physical, enzymatic, or chemical means and which have been shown to have physiological effect of benefit to health by generally accepted scientific evidence to competent authorities.*
- *Synthetic carbohydrate polymers, which have been shown to have physiological effect of benefit to health by generally accepted scientific evidence to competent authorities.*[6,9]

Besides, in the dietary fiber definition, an important aspect to keep in mind is its classification and physiological properties. Conventionally, in the literature, dietary fiber is classified into "soluble" (SDF) and "insoluble" (IDF) in an attempt to explain the different fiber-associated physiological effects. However, several authors and organizations emphasized that water solubility is a physicochemical property dependent on the determination/quantification method used, not being directly related to physiological effects of dietary fiber, whether it be the viscosity, fermentability, or fecal-bulking effect. In this sense, FAO/WHO in 1998 advised that the distinction between IDF and SDF should be phased out. Indeed, several authors consider that the classification of dietary fiber according to its main physiological effects might be more accurate.[6,9]

Another important characteristic that is unclear in the dietary fiber definition is the presence of antioxidant compounds (e.g., phenolic compounds) linked to carbohydrate polymers.[6,10] Several researchers have pointed out the phenolic compounds as a significant contributor to the health functionalities of dietary fiber, usually attributed only to carbohydrates fiber components.[6,10] To clarify the health and technological importance of the phenolic compounds as a structural component of dietary fiber, a new concept of "antioxidant dietary fiber" (ADF) was introduced in 1998.[11] The ADF was defined as a source of dietary fiber with natural linked antioxidants, mainly phenolic compounds, that should meet the following requirements by 1 g: (i) dietary fiber content higher than 0.5 g on a dry basis; (ii) lipid oxidation inhibitory capability equivalent to 200 mg of vitamin E (measured by the thiocyanate procedure) and a free radical scavenging capacity equivalent to at least 50 mg of vitamin E (measured by the DPPH assay). Besides that, the antioxidant capacity of ADF must be an intrinsic property of its natural constituents.[8,11]

The ADF has been recognized as a promising bioactive ingredient providing health benefits for humans due to the joint assimilation and metabolization of dietary fiber and antioxidants in the colon. Furthermore, ADF is also considered a technological/functional ingredient with positive effects on food preservation; for instance, it prevents lipid peroxidation and delays sensory deterioration.[8,12]

6.3 VEGETAL BYPRODUCTS WITH POTENTIAL TO RECOVER ANTIOXIDANT DIETARY FIBER

Vegetal foods contain different amounts of fiber content.[13] The edible part of fruits contains from 0.5% (watermelon) to 3.4% (kiwi) of total dietary fiber (percentage of edible portion). Vegetable's total dietary fiber content varies between 0.6% (cucumbers, peeled) and 7.8% (beetroot). Legumes and pulses have a total dietary fiber content from 1.9% (green beans) to 17.7% (white beans, raw). Usually, grains have higher dietary fiber content, except when refined (when bran was removed), varying between 1.3% (rice) and 17.3% (barley).[13] Nowadays, cereal grains and their byproducts are the main sources of fiber used by the food industry to develop processed foods with high-fiber content.[14] However, overall fiber from vegetables, fruit, and their byproducts present some advantages compared with grains and pulses, namely, lower phytic acid content, higher amount of SDF, and higher antioxidant activity due to the high content of bound phenolic compounds.[7]

Vegetal byproducts include unripe and/or noncompliant fruits and vegetables, peels, stems, leaves, pomaces, and bran (cereals). Often, the valorization of byproducts involves water removal, decreasing the foods' perishability, and consequently increasing the dietary fiber content to 30–90% of dry weight.[15] This transformation of the byproducts usually results in powders/flours that can be applied in diverse food industry sectors, namely bakery, dairy, beverage, and meat/fish industries.[16]

There are many studies on the potential applications of vegetal byproducts. Fiber concentrates from apple byproducts are currently at an industrial scale development state, with patented production processes. They were used to develop several functional foods, for instance, bakery, meat, and dairy food products. Probably, they were one of the most studied byproducts, counting with several reviews.[17,19] Grape pomace has also been intensively studied, mainly due to its content of phenolic compounds, which increased the interest in applications in food, pharmaceutical, and cosmetic industries.[20,21] In foods, phenolic compounds from grape pomace are used as natural antioxidants to prevent oxidative rancidity, retard the development of off-flavors, improve color stability, and increase gel strength in cheeses.

White grape pomace powder increased the DF content, phenolic compounds amount (mainly gallic acid, tyrosol, γ-resorcylic acid, and catechin), and antioxidant activity of biscuits.[22] Grape pomace and its extracts revealed antibacterial effects against *Staphylococcus aureus* and *Escherichia coli* in vegetable soup.[23] Berry pomace is another fruit pomace that has been pointed out as a high DF source. After being dried, it contains about 68% of DF. However, until now, berry pomace has mostly been explored as a source of phenolic compounds.[24] Mango, pineapple, and citrus fruits byproducts are other sources of fiber barely explored. Besides fiber, they contain bioactive compounds such as carotenoids, terpenes and phenolic compounds, and enzymes, as lipolytic enzymes, pectinases, and proteases, with potential to be used in food industry.[25,28]

Regarding vegetables, carrot and tomato byproducts are the most studied due to their content of fiber, phenolic compounds, and carotenoids.[29] The incorporation of tomato byproducts has been tested in several food products (bread, bakery, dairy, meat products, tomato products, extrusion snacks).[30] Potato peels can also be rich in fiber if the peeling procedure is optimized to avoid contamination with starch. This byproduct is also rich in phenolics with antioxidant and antidiabetic capacities.[29] Byproducts from the olive oil industry include olive paste, olive pomace, and olive mill wastewater and are used as fuel and ingredients for compost production or as fertilizer. However, they are rich in bioactive compounds such as phenolics (between 0.5 and 24 g/L).[31] A recent study evaluated the application of olive paste and olive mill wastewater byproducts in bread and pasta. The bread was a better product to reuse the olive oil byproducts. Olive paste offered better performance for food fortification than olive mill wastewater, once the paste improved chemical quality (total polyphenols and antioxidant activity) to a greater extent than olive mill wastewater.[32]

The cereal and pulse processing industry also generate a huge volume of valuable byproducts, such as bran, husk, and germ meal. These byproducts represent a source of phenolic acids (ferulic, gallic, sinapic, gentisic, caffeic, *p*-hydroxybenzoic, syringic, vanillic, *p*-coumaric), dietary fibers (arabinoxylans, β-glucan lignin/lignans, cellulose), and vitamins (thiamin, riboflavin, niacin, pantothenic acid), which are associated with anti-inflammatory, antioxidant, antidiabetic, and anticarcinogenic properties.[33] Furthermore, other cheap and easily acquired byproducts can also result from cereals and pulses processing. Okara, the byproduct of tofu or soya beverage production process, results from ground soybean after the extraction of the water-extractable fraction. Dry okara contains about 50% dietary fiber, 25% protein, and 10% lipid. Other constituents found in okara involve lignans, isoflavones, coumestans, saponins, phytosterols, and phytates.[34] Brewery spent grains (BSG) is the most abundant byproduct from the beer industry and it consists of residual malted grains that remain at the end of the brewing process. The composition depends on the barley varieties and mashing conditions, but generally, this byproduct is rich in fiber (35–60% on a dry basis), proteins (15–26%), phenolic compounds, and vitamins.[35,36]

All the mentioned byproducts and many others have been studied to be used as functional ingredients for the development of new products or to be added to food products that need improvement regarding fiber and protein contents while minimizing caloric value. The high potential of these products comes from their environmental and nutritional benefits, low cost, and high availability.

6.4 FROM FOOD BYPRODUCTS TO ANTIOXIDANT DIETARY FIBER INGREDIENT

Briefly, the recovery of ADF from vegetable byproducts involves water removal and milling. The complexity of the process increases with the purity of fiber when intended (Figure 6.1). For food safety assurance, the extraction of ADF starts with a washing process with water and a stabilization step that can be decontamination with a disinfectant solution followed by rising with water,[37] or in the autoclave (less common).[38] Stabilization procedures where low temperatures are applied minimize the losses of SDF, namely of pectins and pentosans, and the reduction of bioactive compounds.[39] Unlike SDF and bioactive compounds, the remotion of simple sugars during washing can be desirable because it decreases the caloric value of fiber concentrates, the browning during the hot air drying, and stickiness and caking during milling.[7]

The following step is separating the liquid from the solid fractions of the byproducts (this step is not applicable in the case of already solid byproducts such as pomaces).[40] This step has been performed using a juicer machine or a vertical cutter followed by pressing. In processing byproducts containing seeds, a previous step (usually a sieving process) can be carried out to remove them. Seeds can contain some undesirable compounds, for instance, toxic molecules and lipids susceptible to oxidation.[18] However, they also could be valorized due to their bioactive compounds, essential amino acids, and healthy lipids.[41,42]

The solid fraction is then subjected to a drying process that can be performed at temperatures between 30°C[43]–70°C[23] using ovens, dryers or dehydrators,[37,43,44] or freeze-drying.[45] Freeze-drying is the gold standard method to preserve heat-labile compounds and the antioxidant properties of the fiber. However, it is economic and environmental expensive.[7] The most common temperatures in hot drying are under 65°C, minimizing the losses of polyphenols, anthocyanidins, proteins,[46] and pectin.[47] The temperatures lower than 30–40°C are avoided because they imply higher drying time.[47] Besides that, a more aggressive thermal treatment may promote the liberation of bound phenolics, increasing the free phenolics in the powder.[48]

After drying, the solid fraction usually was grounded to powder and sieved to a known particle size (PS) according to the intended use. Grinding may occur using mills (cyclone or hammer mill for example)[32,43] or grinders.[23] The milling methods and PS had a high impact on the fiber's physicochemical, technological, bioactive, and sensory properties. It is well documented that ultra-fine grinding increased the SDF content and decreased IDF amount.[49,50] Regarding technological properties, it was reported that the reduction of PS from 1127 to 550 μm increased the water holding capacity (WHC), water retention capacity (WRC), and water swelling capacity (WSC). However, higher PS reductions decreased these hydration properties.[51] It is important to underline that this effect is different for ADF recovered from different vegetable byproducts mainly due to the different SDF: IDF ratio.[52] In terms of bioactive compounds and bioactivities, PS lower than 500 μm may increase the antioxidant activity of the fibers due to the liberation of the bound phenolics into the free form.[52] Micronization of ADF from peel and calyx persimmon increased phenolic and flavonoid compounds content and antioxidant activity, which resulted in lower lipid oxidation of the final product (pork patties) during storage.[53] Furthermore, fiber with lower PS showed a higher potential to decrease the starch digestion rate and the postprandial serum glucose level. For instance, Liu et al.[49] showed that IDF from orange peels with PS from 0.63 to 158.87 μm had a α-amylase activity inhibition ratio 4.0 times higher than the same fiber with PS from 3.99 to 796.21 μm. Besides, fiber with the lowest PS had a higher capacity to retard the glucose diffusion, indicating a higher ability to reduce glucose absorption in the intestine. Concerning sensorial properties, Liu et al.[49] reported that fiber from orange peels with the lowest PS had a higher lightness and lesser yellow and red color, facilitating their incorporation in some foods. Compared with ultra-fine grinding at room temperature, cryogenic grinding enables us to obtain flours with lower PS and high antioxidant capacity. The grinding at low temperatures increases the brittleness of foods matrices, making it easier to obtain flours with low PS, and improve the preservation of heat-labile compounds.[54]

High hydrostatic pressure is an emerging method to improve the properties of fiber.[55] Yan et al.[50] evaluated the impact of ultra-fine grinding and high hydrostatic pressure on fiber from pear pomace. Compared with control flour (60-mesh), samples submitted to high hydrostatic pressure treatment or ultra-fine grinding showed higher SDF content and higher cholesterol binging capacity. These increases were more significant in high hydrostatic pressure treated flour. Besides that, hydrostatic pressure treatment improved water and oil-holding capacity and swelling of pear pomace flour while ultra-fine grinding impaired these technological properties.[50] Several previous studies have been suggested that high hydrostatic pressure treatment improved the prebiotic activity of fiber.[55]

Given the effect of the processing on the ADF properties, their conditions may be defined according to the final product purpose application, for example, if the ADF is going to be integrated into a product with a mild production process as dairy products, for example, the fiber can be produced with application of higher temperature and inducing lower PS, so the bound phenolics are free (released from the fiber) and thus available to be absorbed in the upper gastrointestinal tract after food product digestion. But, on the other hand, if it is intended to be applied in an extruded product, for instance, the powder must have been produced under as mild as possible conditions to assure the bound compounds are protected (still bound to the fiber) so they can be released during the food product production, instead of being lost (which would happen to those phenolics in the free form).

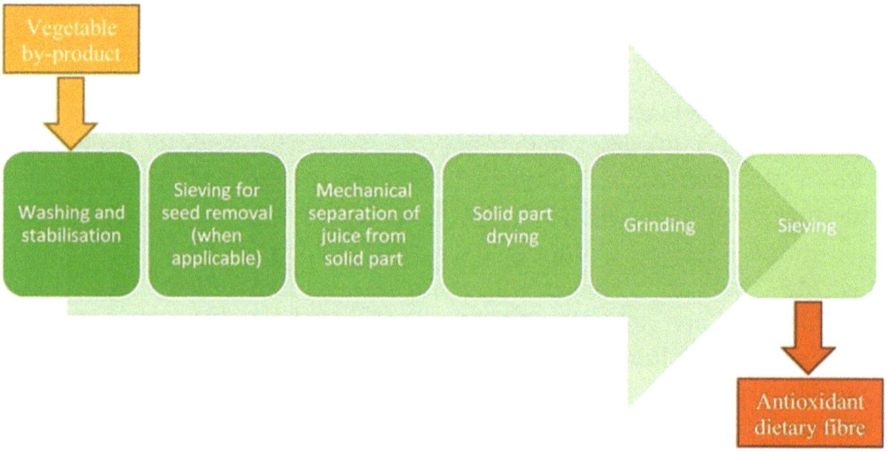

FIGURE 6.1 General procedure for antioxidant dietary fiber recovery.

6.5 TECHNOLOGICAL, NUTRITIONAL AND HEALTH-BENEFICIAL FUNCTIONALITIES OF ANTIOXIDANT DIETARY FIBER

The ADF encompasses a great application potential to be employed as a food ingredient, namely as thickeners, gelling agents, fillers, water retainer agents, as well as in the production of edible films. Additionally, to their technological functionalities, these byproducts present several health benefits related to their content in dietary fiber and bioactive compounds, especially bound phenolics and carotenoids delivered by dietary fiber.[29]

6.5.1 TECHNOLOGICAL FUNCTIONALITY

Fiber obtained from vegetal byproducts presents several technological functionalities related to the soluble and insoluble fiber portions when added to foods. Both SDF and IDF absorb water, but usually, only SDF increases viscosity, presenting technological functionality such as gelling, thickening, filling, production of edible films and coatings,[56] emulsifying,[57] fat replacing, and foam stabilizing.[58] In its turn, IDF usually has a regular structure and can minimize shrinkage retard staling, control moisture and increase food stability.[58]

ADF from different sources has different ratios of SDF: IDF, which results in technological functionalities. For instance, different WHC values from different fruits (blueberry and raspberry pomaces) byproducts flours (presenting 57/2/55% and 52/2/50% of TDF/SDF/IDF, respectively) resulted in different dough structures and consequently different final bakery product textures,[59] the flour with higher WHC (blueberry pomace flour) created a more dense inner structure, higher water loss during baking, resulting in lower cookie thickness, higher spread ratio, more rigid dough structure, and higher cookie hardness. In general, the replacement of wheat flour in bread for fruit byproducts flour increases WHC and oil holding capacity (OHC), consequently increasing the water absorption of dough, increases dough development time but decreases dough stability (fiber slows the rate of hydration and the development of gluten), decreases viscosity (the interaction between fiber and starch limits the swelling of the starch granules), and may or may not increase hardness.[60,61] The use of hard red, soft white, and purple wheat brans into extruded snacks decreased expansion ratios in different proportions (less for purple bran) and affected density and hardness differently among brans.[62]

Other byproducts from grains and pulses rich in dietary fiber could also provide technological features when incorporated into various food matrices. Okara, for example, can be used to improve the quality of pork-based products, especially to replace animal fats and accelerate the formation of an elastic gel network. In pork meat gels, the addiction of okara improves the hydration properties (reducing cooking loss and improving WHC), the texture (increasing the hardness, chewiness, and breaking force), and the color (increasing L* value).[63]

The BSG can also be a functional ingredient to partially replace 1:1 white and whole wheat flour (w/w) in muffins. The batter viscosity increases with the increasing BSG substitution once the high dietary fiber acts as a thickening agent absorbing water. Thus, the idea of incorporation of these byproducts will always find a technological improvement without impacting the consumer's acceptance of the products.[64] However, it is necessary to pay attention to the amount of fiber-rich byproducts incorporated in the development of bakery products. The incorporation of fiber into the dough matrix can interfere with the gluten network formation, competing for water and altering the formation conditions of the protein network, which results in the weakening of the dough. Thus, the addition of these byproducts must be optimized in order not to cause significant changes in dough rheological properties.[65]

Additionally, to the texture-related functionalities, the vegetable byproduct fibers often present a stabilization functionality, since these fibers have antioxidant activity. The addition of tomato peel powder to low-fat sausages decreased hardness and springiness and increased the stability of polyunsaturated/saturated fatty acids ratio during storage.[66] Grape pomace powder and extracts presented antimicrobial effects against *E. coli* and *S. aureus*.[23]

Phenolic compounds present in these fibers can also provide color properties to the final products, carotenoids give yellow, orange, and red colors, and anthocyanins provide red, purple, and blue colors.[67]

The emulsifying properties of ADF byproducts could also be explored in nonfood sectors. Recently, apple pomace and oat bran micronized powders showed useful emulsifying properties to cosmetic formulations, mainly when employed with myritol and rapeseed oil.[68]

6.5.2 NUTRITIONAL AND HEALTH FUNCTIONALITIES

Adequate fiber intake is essential to ensure metabolic health. It has been very well reported that fiber prevents and treats obstipation, stimulates the growth of the beneficial intestinal flora, and increases the production of important microbial metabolites like short-chain fatty acids (SCFAs). Besides, the scientific evidence suggested that fiber consumption is negatively associated with cardiovascular diseases, contributes slightly to weight loss since it reduces the meals frequency and size, improves insulin sensitivity, and decreases chronic inflammation.[69]

The health functionality of ADF is closely related to the interaction between fibers and phytonutrients such as bound phenolics. Fibers play a key role in transporting phenolic compounds through the upper gastrointestinal tract to the colon.[67] Because they are covalently bound to fiber (cellulose, hemicellulose, lignin, pectin), phenolic compounds survive the stomach and small intestine digestions, and can reach the colon, where they are released after fermentation by gut microbiota due to enzymatic activity of esterase and xylanase.[70] It was verified that only 0.5–5% of the ferulic acid is absorbed in the small intestine, which corresponds to the free soluble fraction.[71] Epidemiological studies showed that grain consumption prevents colon cancer and other digestive cancers, which can be explained by the high content of bound phenolic in grains and their higher release and effectiveness in the colon.[72,73] It has been demonstrated for several studies that ferulic acid is mainly released in the colon, which are the probiotic genera *Bifidobacterium* and *Lactobacillus* known as the best ferulic esterase producers.[74]

Nevertheless, studies demonstrated that human cinnamoyl esterases are found in the small intestine, mainly in the mucosa epithelial cells providing a release and absorption of hydroxycinnamic acid.[75] A study that comprised free and bound phenolics concluded that bound phenolics intake provides chemopreventive activity against colon cancer while free and soluble conjugated forms intake provides other health benefits because they are rapidly absorbed in the stomach and small intestine and distributed throughout the body.[75] The consumption of anthocyanin-rich fruits and/or beverages (red wine and red grape juice) was related to coronary heart and cardiovascular diseases' risk reduction.[67] According to an in vitro bioaccessibility assessment study of grape phenolic, these health benefits are associated with the action of the free and bioaccessible phenolics (liquid phase) and bound phenolics that can be partially extracted and released during *in vitro* gastric digestion.[76]

The bound phenolics' content varies among vegetables, and some examples are presented in Table 6.1. The most common phenolic acids in vegetal byproducts dietary fiber are chlorogenic, caffeic, ferulic, *p*-coumaric, sinapic, caftaric, coutaric, and fertaric acids (hydroxycinnamic acids); gallic, *p*-hydroxybenzoic, protocatechuic, protocatechuic, γ-resorcylic, isovanillic, ellagic, and syringic acids (hydroxybenzoic acids).[18,20,22] Fruits are rich in flavanols as catechin, epicatechin, epicatechin gallate, procyanidin B1 and

B2; flavonols as kaempferol, quercetin, quercetin glycosides, hyperin, rutin, reynoutrin, avicularin, quercitrin, isorhamnetin; flavanones as eriodictyol, naringenin, and naringin; flavones as luteolin, apigenin, and apigenin glycosides; coumarins as aesculin;[44] and anthocyanins such as cyanidin, cyanidin-3-O-galactoside, cyanidin 3-O-p-coumaroylglucoside, delphinidin, malvidin, peonidin, and petunidin.[18,20,22,67,77]

TABLE 6.1 Percentage of Bound Phenolics of Total Phenolics (Free Plus Bound) in Several Vegetables or Vegetable Byproducts.[78]

Vegetable/vegetable byproduct	Bound phenolics in fiber
Apple	6.5%
Cranberry pomace	76.3%
Grapefruit	23.6%
Banana	33.1%
Huyou peel	16.65%
Carrot	37.6%
Barley malt	51.9%
Barley	70.1%
Wheat	62–83%
Yellow sweet corn	69.2%
Oat	75–88%

Source: Adapted from [78].

Cereal and pulse byproducts also represent an interesting source of phenolic acids in free, soluble conjugates and bound forms. The majority of them exist in bound form, linked to cell wall components.[79] The BSG contains several bound phenolic compounds, such as catechin, ferulic, *p*-coumaric, 4-hydroxybenzoic, sinapic, syringic, protocatechuic, and caffeic acids.[80] In wheat grains, phenolic acids are mainly found in the cell walls of their outer layers. Thus, the wheat bran contains several compounds of interest like vanillic, syringic, *p*-hydroxybenzoic, *p*-coumaric, ferulic, isoferulic, and salicylic acids. In brief, these byproducts are interesting products to improve health more effectively as they may avoid upper gastrointestinal digestion along with cell wall components and are absorbed into blood plasma during digestion of intestinal microorganism.[79]

Carotenoids are also presented bound to cell wall structures and in several byproduct flours, namely carrot[81] and mango[82] flours. They have a high antioxidant capacity, and their intake has been associated with several benefits, for instance, protection against optical and cardiovascular diseases.[83] The health importance of some carotenoids (the ones having β-ionone end group) is related to the synthesis of vitamin A in the body, once they are the precursors of this vitamin.[67] Rice bran is the most accessible natural source of oryzanol, an oily fraction that contains ferulate, esters of sterols, and triterpene alcohols and has anti-inflammatory and anticarcinogenic properties when applied in low doses.[84, 85] Vitamin E is the generic term for tocols that include tocopherols (α-, β-, γ-, and δ-tocopherol) and tocotrienols (also α-, β-, γ-, and δ-tocotrienol) and have been found in fruit byproduct fibers.[77] Tocols are antioxidants and have been highlighted for their benefits to human health, including the prevention of chronic diseases.[86]

6.6 CONCLUSION

The nutritional, bioactive, and technological properties of DF have been attributed to carbohydrates and the bioactive compounds bound to them. However, the conventional DF definition and its classification into SDF and IDF did not consider the presence of bounded bioactive compounds nor their bioactive properties. Thus, the ADF fiber concept was developed to fill this gap. Although DF from different vegetal byproducts has distinct properties, overall, they are considered excellent sources of ADF, which can be recovered by succinct processes, including three main steps: Washing and stabilization, remotion of water, and milling. The conditions applied in each of these steps significantly impact the bioactive and technological properties of ADF. Several milling and other physico-mechanical conventional methods have been optimized to improve the ADF characteristics in the last few years, standing out micronization, cryogenic grinding, and high hydrostatic pressure that compared with conventional processes showed to improve technological properties (e.g., higher WHC, WRC, WSC) and health benefits (e.g., higher antioxidant, prebiotic, and cholesterol-binding activities) of ADF powders.

Beyond the technological, nutritional, and health beneficial functionalities exhibited by ADF from vegetal byproducts, their exploitation/valorization could bring several advantages from an economic and environmental point of view. The use of byproducts as raw materials of ADF to produce nutrients as fibers and natural antioxidants could represent new revenue streams to the processing industries, while promoting the reduction of industrial wastes and natural resources depletion, fomenting a more sustainable and circular economy.

KEYWORDS

- **antioxidant dietary fiber**
- **food byproducts**
- **health functionality**
- **technological functionality**

REFERENCES

1. ALINORM 09/32/26. Report of the 30th session of the Codex Committee on Nutrition and Foods for Special Dietary Uses. *Codex Aliment. Comm.* **2008,** *30*, 3–7.
2. Quirós-Sauceda, A. E. et al. Dietary Fiber and Phenolic Compounds as Functional Ingredients: Interaction and Possible Effect After Ingestion. *Food Funct.* **2014,** *5* (6), 1063–1072, DOI: 10.1039/C4FO00073K.
3. Anderson, J. W. et al. Health Benefits of Dietary Fiber. *Nutr. Rev.* **2009,** *67* (4), 188–205.
4. European Food Safety Authority. Scientific Opinion on Dietary Reference Values for Carbohydrates and Dietary Fiber. *EFSA J.* **2010,** *8* (3), 1462.
5. US Department of Health and Human Services and Others, *Dietary Guidelines for Americans 2015–2020*. Skyhorse Publishing Inc., 2017.
6. Stephen, A. M. et al. Dietary Fiber in Europe: Current State of Knowledge on Definitions, Sources, Recommendations, Intakes and Relationships to Health. *Nutr. Res. Rev.* **2017,** *30* (2), 149–190. DOI: 10.1017/S095442241700004X.
7. Garcia-Amezquita, L. E.; Tejada-Ortigoza, V.; Serna-Saldivar, S. O.; Welti-Chanes, J. Dietary Fiber Concentrates from Fruit and Vegetable byproducts: Processing, Modification, and Application as Functional Ingredients. *Food Bioprocess Technol.* **2018,** *11* (8), 1439–1463.
8. Xu, J.; Li, Y.; Zhao, Y.; Wang, D.; Wang, W. Influence of Antioxidant Dietary Fiber on Dough Properties and Bread Qualities: A Review. *J. Funct. Foods*, **2021,** *80*, 104434, DOI: 10.1016/j.jff.2021.104434.
9. Johnson, I. T. Fiber. In *Present Knowledge in Nutrition*; Elsevier; 2020, pp 515–529.
10. Liu, S. et al. Removal of Bound Polyphenols and Its Effect on Antioxidant and Prebiotics Properties of Carrot Dietary Fiber. *Food Hydrocoll.* **2019,** *93* (August 2018), 284–292. DOI: 10.1016/j.foodhyd.2019.02.047.
11. Saura-Calixto, F. Antioxidant Dietary Fiber Product: A New Concept and a Potential Food Ingredient. *J. Agric. Food Chem.* **1998,** *46* (10), 4303–4306. DOI: 10.1021/jf9803841.
12. Silva, S.; Oliveira, A. L. S.; Ribeiro, T.; Madureira, A. R.; Pintado, M. M. Chapter 17. Non-extractable Phenolic Compounds as Food Ingredients. In *Non-extractable Polyphenols and Carotenoids: Importance in Human Nutrition and Health*; Fulgencio Saura-Calixto, J. P.-J., Ed.; 2018, pp 345–366.
13. Dhingra, D.; Michael, M.; Rajput, H.; Patil, R. T. Dietary Fiber in Foods: A Review. *J. Food Sci. Technol.* **2012,** *49* (3), 255–266. DOI: 10.1007/s13197-011-0365-5.

14. Tejada-Ortigoza, V. et al. In Vitro Fecal Fermentation of High Pressure-Treated Fruit Peels Used as Dietary Fiber Sources. *Molecules* **2019,** *24* (4), 697.
15. Elleuch, M.; Bedigian, D.; Roiseux, O.; Besbes, S.; Blecker, C.; Attia, H. Dietary Fiber and Fiber-Rich Byproducts of Food Processing: Characterisation, Technological Functionality and Commercial Applications: A Review. *Food Chem.* **2011,** *124* (2), 411–421. DOI: https://doi.org/10.1016/j.foodchem.2010.06.077.
16. Trigo, J. P.; Alexandre, E. M. C.; Saraiva, J. A.; Pintado, M. E. High Value-Added Compounds from Fruit and Vegetable Byproducts–Characterization, Bioactivities, and Application in the Development of Novel Food Products. *Crit. Rev. Food Sci. Nutr.* **2020,** *60* (8), 1388–1416. DOI: 10.1080/10408398.2019.1572588.
17. Piskov, S. et al. Effects of Various Drying Methods on Some Physico-Chemical Properties and the Antioxidant Profile and ACE Inhibition Activity of Oyster Mushrooms (Pleurotus Ostreatus). *Foods* **2020,** *9* (2). DOI: 10.3390/foods9020160.
18. Rabetafika, H. N.; Bchir, B.; Blecker, C.; Richel, A. Fractionation of Apple Byproducts as Source of New Ingredients: Current Situation and Perspectives. *Trends Food Sci. Technol.* **2014,** *40* (1), 99–114. DOI: https://doi.org/10.1016/j.tifs.2014.08.004.
19. Shalini, R.; Gupta, D. K. Utilization of Pomace from Apple Processing Industries: A Review. *J. Food Sci. Technol.* **2010,** *47* (4), 365–371. DOI: 10.1007/s13197-010-0061-x.
20. Fontana, A. R.; Antoniolli, A.; Bottini, R. Grape Pomace as a Sustainable Source of Bioactive Compounds: Extraction, Characterization, and Biotechnological Applications of Phenolics. *J. Agric. Food Chem.* **2013,** *61* (38), 8987–9003. DOI: 10.1021/jf402586f.
21. Yu, J.; Ahmedna, M. Functional Components of Grape Pomace: Their Composition, Biological Properties and Potential Applications. *Int. J. Food Sci. Technol.* **2013,** *48* (2), 221–237. DOI: https://doi.org/10.1111/j.1365-2621.2012.03197.x.
22. Mildner-Szkudlarz, S.; Bajerska, J.; Zawirska-Wojtasiak, R.; Górecka, D. White Grape Pomace as a Source of Dietary Fiber and Polyphenols and Its Effect on Physical and Nutraceutical Characteristics of Wheat Biscuits. *J. Sci. Food Agric.* **2013,** *93* (2), 389–395. DOI: https://doi.org/10.1002/jsfa.5774.
23. Sagdic, O.; Ozturk, I.; Kisi, O. Modeling Antimicrobial Effect of Different Grape Pomace and Extracts on S. Aureus and E. Coli in Vegetable Soup Using Artificial Neural Network and Fuzzy Logic System. *Expert Syst. Appl.* **2012,** *39* (8), 6792–6798.
24. Struck, S.; Plaza, M.; Turner, C.; Rohm, H. Berry Pomace–A Review of Processing and Chemical Analysis of Its Polyphenols. *Int. J. Food Sci. Technol.* **2016,** *51* (6), 1305–1318. DOI: https://doi.org/10.1111/ijfs.13112.
25. Campos, D. A. et al. Optimization of Bromelain Isolation from Pineapple Byproducts by Polysaccharide Complex Formation. *Food Hydrocoll.* **2019,** *87*, 792–804. DOI: https://doi.org/10.1016/j.foodhyd.2018.09.009.
26. Okino Delgado, C. H.; Fleuri, L. F. Orange and Mango Byproducts: Agro-Industrial Waste as Source of Bioactive Compounds and Botanical Versus Commercial Description—A Review. *Food Rev. Int.* **2016,** *32* (1), 1–14.
27. Jahurul, M. H. A. et al. Mango (Mangifera Indica L.) Byproducts and Their Valuable Components: A Review. *Food Chemistry* **2015,** *183*, 173–180. DOI: 10.1016/j.foodchem.2015.03.046.
28. Ledesma-Escobar, C. A.; Luque de Castro, M. D. Towards a Comprehensive Exploitation of Citrus. *Trends Food Sci. Technol.* **2014,** *39* (1), 63–75. DOI: https://doi.org/10.1016/j.tifs.2014.07.002.

29. Gómez, M.; Martinez, M. M. Fruit and Vegetable Byproducts as Novel Ingredients to Improve the Nutritional Quality of Baked Goods. *Crit. Rev. Food Sci. Nutr.* **2018,** *58* (13), 2119–2135.
30. Lu, Z.; Wang, J.; Gao, R.; Ye, F.; Zhao, G. Sustainable Valorisation of Tomato Pomace: A Comprehensive Review. *Trends Food Sci. Technol.* **2019,** *86*, 172–187. DOI: https://doi.org/10.1016/j.tifs.2019.02.020.
31. Obied, H. K.; Allen, M. S.; Bedgood, D. R.; Prenzler, P. D.; Robards, K.; Stockmann, R. Bioactivity and Analysis of Biophenols Recovered from Olive Mill Waste. *J. Agric. Food Chem.* **2005,** *53* (4), 823–837.
32. Cedola, A.; Cardinali, A.; D'Antuono, I.; Conte, A.; Del Nobile, M. A. Cereal Foods Fortified with Byproducts from the Olive Oil Industry. *Food Biosci.* **2020,** *33*, 100490. DOI: https://doi.org/10.1016/j.fbio.2019.100490.
33. Saini, A.; Panwar, D.; Ps, P.; Mb, B. Bioactive Compounds from Cereal and Pulse Processing Byproducts and Their Potential Health Benefits. *Austin J. Nutr. Metab.* **2019,** *6* (2), 1–7.
34. Lu, F.; Liu, Y.; Li, B. Okara Dietary Fiber and Hypoglycemic Effect of Okara Foods. *Bioact. Carbohydrates Diet. Fiber* **2013,** *2* (2), 126–132. DOI: 10.1016/j.bcdf.2013.10.002.
35. Stelick, A.; Sogari, G.; Rodolfi, M.; Dando, R.; Paciulli, M. Impact of Sustainability and Nutritional Messaging on Italian Consumers' Purchase Intent of Cereal Bars Made with Brewery Spent Grains. *J. Food Sci.* **2021,** *86* (2), 531–539. DOI: 10.1111/1750-3841.15601.
36. Puligundla, P.; Mok, C. Recent Advances in Biotechnological Valorization of Brewers' Spent Grain. *Food Sci. Biotechnol.* **2021,** *30* (3), 341–353. DOI: 10.1007/s10068-021-00900-4.
37. do Espírito Santo, A. P. et al. Fibers from Fruit Byproducts Enhance Probiotic Viability and Fatty Acid Profile and Increase Cla Content in Yoghurts. *Int. J. Food Microbiol.* **2012,** *154* (3), 135–144. DOI: https://doi.org/10.1016/j.ijfoodmicro.2011.12.025.
38. Benítez, V.; Mollá, E.; Martín-Cabrejas, M. A.; Aguilera, Y.; López-Andréu, F. J.; Esteban, R. M. Effect of Sterilisation on Dietary Fiber and Physicochemical Properties of Onion Byproducts. *Food Chem.* **2011,** *127* (2), 501–507. DOI: https://doi.org/10.1016/j.foodchem.2011.01.031.
39. Figuerola, F.; Hurtado, M. L.; Estévez, A. M.; Chiffelle, I.; Asenjo, F. Fiber Concentrates from Apple Pomace and Citrus Peel as Potential Fiber Sources for Food Enrichment. *Food Chem.* **2005,** *91* (3), 395–401. DOI: https://doi.org/10.1016/j.foodchem.2004.04.036.
40. Campos, D. A.; Ribeiro, T. B.; Teixeira, J. A.; Pastrana, L.; Pintado, M. M. Integral Valorization of Pineapple (Ananas comosus L.) Byproducts through a Green Chemistry Approach towards Added Value Ingredients. *Foods* **2020,** *9* (1). DOI: 10.3390/foods9010060.
41. Arvanitoyannis, I. S.; Ladas, D.; Mavromatis, A. Potential uses and Applications of Treated Wine Waste: A Review. *Int. J. Food Sci. Technol.* **2006,** *41* (5), 475–487. DOI: https://doi.org/10.1111/j.1365-2621.2005.01111.x.
42. Torres-León, C.; Rojas, R.; Contreras-Esquivel, J. C.; Serna-Cock, L.; Belmares-Cerda, R. E.; Aguilar, C. N. Mango Seed: Functional and Nutritional Properties. *Trends Food Sci. Technol.* **2016,** *55*, 109–117.
43. Lucera, A.; Costa, C.; Marinelli, V.; Saccotelli, M. A.; Del Nobile, M. A.; Conte, A. Fruit and Vegetable Byproducts to Fortify Spreadable Cheese. *Antioxidants* **2018,** *7* (5). DOI: 10.3390/antiox7050061.

44. Gorjanović, S. et al. Evaluation of Apple Pomace Flour Obtained Industrially by Dehydration as a Source of Biomolecules with Antioxidant, Antidiabetic and Antiobesity Effects. *Antioxidants* **2020,** *9* (5), 413.
45. Fuentes-Alventosa, J. M. et al. Effect of Extraction Method on Chemical Composition and Functional Characteristics of High Dietary Fiber Powders Obtained from Asparagus Byproducts. *Food Chem.* **2009,** *113* (2), 665–671.
46. Larrauri, J. A. New Approaches in the Preparation of High Dietary Fiber Powders from Fruit Byproducts. *Trends Food Sci. & Technol.* **1999,** *10* (1), 3–8.
47. Garau, M. C.; Simal, S.; Rossello, C.; Femenia, A. Effect of Air-Drying Temperature on Physico-Chemical Properties of Dietary Fiber and Antioxidant Capacity of Orange (Citrus aurantium v. Canoneta) Byproducts. *Food Chem.* **2007,** *104* (3), 1014–1024.
48. Jung, J.; Cavender, G.; Zhao, Y. Impingement Drying for Preparing Dried Apple Pomace Flour and Its Fortification in Bakery and Meat Products. *J. Food Sci. Technol.* **2015,** *52* (9), 5568–5578. DOI: 10.1007/s13197-014-1680-4.
49. Liu, Y.; Wang, L.; Liu, F.; Pan, S. Effect of Grinding Methods on Structural, Physicochemical, and Functional Properties of Insoluble Dietary Fiber from Orange Peel. *Int. J. Polym. Sci.* **2016,** *2016.*
50. Yan, L.; Li, T.; Liu, C.; Zheng, L. Effects of High Hydrostatic Pressure and Superfine Grinding Treatment on Physicochemical/Functional Properties of Pear Pomace and Chemical Composition of Its Soluble Dietary Fiber. *LWT* **2019,** *107*, 171–177.
51. Raghavendra, S. N.; Ramachandra Swamy, S. R.; Rastogi, N. K.; Raghavarao, K. S. M. S.; Kumar, S.; Tharanathan, R. N. Grinding Characteristics and Hydration Properties of Coconut Residue: A Source of Dietary Fiber. *J. Food Eng.* **2006,** *72* (3), 281–286. DOI: https://doi.org/10.1016/j.jfoodeng.2004.12.008.
52. Esposito, F.; Arlotti, G.; Bonifati, A.; Napolitano, A.; Vitale, D.; Fogliano, V. Antioxidant Activity and Dietary Wbre in Durum Wheat Bran Byproducts. *Food Res. Int.* **2005,** *38*, 1167–1173. DOI: 10.1016/j.foodres.2005.05.002.
53. Ramachandraiah, K.; Chin, K. B. Effect of Particle Size of Persimmon Byproduct Powders on Their Physicochemical Properties and Antioxidant Activities in Porcine Patties. *J. Food Process Eng.* **2018,** *41* (1), e12610. DOI: https://doi.org/10.1111/jfpe.12610.
54. Hemery, Y. et al. Potential of Dry Fractionation of Wheat Bran for the Development of Food Ingredients, Part I: Influence of Ultra-Fine Grinding. *J. Cereal Sci.* **2011,** *53* (1), 1–8.
55. Hurtado-Romero, A.; Del Toro-Barbosa, M.; Garcia-Amezquita, L. E.; García-Cayuela, T. Innovative Technologies for the Production of Food Ingredients with Prebiotic Potential: Modifications, Applications, and Validation Methods. *Trends Food Sci. & Technol.* **2020.**
56. Padam, B. S.; Tin, H. S.; Chye, F. Y.; Abdullah, M. I. Banana Byproducts: An Under-Utilized Renewable Food Biomass with Great Potential. *J. Food Sci. Technol.* **2014,** *51* (12), 3527–3545.
57. Yan, J.; Jia, X.; Feng, L.; Yadav, M.; Li, X.; Yin, L. Rheological and Emulsifying Properties of Arabinoxylans from Various Cereal Brans. *J. Cereal Sci.* **2019,** *90*, 102844. DOI: https://doi.org/10.1016/j.jcs.2019.102844.
58. Föste, M.; Verheyen, C.; Jekle, M.; Becker, T. Fibers of Milling and Fruit Processing Byproducts in Gluten-Free Bread Making: A Review of Hydration Properties, Dough Formation and Quality-Improving Strategies. *Food Chem.* **2020,** *306*, 125451. DOI: https://doi.org/10.1016/j.foodchem.2019.125451.
59. Šarić, B. et al. Fiber Concentrates from Raspberry and Blueberry Pomace in Gluten-Free Cookie Formulation: Effect on Dough Rheology and Cookie Baking Properties. *J. Texture Stud.* **2019,** *50* (2), 124–130.

60. Chareonthaikij, P.; Uan-On, T.; Prinyawiwatkul, W. Effects of Pineapple Pomace Fiber on Physicochemical Properties of Composite Flour and Dough, and Consumer Acceptance of Fiber-Enriched Wheat Bread. *Int. J. food Sci. Technol.* **2016,** *51* (5), 1120–1129.
61. Sudha, M. L.; Baskaran, V.; Leelavathi, K. Apple Pomace as a Source of Dietary Fiber and Polyphenols and its Effect on the Rheological Characteristics and Cake Making. *Food Chem.* **2007,** *104* (2), 686–692. DOI: https://doi.org/10.1016/j.foodchem.2006.12.016.
62. Fleischman, E. F. et al. Physical, Textural, and Antioxidant Properties of Extruded Waxy Wheat Flour Snack Supplemented with Several Varieties of Bran. *J. Food Sci.* **2016,** *81* (11), E2726–E2733.
63. Chang, T.; Wang, C.; Wang, S.; Shi, L.; Yang, H.; Cui, M. Effect of Okara on Textural, Color and Rheological Properties of Pork Meat Gels. *J. Food Qual.* **2014,** *37* (5), 339–348. DOI: 10.1111/jfq.12096.
64. Shih, Y. T.; Wang, W.; Hasenbeck, A.; Stone, D.; Zhao, Y. Investigation of Physicochemical, Nutritional, and Sensory Qualities of Muffins Incorporated with Dried Brewer's Spent Grain Flours as a Source of Dietary Fiber and Protein. *J. Food Sci.* **2020,** *85* (11), 3943–3953. DOI: 10.1111/1750-3841.15483.
65. Amoriello, T.; Mellara, F.; Galli, V.; Amoriello, M.; Ciccoritti, R. Technological Properties and Consumer Acceptability of Bakery Products Enriched with Brewers ⇔ Spent Grains. *Foods* **2020,** *9* (10). DOI: 10.3390/foods9101492.
66. Wang, Q.; Xiong, Z.; Li, G.; Zhao, X.; Wu, H.; Ren, Y. Tomato Peel Powder as Fat Replacement in Low-Fat Sausages: Formulations with Mechanically Crushed Powder Exhibit Higher Stability than Those with Airflow Ultra-Micro Crushed Powder. *Eur. J. Lipid Sci. Technol.* **2016,** *118* (2), 175–184. DOI: https://doi.org/10.1002/ejlt.201400579.
67. Padayachee, A.; Day, L.; Howell, K.; Gidley, M. J. Complexity and Health Functionality of Plant Cell Wall Fibers From Fruits and Vegetables. *Crit. Rev. Food Sci. Nutr.* **2017,** *57* (1), 59–81.
68. Huc-Mathis, D.; Journet, C.; Fayolle, N.; Bosc, V. Emulsifying Properties of Food Byproducts: Valorizing Apple Pomace and Oat Bran. *Colloids Surfaces A Physicochem. Eng. Asp.* **2019,** *568*, 84–91. DOI: 10.1016/j.colsurfa.2019.02.001.
69. Barber, T. M.; Kabisch, S.; Pfeiffer, A. F. H.; Weickert, M. O. The Health Benefits of Dietary Fiber. *Nutrients* **2020,** *12* (10), 3209.
70. Fardet, A. New Hypotheses for the Health-Protective Mechanisms of Whole-Grain Cereals: What is Beyond Fiber? *Nutr. Res. Rev.* **2010,** *23* (1), 65–134. DOI: 10.1017/S0954422410000041.
71. Adam, A. et al. The Bioavailability of Ferulic Acid is Governed Primarily by the Food Matrix Rather than Its Metabolism in Intestine and Liver in Rats. *J. Nutr.* **2002,** *132* (7), 1962–1968.
72. Adom, K. K.; Liu, R. H. Antioxidant Activity of Grains. *J. Agric. Food Chem.* **2002,** *50* (21), 6182–6187.
73. Liu, R. H. Whole Grain Phytochemicals and Health. *J. Cereal Sci.* **2007,** *46* (3), 207–219.
74. Ou, J.; Sun, Z. Feruloylated Oligosaccharides: Structure, Metabolism and Function. *J. Funct. Foods* **2014,** *7*, 90–100. DOI: https://doi.org/10.1016/j.jff.2013.09.028.
75. Acosta-Estrada, B. A.; Gutiérrez-Uribe, J. A.; Serna-Saldívar, S. O. Bound Phenolics in Foods, a Review. *Food Chem.* **2014,** *152*, 46–55.
76. Tagliazucchi, D.; Verzelloni, E.; Bertolini, D.; Conte, A. In Vitro Bio-Accessibility and Antioxidant Activity of Grape Polyphenols. *Food Chem.* **2010,** *120* (2), 599–606. DOI: 10.1016/j.foodchem.2009.10.030.

77. Urquiaga, I. et al. Wine Grape Pomace Flour Improves Blood Pressure, Fasting Glucose and Protein Damage in Humans: A Randomized Controlled Trial. *Biol. Res.* **2015,** *48* (1), 49. DOI: 10.1186/s40659-015-0040-9.
78. Dewanto, V.; Wu, X.; Liu, R. H. Processed Sweet Corn Has Higher Antioxidant Activity. *J. Agric. Food Chem.* **2002,** *50* (17), 4959–4964. DOI: 10.1021/jf0255937.
79. Zhang, J.; DIng, Y.; Dong, H.; Hou, H.; Zhang, X. Distribution of Phenolic Acids and Antioxidant Activities of Different Bran Fractions from Three Pigmented Wheat Varieties. *J. Chem.* **2018,** *2018*. DOI: 10.1155/2018/6459243.
80. Birsan, R. I.; Wilde. P.; Waldron, K. W.; Rai, D. K. Recovery of Polyphenols from Brewer's Spent Grains. *Antioxidants* **2019,** *8* (9), 1–12. DOI: 10.3390/antiox8090380.
81. Kaisangsri, N.; Kowalski, R. J.; Wijesekara, I.; Kerdchoechuen, O.; Laohakunjit, N.; Ganjyal, G. M. Carrot Pomace Enhances the Expansion and Nutritional Quality of Corn Starch Extrudates. *LWT-Food Sci. Technol.* **2016,** *68*, 391–399. DOI: https://doi.org/10.1016/j.lwt.2015.12.016.
82. Ajila, C. M.; Leelavathi, K.; Prasada Rao, U. J. S. Improvement of Dietary Fiber Content and Antioxidant Properties in Soft Dough Biscuits with the Incorporation of Mango Peel Powder. *J. Cereal Sci.* **2008,** *48* (2), 319–326. DOI: 10.1016/j.jcs.2007.10.001.
83. Maria, A. G.; Graziano, R.; Nicolantonio, D. Carotenoids: Potential Allies of Cardiovascular Health?. *Food Nutr. Res.* **2015,** *59* (1), 26762.
84. Srikaeo, K. Chapter 35 - Organic Rice Bran Oils in Health. In Wheat and Rice in Disease Prevention and Health; Watson, R. R.; Preedy, V. R.; Zibadi, S. Eds.; San Diego: Academic Press, 2014, pp. 453–465.
85. Phutthaphadoong, S. et al. Chemopreventive Effect of Fermented Brown Rice and Rice Bran (Fbra) on the Inflammation-Related Colorectal Carcinogenesis in ApcMin/+ Mice. *Oncol. Rep.* **2010,** *23* (1), 53–59.
86. Tiwari, U.; Cummins, E. Nutritional Importance and Effect of Processing on Tocols in Cereals. *Trends Food Sci. Technol.* **2009,** *20* (11), 511–520. DOI: https://doi.org/10.1016/j.tifs.2009.06.001.

PART II

Technological Strategies for Bioactive Compounds Recovery

CHAPTER 7

Biotechnology Strategies to Produce and Recover Bioactive Compounds from Spend Coffee Grounds

KARLA A. LUNA[1,2], CRISTÓBAL N. AGUILAR[1], RODOLFO RAMOS-GONZÁLEZ[3], MAYELA GOVEA-SALAS[2], and MÓNICA L. CHÁVEZ-GONZÁLEZ[1,2]

[1]*Bioprocesses and Bioproducts Research Group, Food Research Department, School of Chemistry, Universidad Autónoma de Coahuila, Saltillo, Mexico*

[2]*Nanobioscience Research Group, Food Research Department, School of Chemistry, Universidad Autónoma de Coahuila, Saltillo, Mexico*

[3]*CONACYT–Universidad Autónoma de Coahuila, Saltillo, Mexico*

ABSTRACT

The coffee industry is an industry that generates a large amount of agro-industrial waste as it is the second-largest commodity in the world. The main residues of this commodity are coffee pulp, which represents 40–45% of the fruit's dry weight, husk coffee, which represents 50% of the weight of the dry fruit, and the spend coffee grounds (SCG) represent 65% in weight of the roasted coffee. This last residue is obtained during the brewing process and generates around 6 million tons per year derived from the consumption of coffee. Its composition is rich in proteins, lipids, mainly fatty acids, cellulose, lignin, tannins, antioxidant compounds of the anthocyanin type, hydroxycinnamic acids, and monosaccharides, such as arabinose, galactose, and mannose. These compounds can be exploited as a source of biological compounds, thus decreasing the environmental pollution generated by these

Food Byproducts Management and their Utilization. Ricardo Gómez-García, Ana A. Vilas-Boas, Débora A. Campos, Maria Manuela Pintado, & Cristóbal Noé Aguilar (Eds.)

residues from the coffee industry. This chapter explores the composition of this residue and the main bio-products that can be obtained from this material using biotechnological strategies.

7.1 INTRODUCTION

This chapter refers to the strategies for recovering bioactive compounds from spend coffee grounds (SCG); coffee is currently the second most consumed product in the world, which generates 6 million tons of coffee spent annually.[1] In Mexico, coffee production is used in more than 500,000 products, such as bakery, beverages, and soaps. Likewise, in 14 states and 480 municipalities, these types of coffee-derived products are produced. Currently, 1.3 kg of coffee is consumed annually per person.[2] The coffee generates significant quantities of wastes, for example, 13,900 tons per year of byproducts that cause damage to the environment, which are deposited on the ground without prior treatment.[3] The different residues produced by coffee process include coffee pulp (CH), coffee husk (CS), and spend coffee grounds (SCG); these byproducts are rich in carbohydrates, proteins, pectins, bioactive compounds as polyphenols, among others are of great importance.[3] The SCG are obtained from the residue that remains in the filter of the coffee machine that leaves the soluble coffee from its preparation.[4] It should be mentioned that the SCG have large areas of opportunity to develop different investigations since they have no commercial value and are generally discarded. It is important to highlight that SCG can be used to extract different value-added products to obtain bioactive compounds with functional characteristics.[5] Its composition is rich in proteins, lipids, mainly fatty acids, cellulose, lignin, tannins, anthocyanin-type antioxidant compounds, as well as monosaccharides, such as arabinose, galactose, and mannose. The diverse chemical composition provides the residue with potential as a source of compounds of commercial interest that can be incorporated into different food products such as pharmaceuticals; however, further research is necessary for such applications. Finally, this chapter seeks to show the richness in the composition of spends coffee grounds and their potential as materials that can be recovered for their nutritional, biological, and industrial value.

7.2 COFFEE OVERVIEW

Coffee belongs to the Rubiaceae family, genus *Coffea*, and it is a crop typical of tropical or subtropical countries. Due to their flavor and aroma, the most important species are *Arabica* (*Coffea arabica*) with approximately 75% of the cultivated area robust (*C. canephora)* with 24%. It should be mentioned that the coffee fruit becomes a 10–15 mm long cherry that contains two seeds, that is, coffee beans. The elliptical seeds are egg shaped that contain endosperm and embryos enclosed in the pericarp layers of the coffee bean.[6] On the other hand, the pericarp is composed of the outer layers of the exocarp, mesocarp, and endocarp of the coffee fruit, as can be shown in Figure 7.1. The skin of coffee fruits is an outer shell made of a cellulose monolayer with a waxy substance. This part of the coffee has a red, yellow cover (color may vary depending on the variety of coffee). The pulp of coffee is fleshy, a soft tissue that lies between the skin and the parchment. It is divided into an internal mesocarp called mucilage and an external mesocarp.[7] Another important layer is the endocarp, also called coffee parchment; it is a layer of cellulosic material that completely envelops the seed. It is made up of different layers of fiber that gives it strength to protect the seed from damage in the process. Also, the perisperm or spermoderm is a thin integument that directly covers the seed. We must remember that the coffee fruit is formed by different kinds of compounds produced by primary and secondary metabolism, which in coffee beans composition is approximately 50% polysaccharide compounds such as cellulose and hemicellulose.

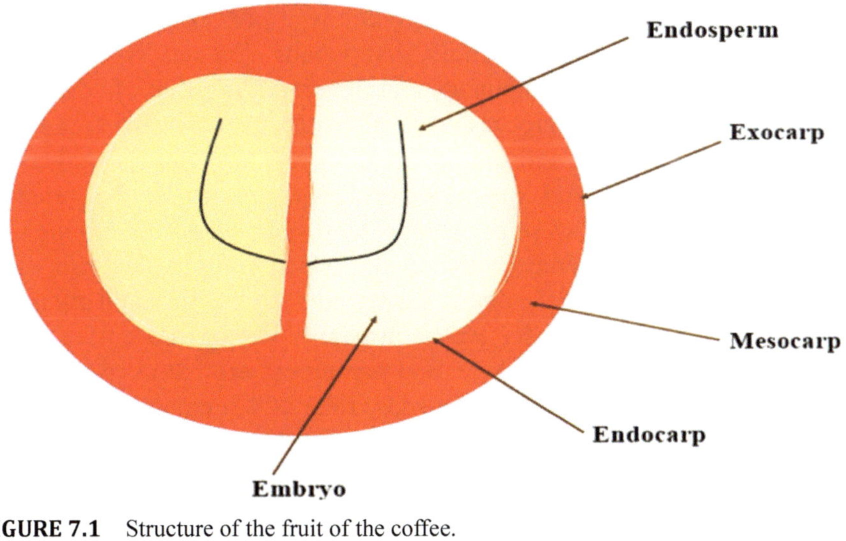

FIGURE 7.1 Structure of the fruit of the coffee.

The coffee fruit is made up of primary and secondary metabolites of the coffee plant that accumulate unevenly in all parts of the fruit. It also contains a significant amount of soluble carbohydrates (fructose, glucose, galactose, arabinose, raffinose stachyose galactose, mannose, arabinose, and glucose polymers). The main component of coffee is sucrose. In addition, oils and proteins, free amino acids, aliphatic acids (volatile and nonvolatile) and minerals are present in coffee.[8] Around the world, the coffee is recognized as one of the most appreciated drinks due to its characteristic sensory properties, such as stimulating the central nervous system and its various benefits on our health.[9] The waste generated from coffee has great health benefits, and there are different methods or technologies to obtain these bioactive compounds to make use of them.[10] The nutritional qualities of coffee byproducts serve as an important source for human consumption, these byproducts represent an interesting alternative to increase its technological potential for the elaboration of new functional foods to attenuate or prevent cardiovascular diseases, cancer, diabetes, Alzheimer's, among others.[11] Coffee residues are an important source of insoluble dietary fiber, antioxidant, and source of proteins, along with other additional ingredients used in the preparation of healthy solid foods. It is also important to mention the bioactive compounds that help the health benefits.[12] Coffee is grown in around 80 countries, of which 90% is grown in developing countries, so, its production is socially and economically important.[13] The most important aspect is that coffee has chemical components of great interest, including organic compounds, such as carbohydrates, lipids, proteins, alkaloids (caffeine), trigonelline, and phenolic compounds. Another aspect is that it also consists of volatile compounds that generate the smell of coffee.[14] As soon as the main coffee processing is subjected to two existing methods, that is, the wet method and the dry method, the first route is called wet processing; in this method, the coffee husk is removed, then the mucilage is removed, and in the last step, coffee is removed and dried.[15] On the other hand, we have the dry way, which is the oldest and simplest process. It consists of the following stages: harvesting, drying, classification, and bagging.[16] Therefore, the dry way consists of drying the coffee cherry. Subsequently, the pulp, mucilage, and parchment are removed utilizing a machine. Drying the cherry requires more time as it is commonly sun-dried compared with the wet drying process of coffee. Dry processing is the treatment given to cherry coffee and consists of dehydrating it, by natural or artificial means, until the grain is preserved.[17]

It may highlight that the coffee making through different production methods, such as wet and dry, generates agro-industrial waste. Both different processes generate a wide variety of wastes. For example, 13,900 tons per year of byproducts, such as CH, CS, SCG cause pollution, which are deposited into the environment without prior treatment and wasting biological compounds of interest. These byproducts are rich in carbohydrates, proteins, pectins, bioactive compounds, such as polyphenols, among others.[3] The content of these byproducts can lead to the obtaining of new products for composting and also ahigh value addition in getting bioactive compounds with functional characteristics.[5] The importance of waste generated by the coffee industry will be described below.

7.3 IMPORTANCE OF THE WASTE GENERATED BY THE COFFEE INDUSTRY

Industrial development implies an increase in waste generation that has become a problem. It involves environmental and economic factors for companies since they must bear the high costs generated by its final disposal.[18] Currently, the industry is looking for new production processes that are more efficient and have a low impact on the environment. Among these new processes the need to reduce the exploitation of natural resources is important by taking advantage of the waste generated in the industry.[19] Coffee is currently the second-largest most used product globally, and its consumption produces an estimated 6 million tons of coffee grounds each year.[1] Coffee is grown in about 80 countries, 90% of which are developing countries, so, its production has relevant social importance.[13] In the coffee cultivation and industrialization process, it is contemplated that less than 5% of the weight of the fruit is used in the production of the beverage; 95% remains in the form of organic residues.[20] These include the leaves, branches, and stems that are generated during the renovation of coffee plantations, fruits that are not suitable for coffee production, the pulp of the fruit, which represents 44% of the weight of the dry fruit, and the chaff which represents 10% of the weight of the dry fruit.[18] SCG are obtained from the residue left in the filter used to brew soluble coffee daily or from the filter of coffee machines.[21] Approximately, 50% of SCG are generated by preparing soluble coffee in the industry and coffee commercialized in stores, coffee shops, and domestic use.[22]

SCG are known to contain various organic compounds, such as fatty acids, lignin, cellulose, hemicellulose, and other polysaccharides, and is a suitable biological source for the production of biodiesel, activated carbon, fertilizers, sugar, and even adsorbents for the removal of heavy metal ions, showing the wide range of possible applications for these wastes. Due to the wide applications of these wastes, it is important to design a careful strategy that seeks to get the most out of these materials.[23] Agro-industrial wastes have a high level of potential for being used to generate benefits for the environment and society, as they can be recycled and used to prevent pollution or recover altered ecosystems. The diversity of agro-industrial waste can be used mainly as an alternative to producing compost.[24]

For this reason, it is essential to take advantage of this type of waste generated by the coffee industry to become an environmental and economic strategy for adequate management of agro-industrial byproducts or waste. When selecting the most appropriate alternative for the use and subsequent treatment of a specific agro-industrial waste, it is necessary to know its composition and the quality of its components, the amount generated, and to take into account that this process can produce a new, more exhausted waste that may have another application or become a waste product.[25]

7.3.1 SPEND COFFEE GROUNDS COMPOSITION AND CHARACTERIZATION

SCG are also called a husk, and in this, it is possible to obtain a significant concentration of polyphenolic compounds.[26] Another important feature of the composition of the SCG samples with a % dry basis is that SCG have compounds of great importance, as given in Table 7.1. Compounds from coffee residues are widely recognized to contain high content of polyphenols, especially chlorogenic acid and flavonoids, which have significant biological functionalities, including antioxidant and antibacterial properties.[27] Approximately, 450 kg of coffee grounds are also obtained from 1000 kg of coffee. Some studies have revealed that boron as bioactive compound in coffee is of great interest that could be used.[28]

The main compounds of SCG are polysaccharides; cellulose and hemicellulose, which combine to constitute approximately 50% of the dry mass of SCG. Mannose, galactose, and arabinose are the major components of hemicellulose sugars, while glucose is the major cellulose component.[29] In addition, SCG are a residue with a high concentration of soluble dietary fiber and proteins and sugars, such as glucose, xylose, galactose, mannose, and arabinose.[30] The various fractions from SCG will are discussed in the following sections.

TABLE 7.1 Chemical Composition of SCG (wt %).

Spend coffee grounds (wt %)	Component	References
Carbohydrates	82	[31]
Cellulose	8.3	[32]
Hemicellulose	36.7	[30].
Moisture	11.69	[32]
Lipids	13.41	[33]
Total fiber	60.45	[34]
Ash	1.6	[35]
Protein	13.6	[30]
Nitrogen	2.3	[30]
Caffeine	0.8	[36]
Tannins	0.02	[37]
Chlorogenic acid	2.3	[38]

7.3.2 IMPORTANT CHEMICAL FRACTIONS OF SCG

7.3.2.1 CELLULOSE

One of the most abundant and renewable natural resources, Cellulose has been extensively studied as an alternative substance for different applications.[39] SCG contain high levels of cellulose and hemicellulose, which can be used to produce bioethanol and biocomponents.[40] Cellulose is the main component of plant cell walls and is defined as a long-chain linear homopolymer with a high degree of polymerization and high molecular weight.[41] It should be noted that cellulose is the most abundant organic polymer on earth, and is a homopolymer of glucose units with β-1,4 linkages as shown in Figure 7.2. It constitutes a major part of the cell wall of growing plant cells.[42] In addition to the above, the presence of three OH^- groups in each of the structural units allows it to react as either alcohol, an ester, or an ether.[43] Cellulose favors the arrangement of polymeric chains in highly crystalline and compact structures, it is insoluble in water and capable of resisting depolymerization.[44] Due to its abundance in the biosphere and its great utility for obtaining fuels and other chemicals, the study of cellulose has proved to be very broad and diverse.[45]

This has made it possible to broaden the range of possible chemical modifications and derivatives, which have been used in the pharmaceutical, food, and textile industries. In recent years, cellulose-derived polymers have gained renewed interest due to their potential for biodegradation, which can be obtained from natural waste such as coffee grounds.[43] The cellulosic content, once in contact, can be hydrolyzed by the action of enzymatic treatment or by the action of microorganisms. While it is true that various microorganisms generate cellulases and use cellulose in a direct way, pretreatment methods are required for large-scale commercial applications to make the process efficient and economically viable.[46]

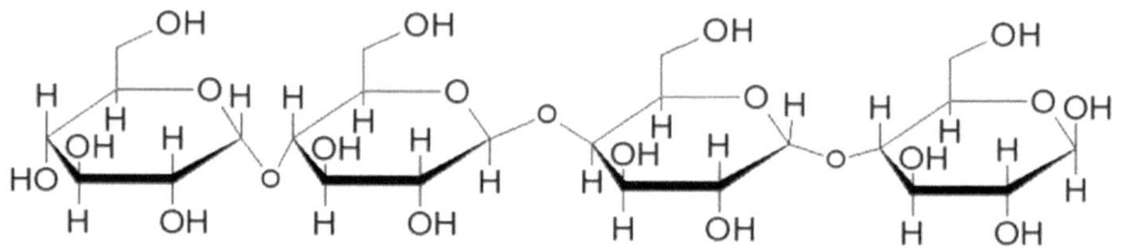

FIGURE 7.2 Cellulose structure.

7.3.2.2 *HEMICELLULOSE*

SCG contain high levels of cellulose and hemicellulose. These materials can be used for bioethanol production.[40] Hemicelluloses are heteropolymers composed of some hexoses (D-glucose, D-galactose, and D-mannose) and some pentoses (L-arabinose and D-xylose). Hemicellulose mainly consists of xylose linking compounds, such as arabinose, glucose, mannose, and other sugars along with acetyl chain[47] Hemicellulose is an important component of plant cell walls and comprises branched chains with a lower degree of polymerization.[41] Hemicellulose can have different configurations, such as monomeric, oligomeric, and polymeric with different degrees of polymerization. Therefore, the properties and industrial applications will differ depending on their configuration and molecular weight distribution. In their monomeric form, hemicelluloses are characterized by their applications in the manufacture of biochemical and energy products.[48,49]

Consequently, the types of hemicelluloses that may be present are mainly glucomannan and galactoglucomannan. Galactomannans are high molecular weight, low-branching polysaccharides composed of a mannose backbone containing individual galactose side groups. Depolymerization of galactomannans occurs during the roasting coffee procedure, increasing their solubility in water.[50] Approximately, 70% by weight of the hemicelluloses glucomannan and galactoglucomannan remain in the matrix of coffee beans after coffee processing or else during the production of soluble coffee.[51] On the other hand, hemicelluloses have different applications, such as in food packaging, hydrogels, and thermoplastics.[52–54]

Different extraction methods can extract high-value hemicelluloses: dilute acid extraction, hot water extraction, steam explosion extraction, alkali extraction, dilute acid steam explosion extraction, organic solvent extraction, and microwave-assisted extraction. The most useful methods are dilute acid extraction and alkaline extraction as hemicelluloses can be extracted under alkaline and acidic conditions.[55]

7.3.2.3 CAFFEINE AND CHLOROGENIC ACIDS COMPONENTS OF SCG

SCG contain high amounts of compounds of great interest, such as caffeine and chlorogenic acids. Caffeine is a more studied compound because of its benefits, and its widely known psychoactive effects and beneficial action on metabolism. This molecule is a water-soluble alkaloid of the xanthine group, an adenosine antagonist, present in coffee.[56] The most important and recognized compound in coffee is caffeine, a central nervous system (CNS) stimulant, which can increase calcitonin concentrations, relax smooth muscles, and stimulate heart rate.[57] It is found in coffee beans and can be extracted during processing on an industrial scale. The coffee used has less caffeine than in coffee preparation.[58] It has been reported that caffeine content is above the concentration of 2.5% in SCG.[1]

Chlorogenic acid is the main component of the phenolic fraction of coffee; it should be noted that it has a potent antioxidant activity in vitro and has excellent health benefits.[59,60] The amount of chlorogenic acid found in SCG residues range about 1.8–5.6 mg/g.[61] It should be mentioned that chlorogenic acids are extracted only in the preparation of coffee beverage. The residues generated from the practice remain a vital source of potentially useful bioactive compounds.[62]

7.3.3 BIOACTIVE COMPOUNDS IN SPEND COFFEE GROUNDS

7.3.3.1 POLYPHENOLIC COMPOUND

Some of the bioactive compounds present in SCG belong to the class of polyphenols, as given in Table 7.2. These compounds are of great interest as they provide a wide range of bioactivities as auxiliaries in inflammatory processes, they show antiviral, antibacterial, and antioxidant activities. Antioxidants are compounds capable of capturing the unpaired electron of the external orbital of free radicals and thus deactivate them, decrease oxidative stress and act on it by inhibiting it to prevent oxidation of proteins, lipids, and DNA.[63] Studies on coffee byproducts such as SGC reveal an excellent profile of bioactive compounds with antioxidant properties. These compounds are related to human health, with antioxidant, antibacterial, antiviral, anti-inflammatory, and anticancer activities.[51] These agro-industrial residues, after valorization, can be used by other industries, such as pharmaceuticals and food, for example.[64] Several studies conducted in recent years have highlighted the antioxidant power of some phenolic compounds present in plants, particularly those found abundantly in coffee byproducts.[65] In addition to their bioactivity, natural agents are generally considered harmless to humans, suitable for use in a wide variety of applications, and are often extracted from reusable sources.[66]

TABLE 7.2 Polyphenol and Caffeine Content in SCG.[67]

Polyphenols	**mg/g sample**
Chlorogenic acids	5.6 ± 0.1
Caffeic acid	0.07 ± 0.0
Caffeine	0.4 ± 0.0
Ferulic acid	0.004 ± 0.0
Gallic acid	2.5 ± 0.06
Quercetin	0.96 ± 0.1
Ellagic acid	± 0.01

Source: Adapted from [67]

7.3.3.2 TANNINS

Tannins are natural water-soluble complex polyphenolic compounds usually found in vascular plants, especially in some leaves and beans such as coffee.[68] The structure and classification of tannins comprise a wide range of substances that possess at least one hydroxyl group (−OH) in one or more phenolic rings.[69] Tannins are classified into two different categories, that is, hydrolyzable and condensed tannins. This classification has been of great economic use for decades due to its potential to replace phenolic synthesis since it is a renewable and inexpensive natural resource available in abundance.[70] It should be mentioned that SCG tannins have immense potential to compete with conventional synthetic phenolics.[71] Studies have reported that the tannin content in the coffee residue is 0.997 mg CE g^{-1} DW.[72] It is relevant to mention that tannins are beneficial to human health including antibacterial effect.[73] Tannins found in SCG are pyrogallol (1,2,3 trihydroxybenzene) and catechol (1,2-dihydroxybenzene) derived from the coffee bean roasting process. However, the hydrolyzable tannins produced from pyrogallol are soluble in solvents such as water, so, they may not be found in coffee grounds at the time of qualitative and quantitative testing of SCG.[74]

It should be mentioned that tannins have cost-effective applications as they are natural biopolymers and excellent candidates for producing biosorbents. Low-cost and environmentally friendly products, such as adhesives, plastics, and polymers can be produced using tannins extracted from SCG. Tannin-added films can be used as environmentally friendly and nontoxic packaging materials for food and pharmaceuticals.[75] Tannin-based sorbents have a natural affinity to absorb heavy metals, dyes, and pharmaceuticals.

7.3.3.3 FLAVONOIDS

Flavonoids are compounds with the highest bioactivity and are considered health promoters and preventive supplements of high nutritional and therapeutic interest. Compounds belonging to the flavonoid family include flavones, isoflavones, flavonols, and anthocyanins.[76] Flavonoids belong to polyphenols and have a 15-carbon skeleton. As shown in Figure 7.3, the flavonoid structure is composed of a C-6 aromatic group, which is identified as an acetate, combined with a phenylpropanoid (C6-C3). Therefore, ring A, acetate and aromatic ring B are linked with a 3-carbon bridge.[77]

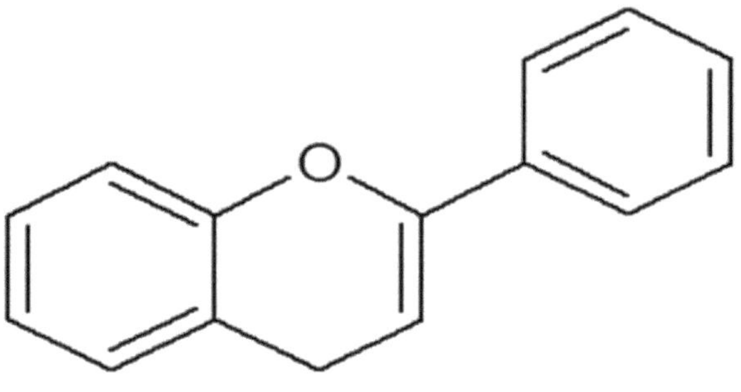

FIGURE 7.3 Chemical structure of flavonoids.

7.3.3.4 ANTHOCYANINS

SCG, within their physicochemical characterization, has compounds of great importance such as anthocyanins which are water-soluble pigments that belong to the flavonoid family (phenolic compounds). They are anthocyanidin glycosides derived from 2-phenylbenzopyrilium (flavilium ion)[78] containing two aromatic rings (A and B) separated by oxygen that forms a six-membered heterocyclic ring (C-ring).[79] Renewed interest in anthocyanins has increased because of their potential health benefits as antioxidant, anticancer, anti-inflammatory agents, and their anti-hypoglycemic effect. Because during the passage from the digestive tract to the bloodstream of mammals, anthocyanins remain intact, hence, they can directly affect human health (Figure 7.4).[31] Indeed anthocyanins have significant benefits as applications, such as incorporation as food dyes. In addition to improving appearance, it has been reported that several studies present scientific evidence that anthocyanin-rich extracts can enhance visual acuity, show antioxidant activity, trap free radicals, and act as protective agents that help to protect healthy tissue from some of the side effects caused by some drugs. So, anthocyanins also play a role in antidiabetic properties, such as lipid control, insulin secretion, and vasoprotective effects, which protect blood vessels.[80]

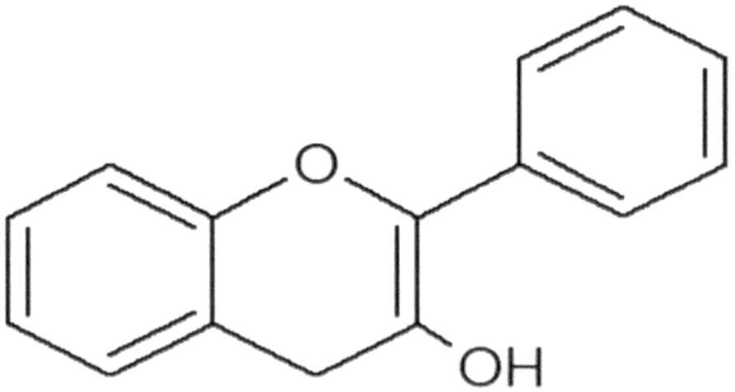

FIGURE 7.4 Chemical structure of anthocyanins.

7.3.4 OILS OF SCG

The oil obtained from SCG is odorless and has a chocolate brown color. The concentrations of extractable lipids from SCG depend on the processing and extraction method. SCG depend on the processing and extraction method.[81] SCG oil has both physiological and biochemical functions and has different organoleptic characteristics; for example, it has a characteristic roasted coffee smell. Its appearance is viscous and shiny liquid.[74] Now, for obtaining the oil extract from coffee grounds, only 1% of the coffee oil is transferred to the extract, thus getting a yield between 28 and 38 g per 250 g of roasted coffee, equivalent to 11.2–15.2%, proving that very little oil is obtained from the beverage. Still, the oils can be recovered from the residues with organic solvents.[74] The most used solvents in oil extraction at the laboratory level are petroleum derivatives, such as hexane and pentane and halogenated solvents.[82] It should be mentioned that coffee oil can be extracted by distillation, as shown in Figure 7.5. On the other hand, SCG oil has physical and chemical characteristics, as represented in Table 7.3.[83] SCG oil generally contains about 75% triacylglycerols, 14% terpenic esters, 5% partial acylglycerols, 1% free fatty acids, 1.5% free sterols, 1% sterol esters, and 1% polar lipids.[84] Coffee oil has special applications as an additive in cosmetic formulations, in biodiesel manufacturing, and bioproduction processes. Of polyhydroxyalkanoates, coffee carbohydrates can be used as food and pharmaceutical additives and as a source of carbon in biotechnology. Bioactives: These compounds are used as supplements and pharmacological products for their antioxidant, antibacterial, antiviral, and even anti-inflammatory, and anticancer properties.[51–84]

TABLE 7.3 Physical and Chemical Properties of SCG Oil.

Physical and chemical properties of SCG oil					
Density (g/cm^3) at 23°C	**Kinematic viscosity (mm^2/s)**	**Dynamic viscosity (Pa s) at 26°C**	**Index of refraction at 23°C**	**Melting point**	**pH at 22°C**
0.923	31.25	0.10	1.478	15°C	4.68

Source: Adapted from ref. [83]

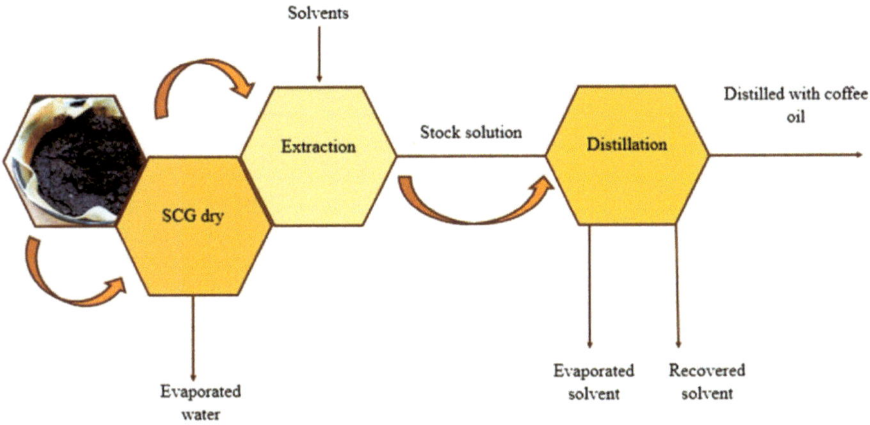

FIGURE 7.5 Flow chart obtaining coffee oil.

7.4 CONCLUSION

The production of the coffee industry generates a large amount of waste that has increased drastically during the transformation of the beverage, causing an environmental impact, and is an essential source of contamination. The most important waste SCG, which have no commercial value, and is therefore, generally discarded without prior treatment. It has proven to be excellent in resource-rich nutrients that can be exploited to recover bioactive compounds, including organic substances, such as carbohydrates, lipids, proteins, caffeine, and chlorogenic acid. Also, one of the most valuable components of this residue is its high content of antioxidants and polysaccharides, among them cellulose and hemicellulose are important, which have a wide range of applications of biological activities, essentially antioxidant and antibacterial activities. They can be used to develop food and pharmaceutical products that contribute satisfactorily to human health. This study concludes that SCG have enormous potential to be used in biotechnological processes thanks to its high potential for reuse as a renewable source and that it can be considered a circular economy strategy to start reusing bioactive compounds, which will support activities to develop new technologies that contribute to the minimization of waste products in the coffee industry.

ACKNOWLEDGMENTS

To CONACYT for the support of the scholarship granted to the student 777738 KAL.

KEYWORDS

- **coffee industry**
- **coffee grounds**
- **residue**
- **spend**
- **bio-products**

REFERENCES

1. Janissen, B.; Huynh, T. Chemical Composition and Value-Adding Applications of Coffee Industry byproducts: A Review. *Resour. Conserv. Recycl.* **2018,** *128*, 110–117.
2. SIAP-SADER. Agrifood and Fisheries Information System, Ministry of Agriculture and Rural Development, Statistical Yearbook of Agricultural Production, 2017. https://nube.siap.gob.mx/cierreagricola/
3. Iriondo, A.; Garcia, N. A.; Fernandez-Gomez, B.; Guisantes, B. E.; Escobar, F. V.; Blanch, G.; Castillo, M. D. Validation of Coffee byproducts as Novel Food Ingredients. *Innov. Food Sci. Emerg. Technol.* **2019,** 194–204.
4. Suárez, C.; Alejandra, J.; Yuleidy, J.; Carriel, G. Extraction and Characterization of Natural Colorant from Coffee Borra. BS Thesis, University of Guayaquil, Faculty of Chemical Engineering, 2017.
5. Caballero-Galván, A. S.; Restrepo-Serna, D. L.; Ortiz-Sánchez, M.; Cardona-Alzate, C. A. Analysis of Extraction Kinetics of Bioactive Compounds from Spent Coffee Grounds (*Coffea arábica*). *Waste Biomass Valorization* **2018,** *9* (12), 2381–2389.
6. Acidri, R.; Sawai, Y.; Sugimoto, Y.; Handa, T.; Sasagawa, D.; Masunaga, T.; ... ; Nishihara, E. Phytochemical Profile and Antioxidant Capacity of Coffee Plant Organs Compared to Green and Roasted Coffee Beans. *Antioxidants* **2020,** *9* (2), 93.
7. Farah, A. Ed.; *Coffee: Production, Quality and Chemistry*; Royal Society of Chemistry, 2019.
8. Esquivel, P.; Jimenez, V. M. Functional Properties of Coffee and Coffee byproducts. *Food Res. Int.* **2012,** *46* (2), 488–495.
9. Peña, J. M.; Murua, B.; Santos, B.; Reynoso, R.; Romero, G.; Vazquez, M. E.; Amaya, S. L. Extraction and Purification of Caffeine and Chlorogenic Acid from Coffee Wet Mill Pulp, 2017; pp 563–569.
10. Cabrera Rodríguez, E.; León Fernández, V.; Montano Pérez, A. D. L. C.; Dopico Ramírez, D. Caracterización de Residuos Agroindustriales Con Vistas a su Aprovechamiento. *Centro Azúcar* **2016,** *43* (4), 27–35.
11. Iriondo M.; Iriondo, A.; Fernández, B.; Miguel, M.; Castillo, M. Optimization of the Formulation and Sensory Quality of New Functional Dairy Products. Food Science Research Institute (CIAL), 2016.
12. Castillo, M.; Martínez, N.; Ullate, M. Food Formulation that Includes Coffee Grounds and Their Applications. Patent 2,489,915, 2015.
13. Dadi, D.; Mengistie, E.; Terefe, G.; Getahun, T.; Haddis, A.; Birke, W.; ... ; Van der Bruggen, B. Assessment of the Effluent Quality of Wet Coffee Processing Wastewater and its Influence on Downstream Water Quality. *Ecohydrol. Hydrobiol.* **2018,** *18* (2), 201–211
14. Toci, A. T.; Boldrin, M. V. Coffee Beverages and Their Aroma Compounds. In *Natural and Artificial Flavoring Agents and Food Dyes*; Academic Press, 2018; pp 397–425.
15. Novita, E. Biodegradability Simulation of Coffee Wastewater Using Instant Coffee. *Agric. Agric. Sci. Procedia* **2016,** 217–229.
16. SAGARPA, Agrifood and Fisheries Information Service (SIAP). Advances in Agricultural Production, 2011
17. Duicela Guambi, L. A.; Martínez Soto, M. E.; Loor Solórzano, R. G.; Morris Díaz, A. T.; Guzmán Cedeño, A. M.; Rodríguez Monroy, C.; Chilán Villafuerte, W. P. Knowledge Management and Organizational Innovation to Reactivate the Production Chain of Robusta Coffee, Ecuador. *Revista Espamciencia.* **2018,** *9* (1), 61–72.

18. Garzón, C.; Marcela, Darío, H.; Gasca, S. Diagnosis of the Environmental Impact of Solid Waste (Pulp and Mucilage), of Coffee Generated by Production Units not Certified in BPA in the Veredas Tabacal and Betania of the Municipality of Pitalito, Department of Huila, 2018.
19. Salamanca, S. Composting of Industrial Waste in Colombia. *Tech. Mag.* **2012,** *28*, 13–18.
20. Rodríguez Martín, J. A.; Álvaro-Fuentes, J.; Gonzalo J.; Gil, C.; Ramos-Miras J. J.; Grau Corbí, J. M.; Boluda, R. Assessment of the Soil Organic Carbon Stock in Spain. *Geoderma* **2016,** *264*, 117–125.
21. Suárez, C.; Alejandra, J.; Yuleidy, J.; Carriel, G. Extraction and Characterization of Natural Colorant from Coffee Borra, BS Thesis, University of Guayaquil, Faculty of Chemical Engineering, 2017.
22. Scully, D. S.; Jalswal, A. K.; Abu-Ghannam, N. An Investigation into Spent Coffee Waste as a Renewable Source of Bioactive Compounds and Industrially Important Sugars. *Bioengineering* **2016,** *3* (4), 33.
23. Pujol, D.; Liu, C.; Gominho, J.; Olivella, M. À.; Fiol, N.; Villaescusa, I.; Pereira, H. The Chemical Composition of Exhausted Coffee Waste. *Ind. Crops Prod.* **2013,** *50*, 423–429.
24. Hernández, A.; Real, N.; Delgado, M.; Bautista, L.; Velasco, J. Agroindustrial Waste with Composting Potential. *Agroproductivity* **2016,** *9* (8), 10–17.
25. Gonzalez, J.; Medina, S.; Rodriguez, A.; Osma, J.; Alméciga, C. J.; Sánchez, O. Production of Trametes Pubescens Laccase Under Submerged and Semi-Solid Culture Conditions on Agro-Industrial Wastes. *PLoS One.* **2013,** *8* (9), e7372.
26. Naranjo, M.; Vélez, L. T.; Rojano, B. A. Antioxidant Activity of Different Grades of Colombian Coffee. *Cuban J. Med. Plants* **2011,** 164–173.
27. López, D. M.; Vázquez, K.; Loarca, M. G.; Campos, R. Spent Coffee Grounds, An Innovative Source of Colonic Fermentable Compounds, Inhibit Inflammatory Mediators In Vitro. *Food Chem.* **2016,** 282–290.
28. Figueroa, G. A.; Homann, T.; Rawel, H. M. Coffee Production Wastes: Potentials and Perspectives. *Austin. Food Sci.* **2016,** *1* (3), 1014.
29. Mussatto, S. I.; Ballesteros, L. F.; Martins, S.; Teixeira, J. A. Extraction of Antioxidant Phenolic Compounds from Spent Coffee Grounds. *Sep. Purif. Technol.* **2011,** *83*,173–179.
30. Zarrinbakhsh, N.; Wang, T.; Rodriguez-Uribe, A.; Misra, M.; Mohanty, A. K. Characterization of Wastes And Coproducts from the Coffee Industry for Composite Material Production. *BioResources* **2016,** *11*(3), 7637–7653.
31. Murthy, P. S.; Naidu, M. M. Recovery of Phenolic Antioxidants and Functional Compounds from Coffee Industry byproducts. *Food Bioproc. Tech.* **2012,** *5* (3), 897–903.
32. Kang, S. B.; Oh, H. Y.; Kim, J. J.; Choi, K. S. Characteristics of Spent Coffee Ground as a Fuel and Combustion Test in a Small Boiler (6.5 kW). *Renew. Energy.* **2017,** *113*, 1208–1214.
33. Urribarrí, A.; Zabala, A.; Sánchez, J.; Arenas, E.; Chandler, C.; Rincón, M.; ... ; Mazzarri, C. A. Evaluation of the Potential of Coffee Grounds as Raw Material for the Production of Biodiesel. *Multisciences* **2014,** *14* (2), 129–139.
34. Ballesteros, Lina F.; Teixeira, José A.; Mussatto, Solange I. Chemical, Functional, and Structural Properties of Spent Coffee Grounds and Coffee Silverskin. *Food and Bioproc. Tech.* **2014,** *7* (12), 3493–3503.
35. Mussatto, S. I.; Carneiro, L. M.; Silva, J. P.; Roberto, I. C.; Teixeira, J. A. A Study on Chemical Constituents and Sugars Extraction from Spent Coffee Grounds. *Carbohydr. Polym.* **2011,** *83* (2), 368–374.

36. Belitz, H.-D.; Grosch, W.; Schieberle, P. Coffee, Tea, Cocoa. *Food Chem.* **2009,** 938–970.
37. Murthy, P. S.; Manjunatha, M. R.; Sulochannama, G.; Naidu, M. M. Extraction, Characterization and Bioactivity of Coffee Anthocyanins. *Euro. J. Biol. Sci.* **2012,** 13–19.
38. Murthy, P. S.; Naidu, M. M. Recovery of Phenolic Antioxidants and Functional Compounds from Coffee Industry byproducts. *Food Bioproc. Technol.* **2012,** 897–903.
39. Braga, N.; Rodrigues, F.; PP Oliveira, M. B. Castanea Sativa byproducts: A Review on Added Value and Sustainable Application. *Nat. Prod. Res.* **2015,** *29* (1), 1–18.
40. Kwon, E. E.; Yi, H.; Jeon, Y. J. Sequential Co-Production of Biodiesel and Bioethanol with Spent Coffee Grounds. *Bioresour. Technol.* **2013,** *136*, 475–480.
41. Basu, P. Biomass Gasification and Pyrolysis: Practical Design and Theory; Academic Press, 2010.
42. Gardner, D. J.; Oporto, G. S.; Mills, R.; Samir, M. A. S. A. Adhesion and Surface Issues in Cellulose and Nanocellulose. *J. Adhes. Sci. Technol.* **2008,** *22* (5–6), 545–567.
43. Müssig, J.; Stevens, C. *Industrial Applications of Natural Fibres: Structure, Properties and Technical Applicationsm*, 1st ed.; Wiley, USA, 2010.
44. Carpita, N. C.; Gibeaut, D. M. Structural Models of Primary Cell Walls in Flowering Plants: Consistency of Molecular Structure with the Physical Properties of the Walls During Growth. *Plant J.* **1993,** *3* (1), 1–30.
45. Payne, C. E.; Wolfrum, E. J. Rapid Analysis of Composition and Reactivity in Cellulosic Biomass Feedstocks with Near-Infrared Spectroscopy. *Biotechnol. Biofuels* **2015,** *8* (1), 1–14.
46. Kumar, A. Pretreatment Methods of Lignocellulosic Materials for Biofuel Production: *A Review. J. Emerg. Trends Eng. Appl. Sci.* **2013,** 181–193.
47. Chandel, A. K.; Singh, O. V.; Venkateswar Rao, L. Biotechnological Applications of Hemicellulosic Derived Sugars: State-of-the-Art. *Sustainable Biotechnology;* Springer: Dordrecht, 2010; pp 63–81.
48. Baptista, S. L.; Cunha, J. T.; Romaní, A.; Domingues, L. Xylitol Production from Lignocellulosic Whole Slurry Corn Cob by Engineered Industrial Saccharomyces Cerevisiae PE-2. *Bioresour. Technol.* **2018,** *267*, 481–491.
49. Luo, Y.; Li, Z.; Li, X.; Liu, X.; Fan, J.; Clark, J. H.; Hu, C. The Production of Furfural Directly from Hemicellulose in Lignocellulosic Biomass: A Review. *Catal. Today* **2019,** *319*, 14–24.
50. Ballesteros, Lina F.; Teixeira, José A.; Mussatto, Solange I. Extraction of Polysaccharides by Autohydrolysis of Spent Coffee Grounds and Evaluation of Their Antioxidant Activity. *Carbohydr. Polym.* **2017,** *157*, 258–266.
51. Zabaniotou, A.; Kamaterou, P. Food Waste Valorization Advocating Circular Bioeconomy-A Critical Review of Potentialities and Perspectives of Spent Coffee Grounds Biorefinery. *J. Clean. Prod.* **2019,** *211*, 1553–1566.
52. Farhat, W.; Venditti, R.; Ayoub, A.; Prochazka, F.; Fernández-de-Alba, C.; Mignard, N.; Becquart, F. Towards Thermoplastic Hemicellulose: Chemistry and Characteristics of Poly-(ε-caprolactone) Grafting onto Hemicellulose Backbones. *Mater. Design* **2018,** *153*, 298–307.
53. García, U.; Delgado, E.; Meléndez, H.; Camacho, T. A.; Esquivel, H.; Gatenholm, P.; Toriz, G. Spruce Xylan/HEMA-SBA15 Hybrid Hydrogels as a Potential Scaffold for Fibroblast Growth and Attachment. *Carbohydr. Polym.* **2018,** *201*, 490–499.
54. Liu, R.; Du, J.; Zhang, Z.; Li, H.; Lu, J.; Cheng, Y.; Wang, H. Preparation of Polyacrylic Acid-Grafted-Acryloyl/Hemicellulose (PAA-g-AH) Hybrid Films with High Oxygen Barrier Performance. *Carbohydr. Polym.* **2019,** 83–88.

55. Fortunati, E.; Luzi, F.; Puglia, D.; Torre, L. Extraction of Lignocellulosic Materials from Waste Products. *Multifunctional Polymeric Nanocomposites Based on Cellulosic Reinforcements*, 2016; pp 1–38.
56. Angeloni, G.; Masella, P.; Guerrini, L.; Innocenti, M.; Bellumori, M.; Parenti, A. Application of a Screening Design to Recover Phytochemicals from Spent Coffee Grounds. *Food Bioprod. Process.* **2019,** *118*, 50–57.
57. Lozano, R. P.; García, Y. A.; Tafalla, D. B.; Albaladejo, M. F. Cafeína: Un Nutriente, Un Fármaco, O Una Droga De Abuso. *Adicciones* **2007,** *19* (3), 225–238.
58. Bravo, J.; Juániz, I.; Monente, C.; Caemmerer, B.; Kroh, L. W.; De Peña, M. P.; Cid, C. Evaluation of Spent Coffee Obtained from the Most Common Coffeemakers as a Source of Hydrophilic Bioactive Compounds. *J. Agric. Food Chem.* **2012,** *60* (51), 12565–12573.
59. Yen, W. J.; Wang, B. S.; Chang, L. W.; Duh, P. D. Antioxidant Properties of Roasted Coffee Residues. *J. Agric. Food Chem.* **2005,** *53* (7), 2658–2663.
60. Brezová, V.; Šlebodová, A.; Staško, A. Coffee as a Source of Antioxidants: An EPR Study. *Food Chem.* **2009,** *114* (3), 859–868.
61. López-Barrera, D. M.; Vázquez-Sánchez, K.; Loarca-Piña, G.; Campos-Vega, R. Spent Coffee Grounds, An Innovative Source of Colonic Fermentable Compounds, Inhibit Inflammatory Mediators In Vitro. *Food Chem.* **2016,** *212*, 282–290.
62. Panusa, A.; Zuorro, A.; Lavecchia, R.; Marrosu, G.; Petrucci, R. Recovery of Natural Antioxidants from Spent Coffee Grounds. *J. Agric. Food Chem.* **2013,** *61* (17), 4162–4168.
63. Conde, C.; Mussatto, S. I. Isolation of Polyphenols from Spent Coffee Grounds and Silverskin by Mild Hydrothermal Pretreatment. *Prep. Biochem. Biotechnol.* **2016,** *46*, 406–409.
64. Braga, N.; Rodrigues, F.; PP Oliveira, M. B. Castanea Sativa byproducts: A Review on Added Value and Sustainable Application. *Nat. Prod. Res.* **2015,** *29* (1), 1–18.
65. Mussatto, S. I.; Machado, E. M. S.; Martins, S.; Teixeira, J. A. Production, Composition and Application of Coffee and its Industrial Residues. *Food Bioproc. Tech.* **2011,** *4*, 661–672.
66. Marto, J.; Gouveia, L. F.; Chiari, B. G.; Paiva, A.; Isaac, V.; Pinto, P.; ... ; Ribeiro, H. M. The Green Generation of Sunscreens: Using Coffee Industrial Sub-Products. *Ind. Crops Prod.* **2016,** *80*, 93–100.
67. López, D. M.; Vázquez, K.; Loarca, M. G.; Campos, R. Spent Coffee Grounds, an Innovative Source of Colonic Fermentable Compounds, Inhibit Inflammatory Mediators In Vitro. *Food Chem.* **2016,** 282–290.
68. Hoong, Y. B.; Pizzi, A.; Tahir, P. M.; Pasch, H. Characterization of Acacia Mangium Polyflavonoid Tannins by MALDI-TOF Mass Spectrometry and CP-MAS 13C NMR. *Eur. Polym. J.* **2010,** 1268–1277.
69. Quideau, S.; Deffieux, D.; Douat-Casassus, C.; Pouységu, L. Plant Polyphenols: Chemical Properties, Biological Activities, and Synthesis. *Angew. Chem. Int. Ed.* **2011,** *50* (3), 586–621.
70. Sanchez-Vazquez, S. A.; Hailes, H. C.; Evans, J. R. G. Hydrophobic Polymers from Food Waste: Resources and Synthesis. *Polym. Rev.* **2013,** *53* (4), 627–694.
71. Low, J. H.; Rahman, W. A. W. A.; Jamaluddin, J. Structural Elucidation of Tannins of Spent Coffee Grounds by CP-MAS 13C NMR and MALDI-TOF MS. *Ind. Crops Prod.* **2015,** 456–461.
72. Saada, M.; Kasmi, M.; Ksouri, R. Towards Coffee Processing byproducts Valorization: Phenolic Compounds and Antioxidant Activity of Spent Coffee Grounds and Coffee Silverskin. *Int. J. Biotech. Trends Technol.* **2019,** *9*, 11–15.

73. Bors, W.; Michel, C.; Stettmaier, K. Electron Paramagnetic Resonance Studies of Radical Species of Proanthocyanidins and Gallate Esters. *Arch. Biochem. Biophys.* **2000,** *374* (2), 347–355.
74. Calle, C. Extraction of Tannins from Coffee Grounds by Soxhlet Leaching, Universidad De Guayaquil Faculty of Chemical Engineering Career of Chemical Engineering, 2017.
75. Missio, A. L.; Mattos, B. D.; Ferreira, D. D. F.; Magalhães, W. L.; Bertuol, D. A.; Gatto, D. A.; Tondi, G. Nanocellulose-Tannin Films: From Trees to Sustainable Active Packaging. *J. Clean. Prod.* **2018,** *184*, 143–151.
76. Moreno- Escamilla, L. A.; Rodrigo, J. O.; Alvarez, P. Phenolic Compounds. In *Postharvest Physiology and Biochemistry of Fruits and Vegetables*; Woodhead Publishing, 2019; pp 253–227.
77. Belwal, T.; Nabavi, S. F.; Nabavi, S. M.; Habtemariam, S. Dietary Anthocyanins and Insulin Resistance: When Food Becomes a Medicine. *Nutrients* **2017,** *9* (10), 1111.
78. Welch, C. R.; Wu, Q.; Simon, J. E. Recent Advances in Anthocyanin Analysis and Characterization. *Curr. Anal. Chem.* **2008,** *4* (2), 75–101.
79. Zhao, C. L.; Yu, Y. Q.; Chen, Z. J.; Wen, G. S.; Wei, F. G.; Zheng, Q.; Xiao, X. L. Stability-Increasing Effects of Anthocyanin Glycosyl Acylation. *Food Chem.* **2017,** *214*, 119–128.
80. Shipp, J.; Abdel-Aal, E. S. M. Food Applications and Physiological Effects of Anthocyanins as Functional Food Ingredients. *Open Food Sci. J.* **2019,** *4* (1).
81. Kovalcik, A.; Obruca, S.; Marova, I. Valorization of Spent Coffee Grounds: A Review. *Food Bioprod. Process.* **2018,** 104–119
82. Adriaans, T. Suitability of Solvent Extraction for Jatropha Curcas. FACT Foundation: Eindhoven, 2006, p 9.
83. Ovalle, L. M. Evaluation of the Extraction Yield and Characterization of the Fixed Oil of Old Genuine Roasted Coffee Type Obtained by the Pressing Process. The University of San Carlos of Guatemala, 2015.
84. Campos-Vega, R.; Loarca-Pina, G.; Vergara-Castaneda, H. A.; Oomah, B. D. Spent Coffee Grounds: A Review on Current Research and Future Prospects. *Trends Food Sci. Technol.* **2015,** *45* (1), 24–36.

CHAPTER 8

Biotechnological Cost-Effective Processes for the Extraction of Bioactive Compounds

MARTINA DE LA ROSA-HERNANDEZ[1], DIANA B. MUÑIZ-MARQUEZ[1], PEDRO AGUILAR-ZARATE[2], MARÍA LUISA CARRILLO INUNGARAY[1], and JORGE E. WONG-PAZ[1]

[1]*Facultad de Estudios Profesionales Zona Huasteca, Universidad Autónoma de San Luis Potosí, Cd. Valles, S.L.P., México*

[2]*Instituto Tecnológico de Ciudad Valles, Tecnológico Nacional de México. Cd. Valles, S.L.P., México*

ABSTRACT

Extraction of bioactive compounds is widely studied due to the great variety of biological properties and potential industrial applications that involve this type of molecules. The need to improve the extraction processes by applying green technologies, such as maceration, ultrasound, microwaves, supercritical fluids, ohmic heating is a current topic of interest. In this matter, biotechnological approaches may help in the reduction of the environmental impact and operational costs generated in the conventional recovery of the bioactive compounds that resulting in the implementation of other effectives and unconventional friendly technique such as the fermentation in solid and liquid states. In this chapter, the economic impact involved in the recovery of bioactive compounds will be analyzed and documented, focusing on the biotechnological processes compared with the conventional process, from the obtaining of the raw materials, the development of the extraction technology processes of bioactive compounds, the continuous use of the equipment, and

Food Byproducts Management and their Utilization. Ricardo Gómez-García, Ana A. Vilas-Boas, Débora A. Campos, Maria Manuela Pintado, & Cristóbal Noé Aguilar (Eds.)

the waste generated. Also, some recommendations will be given considering a global process in the extraction of the bioactive compounds relating their efficacy and market cost for sustainable development. Finally, some advice to improve the process costs will be detailed and highlighted to have realizable alternatives in the present market of the bioactive compounds.

8.1 INTRODUCTION

Today, the possibility of increasing life expectancy as well as the need to achieve human well-being have allowed many phytochemical molecules to be positioned at the top of research aimed at developing products that benefit health and improve the quality of life of individuals.[1] Particularly, the bioactive compounds have been studied in various formulations for the food industry, nutraceutical, pharmaceutical, biomedical, cosmetologically, among others.[2] These molecules generally come from the secondary metabolism of a wide variety of plants, fruits, vegetables, and even microorganisms[3] and have been linked to important biological activities.[4] Consequently, bioactive compounds have a multiple positive health benefits when consumed in the diet, being clinically tested, and documented for the prevention, management or treatment of different diseases or negative health conditions.[5,6]

Therefore, they are highly valued in different areas of research, which has led to the establishment of viable methodologies for the recovery of these metabolites using organic byproducts of the food industry, nutritious plants, including medicinal becoming raw material for extraction processes.[7,8]

Solvent extraction, for example, is part of the conventional techniques in the recovery of secondary metabolites, where hydrodistillation, Soxhlet extraction, and maceration have been the most used for small-scale extraction processes.[9,10] On the other hand, there are other nonconventional techniques recently studied, such as ultrasound-assisted extraction (UAE) and microwave-assisted extraction (MAE), which are also adaptable to laboratory processes.[11] Other unconventional, but more complex methods in instrumentation which include supercritical fluid extraction (SFE), pulsed electric field (PEF) extraction, low polarity water extraction, and press pressure liquid extraction (PLE) are useful in the recovery of large-scale secondary metabolites.[12,13] Nowadays, work is already being done on the development and optimization of new methodologies that reduce extraction time, solvent consumption, and prevent the degradation of thermolabile compounds.[14]

Although all these extraction methods have generated significant scientific advances and influenced the development of food, cosmetic, or pharmaceutical products.[15] The economic expense caused by the recovery of these molecules has not been fully estimated. In addition, there are consequences of the environmental impact on society on the excessive use of solvents and energy.[16] For example, solid-state fermentation (FES) and liquid-state fermentation (FEL), as well as enzyme-assisted extraction[17] are clean, viable, and are considered sustainable biotechnological processes for the extraction of bioactive molecules using microorganisms to metabolize plant material into aromatic compounds, antioxidants, enzymes, organic acids, and other molecules.[18]

Therefore, this chapter will address issues involving the economic impact on the recovery of bioactive compounds in biotechnological processes against classical methodologies, obtaining raw materials, extraction processes, continuous production on a small and medium scale, use of instrumentation, analysis, and purification, as well as the ecological impact of the waste generated. Likewise, recommendations will be established on the effectiveness and cost of sustainable development in practice. In addition, some recommendations are incorporated to improve costs in the process, as well as feasible alternatives within the current market of bioactive compounds.

8.2 BIOACTIVE COMPOUNDS

Bioactive compounds are secondary metabolites of organic origin often produced by plants, fungi, bacteria, and other microorganisms.[19] They are synthesized during the metabolism of these organisms but are not directly involved with their growth, development, and reproduction processes.[20] Particularly in plants and photosynthetic algae, they are formed from the route of shikimic acid and polyacetates[21,22] once active, they are incorporated into defence mechanisms against stressful situations, such as predation, fight against pathogens, and other invasive species.[23]

Humans have been using secondary metabolites as medicine, flavorings, pigments, therapeutics, and recreational drugs.[24,25] Examples of the most studied and representative bioactive compounds (antioxidants, phytochemicals, pigments, nutrients, etc.) and their applications are shown in Figure 8.1, wherein the bioactive capacity has been associated with the presence or absence of specific functional groups, as hydroxyl (OH-), carboxyl (COOH-), carbonyl (CO-), and amino (NH_2-), which make up its structure.[26,27]

Bioactive compounds are in various parts of plants, fruits, and vegetables (Figure 8.2). They can be present as simple or complex molecules.[28] These molecular characteristics determine the viability of their recovery during extraction processes.[29] According to their biosynthetic origin are classified as phenols, terpenes, and alkaloids.[24] The phenols, for example, stand out due to their use in the treatment and prevention of some cardiovascular diseases (CVD), type II diabetes, some types of cancer, and neurodegenerative disorders.[29] Some polyphenols have been highly valued for their multiple biological properties, coupled with the fact that they are highly present in foods, such as fruits and vegetables.[30] Five important groups are distinguished within polyphenols: phenolic acids, flavonoids, stilbenes, coumarins, and tannins.[31]

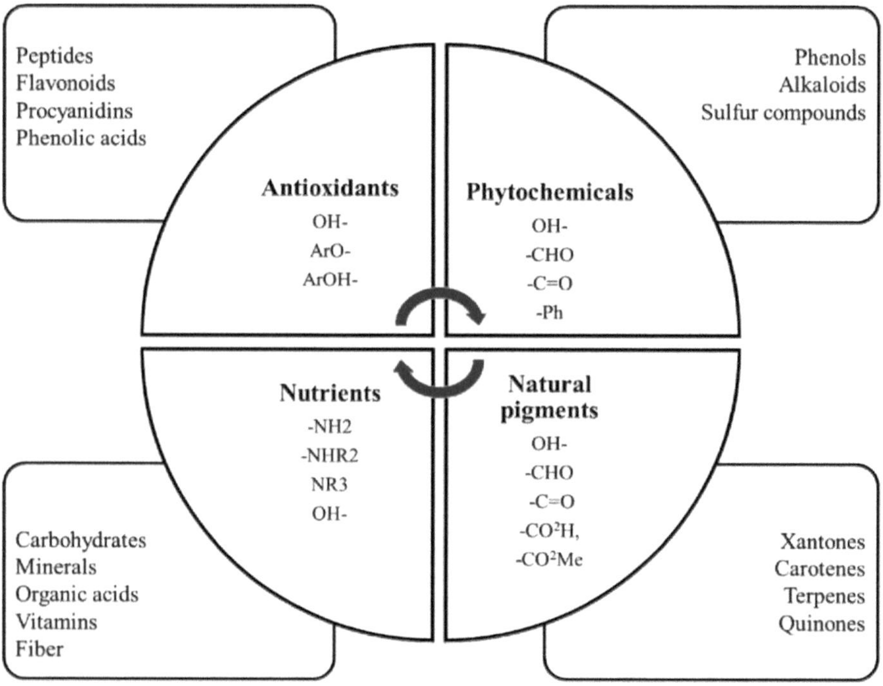

FIGURE 8.1 Representative bioactive compounds and chemical functional group as antioxidants, phytochemicals, pigments, and nutrients.

The last group, can be found in nature as condensed tannins and hydrolyzable tannins, where anthocyanidins and procyanidins are the most interesting compounds for industry and research areas.[32] On the other hand, commonly studied phenolic families, such as simple phenols, phenyl acetic acids, hydroxybenzoic acids, hydroxycinnamic acids, and other organic acids of the second group mentioned above are also known, their structures are usually simpler.[33]

FIGURE 8.2 Example of vegetal material (Tamarindus indica) valued for bioactive compounds extraction in investigation: (a) pulp (1), seeds (2), and peels (3); (b) stem (4), leaves (5), and roots (6).

8.2.1 BIOLOGICAL PROPERTIES

Bioactive compounds can act as alternative therapeutic agents against numerous negative conditions.[34] Alami-Merrouni et al.[17] report the anticancer activity of bioactive compounds from eight plants clinically proven against different types of cancer. In turn, flavonoids contained in citrus fruits, such as red blood, orange, and lemon, have potential antioxidant,[35] antimicrobial,[36] anti-inflammatory,[37] antifungal, and antiviral activities.[21,25,30] Immunomodulatory functions that regulate the expression of pro-inflammatory markers during cellular apoptosis are also reported,[20] including antiallergic properties of flavonoids isolated in food.[34]

It has been studied that saponins and flavonoids isolated from soy have biological, hormonal, immunological, bacteriological, and digestive activities to reduce CVD, liver diseases, obesity, and diabetes.[36] In general, many tannins exert pharmacological effects, antioxidant, antimicrobial, anticancer, cardioprotective, prevent metabolic disorders, and various diseases related to oxidative stress.[38,39]

Examples of bioactive molecules with proven biological properties from in vitro and in vivo experiments are given in Table 8.1. This bioactive capacity of compounds depends on their chemical structure, influencing their storage, extraction processing, bioaccessibility, and bioavailability in

TABLE 8.1 Studied Biological Properties of Extracts Rich in Bioactive Compounds and Their Health-Related Effects.

Compound	Vegetal material	Extraction method	Biological properties	Reference
Hesperidin	Grapefruit (*Citrus paradisi*)	Ethanol/water by soxhlet extraction	Antifungal activity against phytopathogenic fungi.	Salasa et al. [40]
				Castro-Vázquez et al. [41]
		Methanol/accelerated solvent extractor ASE	Antioxidant and pharmacological activity, as an activator of mitochondrial enzymes in the cell.	Tamilselvam et al. [42]
			Neuroprotective effects.	
Procyanidin B1 and B2	Apple (*Malus pumila cv. Fuji*)	Methanol/preparative column (hexane:methanol:ethyl acetate)	Antioxidant activity and anti-inflammatory effects on type 2 diabetes.	Ogura et al. [43]
				Masumoto, et al. [44]
			Prevention of metabolic syndrome.	
Naringin	Grapefruit (*Citrus paradisi*)	Methanol/maceration	Antifungal and antibacterial activity.	Xi et al. [45]
			Antioxidants prevent lipid peroxidation in the body.	Liu et al. [46]
				Li et al. [47]
			Antioxidant activity inhibiting the enzyme Xanthine oxidase (XO) in oxidative damage in tissues.	
Procyanidin A	Sorghum (*Sorghum bicolor (L.)* Grape (*Vitis)*	Ethanol/fraction of column AB-8 resin	Benefit on intestinal metabolism.	Li et al. [48]
			Antioxidant, therapeutic treatment on effects on diabetes, and hypertension. Anticancer activity.	Jara-Palacios et al. [49]
Epicatequin	Apple, grape, cranberries (*Vaccinium Oxycoccus*)	Ethanol/fraction of column	Antioxidant and biomarker of inflammation on animal and human cells. Obesity prevention.	Masumoto et al. [44]

TABLE 8.1 *(Continued)*

Compound	Vegetal material	Extraction method	Biological properties	Reference
Chlorogenic acid	Common wheat (Triticum aestivum)	Water/OHAE	Antioxidant effects.	Aguiar et al. [50]
Procyanidin C1	Cacao (*Theobroma cacao L.)*	Aqueous EtOH/UAE	Therapeutic antiviral activity on HIV virus infection; Inhibits intracellular replication of HIV -1, as well as its binding to host cells.	Hori et al. [51] Feng et al. [52]

food formulations.[53] In foods of plant origin, whether fresh or processed, anticancer phenolic compounds have been reported in green plants, such as *Phyllanthus*, cardioprotective flavonoids in litchi fruits, hypoglycemic agents, such as quercetin in sweet cherries or pomegranate sagittanins, muscle relaxants in watermelon, and fiber-soluble anticholesterol.[46,54]

8.3 CONVENTIONAL, NONCONVENTIONAL, AND INDUSTRIALIZED METHODS

The number of methodologies developed for the extraction process of bioactive compounds and other molecules are very diverse. They range from classical extraction methods such as maceration, Soxhlet extraction, and hydrodistillation[55] to those complex methodologies in instrumentation, such as extraction with supercritical fluids, PEF, and liquid extraction under pressure (PLE).[56] In addition, there are other simple methods to employ but of unconventional use, where the use of solvents is not so required, these are UAE and MAE.[57] Although each technique allows the recovery of metabolites, it is always necessary that the process must be carried out correctly and suitable according to the plant material used. However, in all extraction processes, some considerations will be presented for and against the performance obtained, energy consumption, and expenses related to the material and instrumentation.[58] These characteristics are described in Table 8.2, focusing on the advantages and disadvantages of one extraction method over another for the recovery of secondary metabolites.

8.3.1 CONVENTIONAL EXTRACTION

8.3.1.1 MACERATION

The maceration process is the most traditional extractive technique employed for the recovery of bioactive compounds from plants, fruits, and vegetables. Extraction conditions can occur using temperature or cold. Apparently, cold maceration seems to favor the conservation of the properties of most of the thermolabile compounds recovered from plants, fruits, vegetables, and microorganisms.[59] For example, Paladines-Quezada, et al.,[57] realized a pre-fermentative cold maceration (4 ± 1°C) and traditional maceration (25 ± 1°C) using 50 kg of grapes in 50-L stainless steel tank for 10 days using water and tartaric acid. They obtained a greater extraction of phenolic

TABLE 8.2 Advantages and Disadvantages of Some Extraction Methods of Compounds Most Used in Research and Industry.

Classification	Method	Advantage	Disadvantage
Conventional	**Maceration**	• Simplicity • Low cost • Adaptability on environment • In cold conditions preserve biological properties of compounds	• Low yields • Used of high amounts of organic solvents • Long extraction periods • Its necessary to rotate-evaporate • Produce waste
	Soxhlet extraction	• Extracting good essential oils • Laboratory-scale adaptability simplicity • Low acquisition cost • Processed can process large quantities of sample	• Loss a some properties • Higth cost of organic solvents • Waste generation • Fragility of the team • Long extraction periods
Nonconventional	**Ultrasound-assisted extraction (UAE)**	• No need to use organic solvents • Fast and effective extraction times • Lower energy expenditure • Increased extraction performance	• It is necessary to develop mathematical • methods for its optimization • Expensive extraction equipment
	Microwave-assisted extraction (EAM)	• Time savings • Decrease in solvent volumes • It is reproducible in the laboratory • Overal cost reduction • Good returns • Select method	• Loss of some properties due to heating • Energy expenditure • Automated equipment is expensive • Complexity of the process

TABLE 8.2 *(Continued)*

Classification	Method	Advantage	Disadvantage
Industrialized	**Supercritical fluids extraction (SPE)**	• Higher extraction purity • Lower environmental impact • Replacing the use of solvents • High selectivity of valuable microcomponents • Preservation of thermolabile compounds • Quality of extracts • Massive extraction processes	• Very high operating cost • Low availability of equipment • Energy expenditure
	Extraction by ohmic heating (OHAE)	• Useful for obtaining extracts of microorganisms and hard food • Suitable for the food industry	• Use of high temperatures • Loss of some properties in food • Method complexity • Difficulty scaling
	Pulse electric field extraction (PEF)	• Non.thermal, short-lived, low impact vegetal material • Oganoleptic properties of food are not modified	• Low availability of business units • It must be combined with techniques • Limited use to pumpable products • Electricity conductors • You can't use metal containers
	Means of water at low polarity extraction (PLE)	• Extraction of organic pollutants in food • Ecological • Shorter extraction time • Lower consumption of organic solvents repeatability	• Thermal effect on properties of compounds • Complexity • Economic expenditure

compounds, a higher quantity of total phenols, and greater color intensity in grape extracts and wines.

The maceration technique consists a simple technique where adding a specific amount of plant material in a glass container with solvent at a certain concentration, either in agitation or under static conditions for a certain time and established temperature (Figure 8.3a). This process softens the plant cells by releasing the bioactive compounds from the cells,[60] and the selection of the solvent will be based on the polarity of the metabolite of interest. It is proven that the maceration of fruits using ethanol as a solvent favors the extraction of bioactive and aromatic compounds from plant material.[61,62] Also, maceration has a better impact in the chemical composition of molecules present macerated extracts, hence, this is used in the development of food drinks.[63,64] But, although the maceration method is simple, many volumes of solvents are required to be considered effective. Given this situation, one of the strategies to optimize the extraction of compounds present in fruit peels is the addition of enzymes of microorganism that degrade the cell wall and allow the total disintegration of plant tissue.[65] This is "plant tissue maceration," that consists of the combination of cellulolytic and pectinolytic enzymes (pectinases: hydrolases and lyases and protopectinases) that facilitate the polyphenols release bound to the cell membrane.[66]

8.3.1.2 SOXHLET EXTRACTION

The Soxhlet method is a technique used to extract semi-volatile and nonvolatile compounds from plants on a conventional small scale.[67] The procedure consists of a continuous extraction using a special manual or semi-automated instrumental equipment of glass material until the appropriate phytochemical is obtained (Figure 8.3b). Finally, the recovered extract is concentrated with a rotavapor. The Soxhlet method has been considered more for the extraction of oils and essences from seeds, plants, and fruits, than from polyphenolic compounds.[68] However, Al Juhaimi et al.[69] achieved and tested different cold-pressed extraction systems for bioactive compounds. They showed that the antioxidant activities of different nut oils were improved using the cold temperature. Obtaining the highest contents of total phenols, oleic acid, linoleic acid, and other essential oils. Also, Khan et al.,[70] employed two different extraction methods to study the extract yields, bioactive compounds, and antioxidant activity. The data indicated that with the Soxhlet extraction, higher percentages of yield of compounds and extract, as well as antioxidant activity, were obtained compared with maceration.

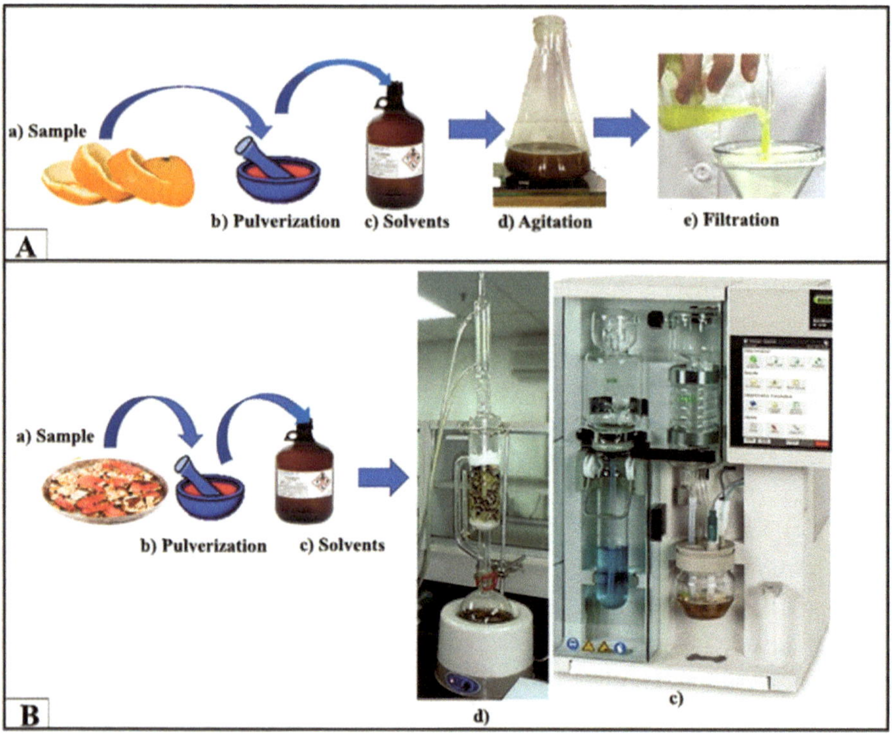

FIGURE 8.3 Extraction procedures: (A) extraction by maceration; (B) soxhlet extraction, where (c) represents the most economical conventional equipment, and (d) the most sophisticated equipment.

8.3.2 UNCONVENTIONAL EXTRACTION

8.3.2.1 ULTRASOUND-ASSISTED EXTRACTION (UAE)

UAE has been an effective methodology in the extraction of active compounds without the use of high amounts of solvents.[71] This technique, like maceration, focuses more on the recovery of secondary metabolites than on the extraction of essential oils. In addition, compared with maceration and Soxhlet extraction, the UAE requires semi-automated instrumental equipment based on ultrasonic wave physics,[72] the ultrasound extraction procedure is shown in Figure 8.4A.

Like the UAE, polyphenols can be recovered from a diversity of plant material. Dalmau et al.[73] made a comparison between conventional aqueous extraction (80 rpm) versus UAE (at 520 and 790 W/L) of orange byproduct biocomposites. Ultrasound-assisted aqueous extraction increased the yields of compounds with increasing temperature and sound power density. Ultrasound assistance allowed higher yields at lower temperatures and shorter times, favoring the phenolic content than by conventional extraction.

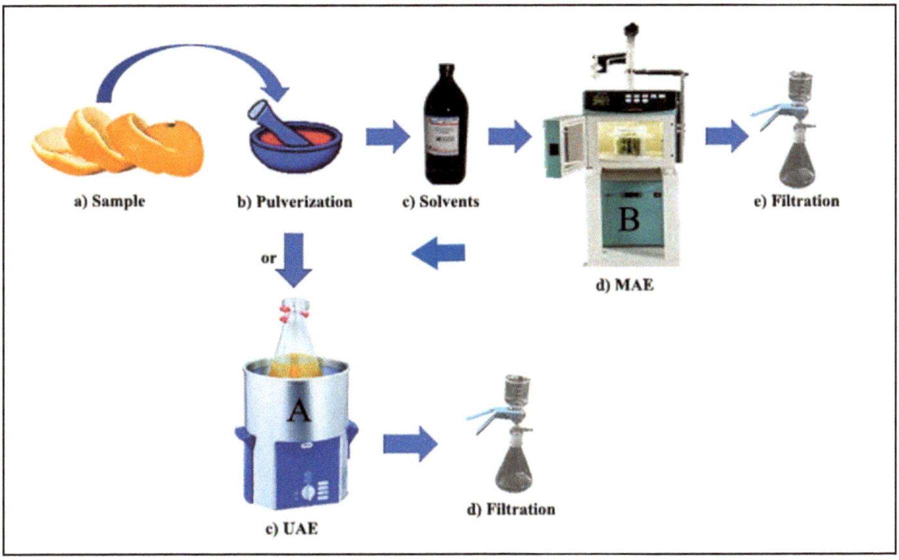

FIGURE 8.4 Extraction procedures: (A) ultrasound-assisted extraction (UAE); (B) microwave-assisted extraction (MAE).

The optimization of the UAE has been carried out based on crucial factors of extraction, such as temperature, sonication, time, and concentration of the solvent, including an extraction by aqueous sonication, is effective. The design of experimental mathematical models is more suited to this type of extraction, for example, Aourabi et al.[74] managed to optimize the extraction process and increase the extraction efficiency of *Zea mays* compounds from the response surface methodology (RSM) from solvent concentration effect, extraction time, and solvent/plant material ratio.

8.3.2.2 MICROWAVE-ASSISTED EXTRACTION (EAM)

MAE is an efficient and modern process for bioactive compounds extraction. In this, aspects, such as solvent type used, current and potential of microwave waves, the time of irradiation exposure, including particle size of vegetal material and of course the temperature, alter the composition of extracts obtained with this technique.[75] The MAE consists of the rapid heating of aqueous samples, solvents such as methanol or a mixture of water for polar compounds and hexane for nonpolar compounds, the solvent absorbs the microwave wave, and transfers it in the form of heat to the sample.[76] The good extraction of the active components of the sample will be mainly influenced by the temperature and nature of the solvent,[77] it is explained in Figure 8.4B.

In Milani et al.,[78] showed that the MAE by 4 min at 105°C increased up to eight times more the obtained phenolic and flavonoids than with the conventional extraction method. In addition, it improved the antioxidant properties of the extract with a significantly increased elimination capacity of DPPH, ABTS, and FRAP. But not only active compounds can be recovered by MAE, a case is the Zghaibi et al.,[79] where they used MAE for the extraction of lipids from *Nannochloropsis* sp. microalgae in the preparation of a functional food. The proposed technique offered better quality lipids containing the highest amount of polyunsaturated fatty acids (PUFA) (44.5%) and omega-3 fatty acids (AG) (43%), achieving a promising method of extraction of high-quality lipids with food application.

8.3.3 OTHER UNCONVENTIONAL INDUSTRIALIZED METHODS

8.3.3.1 SUPERCRITICAL FLUIDS (SFE)

Extraction with supercritical fluids is a more industrialized process to obtain metabolites of interest, where commercially available automated instrumental equipment is required (Figure 8.5a). The process begins with a clean source of fluid, which is a high-pressure cylinder of CO_2 generally, the pressure of the fluid is increased above its critical pressure to perform the extraction using a pump, the extraction pressure will be determined by the density required to dissolve the analytes in the sample.[80] SFE is useful for preparing analytical samples from complex solid samples, such as soils or sediments, biological samples, managing to extract lipids, antioxidants, aromatic compounds, alkaloids, pharmacological excipients, drugs, and diversity of bioactive compounds.[81] Compared with conventional extractions, SFE is useful for thermolabile compounds and can be simultaneously coupled with various chromatographic separation systems, such as supercritical fluid chromatography, gas chromatography, or high-performance liquid chromatography.[82]

For example, Klimek et al.,[83] extracted the bioactive compounds of Polish hops "Marynka" by efficient extraction of supercritical fluids in two steps, achieving greater antibacterial, cytotoxic, and antiproliferative activities in vitro in extracts obtained by SFE than raw extracts by maceration. For its part, Miyazawa et al.[84] obtained a very good elimination of chlorophyll using an extraction of supercritical fluids from a species of green algae (*Chlorella pyrenoidosa*) achieving to enhance the protein content of the food. Therefore, it is also an effective method to eliminate the components present in complex samples.

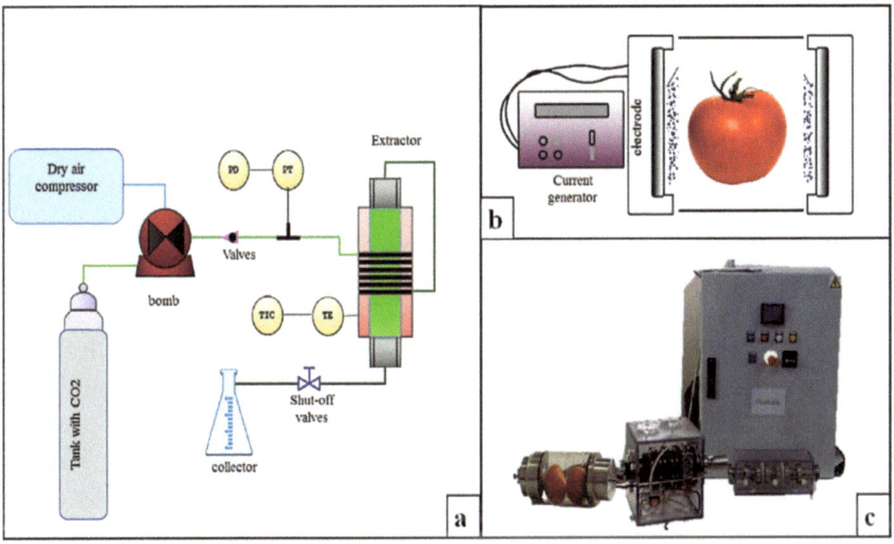

FIGURE 8.5 Extraction procedures: (a) extraction with supercritical fluids (SFE); (b) pulsed electric field (PEF) extraction; and (c) extraction by means of water at low polarity (PLE).

8.3.3.2 EXTRACTION BY OHMIC HEATING (OHAE)

Ohmic heating-assisted extraction (OHAE) has been introduced as a promising emerging technology at laboratory and industrial scales for the extraction of bioactive compounds.[85] This technique is shown in Figure 8.5b. A research work done by Al-Hilphy et al.,[80] which was a pilot-scale OHAE of bioactive wheat bran compounds. It showed that OHAE saved energy consumption, producing extracts with high amount of bioactive compounds (110–460 ppm of total phenolic content) in addition to higher antioxidant activities (antioxidant efficacy of 56–84%) compared with the conventional extraction of heat maceration. Also, Jesus et al.[86] performed an extraction of bioactive compounds using agro-industrial residues from the vine, thought ohmic heating treatment with ethanol–water at 45% (v/v) at 80°C of 20–90 min, the phenolic profile, polyphenolic, and biological properties were evaluated. The extract obtained by OHAE stood out with greater anticancer potential against tumor cell lines, HepG2, MDA-MB-231, MCF-7, and $CacO_2$, the content of polyphenols was also significantly higher as well as the antioxidant activity.[87]

8.3.3.3 PULSED ELECTRIC FIELD EXTRACTION (PEF)

PEF extraction arises from the need to extract all bioactive compounds (nutrients, pigments, and enzymes) safely and efficiently from bacteria, yeasts, fungi, microalgae, and plants applicable in the development of new products in the food and pharmaceutical industry.[88] The procedure is found in Figure 8.5c and consists of a nonthermal technology that allows a better selective release of intracellular compounds as it produces a precise action on the cytoplasmic membrane allowing high extraction yields without thermally damaging the compounds.[89] The effectiveness of PEF will depend on the cell structure, its size, and other factors, such as enzyme activity in microorganisms.[90]

In the food area, PEF has had great relevance due to obtaining different bioactive components, such as vitamins, minerals, phenolic acids, carotenoids, nitrates, ascorbic acids, betalains from fruits and vegetables, which generate a positive effect on human health.[91] For instance, Nowacka et al.[92] obtained a greater increase in the content of betalains, betanin, bulgaxanthin in red beet extract using PEF compared with the control sample extract obtained conventionally. Bobinaite et al.[93] analyzed blueberries treated with PEF, and observed an increase in the concentrations of total anthocyanins, which was reflected in the antioxidant activity, being used in the preparation of a blueberry pressed cake. Another study describes the extraction of C-phycocyanin using biomass of microalgae subjected to PEF and its purity was higher than the sample not treated.[94]

Therefore, the experimental results of various studies have determined PEF as a technique with a greater capacity for extracting bioactive compounds, and is helpful in improving the efficiency of industrial food and pharmaceutical processing.[95,96]

8.3.4 EXTRACTION USING WATER AT LOW POLARITY (PLE)

The PLE method involves using liquid solvents at temperatures above their atmospheric boiling point by applying high pressure to improving solubility and mass transfer compared with traditional methods.[97] Like PEF, PLE is useful in food formulation processes, dietary supplements, and phytopharmaceutical development.[98] The recovery of saponins from cockle seeds of water cow was carried out by means of an extraction with water at low polarity (PLE), against accelerated extractions with solvent (ASE) and ultrasonic extractions (USE), the results showed greater efficacy and purity of saponins in the extracts of PLE.[99]

8.4 ECONOMIC IMPACT OF THE RECOVERY OF BIOACTIVE COMPOUNDS

8.4.1 CONVENTIONAL METHODS

Extraction by maceration is a simple technique, adaptable in the laboratory and low cost in terms of energy consumption, so it is more economical if performed under static conditions.[8] However, maceration requires high consumption of organic solvents to obtain good yields, coupled with the fact that sometimes the extraction kinetics are prolonged, so it becomes an economic contradiction due to the high price of solvents, such as methanol, ethanol, and acetone,[55] evaluated from 50.00–100.00 USD per liter in the market. In addition, when obtaining large quantities of macerated extract, it is necessary to use rotation–evaporation techniques, with equipment evaluated between 2,500.00 and 10,000.00 USD, together with the energy expenditure generated in each manipulation. On the other hand, current research has been responsible for optimizing the extraction process of compounds present in a variety of residues with the addition of pectinase enzymes during maceration, making the process more expensive.[58]

On the other hand, Soxhlet extraction is a process that requires more significant financial capital to extract metabolites compared with maceration.[66] As to achieve the continuous solvent extraction of semi-volatile analytes from solid matrices, an inert glass system is used to extract the compound of interest, its cost can be from 125.00–500.00 USD with certified suppliers. However, the Soxhlet device is easy to install and use. It is made with frosted glass, facilitating disassembly and cleaning, so it is important to consider the manufacturing material to be used to make the process more economical and efficient.[71] In addition, as well as maceration, the use of organic extraction solvents, such as hexane, acetone, or methylene chloride (2500.00–10,000.00 USD) and long extraction cycles on stainless steel heating plates evaluated at 500.00–1000.00 USD is required and also involves a considerable energy expenditure.

8.4.2 UNCONVENTIONAL METHODS

UAE involves having a semi-automated sonication bath. The approximate cost is 1000.00–3000.00 USD, compared with the previous ones; therefore, UAE is economical because the required instrumentation is a long-term investment, coupled with the fact that the use of expensive organic solvents is not required since an ultrasound-assisted aqueous extraction allows the recovery of secondary metabolites using fruit and vegetable byproducts.[67] As for energy expenditure, it is considered minimal; however, some authors reported that the use of organic solvents, measuring temperatures and frequencies of ultrasonic waves make the extraction process a more expensive method.[88]

As for MAE at laboratory scale, it could be considered a complex methodology in the adaptation of equipment, but relatively inexpensive compared with other analytical methods.[90] However, when it comes to a more industrialized process, instrumental equipment is valued at too high a price,[91] coupled with solvents. Therefore, the more sophisticated the equipment, MAE will offer several advantages over more traditional techniques, such as shorter extraction time, lower solvent use, higher compound extraction rates, and lower overall processing cost.[92]

8.4.3 UNCONVENTIONAL INDUSTRIALIZED METHODS

Industrialized unconventional methods are methodologies with great application mainly for the food and pharmaceutical industry, so a significant economic investment is required to solve the purchase and correct maintenance of extraction equipment.[11,12] Therefore, we refer to the development of large-scale extraction techniques, where the most industrialized countries can allow the development and growth of large production plants. Some methods such as ohmic heating extraction (OHAE) can be executed from the laboratory scale, reducing the economic cost that an industrialized process could generate. However, it is necessary to perform a pretreatment with solvents at different molarity on the samples to be analyzed (usually food), also tests of heating time, applied current, up to temperature, which could generate additional costs for the consumption of reagents, energy, and maintenance.[79] So, it is important not to generate higher process costs to prevent extraction from becoming uneconomic.[52]

The large-scale OHAE has generated the development of 19 plants for the extraction of bioactive compounds. Although it is a technology and its initial costs may be high, a very good long-term profitability is generated.[78] In addition, heating is related to operating costs, since 95% of the energy is transformed into heat, allowing greater extraction efficiency, while in microwave heating, it only does so in 70%.[80]

Another method is SFE, commonly used in the food industries, is also called "green solvent technology," which uses as an extracting agent, a supercritical fluid instead of liquids such as conventional extractions.[91] Therefore, based on these properties, the procedure becomes more complex and the economic income required to have access to instrumental equipment of supercritical fluids,[92] its infrastructure cost usually ranges between 20,000.00 and 100,000.00 USD.

Like OHAE, SFE is adaptable to processes of mass extraction of bioactive compounds from plant matrices, mainly seeds, fruits, roots, flowers, rhizomes, and bark carried out in food production, drug production or nutraceutical plants.[90]

PEF is an electroporation technique used in the food industry, the principle of which is to increase the permeability of the cytoplasmic membrane of both plant and microbial cells.[95] Because this treatment is applied to different electrical intensities, the energy cost may be increased, but it is these electrical variations that facilitate the extraction of bioactive compounds, inactivate undesirable microorganisms, and accelerate the process of autolysis of fermenting microorganisms, improving the competitiveness of the extracts stored in the cellars.[94] Due to the very high cost of this equipment and the shortage of suppliers, the instrumentation is valued at approximately 80,000.00 USD, so not all food industries have been able to implement PEF in product development. However, the PEF could be considered a long-term investment for large corporations since it has a maximum processing capacity of 3600 tons of production per fruit season.[93] Currently, PEF equipment has been developed with sufficient power to be installed in medium-sized industries, at much lower costs. Some of the factors that affect the effectiveness of the treatment and cause economic losses during the experiments are technical factors, which relate the equipment and the processing, biological factors (microorganisms), and factors of the environment.[96]

Extraction employing water at low polarity (PLE) is a more practical method than SFE or PEF, and has been successfully applied to extract bioactive compounds since good yields are obtained in addition, a little amount of solvents is required and the process is carried out in less time, thus having greater reproducibility.[86] However, the use of solvents involves a cost of 50.00 USD. In the PLE it has been sophisticated and, in the market, there are devices for the extraction and recovery of components of raw materials from biomass with pressurized water of low polarity. The apparatus is configured with two or more reaction columns, each of which communicates separately with pressurized water sources, pressurized water, and pressure cooling water.[87] During the process, the components are first extracted into the column with water at a certain pressure, then the contents of the column are heated until the water is converted into water at low polarity pressure (PLP). This involves water consumption, the use of heat, and the use of electrical energy.[88]

Figure 8.6 presents the country's economy that allows to solve the creation of massive unconventional industrialized extraction plants around the world, being the European, Western, and North American countries, the pioneers in the development of more than 19 silvers, where this type of processes is frequently carried out.[97] These industrialized methods in the extraction of bioactive compounds have allowed the incorporation of products with adequate microbiological, organoleptic, and nutritional characteristics, since they are carried out under conditions of little fouling and in a minimum space, being able to apply to a wide range of foods and drugs.[84]

8.5 SUSTAINABLE METHODS: FERMENTATION IN SOLID-STATE (FES).

FES is a bioprocess that occurs from the growth of the filamentous fungi (GRAS) on solid material in the absence or almost absence of free water.[98] This process is heterogeneous and three-phase, comprising solid, liquid, and gaseous phases, the objective of which is to benefit the microbial metabolism of the bioprocess and the development of products.[93] The FES has gained significant attention to the need to recover higher yields of compounds and the lower use of electrical energy during the recovery of metabolites of interest.[99] Also, FES is an environmentally friendly technique, since it does not use large amounts of solvents as conventional techniques do, it is much more economical than unconventional and industrialized techniques of new generation[100] in addition to allowing the use and rapid consumption of large quantities of solid agro-industrial waste,[101] which helps the biotransformation, improvement, and increase of biological properties of the extracted compounds.[102].

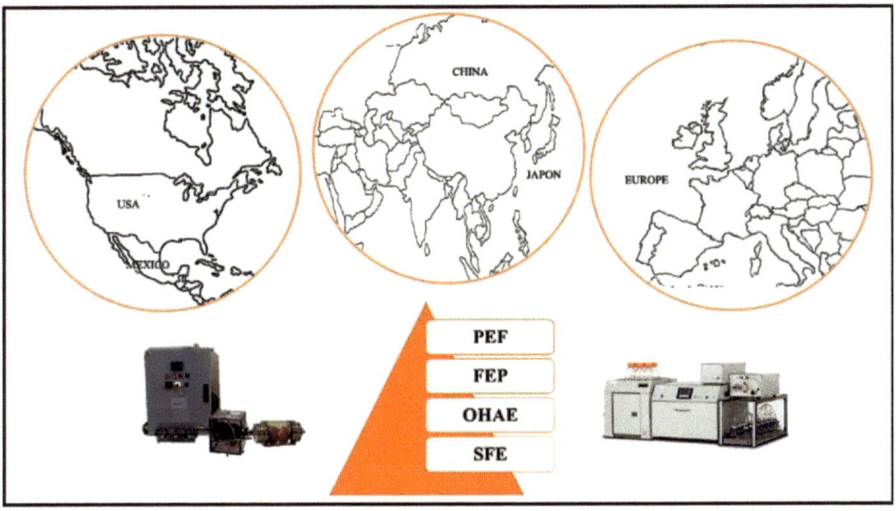

FIGURE 8.6 Countries where new strategies are developed in the recovery of bioactive compounds on an industrial scale focused on the area of food and pharmacy.

Some of the FES applications are shown in Figure 8.7, but the development of pharmaceuticals through this method is a very promising and preferred option in recent years. Therefore, it is considered an alternative in the sustainable development of extraction processes compared with other techniques in the production of bioactive compounds.[103]

8.6 ECONOMIC COST OF WASTE TREATMENT

During the recovery of bioactive compounds of industrial and research interests, from liters to thousands of liters of chemical, physical, and biological wastes have been generated. It can be noted that among the wastes that are mainly generated during the extraction process include organic solvent streams.[68] Unfortunately, the generation of solvents, such as acetone, hexane, methanol, or some other more commonly used ones such as ethanol, which have to be recovered with expensive laboratory equipment. In this situation, waste prevention can be considered the action of generating the least amount of hazardous waste to achieve the reduction of potential damage to the environment. Prevention may include the replacement of toxic solvents by another greener and friendly solvent, where the use of an expensive or health-harmful solvents is avoided, or the recovery of these solvents by vacuum evaporation and their subsequent reutilization.[88] The invention of fugitive losses of one or the other solvent considerably reduces the expense and exaggerated use of this type of reagents.

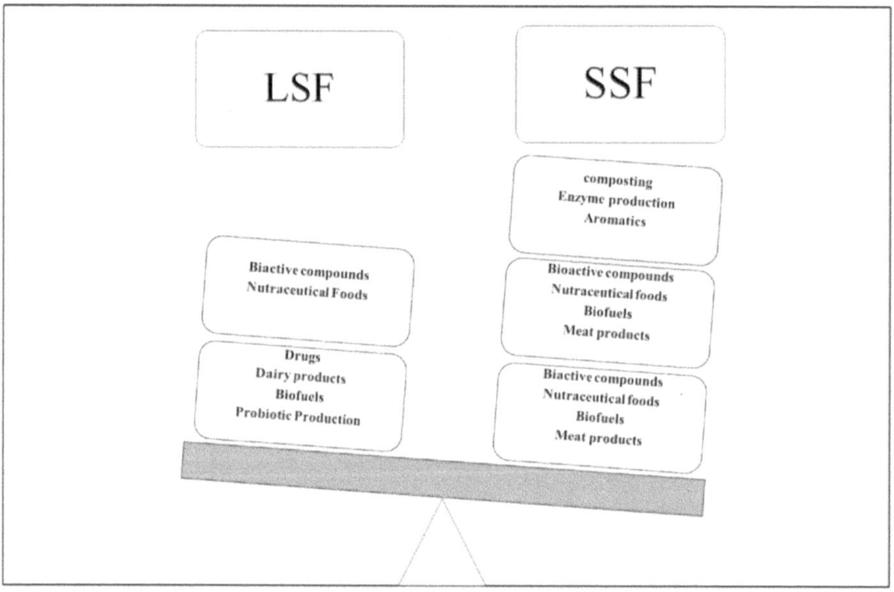

FIGURE 8.7 Main applications of solid-state fermentation (SSF) against liquid fermentation (LSF).

Another strategy for the extraction of compounds on a low and large scale is the action of recovery of the compound, also offering the residue to other companies interested in taking advantage of the recovered solvent.[89] Eliminating the use of chlorinated solvents is an important criterion for achieving the real reduction of the overall impact on the environment.

In economic analysis, where no financial distortions are introduced, any decision to carry out the recovery or regeneration of solvents must be considered as they represent the lowest overall cost and cover the need to avoid harming health and the environment. When estimating the "overall" cost, all factors that can influence the cost balance and are valid must be included.

Some examples are described below:

- Payment for the recovery services of the contracted company, including its profits.
- The disposal of any waste resulting from recovery.
- Transport to and from the waste generator.
- The purchase by the generator of any amount of solvent to compensate for the losses inherent in the reuse process.
- The total of these expenses represents the costs, for the generator to recover its waste so that it can continue with its commercial activity.
- The costs of the logistic strategies to transport waste to the facilities for its treatment.

Also, current strategies for waste treatment include the use of biotechnology (use of microorganism bioremediatory), coking, incineration, as supplementary combustibles, by chemical stabilization, solidification, solvent extraction, and thermal desorption.[104] However, the recommendation is to make agreements with expert companies of public sector and academic research in the treatment of physical and chemical waste. In addition to the use of some byproducts resulting from the vegetal material, they could be economically valued by the private sector to remunerate a minimum percentage of expenditure invested.[105] Finally, the bioactive compounds companies must achieve is the minimization of waste based on exact knowledge of production process, using simple friendly and environment technologies, and knowledge composition that vegetal material can be converted in waste.

8.7 CONCLUSIONS

In this chapter, the most effective extraction methods for the recovery of bioactive compounds, including simple phenolic acids, polyphenols, flavonoids, tannins have been reviewed using conventional, nonconventional, and industrialized technologies. Actually, the panorama of the economic impact of raw materials, instrumental equipment, and waste streams generated are greatly challenging, but recommendations can be established for efficient and economical extraction. The maceration and ultrasound techniques could be the most economical and viable methods for obtaining secondary metabolites. However, it will always depend on the object of study and the budget available to address the cost of the recovery process and environmental impact that could be generated by the development and isolation of active molecules. Therefore, the biotechnological approach requires a long walk between industry professionals, including food, pharmaceutical, cosmetological, biomedical, and experts in environmental care and management.

KEYWORDS

- **bioactive molecules**
- **green extraction**
- **solid-state fermentation**
- **unconventional techniques**
- **bioactive compounds**

REFERENCES

1. Abdel-Salam, A. M. Functional Food: Hopefulness to Good Health. *Am. J. Food Technol.* **2010,** *5* (2), 86–99.
2. Herrera-Chalé, F.; Betancur-Ancona, D.; Segura-Campos, M. Compuestos Bioactivos de la Dieta Con Potencial en la Prevención de Patologías Relacionadas Con Sobrepeso y Obesidad: Péptidos Biológicamente Activos. *Nutr. Hosp.* **2014,** *29* (1), 10–20.
3. Sanguansri, L.; Augustin, M. A. Microencapsulation in Functional Food Product Development. In Functional Food Product Development; Smith, J., Charter, E., Eds.; John Wiley and Sons: New York, NY, USA, 2010; pp 3–23.
4. Le, T. N.; Chiu, C. H.; Hsieh, P. C. Bioactive Compounds and Bioactivities of Brassica Oleracea L. var. Italica Sprouts and Microgreens: An Updated Overview from a Nutraceutical Perspective. *Plants (Basel)* **2020,** *9* (8), 946.
5. Trigo, J. P.; Alexandre, E.; Saraiva, J. A.; Pintado, M. E. High Value-Added Compounds from Fruit and Vegetable byproducts - Characterization, Bioactivities, and Application in the Development of Novel Food Products. *Crit. Rev. Food Sci. Nutr.* **2020,** *60* (8), 1388–1416. https://doi.org/10.1080/10408398.2019.1572588
6. Santos-Buelga, C.; González-Paramás, A. M.; Oludemi, T.; Ayuda-Durán, B.; González-Manzano, S. Plant Phenolics as Functional Food Ingredients. *Adv. Food Nutr. Res.* **2019,** *90*, 183–257. https://doi.org/10.1016/bs.afnr.2019.02.012
7. Martirosyan, D. M.; Singh, J. A New Definition of Functional Food by FFC: What Makes a New Definition Unique? *FFHD* **2015,** *5* (6), 209–223.

8. Altemimi, A.; Lakhssassi, N.; Baharlouei, A.; Watson, D. G.; Lightfoot, D. A. Phytochemicals: Extraction, Isolation, and Identification of Bioactive Compounds from Plant Extracts. *Plants* (*Basel*) **2017,** *6* (4), 42. https://doi.org/10.3390/plants6040042
9. Bonifácio-Lopes, T.; Teixeira, J. A.; Pintado, M. Current Extraction Techniques Towards Bioactive Compounds from Brewer's Spent Grain - A Review. *Crit. Rev. Food* **2020,** *60* (16), 2730–2741. https://doi.org/10.1080/10408398.2019.1655632
10. Shuai, H.; Yong, S.; Shouyao, Z.; Zhongyi, Z. Extraction of Resveratrol and Emondin from Polygonum Cuspidatum by Supercritical CO2 with Different Solubilizers. *Afr. J. Pharmacy Pharmacol.* **2015,** *9* (1), 12–18. http://dx.doi.org/10.5897/AJPP2013.3816
11. Leyva-Jiménez, F. J.; Lozano-Sánchez, J.; Borrás-Linares, I.; Arráez-Román, D.; Segura-Carretero, A. Comparative Study of Conventional and Pressurized Liquid Extraction for Recovering Bioactive Compounds from Lippia Citriodora Leaves. *Food Res. Int.* **2018,** *109*, 213–222. https://doi.org/10.1016/j.foodres.2018.04.035
12. Wen, L.; Zhang, Z.; Sun, D. W.; Sivagnanam, S. P.; Tiwari, B. K. Combination of Emerging Technologies for the Extraction of Bioactive Compounds. *Crit. Rev. Food Sci. Nutr.* **2020,** *60* (11), 1826–1841. https://doi.org/10.1080/10408398.2019.1602823
13. Aires, A. Phenolics in Foods: Extraction, Analysis, and Measurements. In *Phenolic Compounds: Natural Sources, Importance, and Applications*; InTech: London, UK, 2017.
14. Giacometti, J.; Bursac Kovacevic, D.; Putnik,P.; Gabric, D.; Bilusic, T.; Kresic, G.; Stulic, V.; Barba, F. J.; Chemat, F.; Barbosa-Canovas, G.; Jambrak, A. R. Extraction of Bioactive Compounds and Essential Oils from Mediterranean Herbs by Conventional and Green Innovative Techniques: A Review. *Food Res. Int.* **2018,** *113* (2), 245–262.
15. Luna-Castillo, K. P.; Lin, S.; Muñoz-Valle, J. F.; Vizmanos, B.; López-Quintero, A.; Márquez-Sandoval, F. Functional Food and Bioactive Compounds on the Modulation of the Functionality of HDL-C: A Narrative Review. *Nutrients* **2021,** *13* (4), 1165. https://doi.org/10.3390/nu13041165
16. Adeyi, O., Oke, E. O., Okolo, B. I., Adeyi, A. J., Otolorin, J. A., Nwosu-Obieogu, K., Adeyanju, J. A., Dzarma, G. W., Okhale, S., Ogu, D., & Onu, P. N. Process Optimization, Scale-Up Studies, Economic Analysis and Risk Assessment of Phenolic Rich Bioactive Extracts Production from Carica Papaya L. Leaves via Heat-Assisted Extraction Technology. *Heliyon* **2022,** *8* (4), e09216. https://doi.org/10.1016/j.heliyon.2022.e09216.
17. Alami Merrouni, I.; Elachouri, M. Anticancer Medicinal Plants Used by Moroccan People: Ethnobotanical, Preclinical, Phytochemical and Clinical Evidence. *J. Ethnopharmacol.* **2021,** 266. https://doi.org/10.1016/j.jep.2020.113435
18. Yang, F.; Zhang, Y.; Tariq, A.; Jiang, X.; Ahmed, Z.; Zhihao, Z.; Idrees, M.; Azizullah, A.; Adnan, M.; Bussmann, R. W. Food as Medicine: A Possible Preventive Measure Against Coronavirus Disease (COVID-19). *Phytother Res.* **2020,** *34* (12), 3124–3136.
19. Benites-Vílchez, J.; Díaz-García, R.; López-Vivar, J.; Gajardo-Solari, S.; Kusch-Fuschlocher, F.; Rojas-Arredondo, M. Antioxidant and Antibacterial Activity of Six Fruit Peels from the Pica Oasis. *BIOFARBO*, **2010,** *19* (1).
20. More S.; Sathe S.; Sonawane A.; Jadhav A. Citrus Paradisi: An Overview. *IAJPR* **2013,** *3* (10), 8070–8077.
21. Morales-Saldaña, J.; Gómez, N.; Rovira, J.; Abrahams, M. Actividad Larvicida de la Toronja, Citrus Paradisi (Rutaceae) Sobre dos Vectores Del Dengue. Rev. Peru. Biol. **2007,** *14* (2), 297–299.

22. Pan, Y.; Wang, K.; Huang, S.; Wang, H.; Mu, X.; He, C.; Ji, X.; Zhang, J.; Huang, F. Antioxidant Activity of Microwave Assisted Extract of Longan (Dimocarpus Longan Lour.) Peel. *Food Chem.* **2008,** *1* (6), 1264–1270.
23. Seyedrezazadeh, E.; Kolahian, S.; Shahbazfar, A. A.; Ansarin, K.; Pour Moghaddam, M.; Sakhinia, M.; Sakhinia, E.; Vafa, M. Effects of the Flavanone Combination Hesperetin-Naringenin, and Orange and Grapefruit Juices, on Airway Inflammation and Remodeling in a Murine Asthma Model. *Phytother. Res.* **2015,** *29* (4).
24. Liu, J.; Liu, G. Analysis of Secondary Metabolites from Plant Endophytic Fungi. *Methods Mol. Biol.* **2018,** *1848*, 25–38. https://doi.org/10.1007/978-1-4939-8724-5_3
25. Sozanski, K.; Prado, L.; Mularo, A. J.; Sadowski, V. A.; Jones, T. H.; Adams, R. Venom Function of a New Species of Megalomyrmex Forel, (Hymenoptera: Formicidae). *Toxins* **2020,** *12* (11), 679.
26. Zhang, Y.; Han, T.; Ming, Q.; Wu, L.; Rahman, K.; Qin, L. Alkaloids Produced by Endophytic Fungi: A Review. *Nat. Prod. Commun.* **2012,** *7* (7), 963–968.
27. Ngalimat, M. S.; Mohd Hata, E.; Zulperi, D.; Ismail, S. I.; Ismail, M. R.; Mohd Zainudin, N.; Saidi, N. B.; Yusof, M. T. Plant Growth-Promoting Bacteria as an Emerging Tool to Manage Bacterial Rice Pathogens. *Microorganisms* **2021,** *9* (4), 682. https://doi.org/10.3390/microorganisms904068
28. Barrero, A. F.; Herrador, M. M.; Arteaga, P.; Arteaga, J. F.; Arteaga, A. F. Communic Acids: Occurrence, Properties and use as Chirons for the Synthesis of Bioactive Compounds. *Molecules* **2012,** *17* (2), 1448–1467. https://doi.org/10.3390/molecules17021448.
29. Thomas L.; Larroche C.; Pandey A. Current Developments in Solid-State Fermentation (Review). *Biochem. Eng. J.* **2013,** *81*, 46–161.
30. Barros, R.; Andrade, J.; Denadai, M.; Nunes, M. L.; Narain, N. Evaluation of Bioactive Compounds Potential and Antioxidant Activity in Some Brazilian Exotic Fruit Residues. *Food Res. Int.* **2017,** *102*, 84–92.
31. Albuquerque, B. R.; Dias, M. I.; Pereira, C.; Petrović, J.; Soković, M.; Calhelha, R. C.; Oliveira, M.; Ferreira, I.; Barros, L. Valorization of Sicanaodorifera (Vell.) Naudin Epicarp as a Source of Bioactive Compounds: Chemical Characterization and Evaluation of Its Bioactive Properties. *Foods* **2021,** *10* (4), 700. https://doi.org/10.3390/foods10040700
32. Wang, T. Y.; Li, Q.; Bi, K. S. Bioactive Flavonoids in Medicinal Plants: Structure, Activity and Biological Fate. *Asian J. Pharm. Sci.* **2018,** *13* (1), 12–23.
33. Carvalho, E.; Franceschi, P.; Feller, A.; Herrera, L.; Palmieri, L.; Arapotsas, P. Discovery of A-type Procyanidin Dimers in Yellow Raspberries by Untargeted Metabolomics and Correlation-Based Data Analysis. *Metabolomics* **2016,** *12*, 144. DOI: 10.1007/s11306-016-1090.
34. Shashirekha, M. N.; Mallikarjuna, S. E.; Rajarathnam, S. Status of Bioactive Compounds in Foods, with Focus on Fruits and Vegetables. *Crit. Rev. Food Sci. Nutr.* **2015,** *55* (10), 1324–1339. https://doi.org/10.1080/10408398.2012.692736
35. Zhang, B.; Cai, J.; Duan, C.; Reeves, J. M.; He, F. Review of Polyphenolics in Oak Woods. *Int. J. Mol. Sci.* **2015,** *16* (4). DOI:10.3390/ijms16046978.
36. Imai, S. Soybean and Processed Soy Foods Ingredients, and Their Role in Cardiometabolic Risk Prevention. *Recent Pat. food, Nutr. Agric.* **2015,** *7* (2), 75–82.
37. Smeriglio, A.; Barreca, D.; Bellocco, E.; Trombetta, D. Proanthocyanidins and Hydrolysable Tannins: Occurrence, Dietary Intake and Pharmacological Effects. *Br. J. Pharmacol.* **2017,** *174* (11), 1244–1262. https://doi.org/10.1111/bph.13630

38. Liu, R. H. Dietary Bioactive Compounds and Their Health Implications. *J. Food Sci.* **2013,** *1* (78), 18–25. https://doi.org/10.1111/1750-3841.12101
39. Feng, S.; Wang, W. Bioactivities and Structure-Activity Relationships of Natural Tetra-hydroanthraquinone Compounds: A Review. *Front. Pharmacol.* **2020,** *11*, 799. https://doi.org/10.3389/fphar.2020.00799
40. Salasa, M. P.; Céliz, G.; Geronazzo, H.; Daz, M.; Resnik, S. L. Antifungal Activity of Natural and Enzymatically-Modified Flavonoids Isolated from Citrus Species. *Food Chem.* **2011,** *124* (4). DOI: 10.1016/j.foodchem.2010.07.100.
41. Castro-Vázquez, L.; Alañón, M.; Rodríguez-Robledo, V.; Pérez-Coello, M. P.; Hermosín-Gutierrez, I.; Díaz-Maroto, M. C.; Jordán, J.; Galindo, M. F.; Arroyo-Jiménez, M. Bioactive Flavonoids, Antioxidant Behaviour, and Cytoprotective Effects of Dried Grapefruit Peels (Citrus paradisi Macf). *Oxid. Med. Cell Longev.* **2016,** *1* (1). DOI: 10.1155/2016/8915729.
42. Tamilselvam, K.; Braidy, N.; Manivasagam, T. Neuroprotective Effects of Hesperidin, a Plant Flavanone, on Rotenone-Induced Oxidative Stress and Apoptosis in a Cellular Model for Parkinson's Disease. *Oxid. Med. Cell. Longev.* **2013,** Article ID 102741.
43. Ogura, K.; Ogura, M.; Shoji, T.; Sato, Y.; Tahara, Y.; Yamano, G.; Sato, H.; Sugizaki, K.; Fujita, N.; Tatsuoka, H.; Usui, R.; Mukai, E.; Fujimoto, S.; Inagaki, N.; Nagashima, K. Oral Administration of Apple Procyanidins Ameliorates Insulin Resistance via Suppression of Pro-Inflammatory Cytokine Expression in Liver of Diabetic ob/ob Mice. *J. Agric. Food Chem.* **2016,** *64* (46), 8857–8865. DOI: 10.1021/acs.jafc.6b03424.
44. Masumoto, S.; Terao, A.; Yamamoto, Y.; Mukai, T.; Miura, T.; Shoji, T. Non-Absorbable Apple Procyanidins Prevent Obesity Associated with Gut Microbial and Metabolomic Changes. *Sci. Rep.* **2016,** *6,* 31208. DOI:10.1038/srep31208.
45. Xi, W.; Zhang, G.; Jiang, D.; Zhou, Z. Phenolic Compositions and Antioxidant Activities of Grapefruit (Citrus Paradisi Macfadyen) Varieties Cultivated in China. *Int. J. Food Sci. Nutr.* **2015,** *66* (8). DOI: 10.3109/09637486.2015.1095864.
46. Liu, K.; Wang, W.; Guo, B. H.; Gao, H.; Liu, Y.; Liu, X. H.; Yao, H. L.; Cheng, K. Chemical Evidence for Potent Xanthine Oxidase Inhibitory Activity of Ethyl Acetate Extract of Citrus Aurantium L. Dried Immature Fruits. *J. Mol.* **2016,** *21*, 302. DOI: 10.3390/molecules21030302.
47. Li, S.; Liu, Y.; Liu, G.; He, J.; Qin, X.; Yang, H.; Hu, Z.; Lamikanra, O. Effect of the A-Type Linkage on the Pharmacokinetics and Intestinal Metabolism of Litchi Pericarp Oligomeric Procyanidins. *J. Agric. Food Chem.* **2017,** *65* (9), 1893–1899. DOI: 10.1021/acs.jafc.7b00017.
48. Li, W.; Zhaohui, H.; Peiyou, Q.; Yang, Y.; Xianjun, M.; Jianqiu, Z.; Kai, Z.; Guixing, R. Chemical Characterization of a Procyanidin-Rich Extract from Sorghum Bran and Its Effect on Oxidative Stress and Tumor Inhibition in Vivo. *J. Agric. Food Chem.* **2011,** *59* (16), 8609–8615. DOI: 10.1021/jf2015528.
49. Jara-Palacios, M. J.; Hernandez, D.; Escudero-Gilete, M. L; Heredia, F. J. The Use of Grape Seed Byproducts Rich in Flavonoids to Improve the Antioxidant Potential of Red Wines. *Molecules* **2016,** *21* (11), 1526. DOI:10.3390/molecules21111526.
50. Aguiar, J.; Estevinho, N.; Santos, L. Microencapsulation of Natural Antioxidants for Food Application, the Specific Case of Coffee Antioxidants. *Trends Food Sci. Technol.* **2016,** *58*, 21–39.
51. Hori, T.; Barnor, J.; Huu, T. N.; Morinaga, O.; Hamano, A.; Ndzinu, J.; Frimpong, A.; Minta-Asare, K.; Amoa-Bosompem, M.; Brandful, J.; Odoom, J.; Bonney, J.; Tuffour, I.; Owusu, B. A.; Ofosuhene, M.; Atchoglo, P.; Sakyiamah, M.; Adegle, R.; Appiah-Opong, R.; Ampofo, W.; Koram, K.; Nyarko, A.; Okine, L.; Edoh, D.; Appiah, A.; Uto,

T.; Yoshinaka, Y.; Uota, S.; Shoyama, Y.; Yamaoka, S. Procyanidin Trimer C1 Derived from Theobroma Cacao Reactivates Latent Human Immunodeficiency Virus Type 1 Provirus. *Biochem. Biophys. Res. Commun.* **2015,** *459* (2), 288–293. DOI: 10.1016/j.bbrc.2015.02.102.
52. Feng, W. Y.; Tanaka, R.; Inagaki, Y.; Saitoh, Y.; Chang, M. O.; Amet, T.; Yamamoto, N.; Yamaoka, S.; Yoshinaka, Y. Pycnogenol, A Procyanidin-Rich Extract from French Maritime Pine, Inhibits Intracellular Replication of HIV-1 as Well as its Binding to Host Cells. *Jpn. J. Infect. Dis.* **2008,** *61* (4), 279–85.
53. Elshamy, A.; Essa, A. F.; El-Hawary, S. S.; El Gawad, A.; Kubacy, T. M.; El-Khrisy, E.; Younis, I. Y. Prevalence of Diterpenes in Essential Oil of Euphorbia mauritanica L.: Detailed Chemical Profile, Atioxidant, Cytotoxic and Phytotoxic Activities. *Chem. Biodivers.* **2021,** 18 (7), e2100238. Advance online publication. https://doi.org/10.1002/cbdv.202100238.
54. Bimakr, M.; Rahman, R. A.; Taip, F. S.; Adzahan, N. M.; Sarker, M. Z.; Ganjloo, A. Optimization of Ultrasound-Assisted Extraction of Crude Oil from Winter Melon (Benincasa hispida) Seed using Response Surface Methodology and Evaluation of its Antioxidant Activity, Total Phenolic Content and Fatty Acid Composition. *Molecules* **2012,** *17* (10), 11748–11762. https://doi.org/10.3390/molecules171011748.
55. Kusuma, H. S.; Mahfud, M. RETRACTED: Preliminary Study: Kinetics of Oil Extraction from Basil (Ocimum basilicum) by Microwave-Assisted Hydrodistillation and Solvent-Free Microwave Extraction. *S. Afr. J. Chem. Eng.* **2016,** *21* (1), 43–53. https://doi.org/10.1016/j.sajce.2016.06.001.
56. Idris, F. N.; Mohd Nadzir, M. Comparative Studies on Different Extraction Methods of Centella asiatica and Extracts Bioactive Compounds Effects on Antimicrobial Activities. *Antibiotics* **2021,** *10* (4), 457. https://doi.org/10.3390/antibiotics10040457
57. Paladines-Quezada, D. F.; Moreno-Olivares, J. D.; Fernández-Fernández, J. I.; Bleda-Sánchez, J. A.; Martínez-Moreno, A.; Gil-Muñoz, R. Elicitors and Pre-Fermentative Cold Maceration: Effects on Polyphenol Concentration in Monastrell Grapes and Wines. *Biomolecules* **2019,** *9* (11), 671. https://doi.org/10.3390/biom9110671
58. Benucci, I.; Cerreti, M.; Liburdi, K.; Nardi, T.; Vagnoli, P.; Ortiz-Julien, A.; Esti, M. Pre-Fermentative Cold Maceration in Presence of Non-Saccharomyces Strains: Evolution of Chromatic Characteristics of Sangiovese Red Wine Elaborated by Sequential Inoculation. *Food Res. Int.* **2018,** *107*, 257–266. https://doi.org/10.1016/j.foodres.2018.02.029
59. Lerno, L. A.; Panprivech, S.; Ponangi, R.; Brenneman, C. A.; Block, D. E.; Oberholster, A. Impact of Cold Soak Duration on Cabernet Sauvignon Fermentation and Phenolic Composition. *Sci. Food Agric.* **2019,** *99* (2), 805–815. https://doi.org/10.1002/jsfa.9249
60. Hanousek Čiča, K.; Mrvčić, J.; Srečec, S.; Filipan, K.; Blažić, M.; Stanzer, D. Physicochemical and Aromatic Characterization of Carob Macerates Produced by Different Maceration Conditions. *Food Sci. Nutr.* **2020,** *8* (2), 942–954. https://doi.org/10.1002/fsn3.1374
61. Abdulla M. A.; Al-Bayaty F. H.; Younis L. T.; Abu Hassan M. I. Anti-Ulcer Activity of Centella asiatica Leaf Extract Against Ethanol-Induced Gastric Mucosal Injury in Rats. *J. Med. Plants Res.* **2010,** *4*, 1253–1259.
62. Magro L. D.; Goetze D.; Ribeiro C. T.; Paludo N.; Rodrigues E.; Hertz P. F.; Klein M. P.; Rodrigues R. C. Identification of Bioactive Compounds from Vitis labrusca L. Variety Concord Grape Juice Treated with Commercial Enzymes: Improved Yield

and Quality Parameters. *Food Bioprocess Technol.* **2016,** 9, 365–377. DOI: 10.1007/s11947-015-1634-5.

63. Shahidi, F.; Yeo, J. Bioactivities of Phenolics by Focusing on Suppression of Chronic Diseases: A Review. *Int. J. Mol. Sci.* **2018,** *19* (6), 1573.
64. Zhang, Q. W.; Lin, L. G.; Ye, W.C. Techniques for Extraction and Isolation of Natural Products: A Comprehensive Review. *Chin. Med.* **2018,** *13*, 1–26. DOI: 10.1186/s13020-018-0177-x.
65. Zhuoyan, H. Microwave Assisted Extraction: The Effects, Mechanisms and Applications on Selected Plant Materials Thesis. The Hong Kong Polytechnic University; Hong Kong, China, 2011.
66. Harun Al Rashid, M.; Majumder, S.; Mandal, V.; Mandal, S. C.; Thandavarayan, R. A. In Search of Suitable Extraction Technique for Large Scale Commercial Production of Bioactive Fraction for the Treatment of Diabetes: The Case Diospyros melanoxylon Roxb. *J. Tradit. Complement. Med.* **2018,** *9* (2), 106–118.
67. Jemain, S. F. P.; Jamal, P.; Raus, A. R.; Amid, A.; Jaswir, I. Effects of Process Conditions on the Ultrasonic Extraction of Phenolics Scavenger from Curcuma caesia rhizome. *Int. Food Res. J.* **2017,** *24*, 422–427.
68. Al Juhaimi, F.; Özcan, M. M.; Ghafoor, K.; Babiker, E. E.; Hussain, S. Comparison of Cold-Pressing and Soxhlet Extraction Systems for Bioactive Compounds, Antioxidant Properties, Polyphenols, Fatty Acids and Tocopherols in Eight Nut Oils. *J. Food Sci. Technol.* **2018,** *55* (8), 3163–3173. https://doi.org/10.1007/s13197-018-3244-5
69. Khan, W.; Bakht, J.; Nair, M. G.; Uddin, M. N.; Shafi, M. Extraction and Isolation of Important Bioactive Compounds from the Fruit of Physalis ixocarpa. *Pak. J. Pharm. Sci.* **2018,** *31* (6), 2463–2469.
70. González-Centeno, M. R.; Comas-Serra, F.; Femenia, A.; Rosselló, C.; Simal, S. Effect of Power Ultrasound Application on Aqueous Extraction of Phenolic Compounds and Antioxidant Capacity from Grape Pomace (Vitis vinifera L.): Experimental Kinetics and Modeling. Ultrason. sonochem. **2015,** *22*, 506–514.
71. Akhtar I.; Javad S.; Yousaf Z.; Iqbal S.; Jabeen K. Review: Microwave Assisted Extraction of Phytochemicals an Efficient and Modern Approach for Botanicals and Pharmaceuticals. *Pak J Pharm Sci.* **2019,** *32* (1), 223–230.
72. Dalmau, E.; Rosselló, C.; Eim, V.; Ratti, C.; Simal, S. Ultrasound-Assisted Aqueous Extraction of Biocompounds from Orange Byproduct: Experimental Kinetics and Modeling. *Antioxidants* **2020,** *9* (4), 352. https://doi.org/10.3390/antiox9040352
73. Shen, Y.; Liu, A.; Ye, M.; Wang, L.; Chen, J.; Wang, X.; Han, C. Analysis of Biologically Active Constituents in Centella asiatica by Microwave-Assisted Extraction Combined with LC–MS. *Chromatographia* **2009,** *70*, 431–438.
74. Aourabi, S.; Sfaira, M.; Mahjoubi, F. Optimization of Ultrasound-Assisted Extraction of Polyphenol Content from Zea mays Hairs (Waste). *Sci. World J.* **2020,** *2020*, 5072938.
75. Pavlić, B., Kaplan, M., Zeković, Z., Canli, O., Jovičić, N., Bursać Kovačević, D., Bebek Markovinović, A., et al. Kinetics of Microwave-Assisted Extraction Process Applied on Recovery of Peppermint Polyphenols: Experiments and Modeling. *Plants* **2023,** *12* (6), 1391. MDPI AG. http://dx.doi.org/10.3390/plants12061391.
76. Bobinaitė, R.; Pataro, G.; Lamanauskas, N.; Šatkauskas, S.; Viškelis, P.; Ferrari, G. Application of Pulsed Electric Field in the Production of Juice and Extraction of Bioactive Compounds from Blueberry Fruits and their byproducts. *J. Food Sci. Technol.* **2015,** *52* (9), 5898–5905.

77. Milani, G., Curci, F., Cavalluzzi, M.; Crupi, P.; Pisano, I.; Lentini, G.; Clodoveo, M.; Franchini, C., Corbo, F. Optimization of Microwave-Assisted Extraction of Antioxidants from Bamboo Shoots of Phyllostachys pubescens. *Molecules* **2020,** *25* (1), 215.
78. Zghaibi, N.; Omar, R.; Kamal, S.; Biak, D.; Harun, R. Microwave-Assisted Brine Extraction for Enhancement of the Quantity and Quality of Lipid Production from Microalgae Nannochloropsis sp. *Molecules* **2019,** *24* (19), 3581.
79. Uwineza, P. A., Waśkiewicz, A. Recent Advances in Supercritical Fluid Extraction of Natural Bioactive Compounds from Natural Plant Materials. *Molecules* **2020,** *25* (17), 3847. https://doi.org/10.3390/molecules2517384
80. Jesus, M. S.; Ballesteros, L. F.; Pereira, R. N.; Genisheva, Z.; Carvalho, A. C.; Pereira-Wilson, C.; Teixeira, J. A.; Domingues L. Ohmic heating polyphenolic extracts from vine pruning residue with enhanced biological activity. *Food Chem.* **2020,** 30.
81. Carullo, D.; Pataro, G.; Donsì, F.; Ferrari, G. Pulsed Electric Fields-Assisted Extraction of Valuable Compounds From Arthrospira Platensis: Effect of Pulse Polarity and Mild Heating. *Front. Bioeng. Biotechnol.* **2020,** *4* (8).
82. Klimek, K.; Tyśkiewicz, K.; Miazga-Karska, M.; Dębczak, A.; Rój, E.; Ginalska, G. Bioactive Compounds Obtained from Polish “Marynka” Hop Variety Using Efficient Two-Step Supercritical Fluid Extraction and Comparison of Their Antibacterial, Cytotoxic, and Anti-Proliferative Activities In Vitro. *Molecules* **2021,** 26 (8), 2366.
83. Martínez, J. M.; Delso, C.; Álvarez, I.; Raso, J. Pulsed Electric Field-Assisted Extraction of Valuable Compounds from Microorganisms. *Compr. Rev. Food Sci. Food Saf.* **2020,** *19* (2), 530–552.
84. Miyazawa, T.; Higuchi, O.; Sasaki, M.; Ota, M.; Aida, T.; Takekoshi, H.; Inomata, H.; Miyazawa, T. Removal of Chlorophyll and Pheophorbide from Chlorella Pyrenoidosa by Supercritical Fluid Extraction: Potential of Protein Resource. *Biosci. Biotechnol. Biochem.* **2021,** 24
85. Nowacka, M.; Tappi, S.; Wiktor, A.; Rybak, K.; Miszczykowska, A.; Czyzewski, J.; Drozdzal, K.; Witrowa-Rajchert, D.; Tylewicz, U. The Impact of Pulsed Electric Field on the Extraction of Bioactive Compounds from Beetroot. *Foods* **2019,** *8* (7), 244.
86. Jesus, M. S.; Ballesteros, L. F.; Pereira, R. N.; Genisheva, Z.; Carvalho, A. C.; Pereira-Wilson, C.; Teixeira, J. A.; Domingues, L. Ohmic Heating Polyphenolic Extracts from Vine Pruning Residue with Enhanced Biological Activity. *Food Chem.* **2020,** *316*, 126298.
87. Pavez-Guajardo, C.; Ferreira, S. R. S.; Mazzutti, S.; Guerra-Valle, M. E.; Sáez-Trautmann, G.; Moreno, J. Influence of In Vitro Digestion on Antioxidant Activity of Enriched Apple Snacks with Grape Juice. *Foods* **2020,** *9* (11).
88. Cacace, J.; Mazza, G. Pressurized Low-Polarity Water Extraction of Biologically Active Compounds from Plant Products, 2015.
89. King, J. W.; Grabiel, R. D.; Wightman, J. D. In *Subcritical Water Extraction of Anthocyanins from Fruit Berry Substrates*, International Symposium on Supercritical Fluids, 2003; vol 1 (1), pp 409–418.
90. Güçlü-Üstündağ, Ö.; Balsevich, J.; Mazza, G. Pressurized Low Polarity Water Extraction of Saponins from Cow Cockle Seed. *J. Food Eng.* **2007,** *80*, 619–630.
91. Quintero-Ramos, A.; Mascheroni, R.; Rodríguez, M. Extracción con microondas de compuestos antioxidantes en cáscaras de girasol (Helianthus annuus L), 2014.
92. De Melo.; M. M. R.; Silvestre, A. J. D.; Silva, C. M. Supercritical Fluid Extraction of Vegetable Matrices: Applications, Trends and Future Perspectives of a Convincing Green Technology. *J. Supercrit. Fluids* **2014,** *92*, 115–176.

93. Pereira, C. G.; Meireles, M. A. A. Supercritical Fluid Extraction of Bioactive Compounds: Fundamentals, Applications and Economic Perspectives. *Food Bioproc. Tech.* **2010,** *3* (3), 340–372.
94. Cerón-Carrillo, T. G.; Palou, E.; López-Malo, A. Pulsos eléctricos: fundamentos y aplicaciones en alimentos: Temas selectos de Ingenieria de Alimentos. Departamento de Ingenieria Quimica, Alimentos y Ambiental. Acta Universitaria de las Américas. **2010,** *1* (2), 9–26.
95. Sotomayor Carrasco, M. F.; Vargas Arroyo, D. A. Optimización de extracción, microencapsulación y evaluación de la capacidad antioxidante de antocianinas de flor de Jamaica (Hibiscus sabdariffa) mediante secado por aspersión. *Agrícola Panamericana.* **2016,** 44–64.
96. Villacrés Brazales, A. S.; Casas Murillo, D. E. Efectos de pretratamientos en la actividad de polifenol oxidasa extraída de manzana (Malus domestica). *Escuela Agrícola Panamericana.* **2015,** 38–50.
97. Saldaña, G.; Luengo, E.; Puértolas, E.; Álvarez, I.; Raso, J. Pulsed Electric Fields in Wineries: Potential Applications. In *Handbook of Electroporation*; Miklavcic, D., Ed.; Springer International Publishing, 2016; pp 1–18. DOI:10.1007/978-3-319-26779-1_155-1
98. Puértolas, E.; Saldaña, G.; Condón, S.; Álvarez, I.; Raso, J. Evolution of Polyphenolic Compounds in Red Wine from Cabernet Sauvignon Grapes Processed by Pulsed Electric Fields during Aging in Bottle. *Food Chem.* **2010,** *119*, 1063–1070.
99. Feng, H.; Fan, J.; Lin, L.; Liu, Y.; Chai, D.; Yang, J. Immunomodulatory Effects of Phosphorylated Radix Cyathulae officinalis Polysaccharides in Immunosuppressed Mice. *Molecules* **2019,** *24*, 4150. https://doi.org/10.3390/molecules24224150
100. Zhang, Q.; Zhao, C.; Wang, X.; et al. Bioaugmentation by Pediococcus acidilactici AAF1-5 Improves the Bacterial Activity and Diversity of Cereal Vinegar Under Solid-State Fermentation. *Front Microbiol.* **2021,** *11*, 603721.
101. Šelo, G.; Planinić, M.; Tišma, M.; Tomas, S.; Koceva Komlenić, D.; Bucić-Kojić, A. A Comprehensive Review on Valorization of Agro-Food Industrial Residues by Solid-State Fermentation. *Foods* **2021,** *10* (5), 927. DOI: 10.3390/foods10050927
102. Majumder, K.; Paul, B.; Sundas, R. An Analysis of Exo-polygalacturonase Bioprocess in Submerged and Solid-State Fermentation by Pleurotus ostreatus using Pomelo Peel Powder as Carbon Source. *J. Genet. Eng. Biotechnol.* **2020,** *18* (1), 47. DOI:10.1186/s43141-020-00061-7
103. Pandey, A.; Soccol, C. R.; Larroche, C. Current Developments in Solid-state Fermentation, In *Current Developments in Solid-state Fermentation*; Springer Science, 2008; pp 13–25.
104. Pandey, A.; Höfer, R.; Larroche, C.; Taherzadeh, M. J.; Nampoothiri, M. In *Industrial Biorefineries and White Biotechnology*; 1st Ed.; Elsevier: USA, 2015
105. Alkorta, I.; Garbisu, C.; Llama, M.; Serra, J. L. Industrial Applications of Pectic Enzymes: A Review. *Process Biotechem.* **1998,** *33*, 21–28.
106. Carné, M.; Cavalitto, S. F.; Hours, R. A. Maceración de papa con protopectinasa-SE de Geotrichum klebahnii. **2008,** 873–880.

CHAPTER 9

Waste Valorization Through Eco-Innovative Technologies and Yeast Conversion into High-Value Products

MIGUEL ÁNGEL VILLEGAS-MÉNDEZ[1], JAZEL DOMÉNICA SOSA-MARTÍNEZ[1], ELIZABETH XIMENA CASTRILLÓN-DUQUE[1], CRISTIAN SALVADOR COSSIO-CARRILLO[1], JUAN CARLOS CONTRERAS-ESQUIVEL[2], IVÁN SALMERÓN[3], JULIO MONTAÑEZ[1], and LOURDES MORALES-OYERVIDES[1]

[1]*Department of Chemical Engineering, Faculty of Chemical Sciences, Autonomous University of Coahuila, Boulevard Venustiano Carranza SN, Saltillo, Coahuila, México*

[2]*Food Research Department, Faculty of Chemical Sciences, Autonomous University of Coahuila, Boulevard Venustiano Carranza SN, Saltillo, Coahuila, México*

[3]*School of Chemical Science, Autonomous University of Chihuahua, New University Campus, Chihuahua, México*

ABSTRACT

Every year up to 5 billion tons of biowaste (husks, stalks, peels, bagasse, cake) can be generated by agro-industry; hence, the management of these residues is a well-known concern that carries environmental damage by greenhouse emissions, land, and water contamination. Although the accumulation of agro-industrial waste represents a global threat, it is also an opportunity to develop a sustainable, low-cost, and renewable alternative to produce high-value

Food Byproducts Management and their Utilization. Ricardo Gómez-García, Ana A. Vilas-Boas, Débora A. Campos, Maria Manuela Pintado, & Cristóbal Noé Aguilar (Eds.)

bioproducts by microbial fermentation, such as carotenoids, oil, and enzymes. However, the complex composition of agro-industrial waste hinders its utilization as a substrate in submerged fermentation. For that reason, eco-innovative treatments (i.e., ultrasound, subcritical extraction, and enzymatic hydrolysis) can be used to ease the conversion of macromolecules to simple compounds, such as sugars, amino acids, and minerals, which can be easily assimilated by microorganisms. In this respect, yeasts are cell factories that have been exploited in the food industry due to their capacity to assimilate various carbon and nitrogen sources with high cellular yields in short growth cycles. Also, yeast demands neither large cultivation areas nor climacteric conditions; hence, these advantages make yeast a potential alternative source for producing natural value-added products. Therefore, agro-industrial waste valorization as an alternative feedstock in yeast biotechnology reduces the negative environmental effect of waste disposal and raw material expenses, which generally has the most sensitive effect on the bioproduct manufacturing cost. This chapter couples applying innovative green methods and yeast cultivation for agro-industrial waste valorization into high-value bioproducts to contribute to a circular and sustainable bioeconomy.

9.1 INTRODUCTION

The global annual estimation of biomass generated by agro-industry is over 5 billion tons.[1] Therefore, disposal of residues generated from agro-industrial activity to food consumption is a global concern due to its huge accumulation through the years.[2] Agro-industrial residues have a complex composition of proteins, lipids, and carbohydrates, which are commonly incinerated or buried in landfills without restrictions and causing potential contamination in water, soil, and air[3]; however, the strict regulation in some countries has increased the industry investment to ensure adequate disposal.[4,5] On the other hand, agro-industrial and food wastes offer a potential source of nutrients that can be exploited for bioprocesses implementation, such as submerged and solid-state fermentation.[3]

In this respect, oleaginous yeasts are being used as cell factories due to their ability to assimilate a wide variety of substrates to produce high-value biocompounds in submerged fermentation.[6] Even though the nutrients of the waste matrix seem available, microorganisms such as yeast are not able to metabolize the complex matrix. In such cases, physical, chemical, or biological treatments can assist in the release of sugars and amino acids that can be then used to formulate a fermentation culture medium to produce further market valuable ingredients, such as lipids, pigments, enzymes.[6,7] Moreover, considerable efforts are being made toward developing green processes to treat and utilize such wastes, which inquire the application of low-energy processes and mitigation of harmful chemicals such as applying emerging technologies, biocatalyst or the synergic effect of both.[2,8–10] This chapter gives an overview of the recent findings regarding agro-industrial valorization through green alternatives for the bioproduction of lipids, pigments, proteins, and other high-value compounds synthesized by yeast.

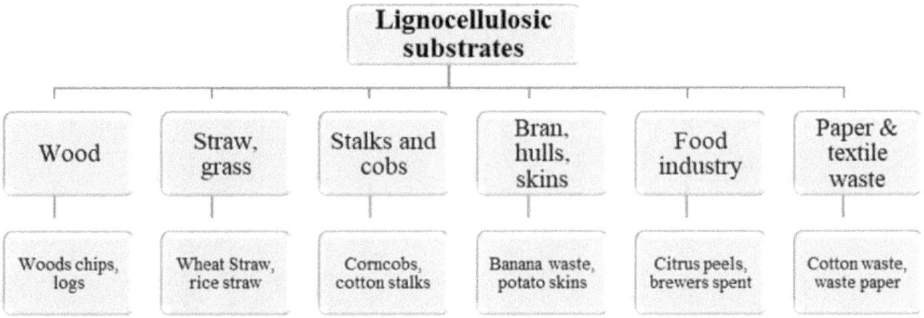

FIGURE 9.1 Main types of waste and examples generated in agro-industrial activities.

9.2 AGRO-WASTE MANAGEMENT AND HANDLING

Waste management consumes an important part of the total treatment cost and waste disposal generated by agricultural and industrial activities has a high cost and generates pollution and risks to human health.[11] Lignocellulosic waste is generated at all stages of the food chain and other products of agricultural origin, from harvest to commercial quality control. Countries, such as India, China, and the United States lead the generation of agro-industrial waste, generated mainly from maize, wheat, rice, sugarcane, soybean. Particularly in the American continent, countries, such as Brazil, Canada, and Argentina also contribute significantly to waste generation.[12] As can be seen in Figure 9.1, different types of materials are generated as waste. Accordingly, their suitable management toward the re-valorization is favorable in terms of the environment and the economy. The most used processes for this purpose are composting, hydrothermal treatments, and anaerobic digestion; however, no longer utilization of them will be accepted as waste management strategy due to several logistical issues and regulatory directives. Some examples are seen in Table 9.1.

TABLE 9.1 Main Processes Used for the Reuse and Re-valorization of Lignocellulosic Wastes.

Waste origin	Waste	Tackled issue	Final use of waste	Reference
Sewage sludge	Strawberry extrudate, fish waste, sewage sludge, and bulking agent	Reducing odorous emission, generation of polluting lixiviate, human health risks	Composting	[19]
Livestock manures	Chicken manure	Sustainably recycling organic waste Waste re-valorization	Composting	[20]
Livestock manures and agricultural waste	Chicken manure Maize straw	Sustainably recycling organic waste Waste re-valorization	Composting	[81]
Waste from microalgae-to-biodiesel manufacturing	Microalgae biomass, glycerol, chicken litter	Recycling or reuse of biomanufacturing residuals	Anaerobic digestion	[14]
Forest and food waste	Leaf litter of neem and vegetable waste	Removal of pollutants	Anaerobic digestion	[82]
Food processing waste	Waste frying oil	Food waste re-valorization	Anaerobic digestion	[83]
Livestock manures	Liquid dairy manure, dissolved air flotation waste and dry poultry manure	Sustainably recycling organic waste Green energy production	Anaerobic digestion	[13]
Agricultural waste, livestock manures	Solid waste from the agricultural and livestock industry	Waste re-valorization as energy sources Reduction of environmental impact	Thermal treatment	[16]
Agro-industrial waste	Jackfruit wastes	Waste re-valorization Green energy production	Thermal treatment	[17]
Pulp and paper waste	Pulp and paper mill waste	Waste re-valorization Green energy production	Thermal treatment	[84]

9.2.1 ANAEROBIC DIGESTION

Anaerobic digestion is the bioconversion of organic matter by bacteria and anaerobic archaeas, producing biogas and methane by their metabolic pathways. This practice is extremely important since it addresses the creation of green energies and the reduction of greenhouse gas emissions.[13] The production of such biofuels has been studied using a wide variety of wastes as substrates (Table 9.1). Using wastes as raw material for anaerobic digestion requires various key features, such as the biodegradability of the substrate, the carbon-to-nitrogen ratio, the structural and nutritional composition, the natural presence of inhibitor compounds of the process.[14] Also, the potential of waste to biogas conversion yield can be predicted by developing databases regarding experimental and theoretical methane yield values.[15]

9.2.2 THERMAL TREATMENT

Direct combustion of solid waste is the most widely used process. In thermal treatments, the generated heat can be used directly or for the production of electricity.[16] It does not contribute to the carbon dioxide concentration increment in the atmosphere, which is advantageous.[17] Various parameters of evaluation, such as fossil fuel equivalence, fuel value index, annual bioenergy yield, potential CO_2 retention, energy content, and thermal behavior provide crucial knowledge about the potential of agricultural residues for direct combustion.[18]

9.2.3 COMPOSTING

It is a simple and low-cost alternative for the re-valorization of waste. The suitability of waste for composting is determined by its respirometric activity and physical–chemical characteristics.[19] The process of matter oxidation results in a humic substances-rich material; in addition, its volume is reduced. It is also important to evaluate the behavior of the microbial communities present in the compost since the presence of certain microorganisms is more beneficial than others.[20]

9.2.4 SOLID-STATE FERMENTATION

This process is not considered among the main ones because it is more complex and requires specialized equipment to be implemented, such as a tray bioreactor, packed column, rotary drum, to mention a few examples. Lignocellulosic materials serve as sources of nutrients and support for microbial biomass. Characteristics, such as swelling capacity, water retention, and absorption index, pH, proximal composition are important to determine if the materials are suitable for this bioprocess.[21,22] High-value compounds can be obtained from this process type, depending on the waste and microbial strain used, such as bioactive peptides, monosaccharides, short-chain fatty acids, enzymes, polyphenolic compounds, growth factors.[23–26] The production of such compounds at a low cost might be possible by using low-cost waste as raw material.

9.3 ECO-INNOVATIVE APPROACHES FOR AGRO-WASTE VALORIZATION AS RENEWABLE FEEDSTOCK

In addition to the foregoing, value-added compounds can be extracted from agro-industrial wastes like those presented in Figure 9.2. Due to its massive accumulation, the economic investment that agro-industrial waste management demands is often underrated.[27] Therefore, economically feasible alternatives through waste valorization have been lately proposed to reduce the number of residues and contribute to a circular bio-economy.[3,28,29] To this end, particular attention is placed on utilizing novel and green technologies (Table 9.2). These green approaches focus on reducing biomass recalcitrance given by the lignin content with technological feasibility due to low energy consumption and reduced processing time. Also, these treatments are suitable to obtain sugars, proteins, and high-value compounds from agro-waste when assisted by biocatalyst.[9,10,27]

9.3.1 SUBCRITICAL/SUPERCRITICAL FLUID TREATMENT

Since most of the generated residues in the agro-industry are hemicellulosic-rich sources, such as bagasse, stalks, husk and straws, among others; subcritical technology could be utilized instead of acid hydrolysis. A substance's critical state is reached when heated and pressurized over its critical temperature and pressure points. Also, the main advantage of the subcritical treatment is the utilization of GRAS (generally recognized as safe) solvents, such as water, $CO_{2,}$ or ethanol to extract sugars, proteins, or phenolic compounds from agro-wastes.[27,30] Subcritical water processing assisted in the obtention of sugars (xylose, glucose) from brewers' spent grain,[27] rice husk,[10] soybean straw and hull.[31] Similarly, supercritical water hydrolysis efficiently converted hemicellulosic and pectin fractions of sugar beet pulp into monosaccharides.[32] Also, supercritical CO_2 treatment has been shown to modify the active site of enzymes, which might increase the enzymatic activity but at the same time can modify the pH and temperature of the system, having a negative effect on the biocatalysis of hemicellulose and cellulose to obtain reducing sugars.[33] Even though subcritical/supercritical process research offers a plausive alternative to waste valorization, its industrial design must be detailed to ensure process efficiency with low energy costs.

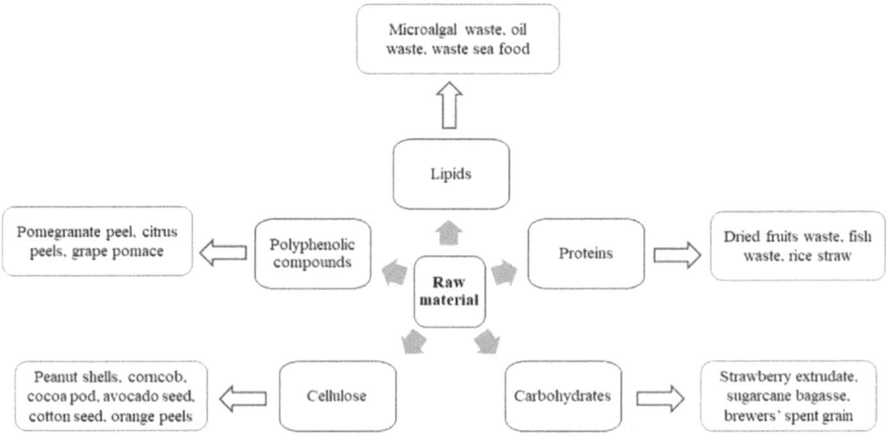

FIGURE 9.2 Main compounds obtained from agro-industrial wastes.

9.3.2 *ELECTRICAL DISCHARGES TREATMENT*

Electrical discharges on biological cells cause electroporation and electrical breakdown, damaging polysaccharides' structure, and facilitating the release of sugars, proteins, and phenolic compounds.[8,34,35] El kantar et al.[34] reported the application of high-voltage electrical discharges (<200 kJ/kg) to orange peels. The main effect of the electric treatment was observed in the reduction of orange peels' particle size, which facilitated the extraction of fermentable sugars and polyphenols.[34] Furthermore, electrical discharges have demonstrated more efficacy in lignin removal than sodium hydroxide in rapeseed straw.[8] The conventional hemicellulose hydrolysis to fermentable sugars is based on a thermochemical method which involves the utilization of high temperatures (≥120°C) and acid consumption.[10,36] However, a proposed electrofluid system by Wu et al.[36] demonstrated that conversion

TABLE 9.2 Parameters and Obtained Products Through Eco-Innovative Approaches for Agro-Waste Valorization.

Eco-innovative approach	Waste	Operational conditions	Product	Product yield	Reference
Subcritical water hydrolysis	Brewers' spent grain	Solvent-to-feed ratio of 64, water flow rate 20 mL/min at 210°C	Hemicellulosic sugars	35 g per 100 g of brewers' spent grain	[27]
Subcritical water hydrolysis	Rice husk	Liquid-to-solid ratio of 7.5 g water per g of husks, flow rate of 15 mL/min at 220°C.	Hemicellulosic sugars	18 g per 100 g of husk	[10]
Supercritical fluids	Sugar beet pulp	Suspended solids of beet pulp were treated by supercritical water at 390°C and 25 MPa with a reaction time of 0.11 s	Reducing sugars	61% w/w	[32]
High-voltage electrical discharges	Orange peels	Liquid to solid ratio of 2.72 w/v treated with 222 kJ/kg for 40 min	Reducing sugars	50% dry matter basis*	[34]
High-voltage electric discharges	Rapeseed straw	Liquid-to-solid ratio of 30 w/v treated with an electrical discharge of 800 kJ/kg for 39 min	Glucose	90% of glucose yield*	[8]
Electrofluid hydrolysis	Corncob	Solid-to-liquid ratio of 1:40 w/v introduced in a electrofluid reactor using a excitation voltage of 120 V, frequency of 40 Hz, 3% H_2SO_4 for 12 h	Reducing sugars	~5 g/L	[36]
Solution plasma process	Cassava starch water	Cassava starch wastewater solution (0.5% w/v) was treated with plasma generated in a reactor voltage of 0.4 kV per pulse of 3 μs frequency of 30 kHz for 300 min	Reducing sugars	Recovery of 99% of reducing sugars (47.9% of glucose)	[37]
Non-thermal plasma	Brewers' spent grain	Suspended biomass in water pretreated in a plasma column with a voltage of 28 kV for 30 min	Reducing sugars	162 mg per g of brewers' spent grain*	[9]
Ultrasound	Brewers' spent grain	Waste mixed with water (17.3% w/v) treated at 20% of amplitude in a bath for 60 min at 26.3°C	Reducing sugars	15 g/L*	[39]

TABLE 9.2 *(Continued)*

Eco-innovative approach	Waste	Operational conditions	Product	Product yield	Reference
Ultrasound	Food waste	Solid-to-liquid (water) ratio 10% w/v, power of 8 W/mL with a 11 mm horn for 5 min at 20°C	Glucose	~75 g/L	[2]
Extrusion coupled with ultrasound	Rice hull	50 g of pre-extruded biomass were suspended in 500 mL of citrate buffer pH 4.8 and put in an ultrasound bath (40 kHz/500 W) for 1.5 h	Reducing sugars	290 mg of glucose and 88 mg of xylose per g of rice hull *	[41]
Crude enzymatic hydrolysis	Confectionery sidestreams	Production of crude enzymatic extract by solid-state fermentation of *Aspergillus awamori*. The crude enzymes were used to hydrolyze the sidestreams at 55°C ~40 h	Nutrient rich hydrolysates	Composed by glucose, 74%; fructose, 10% and galactose, 16%	[47]
Crude enzymatic hydrolysis	Food waste	Production of crude enzymatic extract by solid-state fermentation of *Aspergillus awamori*. The crude enzymes were used to hydrolyze the food waste for 35 h at 55°C	Glucose	97.6% of starch to glucose conversion	[43]

**After enzymatic hydrolysis*

of the hemicellulosic fraction to reducing sugars in corncob was possible at lower energetic costs, which implied that the process is technically feasible, affordable and effective compared with conventional hydrolysis. Apart from this, electroporation processing has been successful, not only in lignocellulosic materials, but also in meat wastes where functional protein extracts could be obtained from waste chicken breasts.[35] These studies demonstrate the wide range of applications of electrical discharge technology and further development aims to reach economic viability.

9.3.3 ATMOSPHERIC PLASMA INDUCTION

A potential sustainable and nontoxic method for lignin removal is the non-thermal plasma treatment which principle is based on the generation of reactive species, such as ozone, hydroxyl, and hydronium ions.[9,37] Indeed, reactive species generated by atmospheric plasma can specifically attack lignin without affecting cellulose and hemicellulose structures in brewers' spent grain[9] and sugar cane bagasse.[38] Interestingly, starch contained in cassava wastewater was degraded to reduce sugars due to the interaction with the hydroxyl radicals in a plasma solution treatment.[37] The application of non-thermal plasma seems a plausible alternative for pretreatment of lignocellulosic materials; however, further research must be focused on the design of atmospheric plasma reactors to develop an affordable and efficient process.

9.3.4 ULTRASOUND-ASSISTED TREATMENT

The ultrasound is based on the cavitation phenomena in which the acoustic energy applied into a liquid causes shear forces, pressure, and high temperature in the system by the continuous formation and burst of microbubbles.[39] The parameters to control ultrasound are the amplitude, frequency, processing time and solid-to-liquid ratio, and equipment design (probe or bath).[40] In general, the effect of ultrasound in residues is increasing the surface area and pore size,[2,39] but lower amplitude (20%) is insufficient to remove lignin in brewers' spent grain.[39] In consideration of low-energy cost, the processing time plays a key role since the longer time elapses, the more energy is consumed; in this respect, 30 min in a sonication bath were sufficient to pretreat pineapple peels to obtain reducing sugars.[40] On the other hand, a sonication pretreatment of 5 min with a sonication probe was efficient to enhance the nutrient recovery of food waste.[2] The ultrasound treatment can be coupled with other unit operations making this technology highly adaptable to industry.[41,42] Extrusion followed by ultrasound treatment assisted in the obtention of 380 mg of reducing sugars per gram of rice hull,[41] whereas the hydrolysis degree in whey protein concentrate was enhanced by the synergic effect of ultrasound and ohmic heating processing.[42] As well as electric fields technology, ultrasonication is a feasible industrial-scale process and its configurations made a versatile process for a wide variety of agro-industrial wastes.

9.3.5 BIOCATALYSIS-ASSISTED EXTRACTION

Enzymatic hydrolysis has been adopted as a green alternative to release fermentable sugars or amino acids from agro-industrial wastes, including food wastes.[43] Previously, it was discussed the importance of lignin removal, which makes the hemicellulosic fraction of materials more accessible for enzymatic action. It can be emphasized that lignin removal around 90% improves the carbohydrases' accessibility to obtain high conversion yields of fermentable sugars.[8,9] Most of the enzymatic hydrolysis performed in lignocellulosic residues exploited the application of commercial mixtures of cellulases, hemicellulases, xylanases, β-glucanases, and arabinases to increase the conversion yields to fermentable sugars.[44–46] On the other hand, crude enzymatic extracts with high amylolytic and proteolytic activities are produced by solid-state fermentation of *Aspergillus* species as an alternative low-cost approach to commercial enzymes' mixtures.[43,47,48] Side streams from wheat milling and confectionery industries,[47] soybean meal,[48] potato peels, and food waste[43] were suitable for the solid-state fermentation process to obtain crude enzymatic extracts, which were then used to yield hydrolysates rich in glucose, free amino nitrogen, and inorganic phosphorous using the same side streams.[43,47,48] This approach is an interesting contribution to a circular biorefinery concept due to using agro-waste as a raw material for consecutive operations.

9.4 VALORIZATION OF RESIDUES IN THE PRODUCTION OF HIGH-VALUE PRODUCTS BY OLEAGINOUS YEAST

Oleaginous yeast is gaining attention in industrial applications due to its ability to synthesize metabolites, such as lipids, carotenoids, and enzymes; in other words, oleaginous yeast can be considered as cell factories of high-value products.[6] Agro-industrial waste valorization allows their utilization as a source of nutrients like carbon and nitrogen for the growth of the oleaginous yeast and simultaneously complements the circular bio-economy concept by waste conversion into high-value products as antioxidants, fatty acids, ergosterol, coenzyme Q10, widely used in pharmaceutical, food, and cosmetic industries.[49–51]

Rhodotorula, *Rhodosporidium*, *Sporobolomyces*, *Cystofilobasidium*, *Sporidiobolus* are the common genera referred in the literature as the feasible yeast for industrial potential.[52,53] Figure 9.3, shows the followed integrative process to convert agro-industrial residues into valuable compounds, which consist of the biowaste pretreatment to release nutrients, fermentation, and recovery of products.[54] Furthermore, Table 9.2 presents the valorization of agro-industrial residues as a substrate on the bioprocessing of high-value compounds, such as oil and carotenoids.

FIGURE 9.3 General integrative process from agro-industrial waste to high-value compounds bioconversion by oleaginous yeast.

9.4.1 SYNTHESIS OF MICROBIAL OIL BY OLEAGINOUS YEAST USING RENEWABLE FEEDSTOCK

Oleaginous yeasts have advantages such as being single-celled organisms with high cell density and accumulation up to 70% w/w of microbial oil; also, they can grow on different carbon sources, such as glucose and xylose obtained from the hydrolysis of hemicellulosic materials, food wastes, and effluents which could lower the manufacturing costs for oil biosynthesis.[45,55,56] Microbial oil obtained from oleaginous yeast bioprocessing can be used to produce biodiesel, oleochemicals, or vegetable oil substitute.[44,57,58]

Wheat straw cellulose enzymatic hydrolysis showed a great performance as a substrate for *Rhodosporidium toruloides* and *Lipomyces starkeyi* bioprocess with 24.95% accumulated oil.[45] Byproducts from the sugar industry processing (molasses, juices, syrups) contain sucrose, glucose, and fructose as the main carbohydrates.[57–59] In this respect, strains of *Rhodosporidium toruloides* and *Rhodotorula glutinis* using molasses can accumulate an average of 50% (w/w)[58] also is an excellent substrate to use as feeding stock in a fed-batch process in which a maximum accumulation of 75% (w/w) can be reached.[57] A two-phase treatment of corn cob consisting of nitric acid pretreatment followed by enzymatic hydrolysis allowed releasing 47 g/L of total sugars (glucose, xylose, arabinose), which was suitable to develop a two-stage batch fermentation with an average oil accumulation of 75% (w/w).[44] In specific processes, supplementation is necessary to complement the carbon source; for instance, glycerol addition to a sugar cane bagasse hydrolysate was required to increase the carbon-to-nitrogen ratio on the fermentation of *Rhodosporidium toruloides.*[60]

9.4.2 CAROTENOIDS OBTAINED THROUGH THE VALORIZATION OF AGRO-WASTE IN YEAST BIOPROCESS

Chemically synthesized food colorants used as food additives have the risk of toxicity and potentially hazardous effects to the consumers, while yeast pigments present higher stability to heat, light exposure, and pH variations, and are highly soluble in water with the advantage of mitigating environmental harm.[52,61] Main carotenoids produced by yeast include β-carotene, astaxanthin, torulene, and torularhodin, which are reported as antioxidants, anticancer agent, enhancers of the immune response protection against bacteria and fungus, and later recognized as production enhancers of eggs in laying hens.[50,51,62,63]

Similar to microbial oil production, hemicellulosic and rich-pectin residues are suitable for biocatalyst, which released products used in carotenoid production by yeast.[46,64,65] Rapeseed meal was a proper matrix for carrying enzymatic hydrolysis when applying a heat pretreatment in the material to improve sugar conversion yield, possibly by assisting in lignin removal, which allowed using the hydrolysate for astaxanthin production by *Xanthophyllomyces dendrorhous.*[64] Glucose and galactose obtained by the enzymatic hydrolysis of sugar beet pulp (rich-pectin residue) were metabolized by *Rhodotorula mucilaginosa* and *Rhodosporidium toruloides* producing 5 mg/L of total carotenoids production.[65] In the valorization of cinnamon waste, it was possible to obtain 12 g/L of total sugars (mainly glucose and fructose) through enzymatic conversion.[46] Although cinnamon hydrolysates contained cinnamic acid and cinnamaldehyde, which are recognized for their antimicrobial and antifungal activity, *Rhodosporidium toruloides* was able to grow and produce 2 mg/L of carotenoids.[46] Residues collected from juices stores and cafeterias were suitable to obtain glucose, xylose, and glycerol after heating treatment. This extract was used to obtain carotenoids by *Rhodotorula* and *Rhodosporidium* strains.[66] Also, enzymatic hydrolysis performed in paddy straw resulted in the obtention of glucose to evaluate carotenoid production in *Rhodosporidium toruloides* and several *Rhodotorula* strains.[66]

9.4.3 OTHER HIGH-VALUE COMPOUNDS SYNTHESIZED BY OLEAGINOUS YEAST

9.4.3.1 SINGLE CELL PROTEIN

Yeast was the first microorganism whose importance as an animal feed supplement was recognized; in addition, yeast utilizes inexpensive feedstock and wastes as carbon and energy sources for microorganism growth.[67] In bioprocess design for yeast protein production, which mainly focuses on the supplementation of food preparation and feed, the most reported strains belong to the genera *Candida*, *Pichia*, *Rhodotorula*, *Sporidiobolus*, and *Saccharomyces*.[67,68]

Regarding waste utilization in single-cell protein manufacture, waste-water from food industry streams has the potential to be used by *Candida utilis* for growth and protein accumulation.[69,70] *Saccharomyces cerevisiae* and red yeast *Sporidiobolus pararoseus*, cultivated in glycerol, have been used to supplement feed in laying hens; this dietary administration improved the egg yolk color, decreased serum, and egg yolk cholesterol levels.[71]

9.4.3.2 ERGOSTEROL

The ergosterol, one of the most important components in fungal membranes, is involved in the regulatory functions of cells; however, it has been used in the development of drugs with potential antitumoral effects.[72] The biosynthesis of ergosterol is derived from the mevalonate pathway like carotenoids[72,73]; therefore, the synthesis of this compound has been reported in yeast belonging to the genera *Rhodotorula* and *Sporobolomyces.*[73,74] Utilization of enzymatically hydrolyzed whey protein, cereal byproducts, and animal waste fat[74] in *Sporobolomyces* sp. could reach up to 8.9 mg/g of ergosterol[73]

9.4.3.3 COENZYME Q10

Coenzyme Q10 or ubiquinone is a compound with health-related beneficial properties against chronic degenerative diseases, such as Parkinson's, Alzheimer and diabetes.[75] Some yeast species, such as *Rhodotorula glutinis, Sporobolomyces shibatanus, Cystofilobasidium macerans* can synthesize ubiquinone due to the common precursor acetyl-CoA in the mevalonate pathway.[74–76] Recent contributions in animal fat waste valorization as substrate in yeast bioprocess emphasize the carbon-to-nitrogen ratio effect and its relation with ubiquinone synthesis.[74,76]

9.5 FUTURE TRENDS IN THE ENVIRONMENTAL IMPACT OF OLEAGINOUS YEAST BIOPROCESS USING RENEWABLE FEEDSTOCK

The strategies in valorizing agro-industrial residues seem a promising short-term solution for the bioconversion of wastes into valuable compounds (Table 9.3). Also, the implementation of biocatalyst to obtain microbial feedstock could contribute to the transformation of chemical processes into green practices in the industrial sector.[3] The implementation of a circular bio-economy based on oleaginous yeasts bioprocess represent an upcoming techno-economic challenge due to the required migration of conventional process to developed green approaches.[54,77]

An effective tool to assess the ecological footprint of the biotechnological process is the life cycle assessment, where it is possible to identify the environmental impacts regarding both the valorization of agro-industrial residues and the obtention of high-value compounds.[78] In general, bioproduction of microbial oil by oleaginous yeast reported the elevated greenhouse gas (GHG) emissions due to the electricity consumption given by bioreactors operation and the energy required to downstream the main products.[79,80] On the other hand, the land utilization for vegetable oils or carotenoids extracted from vegetables is greater than the microbial process.[79] Also, bioprocesses have

TABLE 9.3 Agro-Industrial Waste Valorization as Alternative Feedstock on Yeast Bioprocess.

Source	Valorization technique	Cell factory	High-value synthesized compound	Reference
Molasses	Direct utilization	*Rhodosporidium toruloides*	Microbial oil	[58]
Food waste	Saccharification using crude enzymatic extract of *Aspergillus awamori*	*Rhodosporidium toruloides*	Microbial oil	[43]
Wheat straw	Acid and enzymatic hydrolysis of hemicellulosic fraction	*Lipomyces starkeyi*	Microbial oil	[45]
Corncob	Nitric acid digestion followed by enzymatic hydrolysis to breakdown hemicellulosic fraction	*Rhodotorula glutinis*	Microbial oil	[44]
Sugar cane bagasse	Acid hydrolysis of hemicellulosic fraction and glycerol supplementation	*Rhodosporidium toruloides*	Microbial oil	[60]
Rapeseed meal	Hydrothermal pretreatment followed by enzymatic hydrolysis of hemicellulosic fraction	*Xanthophyllomyces dendrorhous*	Astaxanthin	[64]
Sugar beet pulp	Enzymatic hydrolysis of pectins	*Rhodotorula mucilaginosa*	Mixture of carotenoids	[65]
Cinnamon waste	Enzymatic hydrolysis of hemicellulosic fraction	*Rhodosporidium toruloides*	β-carotene	[46]
Juice store and cafeteria wastes	Hydrothermal extraction	*Rhodosporidium toruloides*	β-carotene	[66]
Paddy straw	Enzymatic hydrolysis of hemicellulosic fraction	*Rhodosporidium toruloides*	β-carotene	[66]
Salad oil manufacturing wastewater	Dilution of wastewater	*Candida utilis*	Single cell protein	[69]
Potato wastewater	Direct use and supplementation with glycerol	*Candida utilis*	Single cell protein	[70]
Animal fat waste	Enzymatic hydrolysis of fat waste	*Sporobolomyces pararoseus*	Ergosterol	[74]
Animal fat waste	Enzymatic hydrolysis of fat waste	*Cystofilobasidium macerans*	Ubiquinone	[76]

shown lower acidification and eutrophication potential than agro-industrial activities, which is evident by the utilization of fertilizers.[80] Nonetheless, the optimal development of process and adequate industrial scaling will generate more data to assess the impacts of the bioprocess.

9.6 CONCLUDING REMARKS

Agro-industrial waste is gaining great potential as an alternative renewable feedstock in bioprocess design, which has the advantage of reducing raw material costs and presents a solution for mitigating the environmental impact by disposal of wastes generated in the agro-industrial chain. The reviewed eco-friendly treatments aim to add sustainability to the circular bioeconomy chain, fulfilling waste conversion to high-value compounds. Following this, yeast biotechnology offers the feasibility of synthesizing lipids, carotenoids, or proteins employed in fuel, supplement, feed, and cosmetics industries. Therefore, the yeast cell factories seem a plausible and sustainable alternative to ameliorate the current disposal of agro-waste due to their valorization and utilization as raw materials. Moreover, the ecological footprint impact must be considered to ensure the sustainability of the developed process. In the overcoming years, research will focus on assessing the environmental impact categories, such as global warming potential, land utilization, and acidification, among the most important.

KEYWORDS

- **agro-industrial waste**
- **biorefinery**
- **high-value bioproducts**
- **sustainability**
- **yeast conversion**

REFERENCES

1. Torres-Valenzuela, L. S.; Ballesteros-Gómez, A.; Rubio, S. Green Solvents for the Extraction of High Added-Value Compounds from Agri-food Waste. *Food Eng. Rev.* **2020,** *12*, 83–100.

2. Li, X.; Mettu, S.; Martin, G. J. O.; Ashokkumar, M.; Lin, C. S. K. Ultrasonic Pretreatment of Food Waste to Accelerate Enzymatic Hydrolysis for Glucose Production. *Ultrason. Sonochem.* **2019,** *53*, 77–82.
3. Sadh, P. K.; Duhan, S.; Duhan, J. S. Agro-Industrial Wastes and Their Utilization Using Solid State Fermentation: A Review. *Bioresour. Bioprocess.* **2018,** *5*, 1–15.
4. Carmona-Cabello, M.; García, I. L.; Sáez-Bastante, J.; Pinzi, S.; Koutinas, A. A.; Dorado, M. P. Food Waste from Restaurant Sector–Characterization for Biorefinery Approach. *Bioresour. Technol.* **2020,** *301*, 1–11.
5. Cheng, S. Y.; Tan, X.; Show, P. L.; Rambabu, K.; Banat, F.; Veeramuthu, A.; Lau, B. F.; Ng, E. P.; Ling, T. C. Incorporating Biowaste into Circular Bioeconomy: A Critical Review of Current Trend and Scaling up Feasibility. *Environ. Technol. Innov.* **2020,** *19*, 1–16.
6. Kot, A. M.; Błażejak, S.; Kurcz, A.; Gientka, I.; Kieliszek, M. *Rhodotorula glutinis*—Potential Source of Lipids, Carotenoids, and Enzymes for Use in Industries. *Appl. Microbiol. Biotechnol.* **2016,** *100*, 6103–6117.
7. Nayak, A.; Bhushan, B. An Overview of the Recent Trends on the Waste Valorization Techniques for Food Wastes. *J. Environ. Manage.* **2019,** *233*, 352–370.
8. Brahim, M.; Checa Fernandez, B. L.; Regnier, O.; Boussetta, N.; Grimi, N.; Sarazin, C.; Husson, E.; Vorobiev, E.; Brosse, N. Impact of Ultrasounds and High Voltage Electrical Discharges on Physico-Chemical Properties of Rapeseed Straw's Lignin and Pulps. *Bioresour. Technol.* **2017,** *237*, 11–19.
9. Ravindran, R.; Sarangapani, C.; Jaiswal, S.; Lu, P.; Cullen, P. J.; Bourke, P.; Jaiswal, A. K. Improving Enzymatic Hydrolysis of Brewer Spent Grain with Nonthermal Plasma. *Bioresour. Technol.* **2019,** *282*, 520–524.
10. Abaide, E. R.; Ugalde, G.; Di Luccio, M.; Moreira, R. de F. P. M.; Tres, M. V.; Zabot, G. L.; Mazutti, M. A. Obtaining Fermentable Sugars and Bioproducts from Rice Husks by Subcritical Water Hydrolysis in a Semi-Continuous Mode. *Bioresour. Technol.* **2019,** *272*, 510–520.
11. Vargas Corredor, Y. A.; Peréz Pérez, L. I. Aprovechamiento de Residuos Agroindustriales en el Mejoramiento de la Calidad Del Ambiente. *Rev. Fac. Ciencias Básicas* **2018,** *1*, 59–72.
12. Gopinath, K. P.; Sankaranarayanan, A. R.; Nivedhitha, L. Platform Chemical Biorefinery and Agroindustrial Waste Management. *Platf. Chem. Biorefinery* **2016,** 379–391.
13. Johannesson, G. H.; Crolla, A.; Lauzon, J. D.; Gilroyed, B. H. Estimation of Biogas Co-production Potential from Liquid Dairy Manure, Dissolved Air Flotation Waste (DAF) and Dry Poultry Manure Using Biochemical Methane Potential (BMP) Assay. *Biocatal. Agric. Biotechnol.* **2020,** *25*, 1–10.
14. Meneses-Reyes, J. C.; Hernández-Eugenio, G.; Huber, D. H.; Balagurusamy, N.; Espinosa-Solares, T. Oil-Extracted *Chlorella vulgaris* Biomass and Glycerol Bioconversion to Methane via Continuous Anaerobic Co-digestion with Chicken Litter. *Renew. Energy* **2018,** *128*, 223–229.
15. Patricio, J.; Kalmykova, Y.; Rosado, L. A Method and Databases for Estimating Detailed Industrial Waste Generation at Different Scales–With Application to Biogas Industry Development. *J. Clean. Prod.* **2020,** *246*, 1–14.
16. Sagastume Gutiérrez, A.; Cabello Eras, J. J.; Hens, L.; Vandecasteele, C. The Energy Potential of Agriculture, Agroindustrial, Livestock, and Slaughterhouse Biomass Wastes Through Direct Combustion and Anaerobic Digestion. The Case of Colombia. *J. Clean. Prod.* **2020,** *269*, 1–17.

17. Alves, J. L. F.; da Silva, J. C. G.; Mumbach, G. D.; Domenico, M. Di; da Silva Filho, V. F.; de Sena, R. F.; Machado, R. A. F.; Marangoni, C. Insights into the Bioenergy Potential of Jackfruit Wastes Considering Their Physicochemical Properties, Bioenergy Indicators, Combustion Behaviors, and Emission Characteristics. *Renew. Energy* **2020,** *155*, 1328–1338.
18. Virmond, E.; De Sena, R. F.; Albrecht, W.; Althoff, C. A.; Moreira, R. F. P. M.; José, H. J. Characterisation of Agroindustrial Solid Residues as Biofuels and Potential Application in Thermochemical Processes. *Waste Manag.* **2012,** *32*, 1952–1961.
19. Gutiérrez, M. C.; Serrano, A.; Siles, J. A.; Chica, A. F.; Martín, M. A. Centralized Management of Sewage Sludge and Agro-Industrial Waste Through Co-composting. *J. Environ. Manage.* **2017,** *196*, 387–393.
20. Bao, Y.; Feng, Y.; Qiu, C.; Zhang, J.; Wang, Y.; Lin, X. Organic Matter- and Temperature-Driven Deterministic Assembly Processes Govern Bacterial Community Composition and Functionality During Manure Composting. *Waste Manag.* **2021,** *131*, 31–40.
21. Orzua, M. C.; Mussatto, S. I.; Contreras-Esquivel, J. C.; Rodriguez, R.; de la Garza, H.; Teixeira, J. A.; Aguilar, C. N. Exploitation of Agro Industrial Wastes as Immobilization Carrier for Solid-State Fermentation. *Ind. Crops Prod.* **2009,** *30*, 24–27.
22. Bilal, M.; Iqbal, H. M. N. Sustainable Bioconversion of Food Waste into High-Value Products by Immobilized Enzymes to Meet Bio-Economy Challenges and Opportunities–A Review. *Food Res. Int.* **2019,** *123*, 226–240.
23. Barcelos, M. C. S.; Ramos, C. L.; Kuddus, M.; Rodriguez-Couto, S.; Srivastava, N.; Ramteke, P. W.; Mishra, P. K.; Molina, G. Enzymatic Potential for the Valorization of Agro-industrial byproducts. *Biotechnol. Lett.* **2020,** *42*, 1799–1827.
24. Patel, A. K.; Singhania, R. R.; Pandey, A. Production, Purification, and Application of Microbial Enzymes. *Biotechnol. Microb. Enzym. Prod. Biocatal. Ind. Appl.* **2017,** 13–41.
25. Zanutto-Elgui, M. R.; Vieira, J. C. S.; Prado, D. Z.; Buzalaf, M. A. R.; Padilha, P. M.; Oliveira, D.; Fleuri, L. F. Production of Milk Peptides with Antimicrobial and Antioxidant Properties Through Fungal Proteases. *Food Chem.* **2019,** *278*, 823–831.
26. Subsamran, K.; Mahakhan, P.; Vichitphan, K.; Vichitphan, S.; Sawaengkaew, J. Potential Use of Vetiver Grass for Cellulolytic Enzyme Production and Bioethanol Production. *Biocatal. Agric. Biotechnol.* **2019,** *17*, 261–268.
27. Torres-Mayanga, P. C.; Azambuja, S. P. H.; Tyufekchiev, M.; Tompsett, G. A.; Timko, M. T.; Goldbeck, R.; Rostagno, M. A.; Forster-Carneiro, T. Subcritical Water Hydrolysis of Brewer's Spent Grains: Selective Production of Hemicellulosic Sugars (C-5 sugars). *J. Supercrit. Fluids* **2019,** *145*, 19–30.
28. Wang, Z.; He, X.; Yan, L.; Wang, J.; Hu, X.; Sun, Q.; Zhang, H. Enhancing Enzymatic Hydrolysis of Corn Stover by Twin-Screw Extrusion Pretreatment. *Ind. Crops Prod.* **2020,** *143*, 1–7.
29. Reis, C. L. B.; Silva, L. M. A. e.; Rodrigues, T. H. S.; Félix, A. K. N.; Santiago-Aguiar, R. S. de; Canuto, K. M.; Rocha, M. V. P. Pretreatment of Cashew Apple Bagasse Using Protic Ionic Liquids: Enhanced Enzymatic Hydrolysis. *Bioresour. Technol.* **2017,** *224*, 694–701.
30. Lachos-Perez, D.; Brown, A. B.; Mudhoo, A.; Martinez, J.; Timko, M. T.; Rostagno, M. A.; Forster-Carneiro, T. Applications of Subcritical and Supercritical Water Conditions for Extraction, Hydrolysis, Gasification, and Carbonization of Biomass: A Critical Review. *Biofuel Res. J.* **2017,** *4*, 611–626.

31. Vedovatto, F.; Ugalde, G.; Bonatto, C.; Bazoti, S. F.; Treichel, H.; Mazutti, M. A.; Zabot, G. L.; Tres, M. V. Subcritical Water Hydrolysis of Soybean Residues for Obtaining Fermentable Sugars. *J. Supercrit. Fluids* **2021,** *167*, 1–12.
32. Martínez, C. M.; Cantero, D. A.; Cocero, M. J. Production of Saccharides from Sugar Beet Pulp by Ultrafast Hydrolysis in Supercritical Water. *J. Clean. Prod.* **2018,** *204*, 888–895.
33. de Carvalho Silvello, M. A.; Martínez, J.; Goldbeck, R. Application of Supercritical CO2 Treatment Enhances Enzymatic Hydrolysis of Sugarcane Bagasse. *Bioenergy Res.* **2020,** *13*, 786–796.
34. El Kantar, S.; Boussetta, N.; Rajha, H. N.; Maroun, R. G.; Louka, N.; Vorobiev, E. High Voltage Electrical Discharges Combined with Enzymatic Hydrolysis for Extraction of Polyphenols and Fermentable Sugars from Orange Peels. *Food Res. Int.* **2018,** *107*, 755–762.
35. Ghosh, S.; Gillis, A.; Sheviryov, J.; Levkov, K.; Golberg, A. Towards Waste Meat Biorefinery: Extraction of Proteins from Waste Chicken Meat with Non-thermal Pulsed Electric Fields and Mechanical Pressing. *J. Clean. Prod.* **2019,** *208*, 220–231.
36. Wu, F.; Jin, Y.; Li, D.; Zhou, Y.; Guo, L.; Zhang, M.; Xu, X.; Yang, N. Electrofluid Hydrolysis Enhances the Production of Fermentable Sugars from Corncob via in/ Reverse-Phase Induced Voltage. *Bioresour. Technol.* **2017,** *234*, 158–166.
37. Prasertsung, I.; Aroonraj, K.; Kamwilaisak, K.; Saito, N.; Damrongsakkul, S. Production of Reducing Sugar from Cassava Starch Waste (CSW) Using Solution Plasma Process (SPP). *Carbohydr. Polym.* **2019,** *205*, 472–479.
38. Miranda, F. S.; Rabelo, S. C.; Pradella, J. G. C.; Carli, C. Di; Petraconi, G.; Maciel, H. S.; Pessoa, R. S.; Vieira, L. Plasma in-Liquid Using Non-contact Electrodes: A Method of Pretreatment to Enhance the Enzymatic Hydrolysis of Biomass. *Waste Biomass Valorization* **2020,** *11*, 4921–4931.
39. Hassan, S. S.; Ravindran, R.; Jaiswal, S.; Tiwari, B. K.; Williams, G. A.; Jaiswal, A. K. An Evaluation of Sonication Pretreatment for Enhancing Saccharification of Brewers' Spent Grain. *Waste Manag.* **2020,** *105*, 240–247.
40. Casabar, J. T.; Ramaraj, R.; Tipnee, S.; Unpaprom, Y. Enhancement of Hydrolysis with Trichoderma Harzianum for Bioethanol Production of Sonicated Pineapple Fruit Peel. *Fuel* **2020,** *279*, 1–7.
41. Zhang, Y.; Li, T.; Shen, Y.; Wang, L.; Zhang, H.; Qian, H.; Qi, X. Extrusion Followed by Ultrasound as a Chemical-Free Pretreatment Method to Enhance Enzymatic Hydrolysis of Rice Hull for Fermentable Sugars Production. *Ind. Crops Prod.* **2020,** *149*, 1–10.
42. Alizadeh, O.; Aliakbarlu, J. Effects of Ultrasound and Ohmic Heating Pretreatments on Hydrolysis, Antioxidant and Antibacterial Activities of whey Protein Concentrate and its Fractions. *Lwt* **2020,** *131*, 1–7.
43. Carmona-Cabello, M.; García, I. L.; Papadaki, A.; Tsouko, E.; Koutinas, A.; Dorado, M. P. Biodiesel Production Using Microbial Lipids Derived from Food Waste Discarded by Catering Services. *Bioresour. Technol.* **2021,** *323*, 1–9.
44. Rane, D. V.; Pawar, P. P.; Odaneth, A. A.; Lali, A. M. Microbial Oil Production by the Oleaginous Red Yeast, *Rhodotorula glutinis* NCIM 3168, Using Corncob Hydrolysate. *Biomass Convers. Biorefinery* **2021,** 1–11.
45. Liu, Z.; Feist, A. M.; Dragone, G.; Mussatto, S. I. Lipid and Carotenoid Production from Wheat Straw Hydrolysates by Different Oleaginous Yeasts. *J. Clean. Prod.* **2020,** *249*, 1–10.

46. Bertacchi, S.; Pagliari, S.; Cantù, C.; Bruni, I.; Labra, M.; Branduardi, P. Enzymatic Hydrolysate of Cinnamon Waste Material as Feedstock for the Microbial Production of Carotenoids. *Int. J. Environ. Res. Public Health* **2021,** *18*, 1–13.
47. Tsakona, S.; Papadaki, A.; Kopsahelis, N.; Kachrimanidou, V.; Papanikolaou, S.; Koutinas, A. Development of a Circular Oriented Bioprocess for Microbial Oil Production Using Diversified Mixed Confectionery Side-Streams. *Foods* **2019,** *8*, 1–17.
48. Papadaki, A.; Papapostolou, H.; Alexandri, M.; Kopsahelis, N.; Papanikolaou, S.; de Castro, A. M.; Freire, D. M. G.; Koutinas, A. A. Fumaric Acid Production Using Renewable Resources from Biodiesel and Cane Sugar Production Processes. *Environ. Sci. Pollut. Res.* **2018,** *25*, 35960–35970.
49. Panesar, R.; Kaur, S.; Panesar, P. S. Production of Microbial Pigments Utilizing Agro-Industrial Waste: A Review. *Curr. Opin. Food Sci.* **2015,** *1*, 70–76.
50. Lopes, F. C.; Ligabue-Braun, R. Agro-Industrial Residues: Eco-Friendly and Inexpensive Substrates for Microbial Pigments Production. *Front. Sustain. Food Syst.* **2021,** *5*, 1–16.
51. Rapoport, A.; Guzhova, I.; Bernetti, L.; Buzzini, P.; Kieliszek, M.; Kot, A. M. Carotenoids and Some Other Pigments from Fungi and Yeasts. *Metabolites* **2021,** *11*, 1–17.
52. Beltrán-Ramírez, F.; Orona-Tamayo, D.; Cornejo-Corona, I.; Luz Nicacio González-Cervantes, J.; de Jesús Esparza-Claudio, J.; Quintana-Rodríguez, E. In *Agro-Industrial Waste Revalorization: The Growing Biorefinery*, in *Biomass for Bioenergy-Recent Trends and Future Challenges*, 2019; pp 1–20.
53. Panesar, P. S.; Kennedy, J. F. Biotechnological Approaches for the Value Addition of Whey. *Crit. Rev. Biotechnol.* **2012,** *32*, 327–348.
54. Cristobal-Sarramian, A.; Atzmuller, D. Yeast as a Production Platform in Bioreflneries: Conversion of Agricultural Residues into Value-Added Products. *Agron. Res.* **2018,** *16*, 377–388.
55. Dias, C.; Santos, J.; Reis, A.; Lopes da Silva, T. Yeast and Microalgal Symbiotic Cultures Using Low-Cost Substrates for Lipid Production. *Bioresour. Technol. Rep.* **2019,** *7*, 1–15.
56. Zuccaro, G.; Pirozzi, D.; Yousuf, A. *Lignocellulosic Biomass to Biodiesel*, in *Lignocellulosic Biomass to Liquid Biofuels*, 2019; pp 127–167.
57. Singh, G.; Sinha, S.; Kumar, K. K.; Gaur, N. A.; Bandyopadhyay, K. K.; Paul, D. High Density Cultivation of Oleaginous Yeast Isolates in 'Mandi' Waste for Enhanced Lipid Production Using Sugarcane Molasses as Feed. *Fuel* **2020,** *276*, 1–11.
58. Lakshmidevi, R.; Ramakrishnan, B.; Ratha, S. K.; Bhaskar, S.; Chinnasamy, S. Valorisation of Molasses by Oleaginous Yeasts for Single Cell Oil (SCO) and Carotenoids Production. *Environ. Technol. Innov.* **2021,** *21*, 101281.
59. Gosalawit, C.; Imsoonthornruksa, S.; Gilroyed, B. H.; Mcnea, L.; Boontawan, A.; Ketudat-Cairns, M. The Potential of the Oleaginous Yeast *Rhodotorula paludigena* CM33 to Produce Biolipids. *J. Biotechnol.* **2021,** *329*, 56–64.
60. Bonturi, N.; Crucello, A.; Viana, A. J. C.; Miranda, E. A. Microbial Oil Production in Sugarcane Bagasse Hemicellulosic Hydrolysate Without Nutrient Supplementation by a *Rhodosporidium toruloides* Adapted Strain. *Process Biochem.* **2017,** *57*, 16–25.
61. Saini, R. K.; Keum, Y. S. Carotenoid Extraction Methods: A Review of Recent Developments. *Food Chem.* **2018,** *240*, 90–103.
62. Nabi, F.; Arain, M. A.; Rajput, N.; Alagawany, M.; Soomro, J.; Umer, M.; Soomro, F.; Wang, Z.; Ye, R.; Liu, J. Health Benefits of Carotenoids and Potential Application in Poultry Industry: A Review. *J. Anim. Physiol. Anim. Nutr. (Berl).* **2020,** *104*, 1809–1818.

63. Saini, R. K.; Keum, Y. S. Microbial Platforms to Produce Commercially Vital Carotenoids at Industrial Scale: An Updated Review of Critical Issues. *J. Ind. Microbiol. Biotechnol.* **2019,** *46*, 657–674.
64. Harith, Z. T.; Charalampopoulos, D.; Chatzifragkou, A. Rapeseed Meal Hydrolysate as Substrate for Microbial Astaxanthin Production. *Biochem. Eng. J.* **2019,** *151*, 1–12.
65. Martins, L. C.; Palma, M.; Angelov, A.; Nevoigt, E.; Liebl, W.; Sá-Correia, I. Complete Utilization of the Major Carbon Sources Present in Sugar Beet Pulp Hydrolysates by the Oleaginous Red Yeasts Rhodotorula Toruloides and R. mucilaginosa. *J. Fungi* **2021,** *7*, 1–21.
66. Sinha, S.; Singh, G.; Arora, A.; Paul, D. Carotenoid Production by Red Yeast Isolates Grown in Agricultural and "Mandi" Waste. *Waste Biomass Valorization* **2021,** *12*, 3939–3949.
67. Tibebu Bogale, T. Microbial Protein Production from Agro-industrial Wastes as Food and Feed. *Am. J. Life Sci.* **2020,** *8*, 121–126.
68. Reihani, S. F. S.; Khosravi-Darani, K. Influencing Factors on Single-Cell Protein Production by Submerged Fermentation: A Review. *Electron. J. Biotechnol.* **2019,** *37*, 34–40.
69. Zheng, S.; Yang, M.; Yang, Z. Biomass Production of Yeast Isolate from Salad Oil Manufacturing Wastewater. *Bioresour. Technol.* **2005,** *96*, 1183–1187.
70. Kurcz, A.; Błażejak, S.; Kot, A. M.; Bzducha-Wróbel, A.; Kieliszek, M. Application of Industrial Wastes for the Production of Microbial Single-Cell Protein by Fodder Yeast *Candida utilis*. *Waste Biomass Valorization* **2018,** *9*, 57–64.
71. Tapingkae, W.; Panyachai, K.; Yachai, M.; Doan, H. V. Effects of Dietary Red Yeast (*Sporidiobolus pararoseus*) on Production Performance and Egg Quality of Laying Hens. *J. Anim. Physiol. Anim. Nutr. (Berl)* **2018,** *102*, e337–e344.
72. Liu, J. F.; Xia, J. J.; Nie, K. L.; Wang, F.; Deng, L. Outline of the Biosynthesis and Regulation of Ergosterol in Yeast. *World J. Microbiol. Biotechnol.* **2019,** *35*, 1–8.
73. Marova, I.; Haronikova, A.; Petrik, S.; Dvorakova, T.; Breierova, E. Production of Enriched Biomass by Red Yeasts of *Sporobolomyces* sp. Grown on Waste Substrates. *J. Microbiol. Biotechnol. Food Sci.* **2016,** *2016*, 534–551.
74. Szotkowski, M.; Byrtusova, D.; Haronikova, A.; Vysoka, M.; Rapta, M.; Shapaval, V.; Marova, I. Study of Metabolic Adaptation of Red Yeasts to Waste Animal Fat Substrate. *Microorganisms* **2019,** *7*, 1–28.
75. Shukla, S.; Dubey, K. K. CoQ10 a Super-Vitamin: Review on Application and Biosynthesis. *3 Biotech* **2018,** *8*, 1–11.
76. Marova, I.; Szotkowski, M.; Vanek, M.; Rapta, M.; Byrtusova, D.; Mikheichyk, N.; Haronikova, A.; Certik, M.; Shapaval, V. Utilization of Animal Fat Waste as Carbon Source by Carotenogenic Yeasts – A Screening Study. *EuroBiotech J.* **2017,** *1*, 310–318.
77. Martins, L. C.; Monteiro, C. C.; Semedo, P. M.; Sá-Correia, I. Valorisation of Pectin-Rich Agro-Industrial Residues by Yeasts: Potential and Challenges. *Appl. Microbiol. Biotechnol.* **2020,** *104*, 6527–6547.
78. Parsons, S.; Abeln, F.; McManus, M. C.; Chuck, C. J. Techno-Economic Analysis (TEA) of Microbial Oil Production from Waste Resources as Part of a Biorefinery Concept: Assessment at Multiple Scales Under Uncertainty. *J. Chem. Technol. Biotechnol.* **2019,** *94*, 701–711.
79. Braunwald, T.; French, W. T.; Claupein, W.; Graeff-Hönninger, S. Economic Assessment of Microbial Biodiesel Production Using Heterotrophic Yeasts. *Int. J. Green Energy* **2016,** *13*, 274–282.

80. Bonatsos, N.; Marazioti, C.; Moutousidi, E.; Anagnostou, A.; Koutinas, A.; Kookos, I. K. Techno-Economic Analysis and Life Cycle Assessment of Heterotrophic Yeast-Derived Single Cell Oil Production Process. *Fuel* **2020,** *264*, 1–8.
81. Ajmal, M.; Shi, A.; Awais, M.; Mengqi, Z.; Zihao, X.; Shabbir, A.; Faheem, M.; Wei, W.; Ye, L. Ultra-High Temperature Aerobic Fermentation Pretreatment Composting: Parameters Optimization, Mechanisms and Compost Quality Assessment. *J. Environ. Chem. Eng.* **2021,** *9*, 1–11.
82. Muhammad, M. B.; Chandra, R. Enhancing Biogas and Methane Production from Leaf Litter of Neem by Co-digestion With Vegetable Waste: Focus on the Effect of Tannin. *Biomass Bioenergy* **2021,** *147*, 1–7.
83. Duarte, M. S.; Sinisgalli, E.; Cavaleiro, A. J.; Bertin, L.; Alves, M. M.; Pereira, M. A. Intensification of Methane Production from Waste Frying Oil in a Biogas-Lift Bioreactor. *Renew. Energy* **2021,** *168*, 1141–1148.
84. Valdés, C. F.; Marrugo, G. P.; Chejne, F.; Marin-Jaramillo, A.; Franco-Ocampo, J.; Norena-Marin, L. Co-Gasification and Co-Combustion of Industrial Solid Waste Mixtures and Their Implications on Environmental Emissions, as an Alternative Management. *Waste Manag.* **2020,** *101*, 54–65.

CHAPTER 10

Use of Peel and Seed of Tropical Fruits Processed Through Flash Vacuum Expansion

LETICIA X. LÓPEZ-MARTÍNEZ[1], EBER A. QUINTANA-OBREGÓN[1], AARÓN F. GONZÁLEZ-CÓRDOVA[2], ADRIÁN HERNÁNDEZ-MENDOZA[2], BELINDA VALLEJO-CORDOBA[2], and MANUEL A. VARGAS-ORTIZ[3]

[1]CONACYT-CIAD (Centro de Investigación en Alimentación y Desarrollo A.C), Coordinación de Alimentos de Origen Vegetal, Hermosillo, Sonora, Mexico

[2]Centro de Investigación en Alimentación y Desarrollo, Coordinación de Tecnología de Alimentos de Origen Animal. Laboratorio de Calidad, Autenticidad y Trazabilidad de los Alimentos/Química y Biotecnología de Productos Lácteos, Hermosillo, Sonora, Mexico

[3]CONACYT-CIAD (Centro de Investigación en Alimentación y Desarrollo A.C), Coordinación de Tecnología de Alimentos de Origen Animal. Laboratorio de Calidad, Autenticidad y Trazabilidad de los Alimentos/Química y Biotecnología de Productos Lácteos, Carretera Gustavo Enrique Astiazarán Rosas, Hermosillo, Sonora, Mexico

ABSTRACT

The flash vacuum expansion (FVE) process is a nontraditional technology that generates the disintegration of plant tissues through pressure and temperature differences that promote a change of phase in the constitutive water of plant tissues. Due to its application principle, the ultrafast expansion technology allows the processing of whole fruits without removing the peel or the seed, currently considered as byproducts or waste, allowing full

Food Byproducts Management and their Utilization. Ricardo Gómez-García, Ana A. Vilas-Boas, Débora A. Campos, Maria Manuela Pintado, & Cristóbal Noé Aguilar (Eds.)

use of the fruits so that it is possible to take advantage of the compounds of interest mainly pigments, antioxidants, fiber, vitamins, essential oils, and aromatic compounds that still are present in both byproducts, which could be incorporated into a final food product. It has been reported that the incorporation of peel and seed components to the purees give them physical, biochemical, and sensory properties very different from the purees generated by traditional technologies. On the other hand, there are reports of the extraction of essential oils and aromatic compounds from the peel, which can be used in different industries such as food, cosmetics, or pharmaceuticals. In this chapter, an overview of the integral processing of tropical fruits through the flash vacuum expansion process is presented and also the advantages and disadvantages of incorporating peel and seed components into different final products (puree or juice of the processed fruit).

10.1 INTRODUCTION

According to the FAO,[1] mango, pineapple, papaya, and avocado are the fruits with the highest production worldwide; on the other hand, the main minor tropical fruits (referred to as being less traditional) are guava, longan, lychee, durian, passion fruit, and rambutan which are gaining importance for their contribution to a healthy diet.[2] Mango is one of the fruits most appreciated by the population worldwide. Its industrial processing generates peel and seed residues that, depending on the variety, represent between 25–40% of the weight of each processed fruit.[3] In the case of passion fruit, waste from processing is very high, representing between 45 and 50% of each processed fruit,[4] since in this case, the seed and the peel represent the most significant part of the fruit from which only the juice contained in the arils is used. However, it is a less traditional fruit than mango and avocado; therefore, residues from the industrialization of passion fruit generate up to 500,000 tons per year in Brazil alone.[5] On the other hand, guava is a tropical fruit with fragile skin, where about 30% of each processed fruit is lost.[6] The grape is used mainly for the industrial production of wine, juices, or concentrates and in smaller quantities to be consumed fresh as table grapes. In the case of grapes, between 13 and 20% of each processed fruit becomes waste, mainly composed of the peel and seed.[7] The grape is cultivated in less volume than other fruits; however, the wine industry's commercial value is of economic importance. Avocado is mainly consumed fresh, although some avocado and oil-based products have been developed; this processing generates waste between 21 and 30%[8] of each processed fruit. Citrus fruits include a great variety of species, of which the orange, lemon, grapefruit, and tangerines stand out for their production volume; about 33% of the world's citrus production is processed, generating the peel and seed as waste, which depending on the species can represent up to 50% of each fresh fruit.[9] The high volumes of fruit waste have meant a phytosanitary problem in the places where they are processed because the accumulation of waste serves as a host for pests and fungi that naturally attack the fruit.

On the other hand, various investigations have shown that the peel of the fruits contains attractive proportions of bioactive compounds such as fiber, pigments, vitamins, phenolic compounds, and some enzymes. As mentioned above, mango, avocado, grape, guava, passion fruit, and citrus fruits represent a great economic value, either due to their volume of production and consumption or because of the added value generated by their industrialization. In all these fruits, bioactive compounds have been found in peels and seeds, so it is attractive to develop processes that can promote the migration of these compounds to the final products or even look for comprehensive processing where the peel becomes part of the fruit puree, to take advantage of the benefits that this could represent. A technology that could answer this question is the flash vacuum expansion (FVE) process, which is based on subjecting the plant material to variations in temperature and pressure that will promote phase changes in the part of the constitutive water of plant tissues, generating cellular rupture and therefore the disintegration of the tissues. Because the design of the FVE equipment is independent of the morphology of the fruits (which allows it to be applied to all fruits), several investigations have focused on evaluating the potential use of FVE for the processing of tropical fruits, seeking to take advantage of the qualities of peel and seed. Hence, this chapter shows the application of FVE for the processing of fruits and how the incorporation of peels and seeds affect the properties of the final products.

10.2 COMPOUNDS OF INTEREST IN SOME TROPICAL FRUITS

10.2.1 COMPOUNDS OF INTEREST IN THE PEEL AND SEED OF SOME TROPICAL FRUITS

Evidence shows that tropical fruits can modulate oxidative stress and its effects due to their content of bioactive compounds such as phenolic acids, flavonoids, carotenoids, fatty acids, sterols, and dietary fiber, which have been isolated from pulps, seeds, and peels. The chemical compositions of avocado, guava, and passion fruits seeds and peels are shown in Table 10.1.

TABLE 10.1 Avocado, Guava, and Passion Fruit Are Rich Sources of Nutritional Compounds.

	Avocado		**Guava**		**Passion fruit**	
	Seed[a]	**Peel**[b]	**Seed**[c]	**Peel**[d]	**Seed**[e]	**Peel**[f]
*Protein	0.14-3.4	0.17–8.18	11.19	6.0	13.07c	3.40
*Total lipids	3.2–14.7	2.2–9.14	13.93	1.6	12.31c	4.20
*Carbohydrate	42.5–79.5	43.6-62.03	3.08	58.2	71.7	85.78
*Fiber total	47.8[g]	53.3[g]	63.94	15.2	65.60c	61.16
*Soluble	11.08[g]	NR	0.39	NR	NR	NR
*Insoluble	36.93[g]	NR	63.55	NR	NR	NR
Total phenolics (mg GAE/100 g	75.1–5920[h]	94–17280[h]	27.15		26.3	106.1
Total carotenoids (µg ß-carotene/100 g)			670		660	918.41

**Expressed as g/100 g of dw*

Source: [a]Daiuto et al. [10], [b]Uchôa-Thomaz et al. [11], [c]Bertagnolli et al. [12]2014; [d]dos Reis et al. [13] [e]Barbosa-Martín et al.[14], [f]Salazar-López et al.[15]

Phytochemicals are secondary plant metabolites, which protect them against various biotic and abiotic stressors. Some of them play a role in human health, including phenolic compounds, carotenoids, sterols, and tocopherols, which are associated with beneficial effects on improvements in cardiovascular disease risk factors, inhibition of inflammation, reducing oxidative stress, and preventing or delaying oxidation by scavenging free radicals; among others, also the presence of fatty acid plays an essential key in human health.

10.2.2 COMPOUNDS OF INTEREST IN AVOCADO PEEL AND SEED

Avocado (*Persea americana* Mill.) belongs to the Lauraceae family and the genus *Persea*, with more than 150 known species.[8] Avocado is a tropical fruit native to Central America and Mexico. An average avocado fruit weighs approximately 150–400 g of which seed and peel weigh about 16 and 11%, respectively, although these proportions depend on the cultivars. The seed and peel of "Hass" avocado are about 15 and 14%, respectively. After the processing, peels, seeds, and exhausted pulp are discarded as waste, which causes a large amount of solid residues which represents 30% of the fruit.[16] In avocado seed, conjugated derivatives of caffeic acid such as trans-5-O-caffeoylquinic acid (1.63–5.74 mg/g dw), 1-caffeoylquinic acid (1.12–2.44 mg/g dw), 4-caffeoylquinic acid (0.067–0.104 mg/g dw), and 3-O-caffeoylquinic acid (0.054–0.055 mg/g dw) are the main phenolic compounds reported, along with one conjugated derivative of coumaric acid, 3-O-p-coumaroylquinic acid (0.081–0.36 mg/g dw). In addition, hydroxytyrosol glucoside (0.389 mg/g dw) and vanillic acid tyrosyl glucoside (2.36 mg/g dw) were also reported.[8] Studies from Kosinska et al.[17] showed the presence of different procyanidins, such as procyanidin trimer A, procyanidin trimer A (II), and catechin/epicatechin gallate in the avocado seed. On the other hand, seeds of guava have shown the presence of vanillic acid (3.9 mg/g oil), 4-hydroxybenzaldehyde (0.2 mg/g oil), vanillin 9.6 (mg/g oil), syringaldehyde 1.2 (mg/g oil), coniferylaldehyde 1.5 (mg/g oil), sinapaldehyde 0.3 (mg/g oil), cinnamic acid 2.4 (mg/g oil), and cinnamaldehyde 9.4 (mg/g oil).[18] Furthermore, the main phenolic compounds reported in passion fruit seeds are caffeic acid 69 mg/100 g, coumaric acid 73 mg/100 g, and salicylic acid 516 mg/100 g.[18]

In avocado peel, 5-O-caffeoylquinic acid (0.077–0.082 mg/g dw) and chlorogenic acid (1.59–1.9 mg/g dw) were the most common phenolic compounds reported.[16] Quercetin 3-O-galactoside, quercetin-3,4'-diglucoside, quercetin-3-O-arabinosyl glucoside, quercetin 3-O-rutinoside, quercetin-3-O-arabinoside, quercetin-3-O-glucoside, and quercetin derivatives (I, II, and III) were also reported by Tremocoldi et al.[19] In the analysis of lipids of hexane extract of avocado seeds, Leite et al.[20] found linoleic acid (27.6%), oleic acid (24.1%), palmitic acid (21.3%), palmitoleic acid (1.6%), and stearic acid (2.2%). Acetogenins are polyhydroxylated fatty alcohol derivatives with a chemical structure containing a long-chain fatty acid with a terminal γ-lactone although are characteristic for the Annonaceae family but have also been detected in the Lauraceae family to which avocado belongs.[21] The main acetogenins reported in the seeds were AcO-avocadene, persenone-A, AcO-avocadyne, persin, and persenone-B.[22]

Avocado byproducts contain other minor components such as alkaloids including scopolamine 0.24 mg/100 g dw), atropine (0.46 mg/100 g dw), solanin (0.04 mg/100 g dw), solasodine (0.08 mg/100 g dw), hyoscyamine (0.6 mg/100 g dw), norhyoscyamine (0.04 mg/100 g dw), and solanidine (0.041 mg/100 g dw).[23] Campesterol, stigmasterol, and ß-sitosterol have been reported by Alkhalf et al.,[24] while Barrera-López and Arrubla-Velez[25] described the presence of stigmasterol (19.17 μg/g dw) and 5α-colestane (49.77 μg/g dw) in seeds of avocado.

10.2.3 COMPOUNDS OF INTEREST IN GUAVA PEEL AND SEED

The guava plant (*Psidium guajava* L.) belongs to the Myrtaceae family. *Psidium guajava* is the most important fruit of the genus Psidium, containing around 150 species. Guava is thought to have originated from southern Mexico or through Central America and is widely known and cultivated throughout the world.[26] The seeds of guava are round and pale yellowish-brown in color, and are in the central pulp of the fruits, making up between 6 and 12% of the total weight of the fruit.[27] The fatty acid lipid profile of the guava seed showed a predominance of unsaturated fatty acids, mainly linoleic acid (77.35%) and oleic acid (9.42%), followed by palmitic acid (8.0%), stearic acid (4.48%), linolenic acid (1.2%), gondoic acid (1.4%), arachidic acid (1.2%), myristic acid (1.0%), behenic acid (1.0%), lauric acid (0.7%), and heptadecanoic acid (0.7%). Campesterol 23.9 mg/g oil, *β*-sitosterol from 437.6–1048.9 mg/g oil, stigmasterol 319.139 mg/g oil, and stigmastanol 26.6 mg/g oil have been reported in seeds oils of guava.[11,18] In the same study were reported the values of tocopherols including *a*-tocopherol from 4.6 mg/100, *β*-tocopherol 0.3 mg/100, *δ*-tocopherol 0.5 mg/100 *γ*-tocopherol from 82.6 mg/100 g.

10.2.4 COMPOUNDS OF INTEREST IN PASSION FRUIT PEEL AND SEED

Passion fruit is native to Brazil, whose production is mainly intended to produce juices and beverages. At the same time, its peel is used to make flour and as a functional food ingredient, which is the major component (approximately 64%), followed by its pulp and seeds.[28] Some of its most representative bioactive compounds include vitamin E and C and some carotenoids like lycopene and β-carotene.[13,29] The passion fruit contains fatty acids including myristic acid 0.119%, palmitic acid 13.2%, palmitoleic acid 0.26%, stearic acid 3.69%, oleic acid 21.7%, linoleic acid 57.9%, linolenic acid 0.25%, arachidonic acid 0.21%, behenic 1.1% acid, and erucic acid 1.3%.[30]

For oil from passion fruit seed campesterol 27.7 mg/100 g oil, *β*-sitosterol 138.6 mg/100 g oil, stigmasterol 78.1 mg/100 g oil, stigmastanol 30.2 mg/100 g oil were described by da Silva and Jorge[18] as well as tocopherols such as

δ-tocopherol 1739 mg/100 g and γ-tocopherol 1072 mg/100 g in seeds of passion fruit.

10.2.5 COMPOUNDS OF INTEREST IN MANGO PEEL

It is well known that mango (*Mangifera indica* L.) is one of the fruits with the highest demand globally and should always be available in most markets worldwide. The flavor, color, odor, and texture of its pulp make it very attractive for taste. Furthermore, some compounds have even been formulated that imitate the flavor and smell of mango as additives for various food products, cosmetics, pharmaceuticals, decorations, and others. Mango processing operations generate huge quantities of peel, which has been reported as rich in phenolic compounds, dietary fiber, pigments, antioxidant compounds, aromatic compounds, and others. Mango peel from different varieties could become a biomass of high commercial value if handled correctly and if alternatives to seize the high quantity of compounds were developed. Many phenolic compounds are found in the peel; 98 of these compounds have been reported in extracts from different varieties. In aqueous and alcoholic mango peel extracts, the antioxidant capacity evaluated by DPPH, FRAP, ABTS, and ORAC assays has been mainly attributed to phenolic compounds.[31,32] Mango peel has also been labeled as a potential source of carotenoids, vitamins C and E, dietary fiber, and other bioactive compounds; thus, it is recommended to find processing alternative to seize these benefits, which aims to diminish the peel residue generation from the mango processing industry.[33] The peel might contain between 19 and 23% soluble fiber and 9–16% insoluble fiber depending on the mango variety. Considering that mango peel could represent between 15 and 20% of the overall fruit weight, the peel is a good source of fiber that could be exploited to develop functional foods.[32,34]

10.2.6 COMPOUNDS OF INTEREST IN CITRUS PEEL

Due to their characteristics, the essential oils from citrus fruit peels are of great interest for diverse industries, including food, pharmaceutical, cosmetics, and decorative. Although all citrus fruits are potential sources of essential oils, traditionally and by sheer production volume, the primary species used for the extraction of essential oils are sweet orange (*Citrus sinensis*), mandarin (*Citrus reticulata*), sour orange (*Citrus aurantium*), lemon (*Citrus limon*), and pomelo (*Citrus grandis*). The extraction yield varies between different species, ranging from 0.49 up to 4.62% (w/w).[35] The citrus fruits peels contain interesting quantities of diverse functional compounds; the proportion of each compound varies in function of the species from which it is extracted, and this variability contributes to defining the aroma and flavor of each species. In the essential oils of the main citrus fruits evaluated, almost 70 compounds have been identified, mainly monoterpenes, oxygenated derivatives like aldehydes, alcohols, and esters, the majority of these being volatile compounds (approximately 95%), although the presence of flavonoids and carotenoids have also been detected. In all the evaluated cases, limonene and β-pinene are the major compounds, independently of the species.[36–39] The peel of citrus fruits has two well-defined sections, the interior layer of white color and the exterior layer that acquires a coloration characteristic to each species. In these two layers, the idioblasts are widely distributed, which are secreting cells specialized in producing and storing essential oils. However, the accumulation of essential oils can also occur in cavities and intracellular ducts.[39,40]

10.2.7 COMPOUNDS OF INTEREST IN GRAPE PEEL

The peels of grapes are characterized by a high content of phenolic compounds, which have a direct role in diverse quality parameters of products derived from this fruit. Thus, the extraction efficiency of these compounds is a critical parameter in the grape processing industry. To guarantee an adequate extraction, it is necessary to consider the location of the phenolic compounds in grape peel cells; also, the variety and ripeness can affect the proportion of phenolic compounds. It has been proved that the phenolic compound profile of grape-based products differs from those of the peel from freshly harvested grapes, while the phenolic profile of ground grape peels is similar to a whole freshly harvested grape. This can be due to the phenolic compounds in grape peel cells are mostly located in two different cellular structures, and interact in different ways in each structure. First, we will mention that the phenolic compounds are strongly linked to the cell wall polysaccharides. Therefore, these compounds are not extractable by physical means, so they are not transferred to grape-based products.

For this reason, some enzyme-assisted extraction systems or cell wall polysaccharide degrading microorganisms have been implemented to release these phenolic compounds. Secondly, we will mention those compounds stored in vacuoles, which are easily extracted by physical means, and widely contribute to the sensory properties of products like wine. Also, they are responsible for the antioxidant capacity attributed to grape juices and their derivatives. The main phenolic compounds found in grape peels are gallic acid, epicatechin, cyanidin-3-glucoside, catechin gallate, ferulic acid, rutin, and resveratrol, as well as flavonoids; all of them have bioactive properties reported.[41–43]

10.3 PRINCIPLES AND DESCRIPTION OF THE FLASH VACUUM EXPANSION PROCESS

Water is the only compound that can exist naturally in the three aggregation states under normal ambient conditions of the geographical location where it is found. This is due to the molecule configuration which possesses diverse qualities that might be considered abnormal, considering the behavior of other materials. As it is known, the aggregation state of pure substances, and specifically water, is ruled by factors like pressure and temperature. For example, under average conditions of 14°C and 101.3 kPa, water is liquid; however, it can boil within a wide range of temperatures from 0–347°C because of pressure.[44] On the other hand, it is known that pure substances like water, pressure, and temperature modify the volume. Therefore, in pressures lower than 10 kPa, the volume of water is inversely proportional to pressure and proportional to the substance's temperature. In the particular case of water, when it changes from liquid to gaseous state, its volume increases approximately by a ratio of 1:1000 requiring a higher volume for containment. Therefore, this could lead to a loss of integrity in the container if it is not resistant.

In most plant tissues, water is accumulated in the central vacuoles, where enzymes, pigments, anthocyanins, betalains, carotenoids, and sugars are also stored.[45] In addition, vacuoles can act as repositories for toxic compounds generated by cellular metabolism or xenobiotic compounds that result from the interaction with the environment. Because of this accumulation, vacuoles occupy a significant portion of the volume in the plant cells and have a role in multiple physiological processes. Besides maintaining the cell turgor, the principal function of this organelle is helping in the cell growth by storing water, which promotes osmotic potential and reduces the requirement of the cytosol for increasing the cellular size.[46,47] The flash vacuum expansion process (FVE) bases its operation principles on the information described beforehand, with moderate temperature increments and a slight pressure change, to promote the change from liquid to gas in the water constituting plant tissues in the materials where it is applied. Thus, it is possible that a significant portion of the water contained in the vacuoles expands and lead to cellular burst, promoting disaggregation of tissues and, by consequence, the release of the compounds stored in the vacuole. The FVE process has two well-defined stages. First, the plant material is placed in a chamber at atmosphere pressure (101.3 kPa) and exposed to a stream of water vapor, increasing its temperature until it reaches 60 and 90°C for a time ranging from 1 to 5 min. At this moment, the second stage starts, when the plant material passes quickly into a chamber with vacuum pressure (2–5 kPa). The pressure–temperature conditions promote the phase change (liquid to gas) of a portion of water constituting the plant tissue, which represents an increment of volume and, therefore cells burst, resulting in a disaggregated tissue, similar to a puree.[48] Pieribattesti et al.[49] were the first ones to apply this technology to obtain juices and aromatic compounds from vegetable substrates. Later, Cogat et al.[50] utilized FVE in search of other applications like cooking and distillation. The FVE process allows the possibility to choose three paths for the products to obtain. The first one is the most utilized and is aimed to obtain disaggregated plant material in the form of paste or pure, which serves as raw material for juice, puree, or concentrate production, according to the refinement degree of the FVE product. The second path is to obtain specific products that require a purification process and can be utilized to obtain pectins, pigments, or phenolic compounds. It is aimed at recovering aromatic compounds or essential oils that are dragged by water evaporation or well, released during tissue disaggregation in the second stage of the FVE process.[51–54]

10.4 INTEGRAL TROPICAL FRUIT PROCESSING BY FLASH VACUUM EXPANSION

10.4.1 MANGO PROCESSING BY FLASH VACUUM EXPANSION

The mango fruit is immensely appreciated for its striking color and pleasant flavor; however, it is also appreciated for its unique odor, which can be perceived and differentiated from other fruits. The sensory perception of consumers can initiate by the smell, even without seeing or touching the product; in the case of fruits and their derivatives, the aroma is a decisive factor on purchase intent, so keeping this attribute of vital importance. On the other hand, fruit aromas are widely used in foods and cosmetics, drugs, cleaning products, and decorative products. Higher concentrations and diversity of volatile aromatic compounds and monoterpenes are found in the mango peel, higher than in mango pulp. In the ripe mango fruit, monoterpenes and sesquiterpenes are the most abundant in proportion, followed by esters, aldehydes, aromatic compounds, alcohols, norisoprenoids, and lactone. However, it has been reported that four sesquiterpenes are found exclusively in mango peel, which also contains the highest quantity of alkanes and aldehydes.[55] The mango peel is a biological matrix of high value because of its high content of aromatic compounds, even more than mango pulp. The aroma of mango is generated mainly in the peel, with some green, fruity, floral, and resinous notes. Many of the aromatic compounds in mango peel could be extracted and utilized for incorporation into food products or cosmetics, even use the peel as an ingredient in food product formulation.[56]

When FVE was evaluated for mango processing, it was found that the traditional process produced mango puree with a yield of 60%, while the FVE process had a yield of 59%. This difference could be since FVE promotes the evaporation of approximately 10% of the total weight of the fruit, affecting yield. However, the recovered condensates from the FVE process contained aromatic compounds present in both mango peel and pulp, and by reincorporating them into the puree, the FVE yield increased to 65%.[57] It has been observed that for the "Amelie" mango variety, the mesocarp cells were the most susceptible to FVE processing; the peel was modified to a minor degree, while the seed was not affected due to its high lignin content. The physicochemical parameters like pH and titratable acidity of the puree produced by FVE were similar to conventional methods. The most affected parameter by tissue disaggregation was consistency and viscosity; both increased significantly. It is possible that during the expansion stage of the FVE process, the reduction of oxygen avoids the browning of mango puree. However, it is more feasible that this effect is due to the inactivation of enzymes involved in browning; in consequence, mango puree obtained by FVE has a very intense and bright yellow color.[58]

Moreover, when the FVE process to produce "Manila" mango puree was evaluated, it was observed that the concentration of phenolic compounds increased. The antioxidant capacity evaluated by DPPH and FRAP also increased between 38.5 and 68%, compared to puree produced by traditional methods. The increment in phenolic compounds and antioxidant capacity can be related to a higher release of compounds contained in mango peel and seed.[59] In purees from mangoes with yellow peel, such as "Manila," a more intense color was observed in purees produced by FVE, possibly because of the incorporation of pigments from the peel.

10.4.2 GRAPE PROCESSING BY FLASH VACUUM EXPANSION

The phenolic compounds are a group of chemicals responsible for most of sensory properties of wines, mainly regarding color and astringency. The anthocyanins influence the color of wines in all the grape varieties. Monoglycosides of delphinidin, cyanidin, petunidin, peonidin, malvidin, and their acylated derivatives have been identified, and the proportion of each one varies depending on the grape variety. The anthocyanins are found in the grape peel and are transferred to the must and wine during maceration.[41,60,61] However, it has been reported that during enzymatic maceration of grapes, phenolic compounds from the seed can also be extracted and can be present in the wine.[62]

In the production of red wine, the main aspects to consider are the grape harvest, the maceration of fruits, and alcoholic fermentation since all these stages affect the level of antioxidants in wine.[63] The presence of anthocyanins and proanthocyanidins is important for the quality of wines.[64] Wine consumers have also increased their interest in food products with health-promoting benefits, which has driven wine producers to seek technological processes that can increase the level of antioxidants in wine and guarantee a higher acceptance from a healthy viewpoint.[65] For this reason, diverse technologies to increase the yield and content of antioxidants in wine have been evaluated, mainly by improving the maceration process. Among these technologies, we can mention enzymatic treatments, irradiation, pulsed electric field, or microwave-assisted maceration. However, these processes can also require significant time and costs. The flash vacuum expansion process is a technology that can reduce maceration times and improve the yield and phytochemical content of the wine.[63] In recent years, FVE has been applied and developed mainly for the wine industry because it allows the extraction of more polyphenols from grapes, and it also reduces maceration time and produces wines with more intense colors.[66] The FVE process (also known as flash extraction) has existed for more than 20 years and is utilized mainly in Europe and South America (Brazil). However, the first commercial-scale FVE system was installed in 2009 in the USA. During its first years in Europe, the FVE technology was used to process lower quality grapes due to environmental problems during the harvest. The FVE process can reduce undesirable attributes in low quality grapes (reducing pyrazine and bad odors from fungi). Besides, it allows the processing of grapes fruits with an incomplete ripening. However, the utilization of the technology grew and expanded gradually, and nowadays, quite a few wine producers consider that it can be applied to process all kinds of grapes.[67] As described previously, the FVE process uses vacuum pressures, differing from the traditional thermovinification. Besides, the vacuum system promotes the formation of condensates, where volatile compounds considered otherwise undesirable are recovered. In the case of grapes, the FVE process has a short duration thermal stage (maximum 4 min) where grapes are heated (>80°C) and then are submitted to high vacuum. The pressure difference induces instantaneous vaporization of a portion of constitutive water in the fruit and, consequently, an almost immediate cooling of the macerated fruit. The FVE process induces the mechanical disaggregation of grape cells, especially from the peel, favoring phenolic compounds' extraction. This can help with the formulation of more differentiated wines, for example, with more intense fruity flavors and colors.[66]

Experiments on different grape varieties, applying FVE, have been realized, and the content of phenolic compounds (anthocyanins, flavonols, hydroxycinnamic acids, and flavonoids including catechin and epicatechin monomers, and proanthocyanidins) has been evaluated. In each stage of the wine production process, comparisons were made with traditional methods. The musts obtained by FVE showed higher content of phenolic compounds than those produced by conventional maceration. This reflects that the FVE process promoted a higher quantity of solids from grape transferred into the must.[64]

The cells from grape peels are more easily disrupted, allowing the release of anthocyanins and tannins before the alcoholic fermentation; thus, bitter tannins, especially those from the seeds, are not extracted because seeds have lesser water content and are less susceptible to the FVE process.[67] This helps to improve specific sensory attributes from wines and is not achieved with ease using other maceration methods. The concentration of phenolic compounds in musts produced by FVE is due to a higher extraction of hydroxycinnamic acids from the peels; moreover, due to the inactivation of polyphenol oxidase from grape during the thermal stage of the FVE process. As a result, the proportion of proanthocyanidins is very different in the musts produced by FVE. The proanthocyanidins in these musts contain more units of epigallocatechin and a lesser proportion of galloilads. Given that the first compounds are specific of peels, and the second ones are more prominent in seeds, these differences in composition indicate that FVE treatment increased the extraction of tannins from peels, the wines produced by FVE contain higher proportions of galloilads and fewer units of epigallocatechin, and this also reflects a better extraction of peel compounds, increasing the color intensity of wines.[64]

The maceration by FVE has been compared with enzymatic maceration. Results indicate that the FVE process generated a higher yield in the extraction of juice and anthocyanins than enzymatic maceration. In addition, impedance evaluations on intact tissue and processed tissue of FVE processed grapes, enzymatic and thermal maceration, showed that FVE caused a more pronounced cellular damage than the other methods, especially on peel cells. The higher extraction of anthocyanins by FVE processing is of interest to those winemakers that harvest grape wine with low content of these compounds, for example, Pinot Noir grapes, or to process grapes where anthocyanins are concentrated majorly on the peel.[63,68]

It has been proved that wines produced from the Pinot Noir variety, macerated by FVE, showed a higher content of flavan-3-ols extracted from the peel. These compounds represent between 13 and 30% of the phenolic content of red grapes and are ranked in second place in abundance. Together with proanthocyanidins, these compounds are responsible for wine color stability and delaying enzymatic browning and aroma loss.[63]

The FVE process could be utilized to obtain wines with higher content of phenolic compounds and antioxidant activity because grape maceration occurs under a high vacuum, reducing the presence of oxygen and avoiding the oxidation of sensible compounds. When comparing the antioxidant activity of wines produced by FVE and those produced by traditional methods, it was reported that despite the antioxidant activity of both types of wines was reduced during the different processing stages, the wines produced by FVE were characterized by a higher antioxidant activity (in ABTS and FRAP assays), which was almost twice the activity of traditionally produced wines.[63]

An investigation where the FVE process was applied for the extraction of grape juice and its stability was evaluated showed that the FVE process generated less color stability in the freshly extracted juices; however, the FVE process significantly reduced the formation of 5-hydroxymethylfurfural during an accelerated aging test, also in synergy with a low pH, increased color stability in the same test.[69]

The FVE process promotes a more efficient cellular disruption in grape peel cells significantly increasing the extraction yield of grape phenolic compounds, which represents an increment in antioxidant capacity, improved coloration, and some sensory attributes of wines. During the FVE process, vacuum pressures and the absence of oxygen reduce oxidation of compounds, the inactivation of polyphenol oxidase during the thermal stage help to reduce deteriorative processing of juice. The FVE is a technology with the potential to extract more phytochemicals from plant tissues that are considered wastes or that are partially utilized as the grape peel.

10.4.3 AVOCADO PROCESSING BY FLASH VACUUM EXPANSION

According to several authors, the main sources of phenolic compounds in avocado fruits are the seed and peel, while the pulp contributes with a minor quantity,[70,71] Avocado processing in the food and cosmetic industries produces many peels, which are considered wastes, and can become a problem due to the volume and space they require. However, these peels and seeds could be a source of bioactive compounds, like phenolic compounds, which can achieve high commercial value. In some studies, on avocado peel and seed, four main classes of polyphenolic compounds have been identified, such as flavanol monomers, proanthocyanidins, hydroxycinnamic acids, and flavonol glycosides. In seeds, the majorly found phenolic compounds are 3-O-caffeoylquinic acid, 3-O-p-cumaroylquinic acid, and procyanidin A trimers; while in the peel, the majorly found compounds are derivatives of 5-O-caffeoylquinic acid and quercetin. In most avocado varieties, the peel presents a higher content of phenolic compounds and higher antioxidant capacity than seed.[17] Additionally, it has been reported that in avocado peel and seed important quantities of bioactive compounds such as catechin, epicatechin, procyanidins B1, and B2 trans-5-O-caffeoylquinic acid highlight the high antioxidant and anti-inflammatory activities.[19]

Almost 47 phenolic compounds have been identified, highlighting some high molecular weight compounds in avocado and seed extracts, especially proanthocyanidin dimers and trimer, which confers to their high antioxidant capacity in ABTS and DPPH assays. However, there were significant differences between the phenolic compound content and antioxidant capacity from seed and peel, with peel having a higher content of phenolic compounds. In contrast, a higher antioxidant capacity was found in seed extract, mainly due to the differences in proportions and types of phenolic compounds in each part of the fruit.[72,73]

The avocado fruit represents a challenge in terms of industrialization due to its high polyphenols and fat content, which are compounds quite sensitive to oxidation by enzymatic or chemical pathways.[74] The avocado-based industrial products include guacamole, paste, frozen portions, and oil. Pulp is susceptible to browning reactions and spoilage. Avocado contains enzymes that can produce undesirable effects in the pulp during processing, mainly the change in color due to enzymatic browning. On the other hand, fats can be oxidized and generate bitter flavors or rancidity; these problems have not been addressed thoroughly, which is why avocado-based products are scarce compared to products developed from other fruits.[75] To inactivate enzymes in avocado, it is necessary to apply severe thermal treatments that can affect the sensory quality of the product or promote the degradation of compounds of interest like antioxidants, vitamins, or even fats.[76,77] The challenge of avocado processing is also present in its flavor, which is not as intense as in other fruits; hence, the possibilities for modifying the flavor are minor.[74] To overcome all these problems in avocado processing, food additives are used and are aimed to prevent enzymatic browning; some others are utilized to avoid oxidation of fats, preserve the stability between the aqueous and lipid phases, avoid microbial growth, imitate the texture of avocado pulp or mask the flavor in the final product. Finding appropriate processing methods for avocado is a priority because it is a very perishable fruit, and due to its post-harvest physiology, its shelf life is very short.[78] On the other hand, it is also susceptible to mechanical damage during handling, transportation, and storage. The avocado industry, like many others, also has the problem of waste generation such as peels and seeds, which have been a source of raw materials for a few industries that seek to provide their adequate utilization, especially the avocado seeds.[79]

The results of various studies have revealed that most of the phenolic compounds in the avocado fruit are contained in the peel and seed, while in the pulp, only a minor proportion of these compounds is found.[71] So, finding a process that allows the exploitation of avocado peels and seeds to produce products with high added value is of great interest to food technologists.

One of the characteristics of the FVE process is that it can be labeled as a mild thermal process, as generally the plant material is exposed to a steam flux that promotes an increment of temperature in the plant tissue between 60 and 90°C during periods of 1 to 5 min (at maximum temperature). Another characteristic is that FVE allows the processing of whole fruits without the need to peel them off or ground them to produce puree. This gives the possibility of producing puree of any fruit containing enough moisture content to be susceptible to pressure and temperature change that promote the disaggregation of plant tissue. These intricacies of the FVE process make it an interesting alternative to overcome the challenges of avocado fruit processing.

Research works related to avocado processing by FVE have been realized. These studies have aimed to exploit the whole fruit thanks to the characteristics of the process. In these studies, avocado fruits were submitted to different exposure times to water steam flux during the thermal stage, where final temperatures between 85 and 95°C were reached; when transitioning to the expansion stage, temperatures decreased quickly to 20–25°C. To provide evidence of the effect on the peel, peeled fruits were included for comparison. The physicochemical, sensorial, and nutritional properties of purees were evaluated and compared with puree produced by traditional methods. The difference between peeled and whole fruits was evident since during the thermal stage, peeled fruits heated faster than whole fruits, reaching 80°C for more than 10 min, while in whole fruits, temperatures higher than 80°C lasted for 3 to 8 min, until the transition into the expansion stage where temperatures decreased rapidly to 20–25°C. However, in the expansion stage, tissue disaggregation was higher in peeled fruits than whole fruits, resulting in a more homogenous color in the puree, while whole fruits showed a more heterogeneous appearance due to the presence of pulp and seed fragments in the pulp.[74,80]

When the physicochemical properties of puree were evaluated, polyphenol oxidase was inactivated in all treatments, and lipoxygenase activity was significantly reduced.[74,81] However, one of the biggest differences between treatments was a significant increase in the content of phenolic compounds, especially in whole fruits compared to peeled ones[74]; this was due to the majority of phenolic compounds present in the peel and incorporated into the puree when tissues are disaggregated. The content of phenolic compounds in whole fruits was significantly higher between 1.1 and 0.89 mg GAE/g, while in the treatments with peeled fruits, the content of phenolic compounds was 0.30 mg GAE/g. The control sample only contained 0.15 mg GAE/g. In this case, the positive effect of preserving the peel on the fruit during processing with FVE could be observed.

The avocado purees produced by FVE also showed a significant increase in antioxidant capacity. In the treatments with whole fruits, the antioxidant capacity ranged between 0.326 and 0.322 mM TEAC/g FW, meanwhile the treatments with peeled fruit showed an antioxidant capacity of 0.2477 mM TEAC/g FW, and the control sample showed an antioxidant capacity of 0.1212 mM TEAC/g FW. This increment of antioxidant capacity in avocado purees could be attributed to the extracted phenolic compounds from the peel incorporated into the puree.[74] Regarding the nutritional content of the purees produced by FVE, no significant differences were found in the lipid, protein, moisture, and ash content. However, there was an increment in carbohydrates, possibly attributed to the incorporation of small peel fragments in the puree, increasing dietary fiber content, but this parameter was not evaluated.[80] The composition of the lipid fraction in avocado puree produced by FVE was not significantly modified, although it was observed that the oxidative process of that fraction was reduced compared to the control sample. However, it was impossible to identify an effect due to the inclusion of peel or seed fragments, which could affect this lipid fraction's oxidative process.[81] The evaluation of sensory and texture properties showed significant evidence of the incorporation of peel fragments or compounds present in it, mainly in the visual appearance and puree color, which was affected by the incorporation of small peel fragments. This produces a heterogeneous appearance, with darker colors, besides the perception of bitter flavors in the treatments with whole fruits.

Regarding texture, the incorporation of peel fragments promoted a well-defined granularity in the puree samples, besides a slightly bitter flavor.[82] The studies performed about the application of FVE to produce avocado puree show that whole fruits lead to a significant increment of antioxidant capacity, which represents a potential benefit for consumers, besides avoiding enzymatic browning and reducing oxidative processes of the lipid fraction. However, the sensory attributes of the puree are modified, which changes consumers' acceptance and affirms that even when a product might be healthier, consumers will prefer a better flavor to a higher health benefit.

10.4.4 PASSION FRUIT PROCESSING BY FLASH VACUUM EXPANSION

When the effect of FVE on passion fruit was evaluated to produce puree, a yield of 46% was found after the incorporation of the condensate, which contained some of the aromatic compounds, compared to a yield of only 26% obtained when fruits were processed in a conical extractor. This increment could be attributed to the FVE process, especially during the thermal stage, promoting the softening of the peel, helping the incorporation of a higher quantity of it into the final product, besides the reincorporation of the condensates with high content of aromatic compounds. Therefore, the aroma of puree produced by FVE was more intense than that produced by the conical extractor.[83] The soluble solids and titratable acidity of the purees produced by FVE were lower than the control sample, which could be due to the reincorporation of the condensate, or a higher extraction of juice contained in the arils, given their structure and water accumulation are very susceptible to FVE. A very favorable aspect was the increment of anthocyanins, which increased from 3.91 (mg/L) in the control sample to 45.72 (mg/L) in the puree produced by FVE, besides a more intense red-violet color in the FVE samples, mostly due to the incorporation of a great portion of mesocarp (where the peel pigments accumulate) into the final puree.[83]

On the other hand, it is also mentioned that the FVE process applied in purple passion fruit promotes higher incorporation of cellulose and xylose in comparison with purees produced by traditional extraction methods. This makes it evident that the FVE process affects more fruit components like the mesocarp and the arils,[83] which are the least affected when the fruit is processed by other methods. Brat et al.[83] mentioned that another consequence of incorporating peel components was the increment of consistency and viscosity in the purees produced by FVE, in which viscosity was three times higher than those produced by traditional methods. This could be interesting for the fruit processing industry, considering that FVE increases properties like yield, color, and functional compounds in purple passion fruit. Moreover, the concentration of aromatic compounds decreased significantly in passion fruit puree produced by FVE, possibly being dragged during the expansion stage, but this can be solved by reincorporating the condensates.[53]

10.4.5 GUAVA PROCESSING BY FLASH VACUUM EXPANSION

The reports related to the application of FVE for the processing of guava are scarce; in fact, the only report so far is the one by Brat et al.,[57] who produced guava purees by FVE. In the thermal stage, fruits reached 95°C and then transitioned into the expansion chamber at 3 kPa. These authors reported that the FVE process allowed a yield of 58% in comparison to traditional methods with a yield of only 50%, which is a minimal improvement. However, it was not the only variable positively affected, as increased soluble solids and luminosity were higher in FVE treated samples than those produced with a rotary pulper. The increment in these parameters is interesting as it is attractive for the beverage industry. Having raw material with higher content of soluble solids and luminosity allows for better dilution ratios for a better yield, which could reduce production costs. Moreover, the analysis of insoluble wastes in alcohol confirmed that the FVE process increases the proportion of compounds present in the mesocarp of guava; some of these components might be considered as dietary fiber which could improve its functionality.

10.4.6 CITRUS PEEL PROCESSING BY FLASH VACUUM EXPANSION

Based on the mechanisms of action of the FVE process and the cellular structure, on which the essential oils are stored in citrus peels, Brat[54] observed an area of opportunity for the application of FVE on essential oil extraction. The authors utilized peels of mature lemons (*Citrus limon* L.), sweet oranges (*Citrus sinensis* L.), mandarins (*Citrus reticulata* B.), and grapefruits (*Citrus paradisi*). An FVE system where they could be treated independently was used in this case; 100 kg of peels of each fruit were heated between 85 and

90°C and then transferred to a chamber at 3 kPa. The results showed that the FVE process allowed attractive extraction yields compared to theoretical yield, for example, 44% of extraction for lemon peels and 33% for mandarin peels, while for sweet orange and grapefruit the obtained yields were 29 and 27%, respectively. On the other hand, Brat et al.[53] indicated that the essential oils extracted by FVE showed significant changes in the proportion of some compounds, highlighting an increment of limonene in the oils of the four citrus fruit peels evaluated (lemon, sweet orange, mandarin, and grapefruit). In the case of essential oils extracted from lemon peels by FVE, an increment on monoterpenes was observed, in contrast to the monoterpenic alcohols and their acetates, which were reduced, while aliphatic aldehydes were not affected. The essential oils extracted by traditional methods from sweet oranges showed compounds in proportions in agreement with those reported by other authors. The oils extracted by FVE maintained their proportions of monoterpenic aldehydes and acetates but significantly reduced the proportion of monoterpenic alcohols.

As reported by Brat et al.,[54] the FVE process could have a considerable effect on the proportions of compounds of oils from the mandarin peel because in these oils, besides a reduction of monoterpenic alcohols, an increment in decanal, and the reduction of sesquiterpenic alcohols was also observed. This differs from the average proportions of compounds in mandarin peel oils. The results for the oils extracted from grapefruit peel showed that the FVE process promoted a reduction by half in the content of linalool and α-terpineol in comparison with oils in their natural state, besides a reduction in compounds like β-citronellol, nerol, geraniol, and nootkatone. The authors concluded that, in general, when the FVE process is applied for the extraction of essential oils from the four citrus fruit peels, the proportion of limonene increases, and the oxygenated compounds decrease. By tradition, citrus peels have been a source of essential oils and are appreciated for their aromatic and functional properties, which have placed them as references for applications in cosmetics and pharmaceuticals. These properties could be affected by the modification of the proportions of these compounds in the essential oils, which is why it is recommended to determine if oils extracted from citrus peels by FVE preserve their aromatic and functional properties.

10.5 CONCLUSIONS

Through various investigations, it has been shown that the FVE process promotes the migration of compounds from the shell and seed to the final products, increasing the functional properties of the derivatives of these fruits, mainly the antioxidant capacity and the yield, by incorporating parts of tissue that were not possible through other processes. Also, some sensory properties such as color and aroma are benefited by extracting pigments and aromatic compounds present in the peels. However, some other sensory aspects are disadvantaged since the sensory perception of the final products is affected by the incorporation of these compounds. In this sense, it gains in terms of beneficial bioactivity, but sensory acceptance is lost. According to trends in the availability of resources increased incidence of diseases related to oxidative stress and nutrition, we would have to think about reducing the value of sensory attributes to give way to what is more nutritious and beneficial. As a recommendation, the FVE process for the tropical fruit process, it could be suggested to develop research focused on the thermodynamic processes of heat transfer in plant tissues subjected to FVE, to make the process of extraction of compounds of interest more efficient and avoid overheating of the fruits in the thermal stage of the process. On the other hand, it is recommended to carry out research on fruits, which due to their morphology (distribution and texture of the shell mainly), are candidates to be processed by FVE, for example, blackberry, strawberry, tomato, pear, and others. The global trend in reducing arable areas, the increase in the population, food waste, and the incidence of diseases related to food, force us to seek a comprehensive use of all the biomass available in the fruits.

KEYWORDS

- **bioactive compounds**
- **essential oils**
- **flash vacuum expansion**
- **fruit peel and seed**

REFERENCES

1. FAO. Las Principales Frutas Tropicales Análisis Del Mercado 2018; Roma, 2020.
2. Altendorf, S. Minor Tropical Fruits-Mainstreaming a Niche Market. Food Outlook-Biannual Report on Global Food Markets. Food and Agriculture Organization of the United Nations, 2018.

3. Banerjee, J.; Singh, R.; Vijayaraghavan, R.; MacFarlane, D.; Patti, A. F.; Arora, A. A Hydrocolloid Based Biorefinery Approach to the Valorization of Mango Peel Waste. *Food Hydrocoll.* **2018,** 142–151.
4. Almeida, J. M.; Lima, V. A.; Giloni-Lima, P. C.; Knob, A. Passion Fruit Peel as Novel Substrate for Enhanced β-glucosidases Production by *Penicillium verruculosum*: Potential of the Crude Extract for Biomass Hydrolysis. *Biomass Bioenergy* **2015,** *72*, 216–226.
5. Corrêa, R. C.; Peralta, R. M.; Haminiuk, C. W.; Maciel, G. M.; Bracht, A.; Ferreira, I. C. The Past Decade Findings Related with Nutritional Composition, Bioactive Molecules and Biotechnological Applications of *Passiflora* spp. (Passion Fruit). *Trends Food Sci. Technol.* **2016,** *58*, 79–95.
6. da Silva Lima, R.; Ferreira, S. R. S.; Vitali, L.; Block, J. M. May the Superfruit Red Guava and its Processing Waste be a Potential Ingredient in Functional Foods? *Food Res. Int.* **2019,** *115*, 451–459.
7. Gülcü, M.; Uslu, N.; Özcan, M. M.; Gökmen, F.; Özcan, M. M.; Banjanin, T.; Gezgin, S; Dursun, N; Geçgel, Ü.; Ceylan D. A.; Lemiasheuski, V. The Investigation of Bioactive Compounds of Wine, Grape Juice and Boiled Grape Juice Wastes. *J. Food Process. Preserv.* **2019,** *43*, e13850.
8. Araújo, R. G.; Rodriguez-Jasso, R. M.; Ruiz, H. A.; Pintado, M. M. E.; Aguilar, C. N. Avocado byproducts: Nutritional and Functional Properties. *Trends Food Sci. Technol.* **2018,** *80*, 51–60.
9. Sharma, K.; Mahato, N.; Cho, M. H.; Lee, Y. R. Converting Citrus Wastes into Value-Added Products: Economic and Environmently Friendly Approaches. *Nutrition* **2017,** *34*, 29–46.
10. Daiuto, É. R.; Tremocoldi, M. A.; Alencar, S. M. De.; Vieites, R. L. Composição química e atividade antioxidante da polpa e resíduos de abacate "Hass." *Rev. Bras. Frutic.* **2014,** *36*, 417–424.
11. Uchôa-Thomaz, A. M. A.; Sousa, E. C.; Carioca, J. O. B.; Morais, S. M. D.; Lima, A. D.; Martins, C. G.; Alexandrino, C. D.; Ferreira, P. A. T.; Rodrigues, A. L. M.; Rodrigues, S. P.; Thomaz, J. C.; Rodrigues, L. L. Chemical Composition, Fatty Acid Profile and Bioactive Compounds of Guava Seeds (*Psidium guajava* L.). *Food Sci. Technol.* **2014,** *34*, 485–492.
12. Bertagnolli, S. M. M.; Silveira, M. L. R.; Fogaça, A. D. O.; Umann, L.; Penna, N. G. Bioactive Compounds and Acceptance of Cookies made with Guava Peel Flour. *Food Sci. Technol.* **2014,** *34*, 303–308.
13. dos Reis, L. C. R.; Facco, E. M. P.; Salvador, M.; Flôres, S. H.; de Oliveira Rios, A. Antioxidant Potential and Physicochemical Characterization of Yellow, Purple and Orange Passion Fruit. *J. Food Sci. Technol.* **2018,** *55*, 2679–2691.
14. Barbosa-Martín, E.; Chel-Guerrero, L.; González-Mondragón, E.; Betancur-Ancona, D. Chemical and Technological Properties of Avocado (*Persea americana* Mill.) Seed Fibrous Residues. *Food Bioprod. Process.* **2016,** *100*, 457–463.
15. Salazar-López, N. J.; Domínguez-Avila, J. A.; Yahia, E. M.; Belmonte-Herrera, B. H.; Wall-Medrano, A.; Montalvo-González, E.; González-Aguilar, G. A. Avocado Fruit and Byproducts as Potential Sources of Bioactive Compounds. *Food Res. Int.* **2020,** 109774.
16. López-Cobo, A.; Gómez-Caravaca, A. M.; Pasini, F.; Caboni, M. F.; Segura-Carretero, A.; Fernández-Gutiérrez, A. HPLC-DAD-ESI-QTOF-MS and HPLC-FLD-MS as Valuable Tools for the Determination of Phenolic and Other Polar Compounds in the Edible Part and Byproducts of Avocado. *LWT* **2016,** *73*, 505–513.

17. Kosińska, A.; Karamać, M.; Estrella, I.; Hernández, T.; Bartolomé, B.; Dykes, G. A. Phenolic Compound Profiles and Antioxidant Capacity of *Persea americana* Mill. Peels and Seeds of Two Varieties. *J. Agric. Food Chem.* **2012,** *60*, 4613–4619.
18. da Silva, A. C.; Jorge, N. Bioactive Compounds of the Lipid Fractions of Agro-Industrial Waste. *Food Res. Int.* **2014,** *66*, 493–500.
19. Tremocoldi, M. A.; Rosalen, P. L.; Franchin, M.; Massarioli, A. P.; Denny, C.; Daiuto, É. R.; Paschoal, J. A. R.; Melo, P. S.; Alencar, S. M. D. Exploration of Avocado byproducts as Natural Sources of Bioactive Compounds. *PLoS One* **2018,** *13*, e0192577.
20. Leite, J. J. G.; Brito, É. H. S.; Cordeiro, R. A.; Brilhante, R. S. N.; Sidrim, J. J. C.; Bertini, L. M.; de Morais, S. M.; Rocha, M. F. G. Chemical Composition, Toxicity and Larvicidal and Antifungal Activities of *Persea americana* (Avocado) Seed Extracts. *Rev. Soc. Bras. Med. Trop.* **2009,** *42*, 110–113.
21. Simpson, D.; Amos, S. Other Plant Metabolites. In *Pharmacognosy*; Academic Press, 2017; pp 267–280.
22. Rodríguez-Sánchez, D. G.; Pacheco, A.; Villarreal-Lara, R.; Ramos-González, M. R.; Ramos-Parra, P. A.; Granados-Principal, S.; Díaz de la Garza, R. I.; García-Rivas, G.; Hernández-Brenes, C. Chemical Profile and Safety Assessment of a Food-Grade Acetogenin-Enriched Antimicrobial Extract from Avocado Seed. *Molecules* **2019,** *24*, 2354.
23. Oboh, G.; Odubanjo, V. O.; Bello, F., Ademosun, A. O.; Oyeleye, S. I.; Nwanna, E. E.; Ademiluyi, A. O. Aqueous Extracts of Avocado Pear (*Persea americana* Mill.) Leaves and Seeds Exhibit Anti-Cholinesterases and Antioxidant Activities *In Vitro*. *J. Basic Clin. Physiol. Pharmacol.* **2016,** *27*, 131–140.
24. Alkhalaf, M. I.; Alansari, W. S.; Ibrahim, E. A.; ELhalwagy, M. E. Anti-oxidant, Anti-inflammatory and Anti-cancer Activities of Avocado (*Persea americana*) Fruit and Seed Extract. *J. King Saud Univ. Sci.* **2019,** *31*, 1358–1362.
25. Barrera López, R. S.; Arrubla Vélez, J. P. Análisis de Fitoesteroles en La Semilla de *Persea Americana* Miller (Var. Lorena) por Cromatografía de Gases y Cromatografía Líquida de Alta Eficiencia. *Revista Facultad De Ciencias Básicas* **2017,** *13* (1).
26. 26. Paull, R. E.; Duarte, O. Guava. In *Tropical Fruits*; CAB International: Cambridge, 2012; vol 2, p 91.
27. Rodriguez, N. N.; Valdés-Infante, J.; Velázquez J. B.; Rivero, D.; Sourd, D. G.; Martínez, F.; Tamayo R.; Rodríguez, J. A. Colección Cubana de Germoplasma de guayabo (*Psidium guajava* L.). Establecimiento, caracterización y selección de cultivares. *Citri Frut.* **2010,** *27*, 28–38.
28. Cazarin, C. B. B.; Rodriguez-Nogales, A.; Algieri, F.; Utrilla, M. P.; Rodríguez-Cabezas, M.E.; Garrido-Mesa, J.; Guerra-Hernández, E.; Braga, P. A.; Reyes, F. G. R.; Maróstica, M. R.; Gálvez, J. Intestinal Anti-inflammatory Effects of Passiflora Edulis Peel in the Dextran Sodium Sulphate Model of Mouse Colitis. *J. Funct. Food* **2016,** *26,* 565–576.
29. Rotta, E. M.; Rodrigues, C. A.; Jardim, I. C. S. F.; Maldaner, L.; Visentainer, J. V. Determination of Phenolic Compounds and Antioxidant Activity in Passion Fruit Pulp (*Passiflora* spp.) Using a Modified QuEChERS Method and UHPLC-MS/MS. *LWT* **2019,** *100*, 397–403.
30. Santos, O. V. D.; Vieira, E. L. S.; Soares, S. D.; Conceição, L. R. V. D.; Nascimento, F. D. C. A. D.; Teixeira-Costa, B. E. Utilization of Agroindustrial Residue from Passion Fruit (*Passiflora edulis*) Seeds as a Source of Fatty Acids and Bioactive Substances. *Food Sci. Tech.* **2020,** *41*, 218–225.

31. Peng, D.; Zahid, H. F.; Ajlouni, S.; Dunshea, F. R.; Suleria, H. A. Lc-esi-qtof/ms Profiling of Australian Mango Peel byproduct Polyphenols and Their Potential Antioxidant Activities. *Processes* **2019,** *7*, 764.
32. Quintana-Obregón, E.; San Martín-Hernández, C.; Muy-Rangel, M.; Vargas-Ortiz, M. Valorization of Mango (*Mangifera indica* L.) Pericarp Powders as an Alternative for the Generation of Functional Foods. *TIP Revista Especializada en Ciencias Químico-Biológicas* **2019,** *22*, 1–5.
33. Ajila, C. M.; Bhat, S. G.; Rao, U. P. Valuable Components of Raw and Ripe Peels from Two Indian Mango Varieties. *Food Chem.* **2007,** 1006–1011.
34. Serna-Cock, L.; García-Gonzales, E.; Torres-León, C. Agro-Industrial Potential of the Mango Peel Based on its Nutritional and Functional Properties. *Food Rev. Int.* **2016,** *32*, 364–376.
35. Hosni, K.; Zahed, N.; Chrif, R.; Abid, I.; Medfei, W.; Kallel, M.; Brahim, N. B.; Sebei, H. Composition of Peel Essential Oils from Four Selected Tunisian Citrus Species: Evidence for the Genotypic Influence. *Food Chem.* **2010,** *123* (4), 1098–1104
36. Kamaliroosta, L.; Zolfaghari, M.; Shafiee, S.; Larijani, K.; Zojaji, M. Chemical Identifications of Citrus Peels Essential Oils. *J. Food Biosciences Technol.* **2016,** *6*, 69–76.
37. Yang, Y.; Wang, X.; Zhao, C.; Tian, G.; Zhang, H.; Xiao, H.; He, L.; Zheng, J. Chemical Mapping of Essential Oils, Flavonoids and Carotenoids in Citrus Peels by Raman Microscopy. *J. Food Sci.* **2017,** *82*, 2840–2846.
38. Hosni, K.; Zahed, N.; Chrif, R.; Abid, I.; Medfei, W.; Kallel, M.; Brahim, N. B.; Sebei, H. Composition of Peel Essential Oils from Four Selected Tunisian Citrus Species: Evidence for the Genotypic Influence. *Food Chem.* **2010,** *123*, 1098–1104.
39. Ahmad, M. M.; Rehman, S. U.; Anjum, F. M., Bajwa, E. E. Comparative Physical Examination of Various Citrus Peel Essential Oils. *Int. J. Agric. Biol.* **2006,** 186–90.
40. Baser, K. H. C.; Buchbauer, G. *Handbook of Essential Oils: Science, Technology, and Applications*; 1st Ed.; CRC Press: Boca Raton, 2009.
41. Romero-Cascales, I.; Fernández-Fernández, J. I.; López-Roca, J. M.; Gómez-Plaza, E. The Maceration Process During Winemaking Extraction of Anthocyanins from Grape Skins into Wine. *Eur. Food Res. Technol.* **2005,** *221*, 163–167.
42. Pinelo, M.; Arnous, A.; Meyer, A. S. Upgrading of Grape Skins: Significance of Plant Cell-Wall Structural Components and Extraction Techniques for Phenol Release. *Trends Food Sci. Technol.* **2006,** *17*, 579–590.
43. Tang, G. Y.; Zhao, C. N.; Liu, Q.; Feng, X. L.; Xu, X. Y.; Cao, S. Y.; Meng, X.; Li, S; Gan, R. Y.; Li, H. B. Potential of Grape Wastes as a Natural Source of Bioactive Compounds. *Molecules* **2018,** *23*, 2598.
44. Fine, R. A.; Millero, F. J. Compressibility of Water as a Function of Temperature and Pressure. *J. Chem. Phys.* **1973,** *59* (10), 5529–5536.
45. Raven, J. A. The Role of Vacuoles. *New Phytol.* **1987,** 357–422.
46. Marty, F. Plant Vacuoles. *Plant Cell* **1999,** *11*, 587–599.
47. Krüger, F.; Schumacher, K. Pumping Up the Volume−Vacuole Biogenesis in Arabidopsis Thaliana. *Semin. Cell Dev. Biol.* **2018,** *80*, 106–112.
48. Pieribattesti, J. C.; Hunez, M. Procédé d'extraction de jus de fruits et de légumes et appareillages pour sa mise en oeuvre. Institud National de la Propriété Industrielle, Paris, France. Brevet, n° 8413469, 1984.

49. Pieribattesti, J. C.; Hunez, M.; Cogat, P. O. Technique déxtraction dárômes. Institud National de la Propriété Industrielle, Paris, France. Brevet, n° 8814311, 1984.
50. Cogat, P. O.; Hunez, M; Pieribattesti, J. C. Procédé et matériel pour léxtraction de jus et dárômes á partir de substrats végétaux, possibilités de traitements complémentaires sur le meme matériel. Institud National de la Propriété Industrielle, Paris, France. Brevet, n° 88143111, 1988.
51. Cogat, P. O. Inventor, Aurore Développement, assignee. Installation pour le traitement de matières biologiques hydratées. European Patent Office, Munich, Germany. European patent, EP 0,727,948, 1994.
52. Cogat. P. O. Procédé et equipements dímpegnation de matiéres solides organiques fibreuses, par vaporisation sous vide de léau de cosntitution et injection en contrepression d´une solution. Institud National de la Propriété Industrielle, Paris, France. Brevet, n° 9505014, 1995.
53. Brat, P.; Brillouet, J. M.; Reynes, M.; Cogat, P. O.; Olle, D. Free Volatile Components of Passion Fruit Puree Obtained by Flash Vacuum-Expansion. *J. Agric. Food Chem.* **2000,** *48*, 6210–6214.
54. Brat, P.; Ollé, D.; Gancel, A. L.; Reynes, M.; Brillouet, J. M. Essential Oils Obtained by Flash Vacuum-Expansion of Peels from Lemon, Sweet Orange, Mandarin and Grapefruit. *Fruits* **2001,** *56*, 395–402.
55. Lalel, H. J. D.; Singh, Z.; Tan, S. C. Distribution of Aroma Volatile Compounds in Different Parts of Mango Fruit. *J. Hortic. Sci. Biotechnol.* **2003,** *78*, 131–138.
56. Oliver-Simancas, R.; Muñoz, R.; Díaz-Maroto, M. C.; Pérez-Coello, M. S.; Alañón, M. E. Mango byproducts as a Natural Source of Valuable Odor-Active Compounds. *J. Sci. Food Agric.* **2020,** *100*, 4688–4695.
57. Brat, P.; Olle, D.; Reynes, M.; Alter, P; Cogat, P. O. Preparation of Tropical Fruit Purees by Flash Vacuum-Expansion. *Acta Hortic.* **2002,** *575*, 535–541.
58. Brat, P. Application of Flash-Release, A New Extraction Procedure (Juice, Pulp, Essential Oil). *Fruitrop* **2001,** *85*, 11–13.
59. Marin-Castro, U. R.; Salgado-Cervantes, M. A.; García-Alvarado, M. A.; Rodríguez-Jimenes, G.; Servent, A.; Pallet, D.; Vargaz-Ortiz, M. A. Efecto del proceso de expansión ultra-rápida sobre la obtención de puré de mango manila (*Mangifera i.*). Proceedings of XL encuentro nacional de la AMIDIQ "Retos de la ingeniería química para el desarrollo nacional", Oaxaca, México, May 7–10, 2019.
60. Gawel, R. Red Wine Astringency: A Review. *Aust. J. Grape Wine Res.* **1998,** *4*, 74–95.
61. Sarni-Manchado, P.; Cheynier, V. Phenolic Structure et Astringency. *J. Int. Sci. Vigne du Vin* **2000,** *33*, 111–118.
62. Busse-Valverde, N.; Gomez-Plaza, E.; Lopez-Roca, J. M.; Gil-Munoz, R.; Bautista-Ortin, A. B. The Extraction of Anthocyanins and Proanthocyanidins from Grapes to Wine During Fermentative Maceration is Affected by the Enological Technique. *J. Agric. Food Chem.* **2011,** *59*, 5450–5455.
63. Samoticha, J.; Wojdyło, A.; Chmielewska, J.; Oszmiański, J. The Effects of Flash Release Conditions on the Phenolic Compounds and Antioxidant Activity of Pinot Noir Red Wine. *Eur. Food Res. Technol.* **2017,** *243*, 999–1007.
64. Morel-Salmi, C.; Souquet, J. M.; Bes, M.; Cheynier, V. Effect of Flash Release Treatment on Phenolic Extraction and Wine Composition. *J. Agric. Food Chem.* **2006,** *54*, 4270–4276.

65. du Toit, W.; Oberholster, A. Antioxidants in Wine During Fermentation. In *Processing and Impact on Antioxidants in Beverages*; Preedy, V., Ed.; Academic Press, 2014; pp 59–67.
66. Besnard, É.; Laffargue, F.; Relhié, F.; Froment, H.; Alibert, V. Influence de l'itinéraire de vinification après Flash-détente dans l'élaboration d'une nouvelle gamme de vins du Lot. http://www. vignevin-occitanie. com/wp-content/uploads/2018/08/flash-detente-lot-malbec. pdf (accessed Apr 3, 2021).
67. Laurie, D. Flash Extraction Goes to Work. Wines & Vines, January 2011. https://winesvinesanalytics.com/features/article/82256/Flash-Extraction-Goes-to-Work (accessed Apr 23, 2021).
68. Giovinazzo, G.; Grieco, F. Functional Properties of Grape and Wine Polyphenols. *Plant Foods Hum. Nutr.* **2015,** *70*, 454–462.
69. Ntuli, R. G.; Ponangi, R.; Jeffery, D. W.; Wilkinson, K. L. Impact of Juice Extraction Method (Flash Détente vs. Conventional Must Heating) and Chemical Treatments on Color Stability of Rubired Juice Concentrates under Accelerated Aging Conditions. *Foods* **2020,** *9*, 1270.
70. Soong, Y. Y.; Barlow, P. J. Antioxidant Activity and Phenolic Content of Selected Fruit Seeds. *Food Chem.* **2004,** *88*, 411–417.
71. Wang, W.; Bostic, T. R.; Gu, L. Antioxidant Capacities, Procyanidins and Pigments in Avocados of Different Strains and Cultivar. *Food Chem.* **2010,** *122*, 1193–1198.
72. Figueroa, J. G.; Borrás-Linares, I.; Lozano-Sánchez, J.; Quirantes-Piné, R.; Segura-Carretero, A. Optimization of Drying Process and Pressurized Liquid Extraction for Recovery of Bioactive Compounds from Avocado Peel byproduct. *Electrophoresis* **2018,** *39*, 1908–1916.
73. Rosero, J. C.; Cruz, S.; Osorio, C.; Hurtado, N. Analysis of Phenolic Composition of Byproducts (Seeds and Peels) of Avocado (*Persea americana* Mill.) Cultivated in Colombia. *Molecules* **2019,** *24*, 3209.
74. Vargas-Ortiz, M.; Servent, A.; Rodríguez-Jimenes, G.; Pallet, D.; Salgado-Cervantes, M. Effect of Thermal Stage in the Processing Avocado by Flash Vacuum Expansion: Effect on the Antioxidant Capacity and the Qualitaty of the Mash. *J. Food Process. Preserv.* **2017,** *41*, e13118.
75. Elez-Martinez, P.; Soliva, F. R. C.; Gorinstein, S.; Martin, B. O. Natural Antioxidants Preserve the Lipid Oxidative Stability of Minimally Processed Avocado Puree. *J. Food Sci.* **2005,** *70*, S325–S329.
76. Toledo, L.; Aguirre, C. Enzymatic Browning in Avocado (*Persea americana*) Revisited: History, Advances, and Future Perspectives. *Crit. Rev. Food Sci. Nutr.* **2017,** *57* (18), 3860–3872.
77. Jacobo-Velázquez, D. A.; Hernández-Brenes, C.; Cisneros-Zevallos, L.; Benavides, J. Partial Purification and Enzymatic Characterization of Avocado (*Persea americana* Mill, cv. Hass) Lipoxygenase. *Food Res. Int.* **2010,** *43* (4), 1079–1085.
78. Gómez-López, V. M. Inhibition of Surface Browning, Cut Avocado. *J. Food Qual.* **2002,** *25* (4), 369–379.
79. Páramos, P. R.; Granjo, J. F., Corazza, M. L.; Matos, H. A. Extraction of High Value Products from Avocado Waste Biomass. *J. Supercrit. Fluids* **2020,** *165*, 104988.
80. Vargas-Ortiz, M.; Rodríguez-Jimenes, G.; Salgado-Cervantes, M.; Pallet, D. Minimally Processed Avocado Through Flash Vacuum-Expansion: Its Effect in Major Physicochemical Aspects of the Puree and Stability on Storage. *J. Food Process. Preserv.* **2017a,** *41*, e12988.

81. Salgado-Cervantes, M. A.; Servent, A.; Marabel, I.; Vargas-Ortiz, M; Pallet, D. Flash Vacuum-Expansion Process: Effect on the Sensory, Color and Texture Attributes of Avocado (*Persea americana*) Puree. *Plant Foods Hum.* Nutr. **2019,** *74*, 370–375.
82. Vargas-Ortiz, M.; Servent, A.; Salgado-Cervantes, M.; Pallet, D. Stability of the Lipid Fraction of Avocado Puree Obtained by Flash Vacuum-Expansion process. *Innov. Food Sci. Emerg. Technol.* **2017c,** *41*, 109–116.
83. Brat, P.; Olle, D.; Reynes, M.; Cogat, P. O.; Brillouet, J. M. Preparation of Passion Fruit Puree by Flash Vacuum-Expansion. *J. Food Sci.* **2001,** *66*, 542–547.

CHAPTER 11

Recovery of Macromolecules of Biotechnological and Pharmaceutical Importance from Soybean Hulls

PAOLA CAMISCIA, GUILLERMO PICÓ, and NADIA WOITOVICH VALETTI

Instituto de Procesos Biotecnológicos y Químicos, Consejo Nacional de Investigaciones Científicas y Técnicas de Argentina, Rosario, Argentina

ABSTRACT

Soybean is one of the main crops worldwide with a high planted area distributed mainly in the US, Brazil, and Argentina. The main destinations for these grains are the production of flour, oil, and pellets for animal feed. During the processing of soybeans for industrialization, residues are produced, one of them being the soybean hull (endocarp that covers the grain), which represents 2–8% of the total mass of the grain. Soy hull is made up of molecules that have high commercial value: 45% cellulose, 20% xylan, 15% pectin, 2% lignin, and 7–14% soluble proteins. Within the fraction of soluble proteins, there are a series of macromolecules of high biotechnological value, including soybean peroxidase, PR-5, extensins, and different types of protease inhibitors such as Bowman–Birk type inhibitors and the family of Kunitz inhibitors, both present in different isoforms. Taking into account that, in addition to the named proteins, there are in total more than 4000 known and unknown proteins with functions to be determined; the soybean hull is presented as an alternative source of different types of macromolecules. In recent years, different attempts have been made to recover proteins with biological activity at laboratory level using traditional protein purification methods: salting out, solvent, ion exchange chromatography, etc. This chapter presents different

Food Byproducts Management and their Utilization. Ricardo Gómez-García, Ana A. Vilas-Boas, Débora A. Campos, Maria Manuela Pintado, & Cristóbal Noé Aguilar (Eds.)

methodologies for the extraction and purification of some of the proteins present in the soybean hull using simple and nonpolluting methodologies with the potential to be scaled up.

11.1 INTRODUCTION

The food industry annually generates tons of waste, including those generated by its raw material to the ones produced by its final consumption. Food waste can be defined as, "any removal of food from the food supply chain which is or was at some point fit for human consumption, or which has spoiled or expired, mainly caused by economic behavior, poor stock management or neglect.[1]" Most waste is generated in the primary stage where commodities are produced, more specifically within the farms or fields. These residues, besides generating pollution, lead to a loss of different types of resources such as water, fertilizers, fuels, among others.[2]

Within the chain of waste generated by the food industry, in this chapter, we will focus on the waste produced during agroindustrial activity, that is, in the early processing stages. In this field, cereal and oilseed residues stand out being the soybean hull, an example of the latter.

Soybean (*Glycine max.*), is one of the most developed crops on our planet, with the majority of the producers in recent years being Brazil with 131 mTn, USA with 112 mTn, and Argentina with 50 mTn (average annual data of the last harvests).[3–5] Asian cultures use soybean to produce traditional foods such as soy milk, soy sauce, and tofu. In western cultures, soybeans are processed mainly to obtain soybean meal and seed oil.[6] During soybeans' harvesting and subsequent processing, residues such as stems, leaves, and shells are produced. Within them, we are going to focus on the soybean hull (SBH).

The process of industrialization of soybeans begins with the breakage of the grain in a mechanical way to release their hull or endocarp. The SBH can represent between 2 and 4% of the total grain mass. This percentage may seem small, but if the amount of soybeans produced and industrialized is considered, it represents a large number.[7] SBH waste is made up of particles of variable sizes; within them, the small particles can quickly oxidize in contact with air and become inflamed or cause an explosion. Due to this, they must be continuously eliminated from the place where they are produced, which adds an additional cost to the process. In most cases, companies pellet SBH waste to be used as livestock feed.

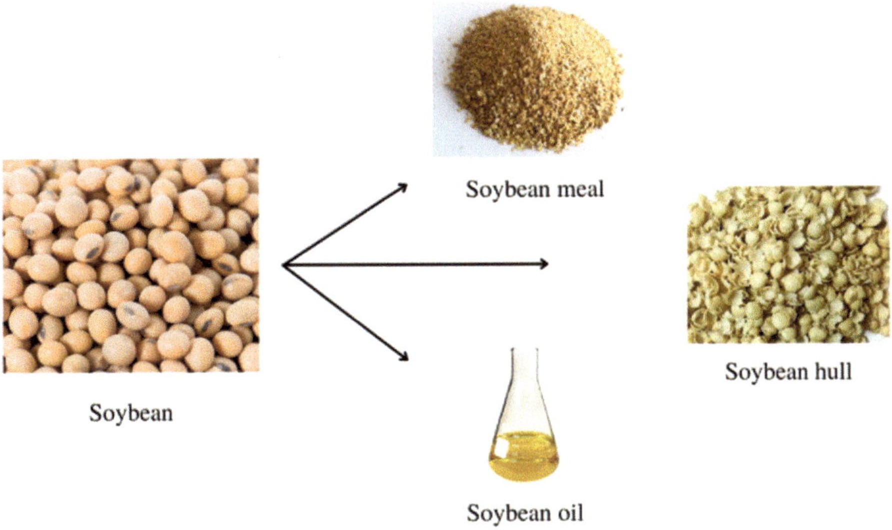

FIGURE 11.1 Main products from soybean.

The selection of SBH is not based only on its massive production. There are different types of hulls from various crops produced in the world. Table 11.1 shows the composition of the different hulls from major crops.

TABLE 11.1 Chemical Compositions of Different Hulls.

Hull	Soybean hull [8–12]	Rice hull [13–16]	Wheat hull [17–18]
Proteins (%)	9–14	2–23	6
Cellulose (%)	29–51	10–30	20–36
Hemicellulose (%)	10–20	10–20	18
Lignin (%)	2–10	2–20	5–16
Inorganic or ash (%)	1–4	5–22	2

As can be seen, the same chemical components are present regardless of their origin. However, the relative composition of each of these components varies between different crops and may even vary within each variant of the cereal. In all cases, cellulose and hemicelluloses are the major components, and together they may represent approximately 80% of the total weight of these wastes. That is why for years, several research groups have focused on reutilizing these residues to obtain cellulose and fermentable sugars for the production of biofuels.[7–11] On the other hand, rice hulls have a large amount of inorganic residues and, within them, a high silicon content (20%) for which most of its applications are focused on obtaining said compound with various applications.[12,13]

The lignin content is high in most plant biomass. This polymer and hemicelluloses will crosslink the cellulose chains present, giving the material strength and hardness.[14] The lignin content is important when selecting the type of biomass, which will be used in any process, since a high content of this compound increases its hardness, and consequently makes it more difficult to grind the material. It also affects the ability to release the other components included in the fiber such as soluble proteins, polyphenols, cellulose, etc. In these cases, to extract compounds of interest that are present in the biomass, extreme working conditions such as strong acids or bases, high temperatures, and a combination of chemical and enzymatic treatments are required.[15–17] Regarding this point, SBH is an excellent biomass to be used as a raw material since it has a low lignin content, which allows working in mild conditions facilitating its handling.[7,18]

Finally, the protein content is variable in each different type of residue, with soybean and rice hulls having the highest content in their composition. A more in-depth analysis of the above allows us to conclude that SBH is an excellent residue to be used as a source of different molecules of interest since it is produced in large quantities around the world. It has low lignin content which favors its processing. It also presents a high percentage of proteins and cellulose that have different industrial applications.

11.2 SOLUBLE PROTEINS FROM SBH

SBH is composed of 9–14% soluble proteins that correspond to more than 4400 proteins, some identified and others still to be identified.[19] In Tables 11.2 and 11.3 part of the identified proteins can be seen:

Among the many proteins present, several stand out in the literature for their potential and applications. Some of them are detailed below.

11.2.1 SOYBEAN PEROXIDASE

Peroxidases are widely versatile enzymes that use hydrogen peroxide to catalyze the oxidation of a broad variety of substrates. They have been isolated and characterized from different plant and animal sources and most contain a protoporphyrin IX (heme) as a cofactor.[20] They are divided into two large families: the superfamily of mammalian peroxidases and plants' superfamily.[21]

TABLE 11.2 Enzymes of Higher Relative Abundance Present in SBH Extract, Identified by LC-MS.

Description	MW (kDa)	pI	Relative abundance (%)
Lipoxygenase	78.7	6.44	3.85093168
Seed linoleate 13S-lipoxygenase-1	94.3	6.39	3.74741201
Lipoxygenase	96.7	6.58	3.58178054
Beta-amylase	56	5.55	1.5942029
Methionine synthase	84.2	6.33	1.36645963
Phospholipase D	91.6	5.69	0.72463768
Fructose-bisphosphate aldolase	38.2	7.55	0.62111801
Malate dehydrogenase	35.4	6.3	0.41407867
Glyceraldehyde-3-phosphate dehydrogenase	36.7	7.24	0.33126294
Beta-galactosidase	91.1	7.28	0.39337474
Pectin acetylesterase	45.8	8.32	0.2484472
Chitinase class I	34.3	7.42	0.49689441
Peroxidase	38.1	5.21	0.1242236

Source: Adapted from Ref. [19].

TABLE 11.3 Protease Inhibitors Found in SBH.

Description	MW [kDa]	pI	Relative abundance (%)
Kunitz trypsin inhibitor	24.1	5.11	0.95238095
Kunitz trypsin protease inhibitor	22.7	5.35	0.14492754
Kunitz-type trypsin inhibitor	22.8	6.55	0.10351967
Putative Kunitz trypsin protease inhibitor	23.6	5.38	0.04140787
Trypsin inhibitor B	20	4.82	0.41407867
Trypsin inhibitor 26 kDa isoform	25.9	8.32	0.22774327
Cysteine proteinase inhibitor	14.2	9.13	0.0621118
Cysteine proteinase inhibitor	10.7	6.09	0.0621118

Source: Adapted from Ref. [19].

In plants, peroxidases are involved in many physiological functions, from germination to senescence. Many isozymes are constitutively expressed, playing an important role in the structure and development of plants, while other isozymes are induced in response to stress. Peroxidases participate in the oxidation of phenolic compounds,[22] defense against pathogens,[23,24] cell elongation,[25] in the metabolism of reactive oxygen and nitrogen species,[24] the metabolism of phytohormones, the crosslinking of cell wall components, and the formation of suberin and lignin.[26]

In industries, peroxidases have been investigated to be used in a wide range of applications such as: wastewater bioremediation,[27–29] biocatalysis,[30,31] for diagnostic tests, and biosensors.[32] Peroxidases have been used in the baking industry to replace potassium bromide[33] and protein crosslinking agents such as formaldehyde.[34]

Despite peroxidase's large number of applications, horseradish is the main source which makes the purification process expensive.[20] For this reason, the need to look for alternative sources arose. Soybean peroxidase (SBP) attracted the attention of researchers due to its high thermal stability even at low pH and the particularity of presenting enzymatic activity in organic solvents.[35] This peroxidase is expressed in the soybean seed coat as a single isoenzyme and, until now, its function in the plant is unknown. SDS-PAGE electrophoresis and mass spectrometry indicate that the isozyme present in SBH has a molecular mass of approximately 38 kDa and a pI of 5.21.[19] On the other hand, the treatment of the enzyme with glycopeptidases together with MALDI-TOF mass spectroscopy studies concluded that the carbohydrates bound to the protein represent an 18.2% of the enzyme mass.[36]

On the other hand, SBP and horseradish peroxidase have a 52% homology. In addition, it was shown that SBP is kinetically more powerful and is much more stable, characteristics that make it a good substitute for commercial peroxidase.[37–39]

Another important characteristic of SBP is that it has been shown to present Lignin peroxidase type activity, which is capable of oxidizing veratryl alcohol as well as methoxybenzenes while remaining stable under the necessary reaction conditions.[40] This increases the potential of SBP as an alternative to the existing lignin peroxidases in the market.

11.2.2 PROTEASE INHIBITORS

Kapoor and Gupta (1978) studied the distribution of Trypsin inhibitors (TI) in soybeans and were able to determine that the highest TI activity resides in its hull.[41] Among the protease inhibitors present in soybeans, the most representative is the Kunitz soybean trypsin inhibitor (KTI) and the Bowman–Birk inhibitor (BBI).[42]

BBI comprises a family of serine protease inhibitors with similar structural features and different isoforms present mainly in legume seeds.[43] BBI has been isolated from both seed and hulls and its main characteristic is that it has the ability to inhibit both trypsin and chymotrypsin.[44] BBI from soybean cotyledons has been isolated, purified, and characterized. It presents a MW of 8 KDa, a pI of 4.2, 71 amino acid residues, and 7 disulfide bridges in its structure.[45] In addition, BBI has been identified in SBH extracts exhibiting characteristics similar to those present in cotyledons.[45] BBI inhibits protease activity by a mechanism similar to that of the substrate enzyme. The inhibitory loop of BBI inserts into the protease active site, blocking it, resulting in a tight complex that lowers its hydrolysis rates.[43]

The importance of the search for new sources for the study, isolation, and purification of this protease inhibitor lies in the potential applications it presents. It has been reported that BBI from soybeans has anticarcinogenic and chemopreventive properties.[46–48] BBI is highly effective in suppressing carcinogenesis in animal systems and in vitro transformation systems. Furthermore, BBI inhibits certain types of proteolytic activities, oncogenic expressions, and reactive oxygen species production by polymorphonuclear leukocytes, among others.[49] There is also epidemiological evidence that soy-derived protease inhibitors, like other protease inhibitors, can reduce cancer death rates.[50]

KTI is one of the main antinutritional factors present in soybeans and is also responsible for inhibiting dietary proteins digestion.[51] Although this inhibitor is mainly isolated from soybeans seeds, its presence in hulls has been demonstrated,[19] identifying different isoforms with MW between 22 and 24 KDa and variable pI (see Table 11.1). Also, other proteins similar to KTI have been identified.[19] Compared to BBI, KTI only has the capacity to inhibit trypsin. Amongst potential applications, KTI showed anticarcino-genic activity, anti-HIV-transcriptase, and immunoregulatory properties.[52] Furthermore, it interacts with cells as a negative modulator and inhibits signaling pathways in ovarian cancer cells that lead to cell invasion.[53]

11.2.3 PATHOGENESIS-RELATED PROTEIN FAMILY

Pathogenesis-related (PR) proteins are a nonhomogeneous family of proteins present in all plants expressed in response to stress, attack by pathogens, and certain abiotic stimuli since they function as the plants defense system.[54] PR proteins are constitutively expressed in different plant tissues and organs during their growth. They are classified into 17 different functional families based on their primary structure, biological activity, and serological criteria.[55] Within these families, the PR-5 stands out due to their properties. It has been shown that overexpression of these proteins in plants could increase their resistance to pathogens.[56] Several PR-5 proteins induced by biotic stress have shown to have antifungal activity, and others showed glucanase activity and binding activity to β-1,3 glucans.[57]

Several members of this protein family have been identified in soybeans. Liu et al. purified and identified an alkaline PR-5 in SBH extracts which presents a MW of 21.6 kDa and a pI of 8.26, a trait that distinguishes it from the rest of PR-5 found in soybeans since they have lower pI.[58] Onishi et al. identified a salt stress-inducible gene that codes for a PR-5 type protein in soybean roots, but the expression in other plant parts has not been found.[57] Graham et al. have identified a homolog of PR-5, named P-21, which is expressed in the leaves of healthy plants and has a MW of 21.48 and a pI of 4.6.[59]

All these identified proteins and those to be identified have potential in various applications, from the design of new transgenic plants with increased resistance to pathogens and environmental stress, or as antifungal and antimicrobial agents.

11.2.4 CLASS I CHITINASE

These are a group of enzymes that catalyze the hydrolysis of β-1,4-N-acetylglucosamine bonds present in chitin. Chitinases are constitutively expressed in plants, although some are expressed in response to stress situations.[60] They are another of the plants' defenses against pathogens since chitin is the primary component of cell walls in fungi.[61]

According to their primary and secondary structure, plant chitinases have been classified into six classes. Enzymes with globular domains are grouped in classes I, II, and IV, while in classes III and V the proteins present 8 α-helices and 8 β-strands. The former the hydrolysis of the β-1,4 glycosidic linkage through an inverting mechanism, and the latter through a retaining mechanism.[62] Class I plant chitinases have been identified as major food and contact allergens and are implicated in latex-fruit syndrome.[63]

Chitinases have been identified in different parts of the soybean plant.[64,65] In particular, a class I chitinase with a MW of 32 kDa has been identified in SBH.[61] In addition to its defense against pathogens, the study and characterization of this family of proteins is important due to its role as an allergen in humans.

11.3 EXTRACTION AND PURIFICATION OF PROTEINS PRESENT IN SBH

As indicated above, residual plant biomass is a highly complex system. Their structure and chemical composition for the same type of biomass usually varies depending on climatic conditions, substrate, humidity, etc., in which the biomass was formed. However, the chemical composition usually varies within a range, which allows the same methodology to be applied many times to extract a desired bioactive. Aqueous extraction, ethanol/water extraction, and leaching with buffers are the most frequent procedures used for extracting soluble proteins from SBH.[19,45,58,66] The procedures for purifying proteins vary according to their characteristics and the degree of purity desired. Below are some of the methodologies used for the purification and extraction of soluble proteins present in SBH.

11.3.1 SBP EXTRACTION AND PURIFICATION

SBP has many potential applications making its commercial production of interest, taking into account that it is a renewable biocatalyst that would not affect animal feed, nor would it require more land or other agricultural resources for its production. Its extraction could be easily incorporated with soybeans crushing operations adding more value to this type of crop.[67] In recent years, interest in SBP has increased notably due to its applications, unusual characteristics, and mainly because a single peroxidase isoenzyme is found in SBH, which facilitates its characterization and purification.

In 1991, Gilking et al. began studying and characterizing SBP. This work showed that the peroxidase activity is located mainly in the soybean's hull. For the isolation and purification of the SBP, they used a leaching process using KH_2PO_4, pH 7.5 as the extractive phase in a 10:1 proportion (10 mL of buffer per gram of SBH). Under these conditions, they achieved the extraction of the enzyme that was subsequently purified using successive chromatographic steps on DEAE Sepharose and A-Sepharose columns, to finally concentrate the sample by ultrafiltration, obtaining a yield of 14% and purifying the enzyme 40.8 times.[68] This work focused on the characterization of the present isoform and was one of the pioneers in the extraction and purification of SBP. Later on, the group continued with the characterization and extraction of this enzyme, which determined its great importance due to its wide range of applications.[35,40] In 1998, Wen et al. purified SBP from an aqueous extract using consecutive treatments of fractional precipitation with ammonium sulfate, ion-exchange chromatography on DEAE-Sephadex A 50, affinity chromatography on a Con A-Sepharose 4B, and gel filtration on Bio-Gel P-60.[69] In addition, they have focused part of their efforts on the characterization and purification of soybean proteins. In 2007, they developed a simple extraction method to obtain SBP from SBH using distilled water and a 24 h incubation time. The purification was later carried out by subjecting the aqueous phase to fractional precipitation with ammonium sulfate and, later, a biphasic extraction using acetone.[70] In this case, they obtained SBP crystals with a yield of 68.5% and high purity. Several works use these extractive and purification techniques due to the high purities obtained. However, the proposed methodologies present low yields and use complex techniques to scale, so the need to develop more straightforward methodologies arises.

In this sense, Bracco et al. have studied the purification of SBP using simpler methodologies and, in addition, analyzed different conditions for the extraction of the enzyme from SBH. They found that, of all the conditions tested for the extraction, a leaching method that uses an aqueous phase of sodium phosphate pH 6.5 with continuous stirring at 25°C was the condition that yielded the best results. To purify the enzyme from the previous extract, they used a dye affinity chromatography with Reactive Blue 4 attached to chitosan minispheres, obtaining an adsorption yield of 96%. Consequently, they had to face with the complication in recovering into the liquid phase, the protein again and finally achieving a recovery of 49.1%, purifying the enzyme 6.9 times.[71]

11.3.2 TI EXTRACTION AND PURIFICATION

The protease inhibitors present in SBH have a high potential to be applied in different fields of research and in medical treatments. However, unlike observed with SBP, few studies extraction and purification of this type of protein present in SBH.

KTI-type inhibitors have been isolated mainly from soybeans using protocols with large number of steps and that use organic solvents due to the grain's lipid composition. Successive ammonium sulfate precipitation steps and various types of chromatography are also used for purification. Despite the lack of a bibliography, SBH is a source of various types of KTI (see Table 11.1), positioning it as an excellent raw material for developing future processes for obtaining and purifying this type of protein.

In the case of BBI, extraction and purification from SBH has been studied. In 2001, Sessa and Wolf identified the presence of BBI in SBH extract.[44]

The developed protocol consists of an extraction stage with various steps that include Soxhlet extraction with hexane:ethanol, extraction with ethanol, and precipitation with acetone. The separation of the proteins obtained is carried out using successive chromatographic steps. Although this protocol allowed to isolate and identify BBI in soybean seed coats, the large number of steps it presents, the use of organic solvents and chromatography makes its application on a large scale difficult. That is why, in 2003, Sessa presented an improved protocol for the extraction of proteins present in SBH. In this work, he compared the extraction of proteins from aqueous extracts using different sources of SBH and concluded that the distribution of BBI does not depend on the species while SBP does. For the purification of BBI he used a chromatographic procedure with a CM Sephadex C-25 column, achieving an acceptable purification of the protein under study.

11.3.3 OPTIMIZATION OF A JOINT SBP AND TI EXTRACTION AND PURIFICATION PROTOCOL

The extraction of the soluble fraction of proteins from SBH has been carried out relatively empirically, leaving a certain amount of biomass in contact with water or buffer. The optimal conditions for this extraction process have not been studied. The two proteins of most significant interest for their use in biotechnology and medicine present in SBH are SBP and the different types of TI. Several authors have studied these types of proteins independently. Given the interest of both macromolecules and the need for a methodology that allows obtaining both proteins simultaneously, the problem was addressed using statistical tools by Camiscia et al.[19] The experimental development was divided into two approaches: in a first stage, they optimized the extraction of proteins from SBH, and in the second stage, they optimized the purification of SBP and TI from the extract obtained. The optimization of any process using statistical tools consists of several steps: (1) Hypothesis plan, determines with the help of bibliography, the potentially significant factors that may influence the response under study; (2) Screening, through a factorial model that determines the significant factors; (3) Optimization, through a response surface model where it is expected to achieve a maximization of the response under study for each property, according to appropriate mathematical models; (4) Validation, where it is expected to verify the robustness of the optimal conditions achieved. Based on these premises, the results obtained are summarized below.

11.3.3.1 LIXIVIATION

The bibliographic analysis shows that the extraction of soluble proteins present in SBH is carried out mainly by extraction in an aqueous phase at different temperatures.[40,67,70] Based on this analysis, Camiscia et al. established a starting condition for the study of the extraction process, a lixiviation with sodium phosphate buffer 50 mM pH 6.0. When screening for the conditions that could affect the SBP and TI extraction process, they selected the following factors: biomass particle size, SBH mass/buffer volume ratio, type of stirring (magnetic or orbital), and incubation time. All aspects were varied at two levels using a complete factorial design. The dependent variables that were quantified were: SBP enzymatic activity, antitrypsin inhibitory capacity, and total proteins.

It was determined that SBH/buffer ratio, particle size, and incubation times were the factors that significantly affected the extraction of SBP. In addition, the interaction between the studied factors were also statistically significant.[19] In the case of TI, differences in the systems' behavior were observed when compared to SBP. The factors that most significantly influenced the extraction were: SBH/buffer ratio and incubation time. The particle size was not as important since it had a slight influence on the process.

Once the factors that influenced the extraction stage were identified, they optimized it using a central compound design. In this case, the previous results were taken into account and the optimization was carried out with the following factors: particle size, SBH/buffer ratio, and incubation time. The optimization of the extraction conditions was analyzed for each protein separately. Then the Derringer desirability function was applied, where different criteria combinations of responses and factors can be optimized simultaneously. The results obtained are presented in Table 11.4.

TABLE 11.4 Optimization of SBP and TI Extraction From SBH.

Condition	Time (min)	Ratio SBH/volume in aqueous phase (g/mL)	Size of SBH
SBP	480	0.066	40
TI	331	0.066	20
SBP + TI	360	0.066	40

The analysis by LC-MS, SDS-PAGE, and measurements of total proteins allowed to verify the presence of more than 4000 proteins in the extract obtained, part of which are presented in Tables 11.2 and 11.3, which demonstrates the optimized method's potential to obtain other proteins present in SBH.[19]

11.3.3.2 PURIFICATION OF SBP AND TI BY ADSORPTION

As detailed above, the methodologies used for the purification of SBP and TI found in the literature require various steps and chromatographic techniques with sophisticated equipment which raises the cost of the process. In this sense, the design of new methodologies for the purification and concentration of proteins is the key. In order to maximize the value of the waste, Camiscia et al. has developed an integrated process for obtaining and purifying SBP and TI using cellulose extracted from SBH that later was chemically modified and used as an adsorbent bed. The process developed is shown in Figure 11.2.

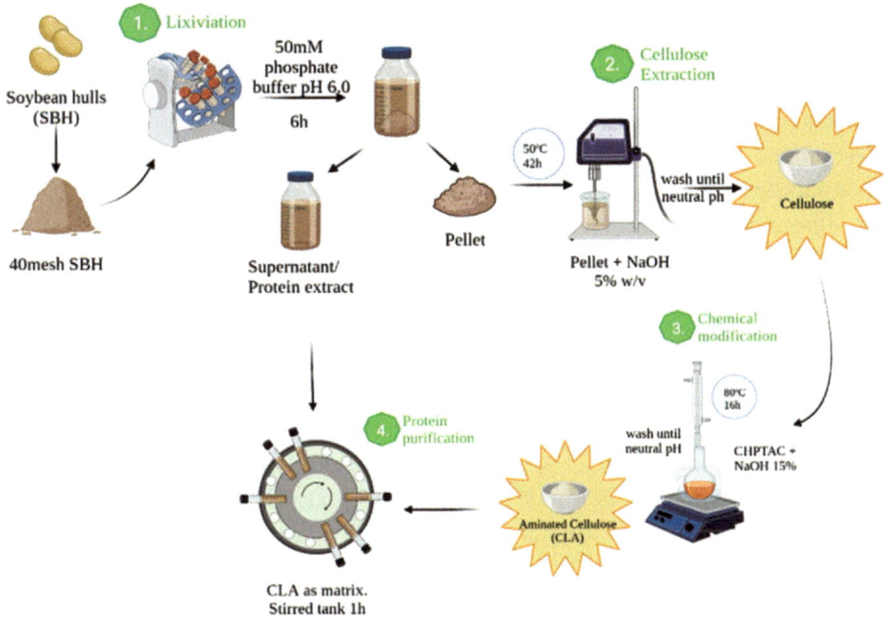

FIGURE 11.2 Optimized protocol for the exploitation of SBH.

As can be seen, the optimization of the extraction process is applied as a first step in exploiting this residue. Subsequently, the residue is subjected to a treatment with NaOH 5% w/v at 50°C for 42 h, to extract its cellulose content. Later, the cellulose is modified by adding quaternary ammonium groups, turning it into an adsorbent bed with ion exchange properties.[7] This bed was studied as a possible adsorbent for the purification of SBP and TI by batch adsorption.

The purification protocol was studied for each protein separately and, again, it was optimized using statistical tools and experimental design. For both proteins, they found that the factors that influenced the adsorption process were pH and the ionic strength of the working medium, which is consistent with an electrostatic mechanism of interaction between the molecules of interest and the bed under study. However, due to the differences in structure and properties of SBP and TI, a differential adsorption behavior was observed. In this sense, the optimal conditions for SBP adsorption using the cationized cellulose bed were: pH 5.75 without adding salts to the medium, yielding a maximum adsorption capacity of 0.0916 SBP/mg matrix. The elution of the enzyme by adding ionic strength to the system resulted in purification of 8.13 with a yield of 66%, more than promising data, comparable with previous protocols developed and with the advantage of being single step purification.

In the case of TI, it is important to note that it is not a single protein but rather a set of proteins with similar characteristics with the same function (Table 11.3). The results obtained for optimizing the adsorption process were the following: pH 8, without the addition of ionic strength and at a temperature of 35C, with an adsorption capacity of 2.47 with a yield of 26%. In this case, the efficiency of the process is much lower than that obtained with SBP, which may be because there is not just one protein present with the same activity but rather a set of similar proteins.

11.4 CONCLUSIONS AND FUTURE PERSPECTIVES

Globalization has been based on a production and consumption model known as a linear economy under the paradigm "take-make-waste" ("extract-manufacture-consume-eliminate"), causing the depletion of natural resources and the generation of tons of residues.[72] The growing concern about the interrelation of the economy and the environment leads to the need to face a change in the development and growth model that allows optimizing the use of available resources, materials, and products in a new paradigm known as the circular economy.[73] Within this paradigm, the use of waste from different industrial activities is framed as a source of molecules with added value. Worlwide, soybean is one of the main crops due to its properties and ease of cultivation. During its industrialization, various wastes with distinctive characteristics are produced making them interesting to use as source of different molecules of industrial interest. In this sense, SBH has proven to be an excellent source of various molecules of interest, additionally containing an enormous protein fraction with potential of discovering new applications and characterization of molecules that have not yet been identified.

Furthermore, the protocols and processes developed could be applied to other waste matrices for the recovery of various molecules of interest, which represents the principal advantage to achieve greater and effective waste management and obtention of diversified value-added natural products.

KEYWORDS

- **antitrypsin**
- **peroxidase**
- **soy hull**
- **waste**
- **soybean**

REFERENCES

1. Kumar, S.; Kushwaha, R.;Verma, M. L. Recovery and Utilization of Bioactives from Food Processing Waste. In *Biotechnological Production of Bioactive Compounds*, Ed.; Elsevier, 2020; pp 37–68.
2. Johnson, L. K.; Dunning, R. D.; Gunter, C. C.; Dara Bloom, J.; Boyette, M. D.;Creamer, N. G. Field Measurement in Vegetable Crops Indicates Need for Reevaluation of On-Farm Food Loss Estimates in North America. *Agric. Syst.* **2018,** *167*, 136–142.
3. World Soybean Production 2020/2021. [Online] 2021. http://www.worldagriculturalproduction.com/crops/soybean.aspx
4. WORLD SOYBEAN PRODUCTION. [Online] 2021. http://www.sopa.org/statistics/world-soybean-production/
5. [Online] 2021. https://capeco.org.py/ranking-mundial-es/
6. Medic, J.; Atkinson, C.; Hurburgh, C. R. Current Knowledge in Soybean Composition. *J. Am. Oil Chem. Soc.* **2014,** *91*, 363–384.
7. Camiscia, P.; Giordano, E. D.; Brassesco, M. E.; Fuciños, P.; Pastrana, L.; Cerqueira, M.; et al. Comparison of Soybean Hull Pre-treatments to Obtain Cellulose and Chemical Derivatives: Physical Chemistry Characterization. *Carbohydr. Polym.* **2018,** *198*, 601–610.
8. Tsai, W. T.; Lee, M. K.; Chang, Y. M. Fast Pyrolysis of Rice Husk: Product Yields and Compositions. *Bioresour. Technol.* **2007,** *98*, 22–28.

9. Mielenz, J. R.; Bardsley, J. S.; Wyman, C. E. Fermentation of Soybean Hulls to Ethanol While Preserving Protein Value. *Bioresour. Technol.* **2009,** *100*, 3532–3539.
10. Sarrouh, B. F.; dos Santos, J. C.; Cunha, M. A. A.; Branco, R. F. Potential Biomass Resources for Cellulosic Ethanol Production in Brazil: Availability, Feedstock Analysis, Feedstock Composition, and Conversion Yields. In *Biofuels in Brazil*, Ed.; Springer, 2014; pp 97–123.
11. Dagnino, E. P.; Chamorro, E.; Romano, S. D.; Felissia, F.; Area, M. C. Optimization of the Acid Pretreatment of Rice Hulls to Obtain Fermentable Sugars for Bioethanol Production. *Ind. Crops Prod.* **2013,** *42*, 363–368.
12. Rahman, I.; Riley, F. The Control of Morphology in Silicon Nitride Powder Prepared from Rice Husk. *J. Eur. Ceram. Soc.* **1989,** *5*, 11–22.
13. Okutani, T.; Ino, S. Synthesis of Metallic Silicon by Reaction of Silica in Rice Hulls with Aluminum and its Hydrochlorination. *J. Solid State Chem.* **2019,** *276*, 93–99.
14. Ghaffar, S. H.; Fan, M. Structural Analysis for Lignin Characteristics in Biomass Straw. *Biomass Bioenergy* **2013,** *57*, 264–279.
15. Chang, V. S.; Nagwani, M.; Kim, C.-H.; Holtzapple, M. T. Oxidative Lime Pretreatment of High-Lignin Biomass. *Appl. Biochem. Biotechnol.* **2001,** *94*, 1–28.
16. Singh, R. D.; Banerjee, J.; Sasmal, S.; Muir, J.; Arora, A. High Xylan Recovery Using Two Stage Alkali Pre-treatment Process from High Lignin Biomass and its Valorisation to Xylooligosaccharides of Low Degree of Polymerisation. *Bioresour. Technol.* **2018,** *256*, 110–117.
17. Gao, L.; Chen, S.; Zhang, D. Advances in Modifying Lignin Structures for Largely Enhancing High-Lignin Biomass Saccharification, *Process Biochem,* **2017,** *57*, 175–180.
18. Barros, P. J. R.; Ascheri, D. P. R.; Santos, M. L. S.; Morais, C. C.; Ascheri, J. L. R.; Signini, R.; et al. Soybean Hulls: Optimization of the Pulping and Bleaching Processes and Carboxymethyl Cellulose Synthesis. *Int. J. Biol. Macromol.* **2020,** *144*, 208–218.
19. Camiscia, P.; Silva, J. M.; Picó, G.; Valetti, N. W. Extraction and Purification of Peroxidase and Trypsin Inhibitor from Soybean Hulls: A Strategy to Revalue a Waste as a Source of Different Types of Molecules of Biotechnological Interest. *Sep. Purif. Technol.* **2020,** *252*, 117456.
20. Welinder, K. G.; Justesen, A. F.; Kjærsgård, I. V.; Jensen, R. B.; Rasmussen, S. K.; Jespersen, H. M.; et al. Structural Diversity and Transcription of Class III Peroxidases from Arabidopsis Thaliana. *Eur. J. Biochem.* **2002,** *269*, 6063–6081.
21. Hiraga, S.; Sasaki, K.; Ito, H.; Ohashi, Y.; Matsui, H. A Large Family of Class III Plant Peroxidases. *Plant Cell Physiol.* **2001,** *42*, 462–468.
22. Fry, S. C. Oxidative Coupling of Tyrosine and Ferulic Acid Residues: Intra- and Extra-Protoplasmic Occurrence, Predominance of Trimers and Larger Products, and Possible Role in Inter-Polymeric Cross-Linking. *Phytochem. Rev.* **2004,** *3*, 97–111.
23. Lehtonen, M. T.; Akita, M.; Kalkkinen, N.; Ahola-Iivarinen, E.; Rönnholm, G.; Somervuo, P.; et al. Quickly-Released Peroxidase of Moss in Defense Against Fungal Invaders. *New Phytol.* **2009,** *183*, 432–443.
24. Almagro, L.; Gómez Ros, L.; Belchi-Navarro, S.; Bru, R.; Ros Barceló, A.; Pedreno, M. Class III Peroxidases in Plant Defence Reactions. *J. Exp. Bot.* **2009,** *60*, 377–390.
25. Liszkay, A.; van der Zalm, E.; Schopfer, P. Production of Reactive Oxygen Intermediates (O2−, H2O2, and OH) by Maize Roots and Their Role in Wall Loosening and Elongation Growth. *Plant Physiol.* **2004,** *136*, 3114–3123.

26. Marjamaa, K.; Kukkola, E. M.; Fagerstedt, K. V. The Role of Xylem Class III Peroxidases in Lignification. *J. Exp. Bot.* **2009,** *60*, 367–376.
27. Darwesh, O. M.; Matter, I. A.; Eida, M. F. Development of Peroxidase Enzyme Immobilized Magnetic Nanoparticles for Bioremediation of Textile Wastewater Dye. *J. Environ. Chem. Eng.* **2019,** *7*, 102805.
28. Bilal, M.; Iqbal, H. M.; Hu, H.; Wang, W.; Zhang, X. Development of Horseradish Peroxidase-Based Cross-Linked Enzyme Aggregates and Their Environmental Exploitation for Bioremediation Purposes. *J. Environ. Manag.* **2017,** *188*, 137–143.
29. Sharma, B.; Dangi, A. K.; Shukla, P. Contemporary Enzyme Based Technologies for Bioremediation: A Review. *J. Environ. Manag.* **2018,** *210*, 10–22.
30. Morsi, R.; Bilal, M.; Iqbal, H. M.; Ashraf, S. S. Laccases and Peroxidases: The Smart, Greener and Futuristic Biocatalytic Tools to Mitigate Recalcitrant Emerging Pollutants. *Sci. Total Environ.* **2020,** *714*, 136572.
31. Ricco, R.; Wied, P.; Nidetzky, B.; Amenitsch, H.; Falcaro, P. Magnetically Responsive Horseradish Peroxidase@ZIF-8 for Biocatalysis. *Chem. Commun.* **2020,** *56*, 5775–5778.
32. Pandey, V.; Awasthi, M.; Singh, S.; Tiwari, S.; Dwivedi, U. A Comprehensive Review on Function and Application of Plant Peroxidases. *Biochem. Anal. Biochem.* **2017,** *6*.
33. Hilhorst, R.; Dunnewind, B.; Orsel, R.; Stegeman, P.; Van Vliet, T.; Gruppen, H.; et al. Baking Performance, Rheology, and Chemical Composition of Wheat Dough and Gluten Affected by Xylanase and Oxidative Enzymes. *J. Food Sci.* **1999,** *64*, 808–813.
34. Heck, T.; Faccio, G.; Richter, M.; Thöny-Meyer, L. Enzyme-Catalyzed Protein Cross-linking. *Appl. Microbiol. Biotechnol.* **2013,** *97*, 461–475.
35. Blinkovsky, A. M.; McEldoon, J. P.; Arnold, J. M.; Dordick, J. S. Peroxidase-Catalyzed Polymerization and Depolymerization of Coal in Organic Solvents. *Appl. Biochem. Biotechnol.* **1994,** *49*, 153–164.
36. Gray, J. S.; Yang, B. Y.; Hull, S. R.; Venzke, D. P.; Montgomery, R. The Glycans of Soybean Peroxidase. *Glycobiology* **1996,** *6*, 23–32.
37. Kamal, J. A.; Behere, D. V. Activity, Stability and Conformational Flexibility of Seed Coat Soybean Peroxidase. *J. Inorg. Biochem.* **2003,** *94*, 236–242.
38. Ryan, B. J.; Carolan, N.; Ó'Fágáin, C. Horseradish and Soybean Peroxidases: Comparable Tools for Alternative Niches? *Trends Biotechnol.* **2006,** *24*, 355–363.
39. Kamal, J. A.; Behere, D. V. Thermal and Conformational Stability of Seed Coat Soybean Peroxidase. *Biochemistry* **2002,** *41*, 9034–9042.
40. McEldoon, J. P.; Pokora, A. R.; Dordick, J. S. Lignin Peroxidase-Type Activity of Soybean Peroxidase. *Enzyme Microb. Technol.* **1995,** *17*, 359–365.
41. Kapoor, A.; Gupta, Y. Trypsin Inhibitor Activity in Soybean Seed as Influenced by Stage of its Development and Different Treatments and the Distribution in its Anatomical Parts [India]. *Indian J. Nutr. Diet.* **1978**.
42. Roosta, H. R.; Javadi, T.; Nazari, F. Isolation and Characterization of Trypsin Inhibitors (Kunitz Soybean Trypsin Inhibitor, Bowman-Birk Inhibitor) in Soybean. *Adv. Environ. Biol.* **2011,** 145–154.
43. Gu, C.; Song, X.; Zhao, L.; Pan, S.; Qin, G. Purification and Characterization of Bowman-Birk Trypsin Inhibitor from Soybean. *J. Food Nutr. Res.* **2014,** *2*, 546–550.
44. Sessa, D. J.; Wolf, W. J. Bowman–Birk Inhibitors in Soybean Seed Coats. *Ind. Crops Products* **2001,** *14*, 73–83.

45. Sessa, D. J. Processing of Soybean Hulls to Enhance the Distribution and Extraction of Value-Added Proteins. *J. Sci. Food Agric.* **2004,** *84*, 75–82.
46. Kennedy, A. Cancer Prevention by Bowman–Birk Inhibitor Concentrate (BBIC). *Cancer Nutrition*; Prasad, K. N., Cole, W. C., Eds.; IOS Press: Amsterdam, 1998; pp 93–97.
47. Kennedy, A. R.;Szuhaj, B. F. Bowman-Birk Inhibitor Product for Use as an Anticarcinogenesis Agent. Ed.; Google Patents, 1994.
48. Maki, P. A.; Paterson, Y.;Kennedy, A. R. Studies Related to the Potential Antigenicity of the Bowman-Birk Inhibitor, An Anticarcinogenic Protease Inhibitor Isolated from Soybeans, 1994.
49. Kennedy, A. R.; Szuhaj, B. F.; Newberne, P. M.; Billings, P. C. Preparation and Production of a Cancer Chemopreventive Agent, Bowman-Birk Inhibitor Concentrate, 1993.
50. Kennedy, A. Overview. Anticarcinogenic Activity of Protease Inhibitors. *Protease Inhibitors as Cancer Chemopreventive Agents*; Troll, W., Kennedy, A. R., Eds.; Plenum Press: New York, 1993; pp 9–64.
51. Zhou, T.; Han, S.; Li, Z.;He, P. Purification and Quantification of Kunitz Trypsin Inhibitor in Soybean Using Two-Dimensional Liquid Chromatography. In *Food Analytical Methods*, 2017; vol 10, pp 3350–3360.
52. Price, S. J.; Pangloli, P.; Krishnan, H. B.;Dia, V. P. Kunitz Trypsin Inhibitor in Addition to Bowman-Birk Inhibitor Influence Stability of Lunasin Against Pepsin-Pancreatin Hydrolysis. *Food Res. Int.* 2016, *90*, 205–215.
53. Inagaki, K.; Kobayashi, H.; Yoshida, R.; Kanada, Y.; Fukuda, Y.; Yagyu, T.; et al. Suppression of Urokinase Expression and Invasion by a Soybean Kunitz Trypsin Inhibitor are Mediated Through Inhibition of Src-Dependent Signaling Pathways. *J. Biol. Chem.* **2005,** *280*, 31428–31437.
54. Hoffmann-Sommergruber, K. Pathogenesis-Related (PR)-Proteins Identified as Allergens. *Biochem. Soc. Trans.* **2002,** *30*, 930–935.
55. Fan, S.; Jiang, L.; Wu, J.; Dong, L.; Cheng, Q.; Xu, P.; et al. A Novel Pathogenesis-Related Class 10 Protein Gly m 4l, Increases Resistance Upon Phytophthora Sojae Infection in Soybean (Glycine max [L.] Merr.). *PLoS One* **2015,** *10*, e0140364.
56. Koiwa, H.; Kato, H.; Nakatsu, T.; Oda, J. i.; Yamada, Y.; Sato, F. Crystal Structure of Tobacco PR-5d Protein at 1.8 Å Resolution Reveals a Conserved Acidic Cleft Structure in Antifungal Thaumatin-Like Proteins. *J. Mol. Biol.* **1999,** *286*, 1137–1145.
57. Onishi, M.; Tachi, H.; Kojima, T.; Shiraiwa, M.;Takahara, H. Molecular Cloning and Characterization of a Novel Salt-Inducible Gene Encoding an Acidic Isoform of PR-5 Protein in Soybean (Glycine max [L.] Merr.). *Plant Physiol. Biochem.* **2006,** *44*, 574–580.
58. Liu, C.; Cheng, F.; Sun, Y.; Ma, H.;Yang, X. Structure–Function Relationship of a Novel PR-5 Protein with Antimicrobial Activity from Soy Hulls. *J. Agric. Food Chem.* **2016,** *64*, 948–959.
59. Graham, J. S.; Burkhart, W.; Xiong, J.;Gillikin, J. W. Complete Amino Acid Sequence of Soybean Leaf P21. *Similarity Thaumatin-Like Polypeptides* **1992,** *98*, 163–165.
60. Hamid, R.; Khan, M. A.; Ahmad, M.; Ahmad, M. M.; Abdin, M. Z.; Musarrat, J.; et al. Chitinases: An Update. *J. Pharm. Bioallied Sci.* **2013,** *5*, 21–29.
61. Gijzen, M.; Kuflu, K.; Qutob, D.; Chernys, J. T. A Class I Chitinase from Soybean Seed Coat. *J. Exp. Bot.* **2001,** *52*, 2283–2289.
62. Fukamizo, T.; Sakai, C.; Tamoi, M. Plant Chitinases: Structure-Function Relationships and Their Physiology. *Foods Food Ingredients J. Jpn.* **2003,** *208*.

63. Breiteneder, H.; Ebner, C. Molecular and Biochemical Classification of Plant-Derived Food Allergens. *J. Allergy Clin. Immunol.* **2000,** *106*, 27–36.
64. Gálusová, T.; Rybanský, Ľ.; Mészáros, P.; Spieß, N.; Piršelová, B.; Kuna, R.; et al. Variable Responses of Soybean Chitinases to Arsenic and Cadmium Stress at the Whole Plant Level. *Plant Growth Regul.* **2015,** *76*, 147–155.
65. Wadsworth, S. A.; Zikakis, J. P. Chitinase from Soybean Seeds: Purification and Some Properties of the Enzyme System. *J. Agric. Food Chem.* **1984,** *32*, 1284–1288.
66. Liu, C.; Cheng, F.-F.; Liu, X.; Ma, H.-Y.; Yang, X.-Q. Improved Extraction of Disulphide-Rich Bioactive Proteins from Soya Hulls: Characterisation of a Novel Aspartic Proteinase. *Int. J. Food Sci. Technol.* **2016,** *51*, 1509–1515.
67. Steevensz, A.; Madur, S.; Al-Ansari, M. M.; Taylor, K. E.; Bewtra, J. K.; Biswas, N. A Simple Lab-Scale Extraction of Soybean Hull Peroxidase Shows Wide Variation Among Cultivars. *Ind. Crops Products* **2013,** *48*, 13–18.
68. Gillikin, J. W.; Graham, J. S. Purification and Developmental Analysis of the Major Anionic Peroxidase from the Seed Coat of Glycine Max. *Plant Physiol.* **1991,** *96*, 214–220.
69. Wen, L.; Jing, F.; Peiji, G. Purification and Characterization of Soybean Hull Peroxidase. *Zhongguo Sheng wu hua xue yu fen zi Sheng wu xue bao= Chinese J. Biochem. Mol. Biol.* **1998,** *14*, 577–582.
70. Liu, J.; Liu, H.; Zhang, Y.; Qiu, L.; Su, F.; Li, F.; et al. A Simple Preparation Method of Crystals of Soybean Hull Peroxidase. *Appl. Microbiol. Biotechnol.* **2007,** *74*, 249–255.
71. Bracco, L. F.; Levin, G. J.; Navarro del Cañizo, A. A.; Wolman, F. J.; Miranda, M. V.; Cascone, O. Simultaneous Purification and Immobilization of Soybean Hull Peroxidase with a Dye Attached to Chitosan Mini-Spheres. *Biocatal. Biotransfor.* **2017,** *35*, 306–314.
72. González Ordaz, G. I.; Vargas-Hernández, J. G. La economía circular como factor de la responsabilidad social. *Economía Coyuntural* **2017,** *2*, 105–130.
73. Sánchez, A. A.; Ocampo, E. D.; Fuentes, N. C. Desarrollo Sostenible & Economía Circular. *Revista de Ciencias Jurídicas* **2021,** *154*, 165–206.

CHAPTER 12

Antimicrobial Activity of Olive Leaf Extract: Bioactives Extraction and Potential as a Food Preservative

MÓNICA SÁNCHEZ-GUTIÉRREZ, ISABEL BASCÓN-VILLEGAS, and ELENA CARRASCO

Food Science and Technology Department, Universidad de Córdoba, Darwin Building, Córdoba, Spain

ABSTRACT

Spain contributes with 45% of worldwide olive oil production and 60% of the European production. As a consequence of this activity, a huge amount of waste is generated, which generally has no industrial application and must be discarded. From an economic point of view, due to the large amount of leaves discarded in the production of olive oil and table olives, the obtention of high value-added products such as polyphenols, constitutes an interesting exploitation via considering the high availability and low price of olive leaves. In this respect, emerging green extraction methods have been developed using different ecological and nontoxic solvents in order to contribute to the sustainable valorization of this byproduct.

Based on scientific evidence, olive leaf is an excellent source of bioactive compounds, including oleuropein, verbascoside, rutin, tyrosol, and hydroxytyrosol with demonstrated beneficial health effects. Their well-known antioxidant activity together with their antimicrobial capacity make these polyphenols highly appreciated by the pharmaceutical, cosmetic, and food industries. In the latter sector, these active compounds would be used as natural additives, revealing the high potential of olive leaf extracts for food applications.

Food Byproducts Management and their Utilization. Ricardo Gómez-García, Ana A. Vilas-Boas, Débora A. Campos, Maria Manuela Pintado, & Cristóbal Noé Aguilar (Eds.)

12.1 INTRODUCTION

Agriculture and agri-food industries generate a huge amount of byproducts and waste with high bioactive potential. In this context, the replacement of the current production model based on a linear economy (take-make-use-dispose) by a more sustainable and eco-friendly production based on a circular economy (grow-make-use-restore), in which the valorization of byproducts plays a key role, would contribute to the reduction of the negative impact of such residues in the environment.[1] Consumers not only have become more environmentally conscious but also more "food conscious." In the last decade, the increasing awareness of diet influences on health has led to a change in consumer preferences, which has resulted in the growing interest of the food industry in the bioactive compounds present in natural extracts obtained from byproducts. Nevertheless, despite their promising applications in the industry, their current valorization is not exploited sufficiently.[2] Polyphenolic compounds are the most abundant secondary plant metabolites. In recent years, the intake of polyphenols has been associated with cardiovascular protection, as well as, their role in the prevention of inflammatory disorders, diabetes, Alzheimer's, or cancer.[3] However, polyphenolic compounds are well known for their antioxidant and antimicrobial activity. This antimicrobial activity has been evaluated against different foodborne and non-foodborne microorganisms, thus adopting a promising role in applications as a natural antimicrobial, thus reducing the use (or abuse) of antibiotics, one of the main global issues to public health.[4] In addition, in the food industry, the reduction or replacement of synthetic preservatives such as nitrates (E-251 and E-252) or nitrites (E-249 and E-250) by extracts rich in polyphenolic compounds from plants byproducts would potentially yield safe products of high quality and nutritional values.[5]

The rich composition of olive leaf in phenolic compounds, the main byproduct of the olive grove, offers a great potential to obtain new high value-added products given their high bioactive effects, with diverse applications in different industries, especially in the food, pharmaceutical, and cosmetic.[6]

12.2 BYPRODUCTS FROM OLIVE TREE: OLIVE LEAF

Olea europaea L., commonly known as olive tree, is a crop that originated in the Mediterranean basin countries, although its cultivation has been extended throughout the world. The main products of the olive grove are table olives and olive oil, two essential foodstuffs of the Mediterranean diet. In 2019, almost 11 M ha were dedicated to olive cultivation, which produced around 20 M tons of olives.[7] Olive oil production generates a large amount of waste and byproducts, such as olive tree leaf and pruning, olive oil pomace and wastewater, olive stone, and olive skin. Olive leaves are one of the main solid wastes in the olive grove, representing roughly 10% of the olives harvested. Olive leaves are obtained from the postharvest pruning of the tree, as well as from the olive mill after olives cleaning. Traditionally, these residues have no industrial application and usually are burned and used as soil fertilizer or as animal feed despite their bitter taste, which generally limits their use. This safe, natural, and low-cost byproduct has been associated with medicinal properties, representing an alternative to conventional antimicrobial drugs.[8]

12.3 BIOACTIVE COMPOUNDS PRESENT IN OLIVE LEAF

As mentioned above, phenolic compounds are the most abundant secondary plant metabolites. Their chemical structure includes one or more hydroxyl groups (polar) bonded to an aromatic ring (nonpolar), often found in plants as esters or glycosides.[9]

The quantitative and qualitative polyphenol composition of olive leaf extracts shows wide differences mainly due to different factors, such as the different olive varieties, the solvent, the methodology used in the extraction, the analytical method, the geographical origin (different climatology, light exposure, and agronomic conditions), the time of harvesting, genotypes, and age of the tree and leaves, among others.[10]

Olive leaves contain a wide variety of phenolic compounds. As can be seen in Figure 12.1, the classification of those mainly present in the olive leaf is based on their molecular structure, such as simple phenols (tyrosol, hydroxytyrosol, p-hydroxybenzoic acid, gallic acid, p-coumaric acid, caffeic acid, and ferulic acid), flavonoids (quercetin, rutin, catechin, gallocatechin, apigenin, luteolin, luteolin-7-glucoside, apigenin-7-glucoside, and verbascoside), and secoiridoids (oleuropein, elenolic acid, and ligstroside).[11] Among the secoiridoids, exclusive to the *Oleaceae* family, oleuropein is the most abundant compound found in olive leaf, with hydroxytyrosol, luteolin-7-glucoside, apigenin-7-glucoside, and verbascoside following. Each compound has a specific bioactivity. For example, oleuropein has shown stronger antimicrobial than antioxidant capacity, comparatively, and usually in combination with hydroxytyrosol, while tyrosol seems to have a marked antioxidant activity in comparison with its antifungal activity.[12]

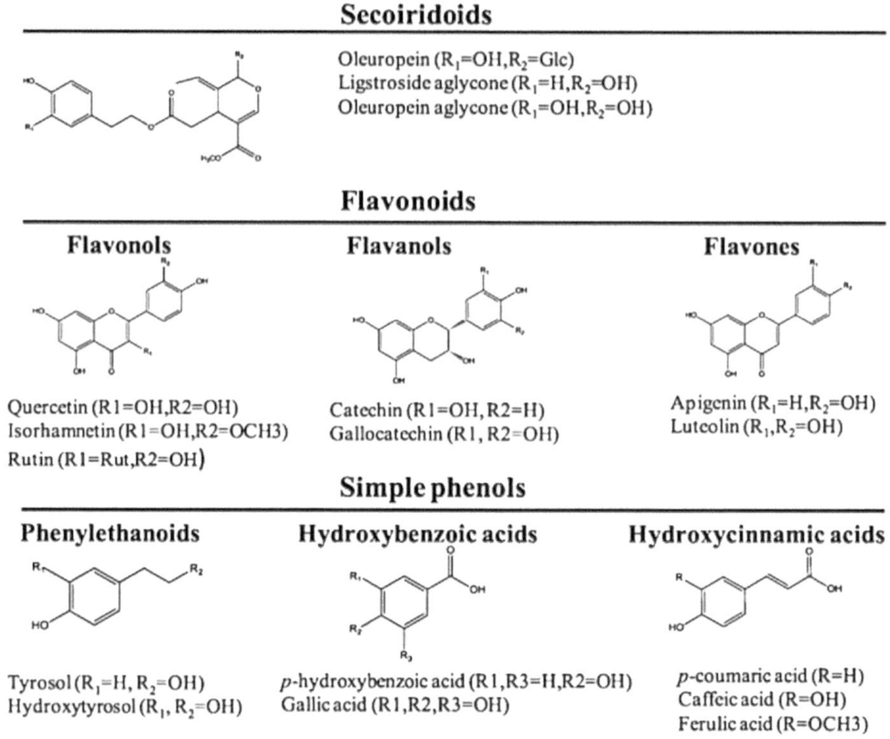

FIGURE 12.1 Chemical structures of phenolic antimicrobial compounds in *Olea europaea*.
Source: Reprinted with permission from Ref. [11]. © 2015 Elsevier.

Within the polyphenolic compounds present in olive leaf, there are some with widely studied antimicrobial effects, such as oleuropein, hydroxytyrosol, tyrosol, and elenolic acid. Traditionally, oleuropein, the major component of olive leaf consisting of hydroxytyrosol esterified with elenolic acid, has received a greater attribution as antimicrobial than hydroxytyrosol.[9,10] However, several authors have claimed that hydroxytyrosol has a stronger antimicrobial activity than oleuropein, as do elenolic acid derivatives.[13] Likewise, the synergistic or even antagonistic effects of the phenolic compounds with recognized antimicrobial activity in olive leaves have been reported. Indeed, it has been observed that in complex mixtures of extracts, certain compounds can interact antagonistically or synergistically with each other or with other substances, resulting in a higher or lower activity compared with the activity of the isolated compounds.[8]

Different action mechanisms for antimicrobial activity have been reported, such as protein denaturation, inhibition of enzymatic reactions required for bacterial growth, and increased cell membrane permeability.[14] Although additional research is needed to establish the antimicrobial effects on Gram-positive and Gram-negative bacteria, it has been demonstrated that the antimicrobial action of oleuropein is associated with the increase of the cell membrane permeability, interfering with the structural and functional properties of the membrane, causing the leakage of cytoplasmic contents.[13]

12.4 EXTRACTION METHODS

The extraction of phenolic compounds from olive leaves deserves special attention as the extraction conditions affect the composition of these compounds. The extraction of polyphenols is a separation process of these compounds from a solid phase (leaf) to a liquid phase (solvent).

Traditionally, conventional extraction methods have been based on the application of heat and/or stir to increase the rate of mass transfer, which requires a long time and results in low extraction efficiency. The potential use of polyphenols within the food industry has prompted research in the chemicals extraction field to find faster, more economical, and environmentally friendly extraction methods. Thus, in recent years, emerging green extraction techniques have been developed, which contribute to a more sustainable valorization, which is characterized by lower energy consumption, extraction time, and use of solvents, improving the recovery of the compounds.[9]

The most common technologies employed to develop bioactive extract of olive leaf associated with conventional technologies were solid–liquid extraction and soxhlet extraction, while in the area of nonconventional extraction, the most commonly applied techniques were ultrasound-assisted extraction,[15,16] microwaved-assisted extraction,[17,18] and supercritical extraction.[19,20]

12.5 IN VITRO ANTIMICROBIAL EVALUATION ASSAYS

Olive leaf extracts have shown variable antimicrobial activity against foodborne and non-foodborne microorganisms. Such variations may be due to the extraction method or the antimicrobial test followed, among other factors.[21]

In order to evaluate the antimicrobial activity of olive leaf extracts, different methods can be employed, being the diffusion methods, disk and well, and the dilution methods, agar and broth, the most commonly used. The latter is used to obtain the minimum inhibitory concentration (MIC) and the minimum bactericidal concentration (MBC).[22]

Several factors, such as inoculum size, type of culture medium, incubation time, and method of inoculum preparation may influence the antimicrobial activity tested, so the use of widely recognized standardized protocols, such as those developed by the Clinical and Laboratory Standard Institute (CLSI) or the European Committee for Antimicrobial Susceptibility Testing (EUCAST), is recommended. In this regard, to evaluate the antimicrobial activity according to CLSI guidelines (Table 12.1), Müller-Hinton agar (diffusion method) or Müller-Hinton broth or agar (dilution method) are used as culture media, being inoculated with a bacterial suspension obtained from previous incubation, once adjusted to the 0.5 McFarland turbidity Standard, as shown in Figure 12.2.[23–25]

12.5.1 DIFFUSION METHOD

12.5.1.1 DISK DIFFUSION METHOD

The evaluation of antimicrobial activity by the disk diffusion method of olive leaf extract involves the observation of the presence (or absence) of an area of inhibition of the growth of bacterial culture on Müller-Hinton agar, caused by the antimicrobial effect of the olive leaf extract. For this purpose, a sterile paper disk (about 6 mm ø) is impregnated with the extract and placed on the inoculated Müller-Hinton agar surface with the tested bacteria in plates. After incubation, a zone of growth inhibition may appear around the disk (inhibition halo). An impregnated disk with the solvent in which the olive leaf extract is dissolved is used as a negative control, while a disk with an antibiotic is used as a positive control. The inhibition halo diameter is used as a reference to compare the effect of different antimicrobial extracts with diverse compositions on different microorganisms.[22–24] The disk diffusion assay offers many advantages, including simplicity, low cost, the possibility to test a large number of antimicrobial agents and microorganisms, and the easiest results interpretation.[26]

12.5.1.2 WELL-DIFFUSION METHOD

This assay uses a Müller-Hinton agar plate, previously inoculated with the microorganism to be tested, in which a series of holes or wells are punched aseptically with a sterile cork borer or a tip, being subsequently filled with a certain volume (20–100 μL) of the solution containing the olive leaf extract

TABLE 12.1 Testing Methods as Recommended by CLSI.

Methods	Microorganism	Growth medium	Final inoculum size	Incubation temperature (°C)	Incubation time (h)
Disk diffusion method	Bacteria	MHA	0.5 McFarland $1–2 \times 10^8$ CFU/mL	35 ± 2	16–18
	Yeast	MHA + GMB[a]	0.5 McFarland $1–5 \times 10^6$ CFU/mL		20–24
	Molds	MHA	0.5 McFarland $0.4–5 \times 10^6$ CFU/mL	–	–
Broth microdilution	Bacteria	MHB	0.5 McFarland 5×10^5 CFU/mL	35 ± 2	20
	Yeast	RPMI 1640[b]	0.5 McFarland $0.5–2.5 \times 10^3$ CFU/mL	35	24–48
	Molds	RPMI 1640[b]	0.5 McFarland $0.4–5 \times 10^4$ CFU/mL	35	48 for most fungi
Broth macrodilution	Bacteria	MHB	0.5 McFarland 5×10^5 CFU/mL	35 ± 2	20
	Yeast	RPMI 1640[b]	0.5 McFarland $0.5–2.5 \times 10^3$ CFU/mL	35	46–50
	Molds	RPMI 1640[b]	0.5 McFarland $0.4–5 \times 10^4$ CFU/mL	35	48 for most fungi
Agar dilution	Bacteria	MHA	0.5 McFarland 10^4 CFU/mL	35 ± 2	16–20

Adapted from Ref. [26].

[a]*GMB, the medium was supplemented with 2% glucose and 0.5 mg/mL methylene blue.*

[b]*RPMI 1460, Roswell Park Memorial Institute medium (with glutamine, without bicarbonate, and with phenol red as a pH indicator) was 1640, buffered to pH 7.0 with MOPS (morpholine sulfonic acid) at 0.165 M.*

is added. As a negative control, the solvent used to obtain the extract is placed in one of the wells, while an antibiotic solution is used as a positive control in another well. After incubation of the plate, the diameter of the inhibition zone around each well is measured.[26,27]

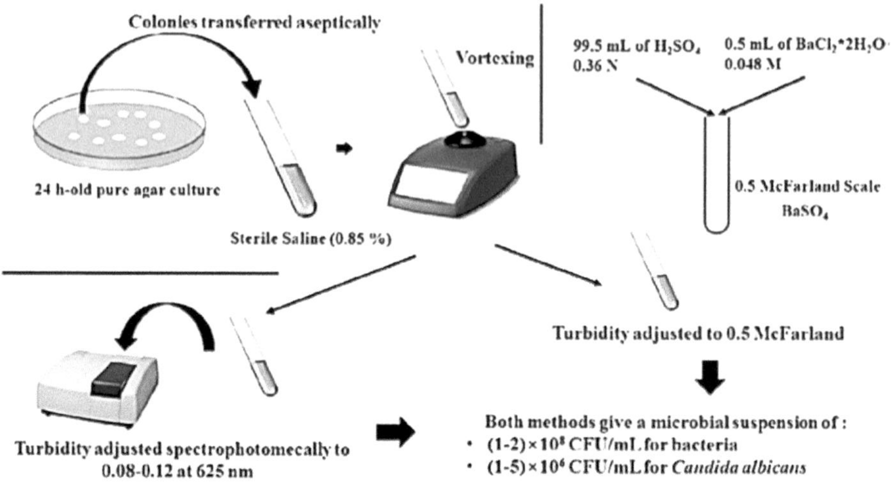

FIGURE 12.2 Establishment of the microbial inoculum level by absorbance measure or by turbidity adjustment to 0.5 McFarland Scale, as recommended by CLSI guidelines.

Source: Reprinted with permission from Ref. [26]. http://creativecommons.org/licenses/by-nc-nd/4.0/

12.5.2 DILUTION METHOD

Dilution methods are the most appropriate methods for estimating MIC values, as it is possible to establish different concentrations of the antimicrobial agent tested in agar (agar dilution) or broth (macrodilution or microdilution).

The term MIC has received different definitions, such as (1) the lowest concentration of a substance that results in the maintenance or reduction of inoculum viability; (2) the lowest concentration required for complete inhibition of the tested organism up to 48 h of incubation; (3) the lowest concentration that inhibits the visible growth of the tested organism; (4) the lowest concentration that results in a significant decrease in inoculum viability (>90%).[9]

The determination of the MBC is the most common measure of the bactericidal effect of a substance. The MBC is defined as the minimum extract concentration at which no bacterial growth is observed and is determined by sub-culturing those negative cultures from the MIC assay, that is, those with no-growth observation. Both MIC and MBC are usually expressed as µg/mL or mg/mL.[26]

12.5.2.1 BROTH DILUTION METHOD

The broth dilution method involves the preparation of serial two-fold dilutions of olive leaf extract dispensed in Müeller-Hinton broth, either in tubes (macrodilution) or in microplate wells (microdilution), inoculated with a standardized bacterial suspension of 5×10^5 CFU/mL (Figure 12.3). After incubation, visible bacterial growth is revealed by turbidity assessment. To facilitate the determination of MIC in microdilution tests, absorbance readers or growth indicators such as resazurin can be used. The main advantages of the microdilution method over the macrodilution method are the small volumes employed (media and reagent) and the subsequent small space required for the test and associated lower cost, and the possibility to run a high number of samples at the same time. The lowest concentration that inhibits the visible growth of the analyzed microorganism is the so-called "minimum inhibitory concentration" (MIC) value.[22,25,26]

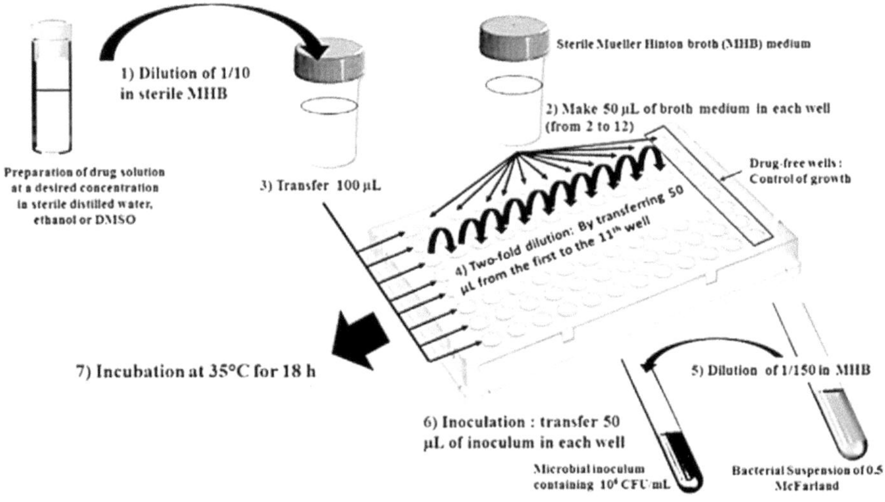

FIGURE 12.3 Broth microdilution for antibacterial testing, as recommended by the CLSI protocol.

Source: Reprinted with permission from Ref. [26]. http://creativecommons.org/licenses/by-nc-nd/4.0/

12.5.3 AGAR DILUTION METHOD

The dilution method can also be performed on Müller-Hinton agar, involving the incorporation of serial two-fold dilutions of olive leaf extract into the agar and then inoculating the tested microorganism onto the agar plate surface.[22,25,26]

12.6 ANTIMICROBIAL ACTIVITY OF OLIVE LEAF EXTRACT

The MIC and MBC of olive leaf extracts have been reported in different publications, showing wide ranges due to several factors, such as the different sensitivity of the strains, different phenolic composition of the extracts, the extraction procedure, the type of solvent, the leaves and trees variety, and the different methods used to evaluate the antimicrobial activity.[8,21]

The influence of the bacterial wall on the sensitivity of microorganisms to olive leaf extracts is also unclear. In this respect, it has been published that Gram-positive bacteria are more sensitive due to the presence of a lipopolysaccharide layer in their wall, absent in Gram-negative bacteria, thereby being more permeable to antimicrobial compounds.[21,28] However, other studies have reported a lower sensitivity of Gram-positive bacteria, thus it is not possible to define a generalized trend in the sensitivity of Gram-negative and Gram-positive bacteria to olive leaf extracts.[8,13]

12.6.1 NON-FOODBORNE MICROORGANISMS

12.6.1.1 PSEUDOMONAS FLUORESCENS

Pseudomonas fluorescens is a Gram-negative, aerobic, motile, non-spore-forming, saprophytic bacteria that can grow from 5–42°C. The microorganism produces a fluorescent pigment, that is, fluorescein, that reacts to ultraviolet light. *P. fluorescens* is found in aquatic and terrestrial (soil) ecosystems. In some cases, it can act as an opportunistic pathogen in the environment under specific temperature and humidity conditions.[29]

As can be observed in Table 12.2, an olive leaf extract obtained by solid–liquid extraction with 80% methanol showed antimicrobial activity against this microorganism, with a MIC value of 5 mg/mL determined by the agar dilution method, and an inhibition halo of 12.5 mm by the well-diffusion method. *Pseudomonas fragi* and *Pseudomonas putida* were also studied in this work, exhibiting a similar antimicrobial activity as *P. fluorescens*.

12.6.1.2 KLEBSIELLA PNEUMONIAE

Klebsiella pneumoniae is a Gram-negative, facultative anaerobic, encapsulated, nonmotile, lactose-fermenting bacterium found in the microbiota of mouth, skin, small, and large intestine. It is considered an emerging nosocomial pathogen. It causes kidney disease, the severity of which depends on certain risk factors, such as the patient's age and health status before infection.[30]

Several studies have reported the antimicrobial activity of olive leaf extract against *K. pneumoniae* (Table 12.2). Diffusion assays have shown inhibition halos between 8 and 27 mm. Using dilution assays, MIC, as well as, MBC values ranged between 32 and 128 μg/mL in some works. In contrast, other authors have reported a lower sensitivity, with a MIC of 15 mg/mL, a MBC of 30 mg/mL, an IC_{25} (extract concentration inhibiting 25% of microbial growth) of 3.13 mg/mL, and even the absence of antimicrobial activity.

12.6.1.3 ENTEROCOCCUS SPP

Enterococcus is a genus that belongs to the *Enterococci* family, whose species are considered potential nosocomial pathogens due to their intrinsic resistance to various antimicrobials and their ability to acquire resistance mechanisms. *Enterococcus* is a member of the group of Gram-positive bacteria.

Within this genus, *E. faecalis* and *E. faecium* have been mostly studied. Although their primary habitat is the gastrointestinal tract of organisms, such as humans and animals, they can also be encountered in the environment (soil and water) and food. *E. faecalis* is responsible for 90% of urinary tract infections. It can also cause bacteremia, endocarditis, and wound infections.[31]

As can be appreciated in Table 12.2, according to the type of solvent employed for extraction, olive leaf extract shows inhibition halos up to 26 mm diameter against *E. faecalis*. For this microorganism, MIC values are reported in the range between 32 μg/mL and 0.63 mg/mL, while MBC values were 32–210 μg/mL. On the contrary, the absence of antimicrobial activity has also been reported. Regarding *E. faecium*, only one author studied the activity of olive leaf extract against this microorganism, whose MIC and MBC values were within the same range as *E. faecalis*.

12.6.1.4 CANDIDA SPP.

Candida spp. are a group of yeasts belonging to the *Saccharomycetaceae* family. The most studied species, *Candida albicans*, is an opportunistic pathogen which takes part in the natural microbiota of the gastrointestinal, respiratory, and genitourinary tract under normal host health conditions. However, when the immune system of the host is impaired, *C. albicans* becomes pathogenic. The disease caused by the pathogenic form of *C. albicans* is known as candidiasis, and is occurring in the vagina, oral cavity, gut, and/or skin.[32]

Based on the antimicrobial diffusion test, olive leaf extract has produced inhibition halos against *C. albicans* and *C. tropicalis* between 8.3 and 24.9 mm. Also, some authors have carried out dilution assays, whose results varied between 20 μg/mL and 1.25 mg/mL for MIC and between 64 μg/mL and 1.5 mg/mL for the minimum fungicidal concentration (MFC). In contrast, other studies found no antimicrobial activity of olive leaf extract against *C. albicans*.

For other species, such as *C. glabrata*, *C. parapsilosis*, and *C. kreusei*, MIC and MFC values were similar to those of *C. albicans*, except for the MIC of *C. kreusei*, which reached up to 2.5 mg/mL. More detailed information is provided in Table 12.2.

12.6.1.5 OTHER MICROORGANISMS

In addition to the microorganisms mentioned above, many others have been reported to be sensitive to olive leaf extract, as illustrated in Table 12.2. Indeed, extract MIC and MBC (or MFC) values up to 250 μg/mL have been found against *Moraxella catarrhalis*, *Streptococcus oralis*, *Trichophyton rubrum*, *Trichophyton mentagrophytes*, *Microsporum canis,* and *Aspergillus niger*. Other microorganisms showed a slightly higher resistance to the extract, with MIC and MBC (or MFC) values up to 1.5 mg/mL, such as *Cryptosporidium neoformans*, *Staphylococcus epidermidis*, *Streptococcus mutans*, *Porphyromonas gingivalis*, *Fusobacterium nucleatum*, *Parvimonas micra,* and *Pasteurella multocida*. *Streptococcus sobrinus* and *Prevotella intermedia* needed MIC and MBC (or MFC) values above 1.5 mg/mL, and *Proteus mirabilis* and *Mycoplasma gallisepticum* were found to be resistant to olive leaf extract.

12.6.2 FOODBORNE MICROORGANISMS

12.6.2.1 ESCHERICHIA COLI

Escherichia coli, a Gram-negative bacteria, is one of the predominant enteric species in the human gut as part of its microbiota, providing benefits to the

TABLE 12.2 Antimicrobial Activity of Olive Leaf Extract Against Non-Foodborne Microorganisms.

Extraction	Compounds	Antimicrobial test method	Microorganisms	Strain number	Antimicrobial activity	References
SLE Hexane, chloroform, dichloromethane, ethyl acetate, acetone, ethanol, methanol, butanol, and water	Flavonoids, reducing sugar, steroids, tannins, terpenes/terpenoids	Microdilution method	*E. faecalis*	ATCC 29212	MIC=0.3–0.63 mg/mL[a]	[33]
			C. albicans	Isolate (Faculty of Health Sciences University of Pretoria)	MIC=0.16–0.63 mg/mL[a]	
			C. neoformans	Isolate (Faculty of Health Sciences University of Pretoria	MIC=0.16–0.47 mg/mL[a]	
SLE Butanol, ethanol, ethyl acetate, hexane, methanol and water	Alkaloids, Tannins, flavones, flavonoids, terpenoids, steroids.	Well-diffusion method Agar dilution method	*E. faecalis*	ATCC 29212	Inhibition halo (mm)= 0–26[a] MIC=62.5–125 μg/mL[a]	[34]
			C. albicans	–	Inhibition halo (mm)=9.4–24.9[a]	
			C. tropicalis	–	Inhibition halo (mm)=8.3–23.7[a]	
			M. catarrhalis	ATCC 25240	Inhibition halo (mm)= 10–16[a] MIC=62.5–125 μg/mL[a]	
			C. neoformans	–	Inhibition halo (mm)=8.6–27.6[a]	
SLE Methanol	Polyphenols (Hydroxytyrosol, tyrosol, 4-hydroxybenzoic	Microdilution method	*E. faecalis*	ATCC29212	MIC=32–64 μg/mL[b] MBC=32–128 μg/mL[b]	[35]

TABLE 12.2 *(Continued)*

Extraction	Compounds	Antimicrobial test method	Microorganisms	Strain number	Antimicrobial activity	References
	acid, rutin, luteolin 7-O-glucoside, apigenin 7-O-glucoside, oleuropein, apigenin, catechin hydrate), flavonoids, chlorophyll.		*E. faecium*	CI234	MIC=32–64 µg/mL[b] MBC=32–128 µg/mL[b]	
			C. albicans	ATCC 90028	MIC=32–125 µg/mL[b] MFC=64 µg/mL[b]	
			C. glabrata	ATCC 90030	MIC=32–64 µg/mL[b] MFC=64 µg/mL[b]	
			C. parapsilosis	ATCC 22019	MIC=32–64 µg/mL[b] MFC=64 µg/mL[b]	
			C. kreusei	ATCC 6258	MIC=32–64 µg/mL[b] MFC=64 µg/mL[b]	
			K. pneumoniae	CI29	MIC=32–128 µg/mL[b] MBC=32–128 µg/mL[b]	
SLE methanol: water (80:20 v/v)	Hydroxytyrosol, catechin, vanillic acid, vanillin, rutin, luteolin-7-glucoside,	Well-diffusion method Agar dilution method	*P. fluorescens*	ATCC 13525	Inhibition halo (mm)=12.5 MIC= 5 mg/mL	[36]
			P. fragi	ATCC 4973	Inhibition halo (mm)=12 MIC= 3 mg/mL	
	verbascoside, oleuropein		*P. putida*	ATCC 17514	Inhibition halo (mm)=13 MIC= 2.5 mg/mL	

TABLE 12.2 *(Continued)*

Extraction	Compounds	Antimicrobial test method	Microorganisms	Strain number	Antimicrobial activity	References
Solvent-free microwaved-assisted extraction	Oleuropein	Microdilution method	*S. epidermidis*	ATCC 12228	MIC=1.25 mg/mL	[37]
			E. faecalis	ATCC 29212	NA	
			K. pneumoniae	ATCC 4352	NA	
			P. mirabilis	ATCC 14153	NA	
			C. albicans	ATCC 10231	NA	
SLE Ethanol: water (70:30 v/v)	Oleuropein Rutin	Disk diffusion method	*K. pneumoniae*	–	NA	[38]
SLE Acetone	60% Oleuropein Hydroxytyrosol	Microdilution method	*S. mutans*	DSM 20523	MIC=1.25 mg/mL	[39]
			S. sobrinus	DSM 20381	MIC=2.5 mg/mL	
			S. oralis	ATCC 35037	MIC=0.07 mg/mL	
			E. faecalis	ATCC 29212	MIC=0.60 mg/mL	
			P. gingivalis	W381	MIC=0.30 mg/mL MBC=0.60 mg/mL	
			P. intermedia	ATCC 25611	MIC=2.50 mg/mL MBC=5 mg/mL	
			F. nucleatum	ATCC 25586	MIC=0.60 mg/mL MBC=1.25 mg/mL	
			P. micra	ATCC 23195	MIC= 0.30 mg/mL MBC=1.25 mg/mL	
			C. albicans	DSM 1386	NA	

TABLE 12.2 *(Continued)*

Extraction	Compounds	Antimicrobial test method	Microorganisms	Strain number	Antimicrobial activity	References
Hydrodistillation Water	–	Microdilution method	*E. faecalis*	ATCC 29212	MIC=70–2500 μg/mL[b]	[40]
			C. glabrata	ATCC 90030	MIC=150–1500 μg/mL[b]	
			C. kreusei	ATCC 6258	MIC= 150–2500 μg/mL[b]	
			C. parapsilosis	ATCC 22019	MIC=310–620 μg/mL[b]	
			C. albicans	ATCC 90028	MIC=70–1250 μg/mL[b]	
Soxhlet Ethyl alcohol, diethyl ether, acetone, water)	Tyrosol, 4-hydroxy benzoic acid, veratric acid, vanillic acid, syringic acid, p-coumaric acid, feluric acid, caffeic acid, oleuropein	Macrodilution method	*E. faecalis*	University of Uludag	MIC=90–105 μg/mL[a] MBC=90–210 μg/mL[a]	[41]
			K. pneumoniae		MIC=25–33 μg/mL[a] MBC=33–52 μg/mL[a]	
SLE Water	–	Macrodilution method	*K. pneumoniae*	–	MBC=0.06 mg/mL	[42]
			T. mentagrophytes		MFC=0.125 mg/mL	
			M. canis		MFC=0.125 mg/mL	
			T. rubrum		MFC=0.125 mg/mL	
			C. albicans		MFC=1.5 mg/mL	
SLE Water	Caffeic acid, verbascoside, oleuropein, luteolin	Macrodilution method	*K. pneumoniae*	ESA 8	IC_{25}=3.13 mg/mL	[43]

TABLE 12.2 *(Continued)*

Extraction	Compounds	Antimicrobial test method	Microorganisms	Strain number	Antimicrobial activity	References
	7-O-glucoside, rutin, apigenin 7-O-glucoside and luteolin 4'-O-glucoside		*C. albicans*	CECT 1394	IC_{25}=0.85 mg/mL	
			C. neoformans	ESA 3	IC_{25}=3.00 mg/mL	
SLE Water	–	Disk diffusion method Agar dilution method	*K. pneumoniae*	–	Inhibition halo (mm)=11.7 MIC=19.03 μg/mL	[44]
			E. cloacae	ATCC 13047	Inhibition halo (mm)=12.5 MIC=21.29 μg/mL	
SLE Dichloromethane Dichloroethane (1:1) mixtures of Chloroform: dichloroethane Chloroform: dichloromethane dichloroethane: dichloromethane ethyl acetate:dichloroethane ethyl acetate: dichloromethane	–	Agar dilution method	*K. pneumoniae*	–	MIC=25–35 μg/mL[a]	[45]
			A. niger		MIC=20–50 μg/mL[a]	
			C. albicans		MIC=20–40 μg/mL[a]	
Soxhlet Ethanol: water (80:20 v/v)	–	Well-diffusion method Macrodilution method	*K. pneumoniae*	Isolate (Al-Yarmuk Hospital, Iraq)	Inhibition halo (mm)=8–27[c] MIC=15 mg/mL MBC=30 mg/mL	[27]

TABLE 12.2 *(Continued)*

Extraction	Compounds	Antimicrobial test method	Microorganisms	Strain number	Antimicrobial activity	References
SLE Ethanol: water (70:30 v/v)	–	Disk diffusion method Macrodilution method	*P. multocida*	Isolate (Microbiology Department, Faculty of Veterinary Medicine, Cairo University, Giza, Egypt)	Inhibition halo (mm)=3.52–9.12[c] MIC=0.625 mg/mL MBC=0.625 mg/mL	[46]
			M. gallisepticum		Inhibition halo (mm)=5.02[c] MIC=NA MBC=NA	

[a] *Range depending on the solvent extraction.*

[b] *Range depending on the different olive tree cultivars.*

[c] *Range depending on the concentration of the extract in the disk.*

IC_{25}, extract concentration which inhibits 25% of microbial growth); MFC, minimum fungicidal concentration; NA, no activity.

host. However, there are groups of pathogenic *E. coli* that can cause severe diarrheal disease in humans.

Six pathogenic *E. coli* groups are recognized: enterotoxigenic *E. coli* (ETEC), enteropathogenic *E. coli* (EPEC), enterohemorrhagic *E. coli* (EHEC), enteroinvasive *E. coli* (EIEC), enteroaggregative *E. coli* (EAEC), and diffusely adherent *E. coli* (DAEC). The transmission of ETEC, EPEC, EHEC, and EIEC is foodborne, with EHEC being the major group implicated in foodborne outbreaks. Infective dose, incubation period, mortality, symptoms, and duration depend on the *E. coli* group.[47]

The antimicrobial activity of olive leaf extract against *E. coli* is variable. Results of different studies are presented in Table 12.3, in which it can be noted a wide range of antimicrobial data, from no effect until a more or less marked antimicrobial effect of the extract, in other words, less or more resistance of *E. coli* to it. In this regard, the wide range of MIC (from 6.97 μg/mL–62.5 mg/mL) and MBC (from 32 μg/mL–60 mg/mL) obtained by dilution assays demonstrates the high spectrum of sensitivity of *E. coli*, which depends on the type of solvent, tree variety, and extraction method. For example, in SLE, Gullón et al.[21] found MIC and MBC values between 30 and 45 mg/mL in 50% ethanolic extracts, while a concentration slightly higher (62.5 mg/mL) of the 80% ethanolic extract was necessary to inhibit 95% growth of *E. coli* O157:H7. Edziri et al.,[35] however, reported substantially lower MIC and MBC values of a methanolic extract of the Tunisian olive tree variety Chetoui, that is, 64 and 32 μg/mL, respectively. Concerning the results obtained by antimicrobial diffusion methods, different authors agree on the size of the inhibition zone, being in the range of 6–22 mm.[27,44,46,48]

12.6.2.2 LISTERIA MONOCYTOGENES

Listeria spp. is a Gram-positive, facultative, motile bacterium, known for being highly ubiquitous and resistant in the environment. The genus *Listeria* includes six species at present, *L. monocytogenes*, *L. ivanovii*, *L. innocua*, *L. welshimeri*, *L. seeligeri*, and *L. gravi*. From these, only *L. monocytogenes* constitutes a food safety issue because of its resistance, the capacity of surviving at temperatures below 1°C and high salt concentrations, and the severe consequences of the disease, listeriosis, especially in immunocompromised persons. There are 13 serotypes of *L. monocytogenes*, being 1/2a, 1/2b, and 4b the serotypes which are associated with foodborne infections.

Although it is a relatively uncommon disease, its presence is one of the leading causes of death from foodborne diseases in infants, children, and the elderly, making it one of the most serious foodborne infections. The incubation period for the gastroenteritis symptoms is usually 2–3 days, and up to 3 months in the severe form of the disease.[47]

Regarding the antilisteria effect of olive leaf extract, controversial results and arguments have been found in the literature. Using diffusion assays, both sensitivity and resistance of the pathogen to olive leaf extract have been reported, while in dilution assays, results from different works seem more homogeneous, with MBC values between 40 and 62.6 mg/mL (Table 12.3). In the case of *L. innocua*, several authors have reported MIC values between 3 and 20 mg/mL, lower than those reported for *L. monocytogenes*.[21,36]

12.6.2.3 SALMONELLA ENTERICA

Salmonella enterica, a Gram-negative, motile, non-spore-forming bacteria of the Enterobacteriaceae family, is one of the most common foodborne pathogens affecting millions of people each year, rarely with fatal consequences. Within the subspecies *enterica*, the target serovars causing foodborne outbreaks are *S. enteritidis*, *S. typhimurium* (including monophasic variants), *S. infantis*, *S. Virchow*, and *S. Hadar*. The source of contamination is the ingestion of contaminated food and/or water, and the incubation period ranges 1–3 weeks, and the infective dose is less than 1000 cells. Gastrointestinal symptoms can remain for 2–4 weeks.[47]

Results from antimicrobial diffusion methods indicate that *S. enterica* show a wide range of inhibition zones between 0 and 24 mm, thus presenting high sensitivity in some studies and no antibacterial activity at all in others. In dilution assays, the MIC values varied between 62.5 μg/mL and 60 mg/mL, while for MBC, values were found to be slightly higher, that is, between 180 μg/mL and 62.5 mg/mL.

As an example, it can be seen in Table 12.3 that Sanchez-Gutierrez et al.[8] found the MIC values of olive leaf extracts against *S. typhimurium* between 40 and 60 mg/mL, in agreement with the results obtained by Liu et al.[28] in ethanol extract. They reported that a concentration of 62.5 mg/mL completely inhibited the growth of *S. enteritidis*. However, significantly lower MIC and/or MBC values were found by other authors when testing the antimicrobial activity of an ethanolic olive leaf extract (5 and 2.5 mg/mL, respectively),[46] with the referenced lowest MIC value established at 62.5 ug/mL.[34]

12.6.2.4 STAPHYLOCOCCUS AUREUS

Staphylococcus aureus is a Gram-positive, nonmotile, non-sporulating, catalase-positive bacteria. It is one of the most resistant human pathogens, being able to survive in low water activity conditions and in a wide range of temperatures and pH. They are ubiquitous in the environment, and could contaminate and grow in foods, even at limiting conditions. *S. aureus* capable of producing the so-called staphylococcal enterotoxin when growing at levels above 100,000 cfu/g. The toxin causes gastrointestinal symptoms after 1–7 h after ingesting the contaminated food, with a short duration (hours), except in severe cases.[47]

S. aureus appears to be one of the most sensitive microorganisms to olive leaf extract, showing inhibition halos by diffusion methods of up to 35 mm. Similarly, MIC values have ranged from 9.88 μg/mL to 25 mg/mL, and MBC values range between 9.12 μg/mL and 30 mg/mL, respectively.

In Table 12.3, the most representative results correspond to the work by Masoko et al.,[33] who studied the influence of the different solvents used in the extraction, with the aqueous extract presenting a MIC value of 2.5 mg/mL, higher than the MIC value of the ethanol extract (0.26 mg/mL), and the study of Brahmi et al.,[40] with MIC ranging from 70 μg/mL to 2.5 mg/mL, depending on the olive tree cultivar studied.

12.6.2.5 YERSINIA ENTEROCOLITICA

Yersinia enterocolitica is a Gram-negative bacteria isolated from animals, soil, and water. It is psychotropic and can survive in freezing for a long period and tolerate a wide pH range (4–10).

It may contaminate foods, causing yersiniosis, a gastrointestinal disease, which can entail autoimmune complications, lasting up to 1 month, except for chronic enterocolitis, which can last for months.[47]

Some researchers have evaluated the antimicrobial activity of the olive leaf extract against *Y. enterocolitica* by the dilution method, showing MIC values between 2 and 20 mg/mL, and MBC values from 2–30 mg/mL, varying according to the type of solvent employed in the extraction.[8,49] Also, inhibition zones of up to 17.68 mm were observed by diffusion methods.[49]

12.6.2.6 PSEUDOMONAS AERUGINOSA

P. aeruginosa is a Gram-negative, aerobic, opportunistic pathogen in humans. It is widely distributed in the environment and is commonly carried by humans due to its rapid multiplication and low nutrient and moisture requirements. It is frequently found in water, dairy products, meat, and plant-based foods.[50]

The evaluation of the susceptibility of *P. aeruginosa* to olive leaf extract has been extensively studied. The different results obtained between studies range from high antimicrobial activity to no antimicrobial activity at all. Dilution methods yielded MIC values between 20 μg/mL and 30 mg/mL and MBC values between 56 μg/mL and 60 mg/mL, while diffusion assays showed inhibition halos from 7 to 28 mm.[21,27,32,34,35,37,38,40–45]

12.6.2.7 BACILLUS CEREUS

Bacillus cereus is a Gram-positive, facultative anaerobic, endospore-forming bacterium that causes food poisoning by producing a toxin. It is found in the environment (soil and vegetation) and its growth conditions range about 4–48°C for temperature, and pH ranges about 4.9–9.3.

Bacillus cereus food poisoning is associated with two toxins (a high molecular weight protein and a low molecular weight non-antigenic peptide), resulting in diarrhea and vomiting, respectively. Pathogenicity arises from the preformed toxin, being necessary for a bacterial concentration between 10^5 and 10^8 cfu/g to be produced. Diarrheal and emetic symptoms usually subside within 24 h of onset.[47]

Using both dilution and diffusion antimicrobial methods, olive leaf extract showed antibacterial activity against *B. cereus*, with inhibition zones up to 23 mm, MIC values ranging from 32 μg/mL–5 mg/mL and MBC values between 64 μg/mL and 2.5 mg/mL.[34,35,41,43,46,49]

12.6.2.8 CAMPYLOBACTER JEJUNI

C. jejuni is recognized as one of the most common foodborne pathogens worldwide. It is a Gram-negative, non-spore-forming, and microaerophilic bacteria, susceptible to desiccation, heat, freezing, and acidity, thus rather sensitive in the environment. The infective dose ranges from 500 to 10,000 cells, depending on the type of food contaminated and the health status of the individual. The incubation period is 2–5 days, resulting in self-limiting gastroenteritis lasting 2–10 days.[47] Regrettably, only one study evaluated the activity of olive leaf extract against this microorganism, showing no antimicrobial effect.[49]

12.6.2.9 OTHER MICROORGANISMS

In addition to the mentioned microorganisms, the effect of olive leaf extract on the inhibition/survival of other microorganisms has been evaluated. Table 12.3 summarizes several microorganisms used in the production of fermented foods, such as *Lactobacillus plantarum*, *Lactobacillus brevis*, *Lactobacillus bulgaricus*, *Streptococcus thermophilus*, *Saccharomyces cerevisiae*, *Leuconostoc mesenteroides*, and *Acetobacter* spp., showing MIC values from 25–46 ug/mL and MBC values from 35 to 92 μg/mL. Similarly, the olive leaf extract shows antibacterial activity against *Brochothrix thermosphacta* with inhibition halos of up to 17 mm and a MIC of 2 mg. High MIC values were found for *Clostridium sporogenes* (3.24 mg/mL), while *Bacillus subtilis* needs concentrations of 4.12 mg/mL of the extract to inhibit 25% of bacterial growth.

12.7 CONCLUSION AND FUTURE RESEARCH

In the last years, the search for new and natural bioactive substances has guided research through different projects in both the public and private sectors, thus increasing the competitiveness through research and innovation. Indeed, plant-based extracts like olive leaf extracts, are currently widely appreciated and recognized for their benefits in preventing inflammatory disorders, cardiovascular protection, and diabetes, and are acting as antioxidant and antimicrobial agents. More recently, the evaluation and exploitation of natural substances have turned into responsibility for researchers and decision-makers, especially in the field of antimicrobials. Antimicrobial resistance is a global issue with severe consequences which may not have yet reached its maximum. Chemical formulae with new antimicrobial substances with different action mechanisms should be investigated, as well as their mode of application and release into the target tissue, organ, or organic system. Also, the design of foods with new preservatives based on plant extracts should be encouraged. For their safe use and applications, safety and toxicity studies of the plant extracts should also be performed as well as the study of their effects on beneficial microbiota. The olive tree constitutes an appreciated heritage, especially in Mediterranean countries, whose byproducts and residues are not always used or valorized. It is the opportunity for the olive leaf to enter into the production cycle, contributing not only to achieve a more sustainable economy or bioeconomy but also to advancing toward a more *one-health*-conscious society.

TABLE 12.3 Antimicrobial Activity of Olive Leaf Extract Against Foodborne Microorganisms.

Extraction	Compounds	Antimicrobial test method	Microorganisms	Strain number	Antimicrobial activity	References
SLE Hexane, chloroform, dichloromethane, ethyl acetate, acetone, ethanol, methanol, butanol, and water	Flavonoids, reducing sugar, steroids, tannins, terpenes/terpenoids	Microdilution method	*E. coli*	ATCC 28922	MIC=0.31–0.63 mg/mL[a]	[33]
			S. aureus	ATCC 29213	MIC=0.26–2.5 mg/mL[a]	
			P. aeruginosa	ATCC 27853	MIC=0.42–2.08 mg/mL[a]	
SLE Water, acetic acid: water (0.5:95.5) Ethanol: water (20:80 and 50:50) Acetone: water (50:50) NaOH: water (1:99) (v/v)	Luteolin diglucoside, loganic acid glucoside, luteolin glucoside (isomer a and b), rutin, apigenin O-rutinoside, chrysoeriol-7-O-glucoside, −)-olivil, 4-O-β-glucopyranoside (Isomer b), luteolin.	Microdilution method	*E. coli*	ATCC 25922	MIC=35 mg/mL MBC=40 mg/mL	[21]
			S. aureus	ATCC 6538	MIC=25 mg/mL MBC=30 mg/mL	
			S. enterica	ATCC 19430).	MIC=30 mg/mL MBC=35 mg/mL	
			L. innocua	NCTC 10528	MIC=20 mg/mL MBC=25 mg/mL	
			P. aeruginosa	ATCC 10145	MIC=30 mg/mL MBC=40 mg/mL	
SLE Butanol, ethanol, ethyl acetate, hexane, methanol, and water	Alkaloids, tannins, flavones, flavonoids, terpenoids, steroids	Well-diffusion method Agar dilution method	*P. aeruginosa*	ATCC 27853	Inhibition halo (mm)=7–19[a] MIC=62.5–125 μg/mL[a]	[34]
			S. typhimurium	ATCC 14208	Inhibition halo (mm)=14–24[a] MIC=62.5–125 μg/mL[a]	
			S. aureus	ATCC 25923	Inhibition halo (mm)=10–28[a] MIC=31.2–62.5 μg/mL[a]	
			B. cereus	ATCC 10876	Inhibition halo (mm)=6-23[a] MIC=62.5–125 μg/mL[a]	

TABLE 12.3 *(Continued)*

Extraction	Compounds	Antimicrobial test method	Microorganisms	Strain number	Antimicrobial activity	References
SLE Methanol: water (80:20 v/v)	Phenols, iridoid, iridoid glycoside, acetophenone glucoside, secoiridoid glycoside, flavonol glycoside, flavonoid, isoflavone glycoside, terpenoid, vanilloid, diterpenoid glycoside, carotenoid.	Disk diffusion method Microdilution method	*E. coli*	ATCC 25922	Inhibition halo (mm)=NA	[49]
			Verotoxigenic *E. coli*	CECT 5947	Inhibition halo (mm)=NA	
			S. enterica	CECT S56	Inhibition halo (mm)=NA	
			S. typhimurium	ATCC 14028	Inhibition halo (mm)=NA	
			Y. enterocolitica	CECT 500	Inhibition halo (mm)=13.76–17.68[b] MIC=2 mg/mL MBC=2 mg/mL	
			L. monocytogenes	CECT 911	Inhibition halo (mm)=NA	
			S. aureus	ATCC 29213	Inhibition halo (mm)=17.20–22.36[b] MIC=0.5 mg/mL MBC=1 mg/mL	

TABLE 12.3 *(Continued)*

Extraction	Compounds	Antimicrobial test method	Microorganisms	Strain number	Antimicrobial activity	References
			B. cereus	CECT 131	Inhibition halo (mm)=10.55–12.14[b]	
					MIC=1 mg/mL MBC=2 mg/mL	
			B.thermosphacta	CECT 847	Inhibition halo (mm)=14.13–17.55[b]	
					MIC=2 mg/mL MBC=4 mg/mL	
			C. jejuni	ATCC 33560	Inhibition halo (mm)=NA	
					MIC=NA MBC=NA	
SLE Methanol	Polyphenols (Hydroxytyrosol, Tyrosol, 4-Hydroxybenzoic acid, Rutin, Luteolin 7-O-Glucoside, Apigenin 7-O-Glucoside, Oleuropein, Apigenin, Catechin hydrate), flavonoids, chlorophyll	Microdilution method	*S. aureus*	ATCC25923	MIC=32–128 μg/mL[c]	[35]
					MBC=32–128 μg/mL[c] [35]	
				MRSA 112,126,234,675	MIC=32–128 μg/mL[c] [35]	
					MBC=32–128 μg/mL[c] [35]	
			E. coli	ATCC 25922	MIC=64 μg/mL	
					MBC=32–128 μg/mL[c]	
				CI423	MIC=64–128 μg/mL[c]	
					MBC=32–128 μg/mL[c]	
			P. aeruginosa	CI122	MIC=64–128 μg/mL[c]	
					MBC=64–128 μg/mL[c]	

TABLE 12.3 *(Continued)*

Extraction	Compounds	Antimicrobial test method	Microorganisms	Strain number	Antimicrobial activity	References
				CI311	MIC=64–128 μg/mL[c] MBC=64–128 μg/mL[c]	
			B. cereus	ATCC 11778	MIC=32–64 μg/mL[c] MBC=64–128 μg/mL[c]	
			B. subtilis	ATCC 14579	MIC=32–64 μg/mL[c] MBC=64–128 μg/mL[c]	
SLE methanol: water (80:20 v/v)	Hydroxytyrosol, catechin, vanillic acid, vanillin, rutin, luteolin-7-glucoside, verbascoside, oleuropein	Well-Diffusion method Agar dilution method	*Listeria innocua*	ATCC 330909	Inhibition halo (mm)=13 MIC= 3 mg/mL	[36]
			B. thermosphacta	ATCC 11509	Inhibition halo (mm)=17 MIC= 2 mg/mL	
			C. sporogenes	ATCC 33090	Inhibition halo (mm)=13 MIC= 3.24 mg/mL	
SLE Methanol: water (25:75 v/v)	Oleuropein	Disk diffusion method	*E. coli*	–	Inhibition halo (mm^2)=227–1017[d]	[51]
			Salmonella spp.		Inhibition halo (mm^2)=201–415[d]	
			L. monocytogenes		Inhibition halo (mm^2)=415–1194[d]	
			S. aureus		Inhibition halo (mm^2)=177–531[d]	
Microwaved-assisted extraction Water, Ethanol: water (50:50 v/v), Glycerol: water (5:95 v/v)	Hydroxytyrosol, luteolin, luteolin-7-O-glucoside, apigenin and apigenin-7-O-glucoside, oleuropein and Verbascoside	Microdilution method	*E. coli*	CECT 8295	MIC=30–60 mg/mL[a] MBC=40–>60 mg/mL[a]	[8]

TABLE 12.3 *(Continued)*

Extraction	Compounds	Antimicrobial test method	Microorganisms	Strain number	Antimicrobial activity	References
			S. typhimurium	CECT 704	MIC=40–60 mg/mL[a] MBC=50–>60 mg/mL[a]	
			L. monocytogenes	CECT 4032	MIC=30–>60 mg/mL[a] MBC=40–>60mg/mL[a]	
			S. aureus	CECT 5193	MIC=2.5–20 mg/mL[a] MBC=5–30mg/mL[a]	
			Y. enterocolitica	CECT 754	MIC=5–20 mg/mL[a] MBC=10–30 mg/mL[a]	
Solvent-free Microwaved-assisted extraction	–	Microdilution method	*S. aureus*	ATCC 29213	MIC=1.25 mg/mL	[37]
			E. coli	ATCC 25922	NA	
			P. aeruginosa	ATCC 27853	NA	
SLE	Oleuropein Rutin	Disk diffusion method	*E. coli*	–	NA	[38]
Ethanol: water			*S. aureus*		Inhibition halo (mm) = 19	
(70:30 v/v)			*P. aeruginosa*		NA	
SLE	60% Oleuropein Hydroxytyrosol	Microdilution method	*E. coli*	ATCC 25922	NA	[39]
Acetone			*S. aureus*	ATCC 25923	MIC= 2.50 mg/mL	
					MBC=5 mg/mL	
SLE	–	Disk diffusion method Film disk	*E. coli*	Clinical strains	NA	[52]
Water			*S. aureus*	ATCC 25923	Inhibition halo (mm)=16.6–26.6[b]	

TABLE 12.3 *(Continued)*

Extraction	Compounds	Antimicrobial test method	Microorganisms	Strain number	Antimicrobial activity	References
Hydrodistillation	–	Microdilution method	*S. aureus*	ATCC 27950	MIC=70–2500 μg/mL[c]	[40]
Water			*E. coli*	ATCC 25922	MIC=70–5000 μg/mL[c]	
			P. aeruginosa	ATCC 27950	MIC= 150–2500 μg/mL[c]	
Soxhlet Ethyl alcohol, diethyl ether, acetone, water	Tyrosol, 4-hydroxy benzoic acid, veratric acid, vanillic acid, syringic acid, p-coumaric acid, feluric acid, caffeic acid, oleuropein	Macrodilution method	*B. cereus*	University of Uludag	MIC=68–75 μg/mL[a]	[41]
					MBC=136–150 μg/mL[a]	
			S. aureus		MIC=50–55 μg/mL[a]	
					MBC=100–110 μg/mL[a]	
			L. plantarum		MIC=25–33 μg/mL[a]	
					MBC=50–66 μg/mL[a]	
			L. brevis		MIC=28–35 μg/mL[a]	
					MBC=56–70 μg/mL[a]	
			L. bulgaricus		MIC=35–42 μg/mL[a]	
					MBC=35–84 μg/mL[a]	
			S. thermophilus		MIC=37–46 μg/mL[a]	
					MBC=74–92 μg/mL[a]	
			S. cerevisiae		MIC=26–35 μg/mL[a]	
					MBC=52–70 μg/mL[a]	
			L. mesenteroides		MIC=30–36 μg/mL[a]	
					MBC=60–72 μg/mL[a]	
			S. typhimurium		MIC=90–110 μg/mL[a]	
					MBC=180–220 μg/mL[a]	

TABLE 12.3 *(Continued)*

Extraction	Compounds	Antimicrobial test method	Microorganisms	Strain number	Antimicrobial activity	References
			S. enteritidis		MIC=170–185 μg/mL[a]	
					MBC=340–370 μg/mL[a]	
			E. coli		MIC=60–72 μg/mL[a]	
					MBC=60–144 μg/mL[a]	
			P. aeruginosa		MIC=28–36 μg/mL[a]	
					MBC=56–72 μg/mL[a]	
			Acetobacter spp.		MIC=35–42 μg/mL[a]	
					MBC=35–84 μg/mL[a]	
SLE Water	–	Macrodilution method	*E. coli*	–	MBC=0.03 mg/mL	[42]
			P. aeruginosa		MBC=0.013 mg/mL	
			S. aureus		MBC=0.06 mg/mL	
			B. subtilis		MBC=0.06 g/mL	
SLE Water	Gallic acid, neochlorogenic acid, p-coumaoylquinic acid, Caffeic acid, vanillin, Ferulic acid, caffeic acid, p-cumaric acid, epicatechin, quercetin-3-O-rutinoside, quercetin-3-O-galactoside, kaempferol, hydrocinnamic	Microdilution method	*L. innocua*	–	MIC=1:10 (w/v)	[53]
			E. coli		MIC=1:10 (w/v)	
			E. coli O157		MIC=1:10 (w/v)	

TABLE 12.3 *(Continued)*

Extraction	Compounds	Antimicrobial test method	Microorganisms	Strain number	Antimicrobial activity	References
	derivatives, verbascoside, luteolin 7-o-glucoside, apigenin 7-o-glucoside, rutin, hydroxytyrosol, tyrosol, oleuropein		*S. enterica*		MIC=1:10 (w/v)	
SLE Water	Caffeic acid, verbascoside, oleuropein, luteolin 7-*O*-glucoside, rutin, apigenin 7-*O*-glucoside, and luteolin 4'-*O*-glucoside	Macrodilution method	*B. cereus*	CECT 148	IC_{25}=0.63 mg/mL	[43]
			B. subtilis	CECT 498	IC_{25}=4.12 mg/mL	
			S. aureus	ESA 7	IC_{25}=2.68 mg/mL	
			E. coli	CECT 101	IC_{25}=1.81 mg/mL	
			P. aeruginosa	CECT 108	IC_{25}=3.22 mg/mL	
SLE Water	–	Disk diffusion method Agar dilution method	*E. coli*	ATCC 25922	Inhibition halo (mm)=13.5 MIC=9.8 µg/mL	[44]
				–	Inhibition halo (mm)=15.3 MIC=6.97 µg/mL	
			P. aeruginosa	ATCC 10145	Inhibition halo (mm)=13.3 MIC=25.01 µg/mL	
			S. aureus	ATCC 6538	Inhibition halo (mm)=9 MIC=9.88 µg/mL	
				ATCC 25923	Inhibition halo (mm) MIC=9.12 µg/mL	

TABLE 12.3 *(Continued)*

Extraction	Compounds	Antimicrobial test method	Microorganisms	Strain number	Antimicrobial activity	References
			B. stearothermophilus	ATCC 11778	Inhibition halo (mm)=6.9 MIC=26.36 µg/mL	
SLE Dichloromethane Dichloroethane (1:1) mixtures of Chloroform: dichloroethane Chloroform: dichloromethane dichloroethane: dichloromethane ethyl acetate:dichloroethane ethyl acetate: dichloromethane	–	Agar dilution method	*E. coli*	–	MIC=20–30 µg/mL[a]	[45]
			P. aeruginosa		MIC=20–30 µg/mL[a]	
			S. aureus		MIC=20–40 µg/mL[a]	
Soxhlet Ethanol: water (80:20 v/v)	–	Well-diffusion method Macrodilution method	*P. aeruginosa*	Isolate (Al-Yarmuk Hospital, Iraq)	Inhibition halo (mm)=10–28[b] MIC=30 mg/mL MBC=60 mg/mL	[27]
			E. coli		Inhibition halo (mm)=6–22[b] MIC=15 mg/mL MBC=30 mg/mL	
			S. aureus		Inhibition halo (mm)=5–35[b] MIC=7.5 mg/mL MBC=15 mg/mL	

TABLE 12.3 *(Continued)*

Extraction	Compounds	Antimicrobial test method	Microorganisms	Strain number	Antimicrobial activity	References
SLE Ethanol: water (70:30 v/v)	–	Disk diffusion method Macrodilution method	*S. aureus*	–	Inhibition halo (mm)=5.5–12.02[b] MIC=0.625 mg/mL MBC=0.625 mg/mL	[46]
			B. cereus		Inhibition halo (mm)=9.1–16.62[b] MIC=5 mg/mL MBC=2.5 mg/mL	
			E. coli		Inhibition halo (mm)=8.5–16.72[b] MIC=2.5 mg/mL MBC=1.25 mg/mL	
			S. enteritidis		Inhibition halo (mm)=9.5–18.02[b] MIC=5 mg/mL MBC=2.5 mg/mL	
SLE Ethanol: water (80:20 v/v)	Luteolin-7-Glucoside Luteolin-4-Glucoside Oleuropein Verbascoside	Microdilution method	*E. coli* O157:H7	Eastern Regional Research Center	62.6 mg/mL=95% bacterial growth inhibition	[28]
			S. enteritidis	(ERRC)	62.6 mg/mL=100% bacterial growth inhibition	
			L. monocytogenes		62.6 mg/mL=100% bacterial growth inhibition	

TABLE 12.3 *(Continued)*

Extraction	Compounds	Antimicrobial test method	Microorganisms	Strain number	Antimicrobial activity	References
SLE Ethanol: water (75:25 v/v)	Oleuropein Rutin	Disk diffusion method Microdilution method	*E. coli*	–	Inhibition halo (mm)=8.33–16.33[c]	[48]
					MIC=1.30–6.25 mg/mL[c]	
					MBC=4.68–15.62 mg/mL[c]	
			S. aureus		Inhibition halo (mm)=8-15[c]	
					MIC=0.78–3.90 mg/mL[c]	
					MBC=1.95–9.37 mg/mL[c]	
			S. Typhimurium		Inhibition halo(mm)=7.66–16[c]	
					MIC=0.78–4.68 mg/mL[c]	
					MBC=1.56–9.37 mg/mL[c]	
SLE Ethanol: water (60:40 v/v)	–	Disk diffusion method	*E. coli* O157:H7	–	NA	[6]
			S. enterica		NA	
			L. monocytogenes		Presence of inhibition halo	

[a] *Range depending on the extraction solvent.*

[b] *Range depending on the concentration of disk extract.*

[c] *Range depending on the different olive leaves cultivars.*

[d] *Range depending on the origin a drying process of the leaf.*

IC_{25}, extract concentration which inhibits 25% of microbial growth; NA, no activity.

KEYWORDS

- **antimicrobial activity**
- **antioxidant activity**
- **foodborne pathogens**
- **microwave-assisted extraction**
- **olive leaf extract**

REFERENCES

1. Bascón-Villegas, I.; Espinosa, E.; Sánchez, R.; Tarrés, Q.; Pérez-Rodríguez, F.;Rodríguez, A. Horticultural Plant Residues as New Source for Lignocellulose Nanofibers Isolation: Application on the Recycling Paperboard Process. *Molecules* **2020,** *25* (14). DOI: 10.3390/molecules25143275.
2. Costa, J. R.; et al. Impact of: In Vitro Gastrointestinal Digestion on the Chemical Composition, Bioactive Properties, and Cytotoxicity of Vitis Vinifera L. cv. Syrah Grape Pomace Extract. *Food Funct.* **2019,** *10* (4), 1856–1869. DOI: 10.1039/c8fo02534g.
3. Panić, M.; et al. Ready-to-Use Green Polyphenolic Extracts from Food byproducts. *Food Chem.* **2018,** *283*, 628–636. DOI: 10.1016/j.foodchem.2019.01.061.
4. Casadey, R.; Challier, C.; Altamirano, M.; Spesia, M. B.; Criado, S. Antioxidant and Antimicrobial Properties of Tyrosol and Derivative-Compounds in the Presence of Vitamin B2. Assays of Synergistic Antioxidant Effect with Commercial Food Additives. *Food Chem.* **2021,** *335*, 127576. DOI: 10.1016/j.foodchem.2020.127576.
5. European Parliament. Regulation (EU) No 1333/2008. **2008,** *1999*, 1–18.
6. Albertos, I.; Avena-bustillos, R. J.; Martín-diana, A. B.; Du, W.; Rico, D.;Mchugh, T. H. Antimicrobial Olive Leaf Gelatin fi lms for Enhancing the Quality of Cold- Smoked Salmon. *Food Packag. Shelf Life* **2017,** *13*, 49–55. DOI: 10.1016/j.fpsl.2017.07.004.
7. FAOSTAT. [Online]. http://www.fao.org/faostat/en/#data/QC/visualize (accessed Jun 15, 2021).
8. Sánchez-Gutiérrez, E.; Bascón-Villegas, M.; Rodríguez, I.; Pérez-Rodríguez, A.; Fernández-Prior, F.; Rosal, Á.; Carrasco, A. Valorisation of Olea Europaea L. Olive Leaves Through the Evaluation of Their Extracts: Antioxidant and Antimicrobial Activity. *Foods* **2021,** *10* (966), 20.
9. Rahmanian, N.; Jafari, S. M.;Wani, T. A. Bioactive Profile, Dehydration, Extraction and Application of the Bioactive Components of Olive Leaves. *Trends Food Sci. Technol.* **2015,** *42* (2), 150–172. DOI: 10.1016/j.tifs.2014.12.009.
10. Gullón, P.; et al. Valorization of byproducts from Olive Oil Industry and Added-Value Applications for Innovative Functional Foods. *Food Res. Int.* **2020,** *137*,. DOI: 10.1016/j.foodres.2020.109683.
11. Talhaoui, N.; Taamalli, A.; Gómez-Caravaca, A. M.; Fernández-Gutiérrez, A.;Segura-Carretero, A. Phenolic Compounds in Olive Leaves: Analytical Determination, Biotic

and Abiotic Influence, and Health Benefits. *Food Res. Int.* **2015,** *77*, 92–108. DOI: 10.1016/j.foodres.2015.09.011.
12. Cejudo Bastante, C.; Casas Cardoso, L.; Fernández Ponce, M. T.; Mantell Serrano, C.;Martínez de la Ossa-Fernández, E. J. Characterization of Olive Leaf Extract Polyphenols Loaded by Supercritical Solvent Impregnation into PET/PP Food Packaging Films. *J. Supercrit. Fluids* **2018,** *140*, 196–206. DOI: 10.1016/j.supflu.2018.06.008.
13. Thielmann, J.; Kohnen, S.; Hauser, C. Antimicrobial Activity of Olea Europaea Linné Extracts and their Applicability as Natural Food Preservative Agents. *Int. J. Food Microbiol.* **2017,** *251*, 48–66. DOI: 10.1016/j.ijfoodmicro.2017.03.019.
14. Gómez-García, R.; Campos, D. A.; Aguilar, C. N.; Madureira, A. R.;Pintado, M. Valorization of Melon Fruit (Cucumis melo L.) byproducts: Phytochemical and Biofunctional Properties with Emphasis on Recent Trends and Advances. Elsevier Ltd.; *Trends Food Sci. Technol.* **2020,** *99* 507–519. DOI: 10.1016/j.tifs.2020.03.033.
15. Martínez-Patiño, J. C.; et al. Optimization of Ultrasound-Assisted Extraction of Biomass from Olive Trees Using Response Surface Methodology. *Ultrason. Sonochem.* **2019,** *51*, 487–495, 2019. DOI: 10.1016/j.ultsonch.2018.05.031.
16. da Rosa, G. S.; Vanga, S. K.; Gariepy, Y.;Raghavan, V. Comparison of Microwave, Ultrasonic and Conventional Techniques for Extraction of Bioactive Compounds from Olive Leaves (Olea europaea L.). *Innov. Food Sci. Emerg. Technol.* **2019,** *58*. DOI: 10.1016/j.ifset.2019.102234.
17. Martiny, T. R.; Raghavan, V.; De Moraes, C. C.; Da Rosa, G. S.; Dotto, G. L. Optimization of Green Extraction for the Recovery of Bioactive Compounds from Brazilian Olive Crops and Evaluation of its Potential as a Natural Preservative. *J. Environ. Chem. Eng.* **2021,** *9* (2), 105130. DOI: 10.1016/J.JECE.2021.105130.
18. da Rosa, G. S.; et al. Eco-friendly Extraction for the Recovery of Bioactive Compounds from Brazilian Olive Leaves. *Sustain. Mater. Technol.* **2021,** *28*, e00276. DOI: 10.1016/J.SUSMAT.2021.E00276.
19. Şahin, S.; Bilgin, M.;Dramur, M. U. Investigation of Oleuropein Content in Olive Leaf Extract Obtained by Supercritical Fluid Extraction and Soxhlet Methods. *Sep. Sci. Technol.* **2011,** *46* (11), 1829–1837. DOI: 10.1080/01496395.2011.573519.
20. Lafka, T. I.; Lazou, A. E.; Sinanoglou, V. J.;Lazos, E. S. Phenolic Extracts from Wild Olive Leaves and Their Potential as Edible Oils Antioxidants. *Foods* **2013,** *2* (1), 18–31. DOI: 10.3390/foods2010018.
21. Gullón, B.; et al. Valorisation of Olive Agro-Industrial byproducts as a Source of Bioactive Compounds. *Sci. Total Environ.* **2018,** *645*, 533–542. DOI: 10.1016/j.scitotenv.2018.07.155.
22. da Silva, A. C.; de F. Santos, P. D.; do P. Silva, J. T.; Leimann, F. V.; Bracht, L.;Gonçalves, O. H. Impact of Curcumin Nanoformulation on its Antimicrobial Activity. *Trends Food Sci. Technol.* **2017,** *72*, 74–82. DOI: 10.1016/j.tifs.2017.12.004.
23. Clinical and Laboratory Standards Institute. *Methods for Antimicrobial Dilution and Disk Susceptibility Testing of Infrequently Isolated or Fastidious Bacteria*, 2015; vol *35* (17).
24. Clinical and Laboratory Standards Institute. Performance Standards for Antimicrobial Disk Susceptibility Tests, 2015; vol *32* (1).
25. Clinical and Laboratory Standards Institute. In *Methods for Dilution Antimicrobial Susceptibility Tests for Bacteria That Grow Aerobically*, 2012; vol *32* (2).
26. Balouiri, M.; Sadiki, M.; Ibnsouda, S. K. Methods for In Vitro Evaluating Antimicrobial Activity: A Review. *J. Pharm. Anal.* **2016,** *6* (2), 71–79. DOI: 10.1016/j.jpha.2015.11.005.

27. Sweedan, E. G.; Al-haidari; Dh, M.; Muayyed, A. Antimicrobial and Antibiofilm Activities of Alcoholic Extract of Olive Leaves (olea europaea) Against Pathogenic Bacteria. *Biosci. Res.* **2019,** *16*, 367–374.
28. Liu, Y.; McKeever, L. C.; Malik, N. S. A. Assessment of the Antimicrobial Activity of Olive Leaf Extract Against Foodborne Bacterial Pathogens. *Front. Microbiol.* **2017**. DOI: 10.3389/fmicb.2017.00113.
29. Pérez Álvarez, G.; Coto Arbelo, S.; Echemendía Pérez, O.; Ávila Quezada, M. Pseudomonas Fluorescens Migula, ¿Control biológico o patógeno?: Pseudomonas Fluorescens, Biological Control or Pathogen?" http://scielo.sld.cu/scielo.php?script=sci_arttext&pid=S1010-27522015000300008 (accessed Jun. 27, 2021).
30. Tártara, S. G. Patógenos Emergentes-Tercera Parte "Klebsiella Pneumoniae Productora De Carbapenemasas (Kpn-Kpc)" Emerging Pathogens-Part Iii. *Transplante* **2013,** *33* (2), 103–109 [Online]. www.renal.org.ar.
31. López, A. I.; Ferrer, I.; Revillo, M. J.; Torres, L.; Sardaña, J.;García-Moya, J. B. Urinary Infection Caused by Enterococcus Faecalis with a Mucoid Morphotype. *Enfermedades infecciosas y microbiología clínica* **2001,** *19* (1), 36–37. DOI: 10.1016/S0213-005X(01)72550-5.
32. Berman, J. Candida Albicans, 2012. DOI: 10.1016/j.tig.2012.01.004.
33. Masoko, P.;Makgapeetja, D. M. Antibacterial, Antifungal and Antioxidant Activity of Olea africana Against Pathogenic Yeast and Nosocomial Pathogens. *BMC Complement. Altern. Med.* **2015,***15* (1), 1–9. DOI: 10.1186/s12906-015-0941-8.
34. Adem, S. R.; Ayangbenro, A. S.;Gopane, R. E. Phytochemical Screening and Antimicrobial Activity of Olea europaea subsp. Africana Against Pathogenic Microorganisms. *Sci. African* **2020,** *10*, e00548. DOI: 10.1016/j.sciaf.2020.e00548.
35. Edziri, H.; et al. A Comparative Study on Chemical Composition, Antibiofilm and Biological Activities of Leaves Extracts of Four Tunisian Olive Cultivars. *Heliyon* 2019, 5 (5), 2019. DOI: 10.1016/j.heliyon.2019.e01604.
36. Testa, B.; Lombardi, S. J.; Macciola, E.; Succi, M.; Tremonte, P.;Iorizzo, M. Efficacy of Olive Leaf Extract (Olea europaea L. cv Gentile di Larino) in Marinated Anchovies (Engraulis encrasicolus, L.) Process. *Heliyon* **2019,** *5* (5), e01727. DOI: 10.1016/j.heliyon.2019.e01727.
37. Şahin, S.; Samli, R.; Birteksöz Tan, A. S.; Barba, F. J.; Chemat, F.; Cravotto, G.; Lorenzo, J. M. Solvent-Free Microwave-Assisted Extraction of Polyphenols from Olive Tree Leaves: Antioxidant and Antimicrobial Properties. *Molecules* **2017,** *22* (7), 1056.
38. Bayçin, O.; Altiok, D.; Ülkü, E.; Bayraktar, S. Adsorption of Olive Leaf Antioxidants on Silk Fibroin. *J. Agric. Food Chem.* **2007,** *55*, 1227–1236. DOI: 10.1021/jf062829o.
39. Karygianni, A.; Cecere, L.; Skaltsounis, M.; Argyropoulou, L.; Hellwig, A.; Aligiannis, E.; Wittmer, N.; Al-Ahmad, A. High-Level Antimicrobial Efficacy of Representative Mediterranean Natural Plant Extract Against Oral Microorganism. *Biomed Res. Int.* **2014,** *2014*, 8. DOI: 10.1155/2014/839019.
40. Brahmi, M.; Flamini, F.; Issaoui, G.; Dhibi, M.; Dabbou, M.; Mastouri, M.; Hammami, M. Chemical Composition and Biological Activities of Volatile Fractions from Three Tunisian Cultivars of Olive Leaves. *Med. Chem. Res.* **2012,** *21*, 2863–2872. DOI: 10.1007/s00044-011-9817-8.
41. Korukluoglu, S.; Sahan, M.; Yigit, Y.; Ozer, A.; Gücer, E.T. Antibacterial Activity and Chemical Constitutions of Olea Europaea L. Leaf Extracts. *J. Food Process. Preserv.* **2010,** *34*, 383–396. DOI: 10.1111/j.1745-4549-2008.00318.x.

42. Markín, D.; Duek, L.;Berdícevsky, I. In Vitro Antimicrobial Activity of Olive Leaves. *Mycoses* **2003,** *46* (3–4), 132–136. DOI: 10.1046/j.1439-0507.2003.00859.x.
43. Pereira, A. P.; et al. Phenolic Compounds and Antimicrobial Activity of Olive (Olea europaea L. Cv. Cobrançosa) Leaves. *Molecules* **2007,** *12* (5), 1153–1162. DOI: 10.3390/12051153.
44. Nora, N. B.; Hamid, K.; Snouci, M.; Boumedien, M.; Abdellah, M. Antibacterial Activity and Phytochemical Screening of Olea Europaea Leaves from Algeria. *Open Conf. Proc. J.* **2014,** *3* (1), 66–69. DOI: 10.2174/1876326x01203020066.
45. Bawadekji, A.; Imran, M.;Randahawa, M. A. Antimicrobial Effects of the Water Immiscible Solvent Extracts of Olive Tree Leaves. *J. Pure Appl. Microbiol.* **2019,** *13* (4), 2189–2194. DOI: 10.22207/JPAM.13.4.31.
46. Hemeg, H. A. et al. Antimicrobial Effect of Different Herbal Plant Extracts Against Different Microbial Population. *Saudi J. Biol. Sci.* **2020,** *27* (12), 3221–3227. DOI: 10.1016/j.sjbs.2020.08.015.
47. Food and Drug Administration. In *Bad Bug Book, Foodborne Pathogenic Microorganisms and Natural Toxins*, Second Ed.; 2012.
48. Khanum, S.; Zahoor, F.; Khan, T.; Asghar, M.; Sablani, M. Antioxidant, Antibacterial and Functional-Food-Packaging Potential of Leaf Extract from Pakistani Olive Cultivars. *Pakistan J. Agric. Sci.* **2019,** *56*. DOI: 10.21162/PAKJAS/19.8313.
49. Moudache, M.; Silva, F.; Nerín, C.; Zaidi, F. Olive Cake and Leaf Extracts as Valuable Sources of Antioxidant and Antimicrobial Compounds: A Comparative Study. *Waste Biomass Valorization* **2021,** *12* (3), 1431–1445. DOI: 10.1007/s12649-020-01080-8.
50. Xu, Z.; et al. Polymicrobial Interaction and Biofilms Between Staphylococcus Aureus and Pseudomonas Aeruginosa: An Underestimated Concern in Food Safety. *Curr. Opin. Food Sci.* **2019,** *26*, 57–64. DOI: 10.1016/j.cofs.2019.03.006.
51. Gulay Baysal, K. G. M. G.; Esra Kasapbas, E.; Yavuz, N.; Hur, Z. Determination of Theoretical Calculations by DFT Method and Investigation of Antioxidant, Antimicrobial Properties of Olive Leaf Extracts from Different Regions, 2021; pp 1909–1917.
52. Ayana, B.; Nazan Turhan, K.;Turhan, K. N. Use of Antimicrobial Methylcellulose Films to Control Staphylococcus Aureus During Storage of Kasar Cheese. *Packag. Technol. Sci. Packag. Technol. Sci.* **2009,** *22*, 461–469. DOI: 10.1002/pts.870.
53. Delgado-Adámez, R.; Bote, J; Parra-Testal, E; Martín, V; Ramírez, M. J. Effect of the Olive Leaf Extracts In Vitro and in Active Packaging of Sliced Iberian Pork Loin. *Packag. Technol. Sci.* **2016,** *29*, 649–660. DOI: 10.1002/pts.2267.

PART III

Recent Applications and Advances on Bioactive Compounds from Food and Byproducts

CHAPTER 13

Biological Activities from Vegetal Products in Latin America

NATALIA MONTELLANO DURAN

Biotechnology, Universidad Católica Boliviana San Pablo, Santa Cruz, Bolivia

ABSTRACT

Plant sources have various potentials due to the variety of ecosystems in which they can be found. Different bioactive molecules with antibacterial, anti-inflammatory, antihyperglycemic, antihypertensive, and anticancer properties, among others, have been found in various Latin American plants. Many of the discoveries of these biological properties are due to the culture of ancestral medicine and the attempt to rescue that knowledge to insert it into modern society. From Mexico to Argentina, ancestral knowledge has been used to create cosmetics, medicines, and foods with biological activities of interest to modern society. Food and traditional medicine are full of bioactive molecules that in modern times have been maintained by tradition, but beyond tradition, pharmacognosy has helped us to know how these foods are improving the quality of life in the region. Each molecule and its variant in Latin American plant sources is rescued by researchers. This chapter identifies these molecules, their mechanisms of action, and their traditional use in Latin American countries.

13.1 INTRODUCTION

A lot of possibilities occur when we refer to the biological functions of biomolecules. Multiple activities are found to be capable of molecular

Food Byproducts Management and their Utilization. Ricardo Gómez-García, Ana A. Vilas-Boas, Débora A. Campos, Maria Manuela Pintado, & Cristóbal Noé Aguilar (Eds.)

interactions in cells, developing important biological activities at the cellular and physiological levels. The interactions between these molecules and the cell receptors or enzymes as structural and modulatory roles are diverse. The biological properties of several molecules have been studied: lipids, peptides, cyclic molecules, neurotransmitters, sugars, from all kinds of sources.[1] Hence, its source is diverse, that is, from microorganisms, fungi, microalgae to higher plants.[2] Chemobiological impact has been designed due to new drug developments from them.[3]

In this chapter, we focus the attention on biomolecules from plants (lower and higher) of Latin American origin that have been studied in regular beginning because of traditional ancient medicine.

Aimed at plant sources, molecules have been identified that can reduce the inflammatory procedure (anti-inflammatory), stop bacterial growth (anti-bacterial), slow down cell oxidation (antioxidants), control blood pressure (antihypertensive), prevent blood clots (anticoagulant), lower blood sugar (antihyperglycemic), slow down the growth of metastatic cells (anticancer) and be necessary for normal cell function, growth and development (vitamins)

At least 60 thousand years ago, humanity has been using natural products from plants, animals, or microorganisms to treat or alleviate diseases.[3] A general panorama of traditional medicine in Latin America has been interested in delimiting the common factors in the region.[1] Latin America traditional medicine has three factors to be considered historically to understand it nowadays: (1) an indigenous medical history, (2) a European cultural influence since the colonization, and (3) an African cultural influence.

According to various studies on biodiversity globally, Latin America is the region with the greatest biodiversity and studies about it.[4,5] The pre-Hispanic communities in the region retain their identity and many of the Native American traditions are preserved. Besides, currently, the genetic and cultural diversity in Latin America is large, so, traditional medicine subsists due to the unexploited biodiversity of the region.[6]

Latin American cultures had been studied since the colonization of America in the XV century. The cultures from the north like Aztecs and Mayans, in Central America like Leka and Takery to the cultures of the south, such as Wayuu, Mapuches, Charrúas, Aymara, Quechuas, Incas, Guaranies, Arhucos, Kogui, and Amazonian tribes, among the hundreds of cultures that remain until today. Each culture has used the plant's natural resources (tree leaves, wild herbs, roots, and fruits) around them, for example, in medicine, gastronomy, and cosmetics. These applications are not only influenced by the cultures but also by the natural resources of each region.[7]

The use of these resources changes accordingly to each country and also it could be employed in different parts of the plant and their application in medicinal treatments may differ. In this regard, the biological activity known is different in every Latin American country, for example, in Bolivia the root of *Acacia farnesiana* (L.) Willd. is used as an antidote to the poisonous seed or a floral tea for dyspepsia, and bark decoction for diarrhea, while in Cuba, its fruit is used for conjunctivitis, dermatitis, and sore throat. Flowers for nervous disorders in Yucatan, root decoction is used for tuberculosis, flowers are used in making unguents for headache, and tea is taken for indigestion in Mexico; in Peru, a floral infusion is used for dyspepsia and the leaf infusion is used for curing fever in Peru.

Ethnopharmacology is an interdisciplinary field of cultural anthropology, biology, and medicine, including field observations, description of the use and preparation of remedies, botanical determination of the material obtained, phytochemical studies to isolate the compounds present in plants, as well as pharmacological studies, so, it is a science field that has gained significant importance around the world because its great relevance between different cultures.[8]

There are references about the endemic species in each country of Latin America: 150 from Argentina,[9] 257 from Colombia,[10] 1676 from Perú,[11] 2402 from Bolivia,[12] 827 from Ecuador,[13] around 700 from Venezuela,[14] 3557 from Brazil,[15] exemplifying the quantity and variety of plants in Latin America. In many countries, endemic species identifications are still being made in each of their regions.

The identification of species and their use in ancestral medicine is closely related, due to the rediscovery of biological properties by researchers in each country of the region. Several researchers screen different plant genera to see biological activities, not only of plants but also of fungi, microorganisms, or animal products traditionally used in some of their forms, drinkable infusions, or topical uses, in traditional medicine.[16]

There has been great interest in using ancestral knowledge and medicine in commercial, industrial, pharmaceutical, and handmade products. This is because Latin American society has powerful cultural beliefs.[17] A business strategy has been increased to associate cultural beliefs with science to gain greater credibility since the last century.[18]

Worldwide, researchers have already studied several molecules that are found in multiple natural plant-based sources with biological activities. Among them, anthocyanins from red or blue fruits, have been studied and they have many biological activities, such as anticarcinogenic, anti-inflammatory, antibacterial, antithrombotic, and antioxidant activities, among others.[19] Several species have aroused gastronomic, pharmaceutical, and medicinal interest throughout the world after measuring biological properties, such as açai (*Euterpe oleracea*) in Brazil, achachairu (*Garcinia humilis*) in Bolivia, lucuma (*Pouteria lucuma*) in Peru, lulo (*Solanum quitoense*) in Colombia and Ecuador, pitaya (*Selenicereus megalanthus*) in Nicaragua and Mexico, among others.

The search for active compounds from different plant species has been an important target since colonization time. It has been taking relevance in recent years due to new techniques and access to experimental laboratories to verify their biological potential. In some countries, such as Brazil and Cuba, a vast pharmacobotanical industry and research have been developed to increase their applicability and mercability. Product development is still very similar to traditional medicine. For example, the use of ointments with antibacterial active compounds as flavonoids are used as natural antiseptics, the use of infusions for antihyperglycemics or antihypertensives.[20]

The action of a bioactive molecule depends on its nature, several active molecules have multiple forms of acting in cellular or/and physiological processes. In this section, the bioactive action of some of these molecules and the ancestral pharmacological presentation in which traditional medicine uses them will be described, including:

- Reducing the inflammatory procedure (anti-inflammatory)
- Stopping bacterial growth (antibacterial)
- Slowing down cell oxidation (antioxidants)
- Controlling blood pressure (antihypertensive)
- Preventing blood clots (anticoagulant)
- Slowing down the growth of metastatic cells (anticancer)
- Controlling blood sugar content (antihyperglycemic)
- Necessary for normal cell function, growth and development (vitamins)

The most studied bioactive molecules known are polyphenols, such as flavonoids, anthocyanidines, tyrosol, cinnamyl alcohol, terpenes, peptides, lipids unsaturated, and polyhydroxyalkaloids, among others.

Some examples of how these molecules interact can be:

- Benzoxazinones, different natural benzoxazinone acetal glucosides, benzoxazinone algucones, and benzoxazolin-2(3H)-ones can be found in some crops, such as *Zea mays*, *Secale cereal*, *Coix lachrymal jobi*, *Triticum aestivum*, *Agropyron*, *Aphelandra tetragona*.[21] Some plant families known for their beneficial biological activities include

Acanthaceae, Poaceae, Ranunculaceae, and Scrophulariaceae. Benzoxazinones are known because they have biological activities that help to protect plant tissues from different pests, they could act as antimicrobial, pesticide, and antifungal agents.

- Molecules that protect higher plants from oxidative damage, such as betalains and anthocyanins have been studied through the years, and researchers have found multiple fruits. The order Caryophyllales is an example. Plant color can have a repelling action against herbivores or UV radiation damage.[22]
- Among the flavonoids, anthocyanins are the largest group of water-soluble pigments in the plant kingdom. They are responsible for most of the red, blue, and purple colors of fruits, vegetables, flowers, and other plant tissues or products. Nearly, 400 forms of anthocyanins are distinguished because of the hydroxyl or methoxyl groups in the flavan ring.

The diversity of natural resources in Latin America has been the focus of the whole world for decades. The researches and results of these research works have been exponential in recent years. It is highlighted that the researchers of these are not always Latin American. Many grains, fruits, and herbs have been brought to other continents due to the interest of scientists and of society for their exploitation.

13.2 ANTI-INFLAMMATORY

The inflammation process is caused by an immune system response to pathogens or injuries. Cells of the immune system recognize certain molecules of microorganisms or any antigen on the outside and trigger inflammation. The cells of the immune system act through response molecules, such as cytokines, histamine, nitric oxide, leukotrienes, and prostaglandins. Each of these molecules has a specific function within the inflammation process. For example, cytokines repair local damage by causing the expression of selectins and integrins. Also, histamine is responsible for vasodilation, and nitric oxide promotes vasodilation. The synthesis of prostaglandins and leukotrienes improves the dilation and patency of blood vessels.[23] Many inflammatory diseases are associated with prostaglandins, a family of 20-carbon fatty acids found widely distributed in almost all living animal cells, tissues, and glands.[24]

13.2.1 TRADITIONAL USES

Among the herbal drugs traditionally used in the treatment of rheumatic diseases, and more recently studied, the following can be mentioned: the bark of devil's claw (*Harpagophytum procumbens* DC. Ex Meisn.), the crystal of aloe (*Aloe vera*), beach sage tincture (*Pluchea carolinensis* (Jacq.) G. Don), eucalyptus leaves (*Eucalyptus globulus* Labill.), aniseed caisimon (*Piper auritum* Kunth), black mustard seeds (*Brassica nigra* (L.) WDJ Koch) and roots of marshmallow (*Althaea officinalis* L.), the latter has been used since ancient times to counteract inflammation of the digestive system, respiratory system, and skin.

Smilax aristolochiifolia Mill, the young shoots, leaves, and tendrils of some pleasantly edible species are used in Argentina, Brazil, and Mexico in infusions of their different plant tissues as an antirheumatic agent; in addition, they act against acne, anorexia, arthritis, dermatosis, fever, gallstones, gout, urticaria, kidney stones, leprosy, muscle weakness, nephrosis, oliguria, rheumatism, burns, cancer, among others.

Solanum tuberosum L., it is consumed directly as tubers, mostly cooked. The tuber is used in different preparations against burns and itches in Bolivia. In addition, the infusion of the peel is used as a diuretic or used against rheumatism in Bolivia. In Peru, slices of raw tuber are used for burns or to treat insect bites; tuber infusion is used to treat kidney stones, as in Venezuela.

Spondias purpurea L., its tender shoots, and infusion of cooked leaves are used in food and traditional medicine. The infusion is used in Peru, Argentina, Bolivia, Haiti, Guatemala, Chile, Costa Rica, and Ecuador for different purposes, that is, to act as anti-inflammatory agent in response to different diseases.

With these aforementioned examples, the most consumed form of plant sources for their anti-inflammatory activity are infusions of different plant tissues. Although, the use of crushed vegetable tissues has also been seen on some external inflammation, either due to disease (such as arthritis) or immediate immune response (such as insect bites).[25]

13.2.2 PLANT FAMILIES WITH ANTI-INFLAMMATORY ACTIVE MOLECULES

Alkaloids are molecules with a basic nitrogenous compound. Plants previously identified with anti-inflammatory activity belong to the families Malvaceae, Buxaceae, Solanaceae, Leguminosae, Apocynaceae, Liliaceae, Papaveraceae, Rutaceae, and Ranunculaceae.

Glycosides and polysaccharides, sugars bind to one another by a glycosidic bond. Plants previously identified with anti-inflammatory properties belong to the families Asteraceae, Rosaceae, Leguminosae, Scrophulariaceae, Polygonaceae, Solanaceae, and Myrsinaceae.

Terpenoids, these compounds are derived from isoprene, a 5-carbon organic compound and polymers of isoprene called terpenes. Plants previously identified for anti-inflammatory properties belong to the families Umbelifereae, Lamiaceae, Taxodiaceae, Capparidaceae, Cucurbitaceae, Burseraceae, and Asteraceae. Resins, multiple polymers like terpenes are found in Burseraceae.

Essential oils are active volatile substances that are embedded in a concentrated hydrophobic liquid. Plants previously identified for anti-inflammatory activities are in the families Boraginaceae, Ericaceae, Labiatae.

Flavonoids are polyphenolic secondary metabolites with a 15-carbon backbone, usually with two rings: a phenyl and a heterocycle. They can be classified into bioflavonoids, isoflavonoids, and neoflavonoids. Plants previously identified for anti-inflammatory properties are in the families Fagaceae, Leguminosae, Lamiaceae, Ginkgoaceae, and Gentianaceae.

Cannabinoids are phenolic compounds, with one or more hydroxyl groups attached to an aromatic hydrocarbon group. They are classified as simple phenols or polyphenols. Plants previously identified for anti-inflammatory activities belong to the families Alliaceae, Caprifoliaceae, Asteraceae, Piperaceae, Bignoniaceae, and Oleaceae.

Steroids are compounds with four rings arranged in a specific configuration, four fused rings of 17 carbons, 3 cyclohexanes, and 1 cyclopentane. Plants previously identified to possess anti-inflammatory properties belong to the family Ganodermataceae.

Fatty acids are composed of a saturated or unsaturated carboxylic acid with a long aliphatic chain. Plants previously identified as having anti-inflammatory activities belong to the family Onagraceae.

Plant glycoproteins are proteins with a sugar molecule attached. Plants previously identified to have anti-inflammatory activity are in the families Anacardiaceae and Rubiaceae.[26]

13.2.3 FUNCTION OF MOLECULES WITH ANTI-INFLAMMATORY ACTIVITY

In each Latin American country, different plant extracts have been used to reduce the inflammatory process and all the molecules involved in the process. Recently, terpenes have been corroborated in plant extracts such as *Tripterygium wilfordii Hook. F.* (Celastraceae) that markedly inhibited mRNA synthesis and the expression of metalloproteinase proteins. In human primary osteoarthritis, an inflammatory disease, these proteins are those induced by pro-inflammatory cytokines.

Scientific research has been supporting the bioactives of the vegetal extracts, which is increasing the interest of pharmaceutical companies. For example, the extract of *Tripterygium wilfordii* Hook. f., in the form of tripterin, has been suggested in antiarthritic medications and has protective role for cartilaginous tissue. Another example, the bioactive tingenome of *Maytenus canariensis* (Loes.) Kunk has been isolated. et Sund. (Celastraceae), since its role in the production of stimulated monocytes has been proven. Glycosylated iridoids, such as aucubin and catalpol, common in oriental medicine, have been isolated from the group of cyclopentane monoterpenoids.

Anti-inflammatory activity has been found in ginsenosides, tetracyclic triterpenic saponins, found in the genus *Panax* (Araliaceae). This plant is known because of its function as inhibitor of the induction of nitric oxide and some cytokines production in microglial cells, possibly by inhibiting the phosphorylation of kinases at the terminal amino group and an inhibitory effect on the activation of nuclear factor kappa-light-chain-enhancer of activated B cells.

Another type of compound in which anti-inflammatory activity has been found is flavonoids. Inhibitory activity of tumor necrosis factor-alpha has been reported in macrophages and the isoflavonoid genistin.

Bioactive mixtures with flavonoids from *Silybum marianum* (Asteraceae) that have inhibited the activity of induced tumor necrosis factor-alpha and nuclear factor kappa-light-chain-enhancer of activated B cells have been isolated in human histocytic lymphoma cells and also, an effect on the increase of both clusters of differentiation-type thyroid lymphocytes and a decrease in the expression of interleukins.

Likewise, antioxidant flavonoids have been isolated from *Scutellaria baicalensis* Georgi (Lamiaceae), traditionally used for the treatment of bronchitis, nephritis, hepatitis, asthma, and atypical dermatitis, demonstrating to inhibit NO production, between other immune responses.

Other extracts with biological activity are those of *Ginkgo biloba* L. (Ginkgoaceae), rhizomes of *Coptis, Makino japonica* (Ranunculaceae), *Distichoselinum tenuifolium*, *Podocarpum desmodium*, *Ajuga bracteosa*, *Ananas comosus*, among others. All the extracts have shown some type of inhibition in the inflammatory process.[27]

As has been seen previously, depending on the type of bioactive present in plant extracts, they will have an inhibitory interaction of various factors that intervene in the inflammatory process, such as pro-inflammatory cytokines, adhesion molecules, nuclear factors, and other inflammatory enzymes, such as iNOS and COX-2. For example, flavonoids have shown anti-inflammatory activity in animal models. There are previous reports about flavone/flavonols, quercetin, myricetin, fisetin, because they possess cyclooxygenase (COX) inhibitory activity. Cyclooxygenase, a membrane-associated enzyme with two isoforms, synthesizes prostaglandins, responsible for the inflammatory response.[28]

13.3 ANTIBACTERIAL

Antibacterial activity has been of great interest in recent years, as antibiotics use from the 18th century has become useless due to the resistance developed by microorganisms. Therefore, the search for antibacterial is in the interest of world public health. Resistances can be innate or acquired by plasmids. Molecules with antibacterial bioactives, such as quinine and coumarin, and their derivatives, have recently been found.[29] Antimicrobial activity is the ability of a molecule to inhibit the growth of a bacterial population, either by stopping population growth or by eliminating the population. Among antimicrobial substances, their effect is distinguished by the concentration necessary to inhibit growth or eliminate bacteria. Taking this into account, they are distinguished between bactericidal or bacteriostatic. In addition, when testing a bioactive, the minimum inhibitory concentration and the minimum bactericidal concentration are measured.

13.3.1 TRADITIONALLY USED SPECIES FOR ANTIBACTERIAL

Here are some examples of plants that are used as antiseptics in order to control bacterial infections in wounds.

Achyrocline satureioides is used as choleretic, hepatoprotective, digestive, and antimicrobial, its most common form of use is in the form of an infusion and many times, the crushed leaves are placed on the lesions after cleaning the wound. It is used in Argentina, Bolivia, Brazil, Uruguay, Peru, and Colombia.

Gailliardia megapotamic var. Scabiosioides, from the same Asteraceae family, is used as an analgesic, antiseptic, antineuralgic, for influenzas, as a decongestant and to treat acne.

Protousnea poeppigii, from the Polytrichaceae family, is used as a curative and antibiotic in the form of a natural extract and infusion.

In different Latin American countries, experiments have been carried out to verify the curative effect of different plants. Among the species most used as antimicrobial or antiseptic are *Thymus vulgaris* L., *Rosmarinus officinalis* L., *Melissa officinalis* L., *Salvia officinalis* L., *Ocimum basilicum* L., *Achillea millefolium* L., *Syzygium aromaticum*, *Caryophyllus aromaticus* L. *Punica granatum* L., *Syzygium cumini*, *Psidium guajava* L.,[30] *Ammi majus* L, *Vinca minor* L, *Solidago chilensis* Meyen, *Leonurus sibiricus* L, *Schkuhria pinnata* Kuntze ex Thel, *Psittacanthus cuneifolius*, *Tripodanthus acutifolius*, *Passiflora edulis* Sims, *Colletia spinosissima* J. Gmelin, *Jodina rhombifolia* Hook. et Arn., *Nicotiana glauca* Graham, T*ropaeolum majus* L., *Achyrocline satureioides*, *Baccharis crispa*, *Gailliardia megapotamica* var.radiat a, *Gailliardia megapotámica* var., *Gailliardia cabrerae*, *Picrosia longifolia*, *Pterocauulon polystachyum*, *Macfadyena unguiscati*, *Terminalia australis*, *Equisetum giganteum*, *Adesmia bicolor*, *Gentianella achalensis*, *Gentianella parviflora*, *Pellaea flavens*, *Lippia turbinata*, *Lippis germinate*, and *Protousnea poeppigii.* All these species have their traditional use not only as antibacterial but also as a diuretic, astringent, antirheumatic, digestive, menstrual regulator, among other uses.[31]

Chamaesyce hirta (L.) Millsp., used in Peru, has antidiarrheal activity and antibacterial, antiamoebic, and antispasmodic action; its extract is used with oral doses. *Coriaria ruscifolia* L. is used in Colombia, Peru, Ecuador, and Bolivia, and its methanolic powdered fruit extract has a slightly antibacterial and cytotoxic activity. *Croton lechleri* Müll. arg., known as "dragon's blood croton." It is used in Brazil for bacterial infections, ulcers, tumors, injuries; in Ecuador it is used for cancer, inflammations and as antiseptic; and in Peru, it is used for stopping bleeding, bug bites, cancer, *Candida* infection, diabetes treatment, diarrhea, eczema, fractures, different infections in the gastrointestinal system as gastroenterosis, gingivosis, hemorrhoid, laryngosis, pharyngosis, toothache and ulcers.

The most common forms are in infusions both to ingest and topical uses (leaves or roots). Essential oil therapies have also been developed as a way to preserve the extract for longer time. Ripe leaves are mostly used for traditional medicine.

13.3.2 INTERACTION OF BIOACTIVE MOLECULES

Depending on the mechanism of each of these active molecules, it will be bactericidal or bacteriostatic. Some molecules interact with the systemic mechanism of the bacterial cell wall, considering the difference between gram-positive and gram-negative. There are different mechanisms that could interact with the cell wall, other mechanisms that have a role in the transport of precursors or the structural organization of the bacterial cell wall have also been observed. Another mechanism of interaction is the inhibition of some processes of bacterial protein synthesis in any of its phases, activation, initiation, formation of complete tRNA-ribosome, elongation. Additional important mechanism is the synthesis of bacterial nucleic acids that can undergo inhibition by RNA polymerase or DNA compaction. Finally, there are molecules capable of inactivating the resistance mechanisms developed by some microorganisms.[32]

Flavonoids have antimicrobial (antibacterial, antifungal, and antiviral) activity as esters of phenolic acids. Quercetin is one of the most studied flavonoids as antibacterial agent against *Staphylococcus aureus*; also, proteins and lipids are affected in their synthesis. Researchers suggest that ring B plays an important role in hydrogen bonding necessary for nucleic acid synthesis.[33] Flavanones without sugars presented antibacterial activity.[34] General structures of various flavonoids can change their biological activity according to their substitution patterns of series of flavonoids: flavones, flavonol, flavanone, flavanonol, isoflavone, flavan-3-ols, and flavylium salts. Also, flavonoids present the inhibition of cytoplasmic membrane function and inhibition of energy metabolism.[33]

Various peptides can also act as antimicrobials, and their mechanism requires an amphipathic property that causes molecules to bind favorably at the interface of lipid bilayers. This binding of the peptide to the bacterial lipid bilayer causes a local expansion of the area in the membrane, causing the formation of pores.[35]

13.4 ANTIOXIDANT

A molecule capable of inhibiting the oxidation process of other molecules is called an antioxidant. Oxidative stress in biological systems is a complex process that is characterized by an imbalance between the production of free radicals (FR) and the ability of the body to eliminate these reactive species through the use of endogenous and exogenous antioxidants. During the metabolic processes, a great variety of reactions take place, where the promoters are the reactive oxygen species, such as hydrogen peroxide. The reactive oxygen species include Superoxide (O_2^-), hydroxyl ($HO^•$), peroxyl ($ROO^•$), alkoxyl ($RO^•$), and nitric oxide ($NO^•$). This reactive species induces oxidative damage through other biomolecules, such as lipids, nucleic acids, proteins, and carbohydrates. The most known damage is aging, cancer, and many other diseases. In addition, ROS are involved in multiple diseases, through different complications, diseases, such as diabetes, malaria, cancer, acquired immunodeficiency syndrome, many heart or cerebrovascular diseases, and even arteriosclerosis.[36]

13.4.1 TRADITIONAL USES

In food, fruits, and vegetables are closely associated with antioxidant substances. This is because there are simple phenolic substances in them, such as monophenols with a single benzene ring (e.g., 3-ethylphenol or 3,4-dimethylphenol), the hydroxycinnamic acid group (e.g., caffeic or ferulic acid), and flavonoids and their glycosides (e.g., catechins, proanthocyanins, anthocyanidins, and flavanols). They also contain some tannins, a complex group of water-soluble phenols. Here there are some examples:

Phyllanthus niruri L., used as antioxidant, has been indicated for alopecia, amenorrhea, cancer, colic, colitis, constipation, diabetes, dysentery, dyspepsia, edema, enterosis, fever, flu, gallstones, gas, kidney stones, malaria, nephrosis, oliguria, pain, spasms, stomachache, vaginitis, and worms. It is traditionally used in the form of an infusion, although currently there are already tablets and drops with active ingredients.

Hibiscus tiliaceus, its methanolic flower extract has antioxidant and antimutagenic activity. The traditional ways of using it vary from country to country. For example, Brazilians use the flowers and leaves directly on sores and chronic wounds. In Cuba, the infusion of the leaves is used. In Nicaragua, the infusion is used, the bark and the leaves are used in childbirth and also against constipation and fever. On the other hand, in Panama, the flowers boiled in milk are used especially for earache.[25]

In this type of activity, high concentrations are needed, so, the most commonly used traditional form includes the leaves, fruits, stems or roots crushed and used topically on the wound or infection.

13.4.2 MOLECULES RESPONSIBLE FOR THIS ACTIVITY

In nature, there are multiple molecules responsible for stopping natural oxidation in cellular processes, among the best known are flavonoids, phenols, vitamin C, and carotenoids.[37]

Phenolic acids compounds are present in food as secondary metabolites, acting as antioxidant protecting food from oxidative rancidity. The potential antioxidant activity depends on the number and arrangement of the hydroxyl groups. Among the natural antioxidants as phenolic acids there are hydroxy-benzoic acid derivates and hydroxycinnamic acid derivates.[38]

Flavonoids are found in large quantities in plants, different types of flavonoids with different antioxidant and metal chelating activities.[34] In the flavonoids groups there are multiple active molecules as flavonols, flavononols, flavones, flavanones, flavanols, antocyanidins, and isoflavonoids.[33]

Carotenoids, as lycopene, α-carotene, β-carotene, β-cryptoxanthin, lutein, zeaxanthin, and astaxanthin, are natural colorants and lipid-soluble pigments exclusive in plants. Their basic structure has 40 carbons with multiple double-bond system. Carotenoids have important roles in cell metabolism as junction communication, modulation of growth factors, stimulation of immune response, cell differentiation, regulation of cell cycle, and important role as antioxidant agents. The antioxidant effect is due to the extinction of the O_2 resulting from the excitation of the carotenoids that end up dissipating the energy by a series of interactions with the medium, rotational, and vibratory in the solvent.[39]

13.4.3 INTERACTIONS

The mechanism of action as an antioxidant of phenolic acids or flavonoids has to do with these substituents on the rings that compose it. Generally, it is a two-step reaction where the antioxidant donates the proton to the medium and an electron, so that later the free radical in the cell medium takes this electron and hydrogen and is reduced.[40]

Electron donation reactions are favored by the electron density in the end rings, in flavonoids.

The reaction is known as HAT-based, which has to do with the ability to quench free radicals by donating hydrogen. Also, it is possible to detect the potential ability to transfer an electron by the SET-based method. When it comes to a sequence followed the mechanism is called sequential proton loss electron transfer.[36]

13.5 ANTIHYPERTENSIVE

Hypertension is the high blood pressure, it increases the heart disease risk, stroke, aneurysm, heart failure and can lead to hypertension. Some researchers studied this disease suggesting that molecular mechanisms are involved in the endothelial dysfunction. Certain molecules as nitric oxide and prostacyclin are associated with the production of vasoactive mediators. Overexpression of some vasoconstrictors like endothelin-1 also promotes vascular remodeling. The phramacological molecules can control hypertension disease.[41]

The leading cause of death in Latin America is the cardiovascular disease. In some countries, researchers have found that patients had hypertension (42.5%) and pre-hypertension (32.5%). Risk factors found in Latin America population for cardiovascular disease were hypertension and obesity. Studies in 20 countries in Latin America found that the most frequent cardiovascular risk factors include arterial hypertension, overweight/obesity, diabetes, and smoking. A glucagon–growth hormone-releasing factor vasoactive intestinal peptide secreted inhibits platelet activation and the proliferation of vascular smooth muscle cells are acting as a potent pulmonary vasodilator.[42]

Previous researches have shown that flavonoids can prevent cardiovascular diseases including hypertension and atherosclerosis, reducing arterial pressure in rats enhancing the vasorelaxant process.

13.5.1 TRADITIONAL USES

In this section, we review some Latin American plants with hypertensive activity and their traditional uses as an example.

Agave americana L. presents antihypertensive, antiallergenic and anti-inflammatory activity, between others. The traditional uses were as food for indigenous in America. Also, they consumed it in fermented form or cooked flowers are consumed. Likewise, in Mexico, they consume the sap of the fermented stem, pulque. Even the distillation of the fermented sap is consumed as mescal or tequila. Chicha and vinegar are made in Peru. In the Andes, it dissolves in water to be used as a pesticide on crops. In Ecuador, the crushed roots are used as shampoo. It has been proven effective in treating hypertension.

Bidens pilosa L. In several Latin American countries, it is used to treat angina, chills, diabetes, dysentery, dysmenorrhea, edema, headache, hepatosis, jaundice, laryngitis, malaria, ophthalmia, parasites, sore throat, stomach pain, stomatosis, toothache, wounds, adenopathy, bleeding, breast engorgement, insect bites, cough, diabetes, dysentery, dysuria, fever, gonorrhea, hemorrhoids, hepatosis, inflammation, malaria, mycosis, parasites, pharyngitis, pulmonosis, rheumatism, sclerosis, toothache, urinary tract infections, vaginosis, and other diseases. In its most rudimentary forms, they are used in traditional communities, such as Panama, where the juice is used directly to treat dermatosis, sores and wounds, and is also used for cancer, fever, gas, itching, oliguria, rashes, and sores. In Panama, it is taking it to treat colds, enterosis, headaches, prostate tumors, and rheumatism. In Peru, it is used in the form of an infusion with lemon, for dropsy, hepatitis, and water retention. It has been experimentally proven that the antihypertensive, vasodilator, and vasorelaxant leaf extract.

Carica papaya L., it is traditionally consumed as raw ripe fruit. In its unripe, green form, it is cooked, pickled for consumption. The seeds are often used as a condiment in salads. The flowers and leaves can be eaten steamed and used in conjunction with meats to tenderize them. Stems and marrows are also eaten cooked.

Cecropia peltata L., it is eaten as grass, leaves, and flowers are consumed in the form of broth. The leaves were used in the form of tobacco. Its fruit is consumed fresh for desserts. Also, in Peru, they have been combined with the consumption of coca leaves. The infusions are used for different purposes in Argentina, Barbados, Belize, and Colombia. It is beneficial against diabetes and also used as diuretics, to treat infections and even Parkinson's disease.

Ipomoea batatas (L.) Lam., in Haiti, they mix the roots with honey and sulfur for coughs; also, they use the root with olive oil against shellfish poisoning, or use the roots directly on wounds. In Nicaragua, they use the leaves topically on cuts or hemorrhages. In Peru, bruised leaves are used for dermatological problems or tubercles for rheumatism or varicose veins.

Morinda citrifolia L., ripe fruits are traditionally consumed directly or in syrup. Immature fruits are eaten cooked or pickled. Also, the tender leaves and blanched shoots are eaten raw.

Cephalanthus occidentalis L., *Uncaria tomentosa* (willd. ex schult.) DC, as many others, the most common way to consume it is in the form of an infusion.

Zea mays L., in Peru, they burn dried corn and inhale the smoke for sinusitis and flu. In Argentina, they use cooked corn water to remove the breast from babies. In Bolivia, the water from boiling corn is drunk to stop diarrhea, cystitis, gout, sciatica, and dysentery. Similarly, in Costa Rica, Haiti, they use boiled corn water as an anti-inflammatory, diuretic agent or to prevent high blood pressure.

The antihypertensive effect is seen with an ingestion of infusions in general and must be regular in order to control blood pressure.

13.5.2 ACTION OF ANTIHYPERTENSIVES MOLECULES

The antihypertensive molecules most seen in nature have the function of inhibiting the enzymes that convert angiotensin, that is, they block angiotensin receptors or inhibit angiotensin-converting enzymes. Angiotensin is a peptide molecule that narrows the blood vessels and increases the blood pressure usually by the action of a flow of calcium. The molecules that regulate blood pressure or intervene with angiotensin giving an antihypertensive effect include alkaloids, diterpenes, flavonoids, and peptides from natural plant sources.[43] The mechanism of action of alkaloids is to modulate the function of calcium ion channels. Other system to lower blood pressure is by relaxing the vascular smooth muscle through diterpenes. Norepinephrine contractions are the result of potassium cation spasms that can be lessened by coumarins. Compounds that show great inhibitory activity against angiotensin-converting enzyme are flavonoids.

In order to reduce blood pressure, cardiac output must be reduced, the baroreflex system is activated by the beta-blockers. In response, total peripheral resistance increases and there is a readjustment of the baroreceptors. The sympathetic activity of the nervous system decreases, so vasomotor tone is reduced. The release of norepinephrine is reduced by the effect of beta receptor pre-binding.

The two classes of diuretics, thiazine and loop, are different pharmacologically as well as their mechanisms of action. The former, loop diuretics, act on the nephron in the apical membrane of the thick ascending limb of the loop of Henle. The reabsorption of Na^+ and chloride (Cl^-) in the $Na^+/K^+/2Cl^-$ co-transporter is inhibited by competition with Cl^-. There is a sigmoid relationship between the excision of Na^+ and loop diuretics, it is that of a sigmoid curve.

On the other hand, thiazide diuretics have an effect on the nephron in the apical membrane of the distal convoluted tubule, where they inhibit the coupled reabsorption of Na^+ and Cl^-. The effect is less than that of loop diuretics. Thiazide diuretics alter urine dilution capacity.

The angiotensin-converting enzyme inhibitors have the ability to inhibit the angiotensin-converting enzyme is given by the transformation of angiotensin I into angiotensin II acting on a pluripotent zinc metalloprotein. This enzyme is present in the endothelial cells of capillary blood vessels and venules, and even in the endothelial cells of the lung.

Another antihypertensive mechanism is the angiotensin II receptor blocker, which antagonizes angiotensin II at the receptor level in cells of the heart, kidney, smooth muscle, and the aorta.

The action of calcium channel blockers is as its name implies to block the depolarizing effect on vascular smooth muscle cells, cardiac myocytes, and cardiac nodal tissue, since they depend on the amount of Ca^{2+} that enters the cell.[44]

13.6 ANTICOAGULANT

One of the most biological activities studied is the anticoagulant activity, because of the increasing blood-contacting coagulation problems in artificial organs or medical procedures. Thrombosis is one of the most common complications that leads to death.[45]

Anticoagulants are used in medical procedures, clinical trials, and science. Heparin is the best example of anticoagulant, the molecule is a linear and highly sulfated glycosaminoglycan, which is a repeating monomeric disaccharide of uronic acid and glucosamine.[46] Heparin, as an anticoagulant, acts in binding fibrin, as an antithrombin, factor Xa, factor IXa, thrombin, heparin cofactor II, and protein C inhibitor.

Multiple saccharides have been used in trials due to their anticoagulant effect. Low-cost alginate sulfated,[45] chitosan sulfated,[47] polysaccharides extracted from natural algae,[48] synthetic peptides,[49] between others.

13.6.1 TRADITIONAL USES

The treatment for anticoagulant activity is systematic, so, most of the vegetable sources with this activity should be consumed regularly in solid foods or infusions. Here are some examples used traditionally:

Achillea millefolium L., it is consumed as an infusion. The fresh leaves are used in salads, sauces, and soups. Also, they are cooked like vegetables, many times they are used to prepare alcoholic beverages. In Latin America, Argentina, and Peru, the indicated infusion is drunk to treat amenorrhea, hemorrhagic diarrhea, hemorrhoids, colds, tuberculosis, and improve blood circulation.

Ananas comosus (L.) Merr., the fruit is consumed raw and ripe, or it is used in natural drinks such as juices or even cocktails. Unripe fruit is cooked for soups or sauces. The Latin American countries that use it in medicine and traditional food are Bolivia, Haiti, Peru, and Venezuela. In Bolivia, it is used with honey for respiratory diseases. In Haiti, it is used as a digestive, diuretic or to treat angina. In Peru, juices are used for anemia, arteriosclerosis, arthritis, colds, gastritis. In Venezuela, it is used for hepatosis, among others.

Bauhinia forficata Link: In Brazil, it has many uses, such as astringent, depurative, diuretic and tonic, cystitis, diabetes, diarrhea, elephantiasis, hyperglycemia, kidney stones, obesity, and polyuria.

Carica papaya L. is eaten as ripe and fresh fruit, but in its unripe form, it is used to tenderize meats. Its anticoagulant activity has been proven and in other parts of the world, such as Asia, it is used to treat diseases of the circulatory system.

Jatropha curcas L., the young leaves are cooked to season foods. The seeds are eaten roasted. It is used in Belize, Colombia, Cuba, and the Dominican Republic. They are used as an infusion to treat different conditions, such as stomatitis, dysentery, and gonorrhea. In addition, the bruised leaves or roots are used on wounds, dermatosis, tooth pain, and wounds.

Tabebuia spp., it has been traditionally used to treat anemia, arthritis, boils, cancer, cold, colitis, cough, fever, flu, fungus, infections, leishmania, rheumatism, snakebite, sores, and urinary infections. Used in Argentina, Bolivia, Brazil, Costa Rica, and Mexico, always in the form of an infusion with additives.

Uncaria tomentosa (willd. ex schult.) DC. The crushed leaves are used for wounds or topical treatments and instead, the infusions are used to treat infections, bleeding or blood disorders, including systematic diseases, such as cancer, cirrhosis, or diabetes.

13.6.2 MOLECULES INVOLVED IN THE ANTICOAGULANT ACTIVITY

Proteins and peptides have been found to have anticoagulant activity. A 65 amino acid peptide is capable of inhibiting thrombin. Another interesting compound is a 20-residue peptide, which is Bivalirudin, a direct thrombin inhibitor. It was isolated and commercialized under the name ANGIOMAX, it is used for surgeries. Not only have natural peptides been isolated, but proteins have also been seen to become active peptides due to proteases in the digestive tract. In addition, many bioactive peptides have other biological activities as previously discussed, as antimicrobials, antioxidants, antihypertensives, and anticancer. One type of anticoagulant bioactive peptides is those that inhibit FXa factors, these play an important role in coagulation, precisely in the union between intrinsic and extrinsic coagulation, they can be direct or indirect. This factor manages to cleave prothrombin, generating thrombin. When they are direct inhibitors, they bind directly to factor or to the prothrombin complex. If they are indirect, they bind to the factor but with an antithrombin intermediate.

FXIa inhibitory peptides have been found. In the human coagulation factor cascade, FXIa factor is zymogenic to factor IX by means of a serine protease. This zymogen is activated by other factors or by the presence of negatively charged surfaces. Although they are identified, it has not been possible to isolate this type of active peptides. There are some peptides with experimentally proven anticoagulant activity but with no known interaction of *Amaranthus mantegazzianus* and *Porphyra yezoensis*, between others.[50]

13.7 ANTICANCER

Anticancer drugs are approved in the time frame of about 1940–2002, approximately, 54% were derived natural products or drugs inspired from knowledge related to it.[3]

In Latin America, cancer mortality decreases since the 90s in both sexes, with some exceptions. But some specific cancer like colorectal and pancreatic in both sexes, and lung and breast cancer in women are increasing the number of patients. Nevertheless, uterus cancer mortality is extremally high. The number of patients and mortality rate changes between countries in Latin America.[51]

Natural sources have played an important role in cancer therapy by affording molecules capable of inhibited cancer developing and providing molecules for pharma against cancer cells.[52]

13.7.1 TRADITIONAL USES

The anticancer effect of plant sources has spread culturally through its systematic use and in different ways. Today, several of them have scientific verification of this activity, either by reducing the size of tumors or reducing symptoms. Here are some examples:

Vanilla planifolia Andrews, the leaves and tender shoots are consumed both cooked and raw in salads. In Brazil, they use the infusion to treat tuberculosis or lung conditions.

Ximenia americana L., the fruits are eaten raw, although they are also fermented or used in preserves. In Brazil, they boil the bark to treat menorrhea or disinfect wounds.

Acnistus arborescens, in Argentina, Peru and the Dominican Republic the boiled leaves are used to treat hemorrhoids, rheumatic pain and fever. In Costa Rica and the Dominican Republic, the infusion is used to treat sore throats, coughs and colds. In Peru, they use the leaves soaked in butter for rheumatic pain.

Anacardium occidentale, the fruits can be consumed as such and their seeds are roasted and eaten. It is considered a diuretic, expectorant, laxative, among other benefits. Its anticancer capacity has been proven in the extract of its leaves; the juice contains three antitumor compounds. It is known in Brazil, Colombia, Cuba, Mexico, Panama, and Venezuela.

Annona reticulata, the fruit is consumed as such. They also make drinks, desserts, ice cream, smoothies with fresh fruit. Traditionally, the crushed leaves were applied topically on tumors, and the boiled leaves on vaginal sores or inflammations. Instead, the boiled green fruits and bark is used to treat diarrhea, dysentery, or parasites. The ripe pulp was applied to abscesses, boils, sores, bile, or nausea. In Panama, they even use the root to treat epilepsy.

Argemone mexicana, it has different applications depending on the country and its use. The sap is used to treat conjunctivitis, warts, and dermatological problems in Bolivia and Haiti. Particularly in Bolivia, the petals are used to treat asthma and whooping cough. Latex is used in Cuba for pimples on the lips, herpes, and warts. In Haiti, it has various uses, for example, the crushed leaves are used to treat rheumatism, the infusion of leaves to treat anorexia, cough, croup, hepatosis, and the crushed seeds as a purgative. The infusions have multiple uses in Venezuela against cancer and epilepsy, among others.

Bertholletia excelsa Bonpl, about three nuts are recommended daily to prevent colon, lung, and prostate cancer. In addition, the infusion is used in the Amazon to treat liver diseases. And particularly in Brazil, the stem bark, fruit tea, and juice are used to treat malaria, gastralgia, and hepatitis, respectively.

Bidens pilosa, the leaves are used for food. It is recommended in the Amazon for angina, chills, diabetes, dysentery, dysmenorrhea, headache, hepatosis, jaundice, laryngitis, malaria, ophthalmia, parasites, sore throat, stomach pain, stomatosis, toothache, worms, and wounds. In several countries, it is used to treat infections, such as Brazil, Peru, and Colombia. In Latin countries, it has multiple uses. The anticancer effect has been verified in extracts with hexane and methanol.

Brosimum acutifolium Huber, the bark is widely used in Brazil, Colombia, and Peru to treat arthritis and rheumatism, it is also considered anti-inflammatory and antirheumatic drug. Four cytotoxic compounds have been identified to treat leukemia, colon, and lung cancer.

Calophyllum brasiliense Cambess., its fruit is not edible. In general, in Brazil, Dominican Republic, Haiti, and Peru, the leaves and stem are used to treat topically from sores or wounds to tumors. Its antitumor activity shows a reduction of up to 83% with a plant extract.

Other plant species with anticancer activity proven experimentally and traditionally used include *Capraria biflora*, *Capsicum* spp., *Cayaponia tayuya* (Vell.), *Chromolaena odorata*, *Chrysobalanus icaco*, *Cissampelos pareira*, *Deguelia utilis*, *Hyptis verticillata* Jacq., *Jacaranda caucana* Pittier, *Larrea tridentate*, *Maytenus boaria* Molina, *Myrciaria dubia* (Kunth) McVaugh, *Phaseolus vulgaris* L., *Quassia amara* L., within several plants that are used in Latin America.

13.7.2 INTERACTIONS AND PRODUCTS

Some active components in medicinal plants include asarone, eugenol, palmitic acid, champhene, flavonoids, tannin, triterpene, coumarin, leucine, napthoquinolines, quercetin, xanthone, benzophenone, polyphenols, carotene, among others found. The anticancer activity can result from several mechanisms of action of the different molecular causes that cause tumors or cancer in any tissue. For example, actions on cytoskeletal proteins that participate in the cell cycle. Also, all these proteins participate specifically in cell division, DNA replication with topoisomerases. Although, other affected mechanisms also have to do with the immune system and the immune response through its stimulation.

Considering all polyphenolic compounds, all those known from natural sources are considered anticancer compounds. For example, resveratrol is found in peanuts, grapes, and red wine, and gallocatechins in green tea. Polyphenols have demonstrated cytotoxic activity, in addition to their antioxidant properties. The mechanism of action of polyphenols could be associated with the regulation of apoptosis by means of copper ions that bind to chromatin. Even inhibitory activity has been seen on proteins present in cancer cells.

Among the polyphenolic compounds, we find the flavonoids, present in various plants and with strongly associated biological activities. Currently, several flavonoid compounds have been identified that have cytotoxic and reducing activity. Activity on lung, cervix, breast, liver cancer, and leukemia has been proven. Some experiments have shown activity on proteins in cancer cells.

One of the compounds of natural origin, brassinosteroid is found in plants; it has various functions from hormonal control to signaling for their growth. They are among the most important natural compounds with anticancer activity. Its action on cell apoptosis and cancer cell inhibition have been proven. Its activity is proven in low concentrations on lung and cervical cancer, lymphoblastic leukemia, and osteosarcoma. Specifically, in breast and prostate cancer, it was found that it acts on the hormone receptor and growth factors, binding and preventing their activity on cyclins. Within anticancer treatments, it is required that the cytotoxic action is not for normal cells.

Most of the natural isolates for cancer treatments are usually nontoxic to normal cells, which is why it is beneficial when taking them to clinical trials and administering them to patients in large quantities. There are some that are toxic, such as glycosides, lectins, saponins, lignans, lectins, and some taxanes. In general, activities can be classified into:

- Methyltransferase inhibitors,
- Preventing DNA damage or antioxidants,
- Histone deacetylases inhibitors and mitotic disruptors (sulforaphane, isothiocyanates, isoflavones, and pomiferin).[53]

13.8 ANTIHYPERGLYCEMIC

In 2014, 422 million patients were identified as diabetics for the World Health Organization (WHO).[54] In 2021, because of COVID-19 pandemic, diabetes patients were in the focus of health staff. Patients with diabetes in Latin America were around 5–10% of the total population.[55]

There is an association between the resistance of peripheral tissues with diabetes and human obesity. Several studies have shown that resistance to the insulin is a progressive disease, hence drugs involved in reversing the resistance are important in medical treatments. One of the molecules studied is thiazolidinediones, an insulin-sensitizing agent, that improves the glucose metabolism.[56] The number of patients with diabetes are increasing worldwide, the lifestyle in Latin America is a risk factor for those numbers in Latin people because it is a chronic disease. The metabolism of biomolecules is altered by hyperglycemia due to the problems with insulin resistance or inactivity, related to disfunction of the beta cells of pancreas.[57] The relationship with some proteins function as tyrosine phosphatases are important in the insulin resistance, there are some findings in the increasing activity of those proteins for insulin resistance, diabetes, and obesity.[58]

Multiple molecules have been studied to treat hyperglycemia. These include Biguanides,[59] sulfonylurea, insulin, thiazolidinediones,[60] incretin mimetics, glucose-dependent insulinotropic polypeptide, glucagon-like peptide-1, and glucagon-like peptide-2, dipeptidyl peptidase-4 inhibitors. The functionality for each molecule is particular, even some researchers have studied synthetic similar molecules.[54]

13.8.1 TRADITIONAL USES

The recommended uses in case of early detection of diabetes or already having this disease, is regular and consistent. Hyperglycemia is systematic and progressive. Here are some examples of infusions given in the case of hyperglycemia, mostly recommended among other known traditional uses:

Achyrocline satureioides, usually is ingested in the form of infusion or commercial capsules available today. It has several uses in Latin America, for example, in Argentina, is used for asthma, diabetes, dysmenorrhea, and dyspepsia. In Bolivia, it is used as an anticonvulsant or antipyretic. In Brazil, they use it to treat colds, anorexia, gastrosis, flu, or enterosis. It is also used in Colombia, Paraguay, Peru, and Uruguay for multiple purposes. Experimentally, extract isolates have significantly lowered blood sugar levels.

Neurolaena lobata is used for different purposes in several Latin American countries, such as Belize, Cuba, Dominican Republic, Nicaragua, in addition to a Latin tradition with the leaves in infusions and topical applications. The ethanol extract has shown antihyperglycemic activity in mice.

Ptychopetalum olacoides Benth, with a long tradition in Brazil, it is not considered food, but is used as an appetite stimulant or even an aphrodisiac. Also, it is used for different ailments, among which are alopecia, dysmenorrhea, heart disease, depression, diarrhea, dysentery, hysteria, neuralgia, neurosis, polio, among the many examples.

Rhizophora mangle, the leaves are used in infusion for many diseases, such as gonorrhea, tuberculosis, asthma, leprosy, among many others throughout Latin America. Antihyperglycemic activity has been found in rabbits, lowering the glycemic peak.

Smallanthus sonchifolius, its roots are edible, both raw and cooked. In Bolivia, it is used as a diuretic, against liver problems and nephrosis. Meanwhile in Peru, its hot leaves are used against myalgia and rheumatism.

Tecoma stans, the roots are used in Guatemala to make beer, in Argentina, the same crushed ones are used as a diuretic. The infusions are used in several countries (Bolivia, Cuba, Mexico, Peru) to treat gastritis, diarrhea, dysentery, diabetes, or even as diuretic or sweaty. Also, the flowers are used in Peru for dermal diseases.[11,61]

These are some of the plants that are traditionally used in Latin America and that experimentally have shown antihyperglycemic activity or have lowered blood sugar levels in experimental models.

13.8.2 BIOACTIVE MOLECULES AND THEIR INTERACTIONS

The best-known molecules with antihyperglycemic activity from natural plant sources are alkaloids, polyphenols, polysaccharides, flavonoids, carotenoids, saponins, and fatty acids.

There are several antidiabetic effects found in plants, such as increase insulin secretion, lower glucose absorption, increase peripheral glucose intake, lower complications associated with diabetes, lower hepatic glucose production, and lower epigenetic effects.[61,62]

Bioactive molecules often have insulinomimetic and insulin secretagogue activity. Charantia, momordicin, p-insulin, and vicin are examples. Charantia is a mixture of steroid saponins. The alkaloid momordicin promotes the regeneration of pancreatic β cells. P-insulin is an insulin-like peptide that mimics its action. The action significantly decreased glucose levels and increased glucose uptake by myocytes and adipocytes in vitro.[64]

Polyphenols, including flavonoids, important in nature, such as quercetin, ellagic, and gallic acid, have activity on the inhibition of the protein glycation. In addition, they have important antioxidant activities as we previously reviewed. Among the bioactives studied, we have ellagic acid that reduces blood glucose levels, because it improves the glucose/insulin balance and also improves insulin signaling. It has also been found as an important antihyperglycemic activity of gallic acid.[64,65]

Within the antihyperglycemic molecules are also free fatty acids. These act by being released during lipolysis from adipocytes. Due to their presence, the secretion profile of adipokines changes, which are usually responsible for inflammation and insulin resistance. In liver and muscles, the high concentration of free fatty acids prevents the binding of insulin and its action.[39]

Polysaccharides from plant extracts are associated with an antihyperglycemic activity, lower blood glucose, lower total cholesterol, and even some triglycines in animal models. This activity has been associated with the stimulus, it generates to secrete insulin in beta cells. Epigenetic activity has also been seen by stimulating insulin receptor mRNA expression in hepatocytes and myocytes. Another polysaccharide has been associated with pancreatic tissue, improving the hypoglycemic effect.[67]

13.9 VITAMINS

Vitamins are complex organic molecules present in foodstuff that interact in metabolism, their absence can cause metabolic disorders.[68] They have an importance in physiologic functions, highly specificity. Activated form of vitamins comes from the metabolic activation, small amounts are needed.[69] Researchers have characterized 13 vitamins due to their solubility. There are four soluble in fat (Vitamin: A, E, K, and D) and 10 insoluble in water (Vitamin B1-6, 9, 12, and Biotin).[70]

13.9.1 TRADITIONAL USES

Besides common daily foods in diet, we can obtain vitamin from plants less known as these examples below:

Manihot esculenta Crantz, the vitamin A content has been studied before. The traditional uses are diverse. In Argentina, starch is used directly in hot milk for the lactation period when the breasts become inflamed. In Bolivia, the leaves are used as an infusion to treat gout and rheumatism. The same infusion is used in Haiti for headaches, allergies, and as an antiseptic; in Peru, it is used to treat inflammations, measles, and smallpox. Even in Peru, a drop of latex is prescribed to treat conjunctivitis. In Venezuela, a powdered root preparation is used for angina or diarrhea.

Mauritia flexuosa, it contains vitamin C and vitamin E, two important antioxidants. It is traditionally used in Brazil and Colombia. They use leaves and fruits in various preparations.

Myrciaria dubia, it contains a high content of vitamin A, C, and E. It is used for various purposes. Many of which we have seen in this chapter as anticancer. It is widely used in Peru, where various artisanal and commercial cosmetic and pharmaceutical products have already been commercialized.

Portulaca oleracea, known throughout the American continent from Alaska to the Amazon, is rich in pre-vitamin A, vitamin C, E, and omega-3 fatty acids. It is known in Latin America for its beneficial properties for the eyes; it is used for these purposes in Bolivia and Brazil, and for other purposes in the Dominican Republic, Haiti, Peru, and Venezuela.

Solanum sessiliflorum, its fruit is rich in vitamin C, showing high antioxidant power, as well as being delicious and diuretic. In Peru, it is used by children, pregnant women, the elderly and people in recovery to relieve pain.

Each type of vitamin will act on its different target molecules in the cell. Vitamins are obtained practically by eating vegetables, meats, and grains in the diet. They are known for their need, since they mostly act as important cofactors in the enzymes involved in human, animal, and plant metabolism.[1,25]

13.9.2 MOLECULES AND THEIR INTERACTIONS

Most known vitamins are cofactors enzymes as vitamins: A, K, and C, thiamin, niacin, riboflavin, vitamin B6, biotin, pantothenic acid, folate, and B12. Some vitamins have antioxidant activity as vitamins E and C. They also act as cofactors of oxide reduction reactions, vitamins E, K, C, niacin, riboflavin, and pantothenic acid. Vitamins A and D act as hormones. Vitamin A act as a photoreceptive cofactor in vision.[68,70]

Retinol is the transported form for vitamin A; its metabolism is through esterification, conjugation, oxidation, and isomerization. Multiple forms can be metabolized in the body: retinol, retinal, retinoic acid. Each form has different function:

- Retinol, as transport or reproduction in mammals.
- Retinyl esters, as storage.
- Retinal, as vision process (found in rod cells).
- Retinoic acid, as epithelial differentiation, gene transcription, reproduction.

Multiple hydroxylated forms of vitamin D are their active forms. Calcidiol is one of the products of hydroxylation of vitamin D, the major circulating form. Each one of the possible hydroxylated forms has a physiological function. The maintenance of calcium and phosphate homeostasis is one of the most important roles of vitamin D. The lack of calcium or phosphate produces diseases, such as rickets and osteomalacia. Other forms are involved in the induction of cell differentiation and proliferation, energy metabolism, hormonal signaling, mineral homeostasis, oncogenes, and chromosomal proteins, for example. Some researchers have evidence for other functions also, such as cell differentiation, membrane structure, mitochondrial metabolism, muscular function, pancreatic function, immunity, neural function, skin protection, parathyroid function, between the other roles in gene regulation.[72]

Vitamin E is metabolized mostly as tocopherols. Most of the molecules are reacting in metabolism as antioxidants: membrane antioxidants, soluble antioxidants, and antioxidant enzymes. Multiple diseases are associated with oxidative damage as: aging, cancer, cardiovascular disease, cataracts, diabetes, exercise, hemodialysis, immunity and infection, inflammatory disease, ischemia-reperfusion injury, lung disease, neurodegenerative diseases, preeclampsia, skin disorders and smoking. The molecular antioxidant path is acting with the enzymes.[72,73]

Vitamin K, it is usually ingested as phylloquinone, to be converted to menaquinone by alkylation. It is the cofactor of microsome carboxylase to oxygenate it and drive the γ-carboxylation of glutamic acid into peptides. It is mainly found in the liver, but is present in other tissues. Also, it can act as a reducer and cleave a C-H bond, bonding by peptide bond and undertaking a carboxylation.[69,74,75]

Vitamin B1, known as thiamin, it is an essential cofactor in carbohydrate metabolism and neural function. In general, its function is to bind to proteins and mainly to thiamine-specific binding protein. It is phosphorylated by two enzymes: thiamine pyrophosphokinase and a phosphoryl transferase. In animals, it can convert pyruvate to acetyl-CoA with pyruvate dehydrogenase, it can convert α-ketoglutarate to succinyl-CoA with α-ketoglutaric dehydrogenase and branched chain α-ketoacids into the corresponding acyl-CoAs with α-ketoacid dehydrogenase.[68,69]

Vitamin B2, known as riboflavin, is associated with several metabolic enzymes in the catabolism of carbohydrates, amino acids, and lipids in addition to having action in cellular antioxidant protection. Its best-known forms are flavin mononucleotide (FMN) and flavin adenine dinucleotide (FAD). Its mechanism of action has two steps. First, it is converted to FMN by phosphorylation. Then, it is converted to FAD by FAD-synthase.[69]

Vitamin B3, known as niacin, it is the precursor of the pyridine nucleotides NAD(H) and NADP(H). It is synthesized from tryptophan. Metabolically its active form in enzymes is NAD(H) and NADP(H), used as co-substrate for hydrogen transfers in hundreds of catabolic reactions of carbohydrates, fatty acids, and amino acids.[68,76]

Vitamin B6 comprises molecules, such as pyridoxal, pyridoxine, or pyridoxamine to carry out the active coenzyme pyridoxal phosphate. It plays an important role in growth, cognitive development, depression, immune function, fatigue, and steroid hormone activity. Vitamin B6 is activated by phosphorylation/dephosphorylation, oxidation/reduction, and amination/deamination processes. The phosphorylated form is the one that is present intracellularly by the action of three enzymes: pyridoxal kinase, alkaline phosphatases, pyridoxamine phosphate oxidase, and pyridoxal dehydrogenase.[69]

Vitamin C, it is present as ascorbic or dehydroascorbic acid. The ascorbic acid form can lose up to four electrons, the first two are successive and reversible, and the second two electrons are successive irreversible. Ascorbic acid can be regenerated in its reversible form by three processes: using an ascorbic acid–dehydroascorbic acid pair, due to reduction of dehydroascorbic acid, and due to ascorbic radical enzymatic reduction. As a result of the easy process that ascorbic acid has to move electrons, it is involved in various biological processes, such as collagen synthesis, 4-hydroxyphenylpyruvate degradation, norepinephrine synthesis, and fatty acid desaturation.[69,77]

On its free form, Biotin binds to lysine residues by the biotin holoenzyme synthetase in a specific region of the same. The action of biotin is to transfer carbons in their most oxidized form; Biotin is useful in carboxylases, transcarboxylases, and decarboxylases. In addition, it regulates gene expression and the cell cycle. Furthermore, it is a substrate in the modification of proteins by biotinylation.[70]

13.10 CONCLUSIONS

In recent years, scientific progress has made possible the re-discovery of biological potentials from plant sources, using ancestral knowledge and scientific experimentation. Access to technology around the world makes it possible for nontraditional plant sources to become potential pharmaceuticals, cosmetics, food, and medicine. They also become a source of sustainability for indigenous communities, transcends the culture and provides an opportunity for the Latin American market and science. Several molecules present in plant sources have many ways of acting at the cellular level, improving the quality of life (antioxidants, antihypertensives) or reducing environmental dangers (antibacterials, anticancerous).

KEYWORDS

- **Latin American plants**
- **functional properties**
- **bioactives**
- **ethnopharmacognosy**
- **native flora**

REFERENCES

1. Zolla, C. Traditional Medicine in Latin America, With Particular Reference to Mexico. *J. Ethnopharmacol.* **1980,** *2*, 37–41.
2. Tringali, C. *Bioactive Compounds from Natural Sources*; 2011.
3. Yuan, H.; Ma, Q.; Ye, L.; Piao, G. The Traditional Medicine and Modern Medicine from Natural Products. *Molecules* **2016,** *21* (5), 559. DOI: 10.3390/molecules21050559.

4. Raven, P. H.; Gereau, R. E.; Phillipson, P. B.; Chatelain, C.; Jenkins, C. N.; Ulloa, C. U. The Distribution of Biodiversity Richness in the Tropics. *Sci. Adv.* **2020,** *6* (37), 5–10. DOI: 10.1126/sciadv.abc6228.
5. Quijas, S. et al. Linking Biodiversity, Ecosystem Services, and Beneficiaries of Tropical Dry Forests of Latin America: Review and New Perspectives. *Ecosyst. Serv.* **2019,** *36*, DOI: 10.1016/j.ecoser.2019.100909.
6. Belbin, G. M.; Nieves-Colón, M. A.; Kenny, E. E.; Moreno-Estrada, A.; Gignoux, C. R. Genetic Diversity in Populations Across Latin America: Implications for Population and Medical Genetic Studies. *Curr. Opin. Genet. Dev.* **2018,** *53*, 98–104. DOI: 10.1016/j.gde.2018.07.006.
7. Montenegro, R. A.; Stephens, C. Indigenous Health in Latin America and the Caribbean. *Lancet* **2006,** *367* (9525), 1859–1869. DOI: 10.1016/S0140-6736(06)68808-9.
8. Corrigan, D. Ethno-Botany and The Search for New Drugs **1981,** *4*, 343–345.
9. Cantero, J. J.; Palchetti, V.; Núñez, C. O.; Barboza, G. E. *Halophytic Flora of Argentina: A Checklist and an Analysis of its Diversity*, Vol. V; 2016.
10. Ávila, F. A.; Funk, V. A. *Catálogo de Plantas y Líquenes*; 2016.
11. Rojas, R.; Bustamante, B.; Bauer, J.; Fernández, I.; Albán, J.; Lock, O. Antimicrobial Activity of Selected Peruvian Medicinal Plants. *J. Ethnopharmacol.* **2003,** *88* (2–3), 199–204. DOI: 10.1016/S0378-8741(03)00212-5.
12. Moraes, M. R.; Maldonado, C.; Zenteno-Ruiz, F. S. Endemic Plant Species of Bolivia and Their Relationships with Vegetation. *Endem. Species* 2019. DOI: 10.5772/INTECHOPEN.82776.
13. Borchsenius, F. Patterns of Plant Species Endemism in Ecuador. *Biodivers. Conserv.* **1997,** *6* (3), 379–399. DOI: 10.1023/A:1018312724137.
14. Huber, O.; García, P. The Venezuelan Guayana Region and the Study Areas: Geo-ecological Characteristics. In *The Venezuelan Guayana Region and the Study Areas: Geo-ecological Characteristics*; 2011; pp 29–89. DOI: 10.1007/978-3-642-20138-7_3.
15. Giulietti, A. M.; Harley, R. M.; De Queiroz, L. P.; Wanderley, M. D. G. L.; Van Den Berg, C. Biodiversity and Conservation of Plants in Brazil. In *Conservation Biology*, Vol. 19, no. 3. John Wiley & Sons, Ltd, 2005; pp 632–639. DOI: 10.1111/j.1523-1739.2005.00704.x.
16. Jiménez, G. et al. Biological Screening of Plants of the Venezuelan Amazons. *J. Ethnopharmacol.* **2001,** *77* (1), 77–83. DOI: 10.1016/S0378-8741(01)00271-9.
17. Lajolo, F. M. Functional foods: Latin American Perspectives. *Br. J. Nutr.* **2002,** *88* (S2), S145–S150. DOI: 10.1079/bjn2002679.
18. Suárez-Torres, J.; Suárez-López, J. R.; López-Paredes, D.; Morocho, H.; Cachiguango-Cachiguango, L. E.; Dellai, W. Agroecology and Health: Lessons from Indigenous Populations. *Curr. Environ. Heal. Reports* **2017,** *4* (2), 244–251. DOI: 10.1007/s40572-017-0146-z.
19. Mazza, G.; Cacace, J. E.; Kay, C. D. Methods of Analysis for Anthocyanins in Plants and Biological Fluids. *J. AOAC Int.* **2004,** *87* (1), 129–145. DOI: 10.1093/jaoac/87.1.129.
20. Caceres Guido, P.; Ribas, A.; Gaioli, M.; Quattrone, F.; Macchi, A. The State of the Integrative Medicine in Latin America: The Long Road to Include Complementary, Natural, and Traditional Practices in Formal Health Systems. *Eur. J. Integr. Med.* **2015,** *7* (1), 5–12. DOI: 10.1016/j.eujim.2014.06.010.
21. Sicker, D.; Schulz, M. Benzoxazinones in Plants: Occurrence, Synthetic Access, and Biological Activity. *Stud. Nat. Prod. Chem.* **2002,** *27*, 185–232.

22. Stintzing, F. C.; Carle, R. Functional Properties of Anthocyanins and Betalains in Plants, Food, and in Human Nutrition. *Trends Food Sci. Technol.* **2004,** *15* (1), 19–38. DOI: 10.1016/j.tifs.2003.07.004.
23. Dvorakova, M.; Landa, P. Anti-inflammatory Activity of Natural Stilbenoids: A Review. *Pharmacol. Res.* **2017,** *124*, 126–145. DOI: 10.1016/j.phrs.2017.08.002.
24. Shale, T. L.; Stirk, W. A.; Van Staden, J. Screening of Medicinal Plants Used in Lesotho for Anti-bacterial and Anti-inflammatory Activity. *J. Ethnopharmacol.* **1999,** *67* (3), 347–354. DOI: 10.1016/S0378-8741(99)00035-5.
25. Duke, J. A.; Bogenschutz-Godwin, M. J.; Ottesen, A. R. *Medicinal Plants of Latin America*, Vol. 64, no. 1. 2009.
26. Beg, S.; Swain, S.; Hasan, H.; Barkat, M. A.; Hussain, M. S. Systematic Review of Herbals As Potential Anti-inflammatory Agents: Recent Advances, Current Clinical Status and Future Perspectives. *Pharmacogn. Rev.* **2011,** *5* (10), 120–137. DOI: 10.4103/0973-7847.91102.
27. Gómez Estrada, H. A.; González Ruiz, K. N.; Medina, J. D. Actividad Antiinflamatoria De Productos Naturales. *Bol. Latinoam. y del Caribe Plantas Med. y Aromat.* **2011,** *10* (3), 182–217.
28. Knights, K. M.; Mangoni, A. A.; Miners, J. O. Defining the COX Inhibitor Selectivity of NSAIDs: Implications for Understanding Toxicity. *Expert Review of Clinical Pharmacology*, Vol. 3, no. 6; Taylor & Francis, 2010; pp 769–776. DOI: 10.1586/ecp.10.120.
29. Farhadi, F.; Khameneh, B.; Iranshahi, M.; Iranshahy, M. Antibacterial Activity of Flavonoids and Their Structure–Activity Relationship: An Update Review. *Phyther. Res.* **2019,** *33* (1), 13–40. DOI: 10.1002/ptr.6208.
30. Nascimento, G. G. F.; Locatelli, J.; Freitas, P. C.; Silva, G. L. Antibacterial Activity of Plant Extracts and Phytochemicals on Antibiotic-resistant Bacteria. *Brazilian J. Microbiol.* **2000,** *31* (4), 247–256. DOI: 10.1590/S1517-83822000000400003.
31. Haag, G.; Marín, G. H.; Brignoles, P.; Del Valle, M. E.; Magarinos, M. C.; Debenedetti, S. L. Antimicrobial Activity of Latin American Medicinal Plant Extracts. *J. Sci.* **2014,** *4* (2), 128–131.
32. Kushagri Singh, K. S. Antibacterial Activity - An Overview | ScienceDirect Topics. *Inorganic Frameworks as Smart Nanomedicines*, 2018. https://www.sciencedirect.com/topics/pharmacology-toxicology-and-pharmaceutical-science/antibacterial-activity (accessed Aug. 22, 2021).
33. Cushnie, T. P. T.; Lamb, A. J. Antimicrobial Activity of Flavonoids. *Int. J. Antimicrob. Agents* **2005,** *26* (5), 343–356. DOI: 10.1016/j.ijantimicag.2005.09.002.
34. Lin, J. K.; Weng, M. S. Flavonoids as Nutraceuticals. *Sci. Flavonoids* **2006,** *7* (September), 213–238. DOI: 10.1007/978-0-387-28822-2_8.
35. Huang, H. W. Molecular Mechanism of Antimicrobial Peptides: The Origin of Cooperativity. *Biochim. Biophys. Acta - Biomembr.* **2006,** *1758* (9), 1292–1302. DOI: 10.1016/j.bbamem.2006.02.001.
36. Gulcin, İ.; *Antioxidants and Antioxidant Methods: An Updated Overview* **2020,** *94* (3).
37. Ríos, J. L.; Recio, M. C. Medicinal Plants and Antimicrobial Activity. *J. Ethnopharmacol.* **2005,** *100* (1–2), 80–84. DOI: 10.1016/j.jep.2005.04.025.
38. Oldoni, T. L. C. et al. Chemical Characterization and Optimization of the Extraction Process of Bioactive Compounds from Propolis Produced by Selected Bees Apis Mellifera. *J. Braz. Chem. Soc.* **2015,** *26* (10), 2054–2062. DOI: 10.5935/0103-5053.20150186.

39. Sayahi, M.; Shirali, S. The Antidiabetic and Antioxidant Effects of Carotenoids: A Review. *Asian J. Pharm. Res. Heal. Care* **2017,** *9* (4), 186–191. DOI: 10.18311/ajprhc/2017/7689.
40. Nina, N. et al. Antibacterial Activity, Antioxidant Effect and Chemical Composition of Propolis from the Región del Maule, Central Chile. *Molecules* **2015,** *20* (10), 18144–18167. DOI: 10.3390/molecules201018144.
41. Tamura, Y.; Kimura, M. Treatment of Pulmonary Arterial Hypertension. *Japanese J. Chest Dis.* **2015,** *74* (3), 286–294. DOI: 10.1056/nejmra040291.
42. Ruilope, L. M. et al. Hipertensión en América Latina: Perspectivas Actuales De Las Tendencias y Características. *Hipertens. y Riesgo Vasc.* **2017,** *34* (1), 50–56. DOI: 10.1016/j.hipert.2016.11.005.
43. Bai, R. R.; Wu, X. M.; Xu, J. Y. Current Natural Products with Antihypertensive Activity. *Chin. J. Nat. Med.* **2015,** *13* (10), 721–729. DOI: 10.1016/S1875-5364(15)30072-8.
44. Laurent, S. Antihypertensive Drugs. *Pharmacol. Res.* **2017,** *124*, 116–125. DOI: 10.1016/j.phrs.2017.07.026.
45. Ma, L. et al. In Vitro and In Vivo Anticoagulant Activity of Heparin-like Biomacromolecules and the Mechanism Analysis for Heparin-mimicking Activity. *Int. J. Biol. Macromol.* **2019,** *122*, 784–792. DOI: 10.1016/j.ijbiomac.2018.11.011.
46. Paluck, S. J.; Nguyen, T. H.; Maynard, H. D. Heparin-Mimicking Polymers: Synthesis and Biological Applications. *Biomacromolecules* **2016,** *17* (11), 3417–3440. DOI: 10.1021/acs.biomac.6b01147.
47. Vongchan, P.; Sajomsang, W.; Subyen, D.; Kongtawelert, P. Anticoagulant Activity of a Sulfated Chitosan. *Carbohydr. Res.* **2002,** *337* (13), 1239–1242. DOI: 10.1016/s0008-6215(02)00098-8.
48. Athukorala, Y.; Lee, K. W.; Kim, S. K.; Jeon, Y. J. Anticoagulant Activity of Marine Green and Brown Algae Collected from Jeju Island in Korea. *Bioresour. Technol.* **2007,** *98* (9), 1711–1716. DOI: 10.1016/j.biortech.2006.07.034.
49. Maraganore, J. M.; Chao, B.; Joseph, M. L.; Jablonski, J.; Ramachandran, K. L. Anticoagulant Activity of Synthetic Hirudin Peptides. *J. Biol. Chem.* **1989,** *264* (15), 8692–8698. DOI: 10.1016/s0021-9258(18)81848-8.
50. Syed, A. A.; Mehta, A. Target Specific Anticoagulant Peptides: A Review. *Int. J. Pept. Res. Ther.* **2018,** *24* (1), 1–12. DOI: 10.1007/s10989-018-9682-0.
51. Carioli, G. et al. Cancer Mortality Predictions for 2019 in Latin America. *Int. J. Cancer* **2020,** *147* (3), 619–632. DOI: 10.1002/ijc.32749.
52. Cragg, G. M.; Newman, D. J. Plants as a Source of Anti-cancer Agents. *J. Ethnopharmacol.* **2005,** *100* (1–2), 72–79. DOI: 10.1016/j.jep.2005.05.011.
53. Greenwell, M.; Rahman, P. K. S. M. Medicinal Plants: Their Use in Anticancer Treatment. *Int. J. Pharm. Sci. Res.* **2015,** *6* (10), 4103–4112. DOI: 10.13040/ijpsr.0975-8232.6(10).4103-12.
54. Kumar, A.; Bharti, S. K.; Kumar, A. Therapeutic Molecules Against Type 2 Diabetes: What We Have and What Are We Expecting? *Pharmacol. Reports* **2017,** *69* (5), 959–970. DOI: 10.1016/j.pharep.2017.04.003.
55. Flores-Ramírez, R.; Berumen-Rodríguez, A. A.; Martínez-Castillo, M. A.; Alcántara-Quintana, L. E.; Díaz-Barriga, F.; Díaz de León-Martínez, L. A Review of Environmental Risks and Vulnerability Factors of Indigenous Populations from Latin America and the Caribbean in the Face of the COVID-19. In *Global Public Health*, Vol 16, no. 7; Taylor & Francis, 2021; pp 975–999. DOI: 10.1080/17441692.2021.1923777.

56. Willson, T. M. et al. The structure-Activity Relationship Between Peroxisome Proliferator-activated Receptor γ Agonism and the Antihyperglycemic Activity of Thiazolidinediones. *J. Med. Chem.* **1996,** *39* (3), 665–668. DOI: 10.1021/jm950395a.
57. Nayak, B. S.; Roberts, L. Relationship between Inflammatory Markers, Metabolic and Anthropometric Variables in the Caribbean Type 2 Diabetic Patients With and Without Microvascular Complications. *J. Inflamm.* **2006,** *3*, 1–7. DOI: 10.1186/1476-9255-3-17.
58. Tamrakar, A. K.; Yadav, P. P.; Tiwari, P.; Maurya, R.; Srivastava, A. K. Identification of Pongamol and Karanjin as Lead Compounds with Antihyperglycemic Activity from Pongamia Pinnata Fruits. *J. Ethnopharmacol.* **2008,** *118* (3), 435–439. DOI: 10.1016/j.jep.2008.05.008.
59. Control, T. D. 7. Approaches to Glycemic Treatment. *Diabetes Care* **2016,** *39* (January), S52–S59, DOI: 10.2337/dc16-S010.
60. Gelman, L.; Feige, J. N.; Desvergne, B. Molecular Basis of Selective PPARγ Modulation for the Treatment of Type 2 Diabetes. *Biochim. Biophys. Acta - Mol. Cell Biol. Lipids* **2007,** *1771* (8), 1094–1107. DOI: 10.1016/j.bbalip.2007.03.004.
61. Mignone, J.; Bartlett, J.; O'Neil, J.; Orchard, T. Best Practices in Intercultural Health: Five Case Studies in Latin America. *J. Ethnobiol. Ethnomed.* **2007,** *3*, 1–11. DOI: 10.1186/1746-4269-3-31.
62. Shanak, S.; Saad, B.; Zaid, H.; Carvalho, J. C. T. Carvalho. Metabolic and Epigenetic Action Mechanisms of Antidiabetic Medicinal Plants. *Evidence-based Complement. Altern. Med.* **2019,** *2019*, DOI: 10.1155/2019/3583067.
63. Han, D. G.; Cho, S. S.; Kwak, J. H.; Yoon, I. S. Medicinal Plants and Phytochemicals for Diabetes Mellitus: Pharmacokinetic Characteristics and Herb-drug Interactions. *J. Pharm. Investig.* **2019,** *49* (6), 603–612. DOI: 10.1007/s40005-019-00440-4.
64. Lankatillake, C.; Huynh, T.; Dias, D. A. Understanding Glycaemic Control and Current Approaches for Screening Antidiabetic Natural Products from Evidence-based Medicinal Plants. *Plant Methods* **2019,** *15* (1), 1–35. DOI: 10.1186/s13007-019-0487-8.
65. Manukumar, H. M.; Shiva Kumar, J.; Chandrasekhar, B.; Raghava, S.; Umesha, S. Evidences for Diabetes and Insulin Mimetic Activity of Medicinal Plants: Present Status and Future Prospects. *Crit. Rev. Food Sci. Nutr.* **2017,** *57* (12), 2712–2729. DOI: 10.1080/10408398.2016.1143446.
66. Wojdyło, A.; Nowicka, P.; Bąbelewski, P. Phenolic and Carotenoid Profile of New Goji Cultivars and Their anti-Hyperglycemic, Anti-aging and Antioxidant Properties. *J. Funct. Foods* **2018**, *48* (April), 632–642. DOI: 10.1016/j.jff.2018.07.061.
67. Xie, J. H. et al. Advances on Bioactive Polysaccharides from Medicinal Plants. *Crit. Rev. Food Sci. Nutr.* **2016,** *56*, S60–S84. DOI: 10.1080/10408398.2015.1069255.
68. Marshall, C. V. Vitamins and Minerals: Help or Harm?: Clinical Nutrition Insight. *Clinical Nutrition Insight*, 1987. https://journals.lww.com/clinnutrinsight/Citation/1987/07000/Vitamins_and_Minerals__Help_or_Harm_.12.aspx (accessed Aug. 09, 2021).
69. Combs, Gerald F. Jr.; McClung, James P. In *The Vitamins: Fundamental Aspects in Nutrition and Health*; Google Libros, 2017. https://books.google.es/books?id=UEy0DAAAQBAJ&printsec=frontcover&hl=es&source=gbs_ge_summary_r&cad=0#v=onepage&q&f=false (accessed Aug. 08, 2021).
70. Maqbool, M. A.; Aslam, M.; Akbar, W.; Iqbal, Z. Biological Importance of Vitamins for Human Health: A Review Fat Soluble Vitamins: Vitamin D: Vitamin A. *Agric. Basic Sci.* **2017,** *2* (3), 50–58.

71. Huang, Z.; Liu, Y.; Qi, G.; Brand, D.; Zheng, S. Role of Vitamin A in the Immune System. *J. Clin. Med.* **2018,** *7* (9), 258. DOI: 10.3390/jcm7090258.
72. Sassi, F.; Tamone, C.; D'amelio, P. Vitamin D: Nutrient, Hormone, and Immunomodulator. *Nutrients* **2018,** *10* (11), 1–14. DOI: 10.3390/nu10111656.
73. Miyazawa, T.; Burdeos, G. C.; Itaya, M.; Nakagawa, K.; Miyazawa, T. Vitamin E: Regulatory Redox Interactions. *IUBMB Life* **2019,** *71* (4), 430–441. DOI: 10.1002/iub.2008.
74. Lee, G. Y.; Han, S. N. The Role of Vitamin E in Immunity. *Nutrients* **2018,** *10* (11), 1–18. DOI: 10.3390/nu10111614.
75. Tsugawa, N.; Shiraki, M. Vitamin K Nutrition and Bone Health. *Nutrients* **2020,** *12* (7), 1–16. DOI: 10.3390/nu12071909.
76. Cozzolino, M.; Mangano, M.; Galassi, A.; Ciceri, P.; Messa, P.; Nigwekar, S. Vitamin K in Chronic Kidney Disease. *Nutrients* **2019,** *11* (1), 1–11. DOI: 10.3390/nu11010168.
77. Heckman, J. J.; Pinto, R.; Savelyev, P. A. *The Vitamins: Fundamental Aspects in Nutrition and Health*; 1967.
78. Carr, A. C.; Maggini, S. Vitamin C and immune function. *Nutrients* **2017,** *9* (11), 1–25. DOI: 10.3390/nu9111211.

CHAPTER 14

Encapsulation of Procyanidins Compounds from Natural Sources

MARÍA DE LOS ÁNGELES VÁZQUEZ-NÚÑEZ[1], MAYRA AGUILAR-ZÁRATE[2], PEDRO AGUILAR-ZÁRATE[1], DIANA B. MUÑIZ-MÁRQUEZ[1], FABIOLA VEANA[1], JORGE E. WONG-PAZ[1], and MARIELA R. MICHEL[1]

[1]*Engineering Department, Tecnológico Nacional de México/Instituto Tecnológico de Ciudad Valles, San Luis Potosí, México*

[2]*Facultad de Ciencias Químicas-CIEP, Universidad Autónoma de San Luis Potosí, San Luis Potosí, Mexico*

ABSTRACT

Recent research and applications of polyphenols have generated great interest in the functional food, nutraceutical, and pharmaceutical industries because of the health benefits that they offer to humans. The application of polyphenols is limited by their unpleasant taste. In addition, the bioactive effectiveness of polyphenols involves preservation, stability, bioactivity, and bioavailability of the active ingredients. Procyanidins are a complex family of polyphenols that are widespread and powerful antioxidants. Due to their strong antioxidant capacity, procyanidins can be easily denaturalized by light, oxygen, and other factors, and they are also difficult to store and transport. Different techniques are used to maintain the stability of these compounds. Micro- and nanoencapsulation are techniques that have been used to protect these bioactive compounds. The main objective of these encapsulation techniques is to protect the bioactive compounds from environmental conditions (effects of light, humidity, and oxygen), allowing them to keep their biological activity and improve their stability, promoting a controlled release

Food Byproducts Management and their Utilization. Ricardo Gómez-García, Ana A. Vilas-Boas, Débora A. Campos, Maria Manuela Pintado, & Cristóbal Noé Aguilar (Eds.)

of the encapsulated compound. Procyanidins encapsulation technologies are discussed in this chapter. It includes the wall materials tested and drying techniques such as spray drying, and freeze-drying, among others.

14.1 POLYPHENOLS OVERVIEW

Polyphenols are secondary metabolites found in plants, fruits, and vegetables involved in the defense mechanism for their self-protection against ultraviolet radiation, fungal parasites, and herbivores, and they are involved in the growth and development of plants.[1] Polyphenols are bioactive compounds with various biological properties for example, the protection of certain cancers, gastrointestinal disorders, anti-inflammatory effects, cardiovascular problems, and neurodegenerative diseases.[2] These compounds can be consumed as part of the human diet from plants in infusions, fruits, vegetables, cereals, and coffee. They are widely distributed in plants especially in the form of glycosides and present different sugar monomers and acylated sugars at various positions of the polyphenols skeletons and are represented by at least two phenyl rings and hydroxyl groups.[3] They are a collective term for various sub-groups of phenolic molecules that constitute one of the most numerous classes of natural compounds in the plant kingdom. Currently, more than 8000 phenolic compounds have been reported of which over 4000 flavonoids are known over 4000 flavonoids have been reported.[4]

14.1.1 CHEMICAL CLASSIFICATION

Principal distribution of polyphenols found in plant-based materials (Figure 14.1). According to Tsao,[4] polyphenols are classified as phenolic acids that include benzoic and cinnamic acids; flavonoids such as flavones, neoflavonoids, chalcones, flavones, flavonols, flavanones, flavanonols, flavan-3-ols, proanthocyanidins, and anthocyanidins; polyphenolic amides that include avenanthramides and capsaicinoids; and other polyphenols such as resveratrol, ellagic acid, curcumin, rosmarinic acid among others.

14.1.2 PROCYANIDINS

Polyphenols are the largest group of secondary metabolites found in plants.[5] The most representative flavanols in foods are of the flavan-3-ol type, and these

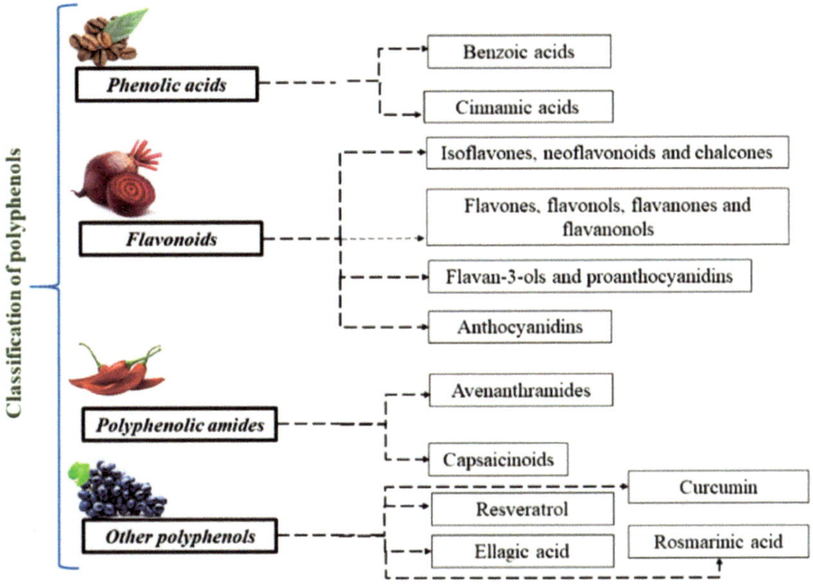

FIGURE 14.1 Principal distribution of polyphenols found in plants.

can appear as monomers (catechins), dimers condensed together, and oligomers (procyanidins), or polymers (proanthocyanidins or condensed tannins).[6] Within polyphenols, procyanidins are derived from proanthocyanidins, also known as condensed tannins,[7] which are derived from a group of flavonoids called flavanols. All of these compounds are known for their remarkable antioxidant activity.[8] The name procyanidins given to flavanol polymers has its origin in the property they present in a strongly acid medium, giving rise to cyanidin by hydrolysis. However, it has come to be considered more correct to speak of proanthocyanidins.[9] Proanthocyanidins are classified according to their bonds and are present in homopolymers and heteropolymers. They are divided into three subgroups: procyanidins, prodelphinidins, and propelargonidins, which yield cyanidins, delphinidins, and pelargonidins, respectively, and these are divided into their elemental units, (epi)catechin, (epi)gallocatechin, and (epi) afzelechin.[10] However, procyanidins are the most common type of proanthocyanidins[8] and these in turn can be classified into types A and B procyanidins.

14.1.2.1 SOURCES

Procyanidins and proanthocyanidins are widely distributed naturally throughout the plant kingdom.[5,11] Commonly consumed foods such as fruits, vegetables, legumes, grains, and nuts,[12–14] as well as cosmetics,[15] and pharmaceuticals[7,11] containing plant materials, are common sources where procyanidins can be found in varied contractions.[5]

As mentioned above, procyanidins are classified according to their bonds, type B procyanidins are predominant in many foods, such as grapes,[16] berries, cereals, nuts, legumes, tea, chocolate, wine[17,18] coffee,[19,20] and litchi.[21] Type A exists mainly in limited foods, such as peanuts, plums, blueberries,[22] and avocados. However, countless research studies mention different foods containing procyanidins in varying concentrations.[13]

14.1.2.2 BIOACTIVITIES

Numerous studies have demonstrated the potential of flavonoids in general and procyanidins in particular. A large part of these studies demonstrates their important antioxidant capacity since procyanidins have an antioxidant power 50 times greater than vitamin C[18] and 20 times greater than that of vitamin E.[23] However, it has now been evidenced that procyanidins and their metabolic derivatives, independently of their antioxidant activity[18] confer important health benefits.[14] One of the applications is in cases where the cardiac function is impaired, proanthocyanidins alleviate cardiovascular diseases by relaxing vessels and inhibiting low-density lipoprotein (LDL) oxidation.[24] In other studies on cancer, which can be directly induced by oxidative stress, it has been studied that it can be eventually controlled using a diet rich in compounds with antioxidant properties such as procyanidins[14] There is evidence for many other beneficial health effects associated with procyanidins, including anti-inflammatory, anti-infective,[11] antibacterial, antiviral, anti-allergic, vasodilatory,[14,25] anti-mutagenic, anti-hyperglycemic, and wound healing effects.[5] In addition, proanthocyanidins interact with numerous cellular proteins, being more important than their interaction with signal transduction factors, nuclear receptors, and transcription factors, thereby modifying cellular metabolism in various tissues, in a beneficial sense.[26] Other studies have shown that they help to protect against sun damage, improve vision and flexibility of joints, arteries, and body tissues such as the heart, and also improve blood circulation.[14,18,23] However, the mechanisms underlying their biological properties are not yet fully understood,[27] and although several studies have been conducted on their health-related benefits, the exact mechanisms underlying their biological activities have yet to be extensively explained. Still, it has been suggested that they lie in their degradation by gut colonic microbiota into low molecular weight compounds.[6,27]

14.1.2.3 CHEMICAL STRUCTURE

Procyanidins are composed of flavan-3-ol units such as (+)-catechin and (−)-epicatechin,[20,28] and possess two types of linkage between the monomeric units,[29] which differentiates them into types A and B procyanidins. The subunit located at the end of the chain is called the terminal unit and the remaining subunits are called extension units.[28] B-type procyanidins (Figure 14.2) are characterized by a bond between the flavan-3-ol units mainly from carbon 4 to carbon 8 of the next monomer or less frequently from carbon 4 to carbon 6. B-type procyanidins are the most abundant, with procyanidins B1, B2, B3, and B4 being the most frequent.[5,8,30] Procyanidins can exist in oligomeric and polymeric forms, and when B-type procyanidins undergo an oxidation process under alkaline conditions (pH 8.0) they are partially converted to A-type procyanidins.[20,31] In A-type procyanidins, structural variations can occur due to the formation of a second bond by oxidative coupling[32] as they have not only an interflavanic bond but also an additional ether bond between the hydroxyl group of the A-ring and carbon 2 of the A-ring (Figure 14.2).[5,30] The most common type A compounds that have been reported are A1 and A2.[5,8]

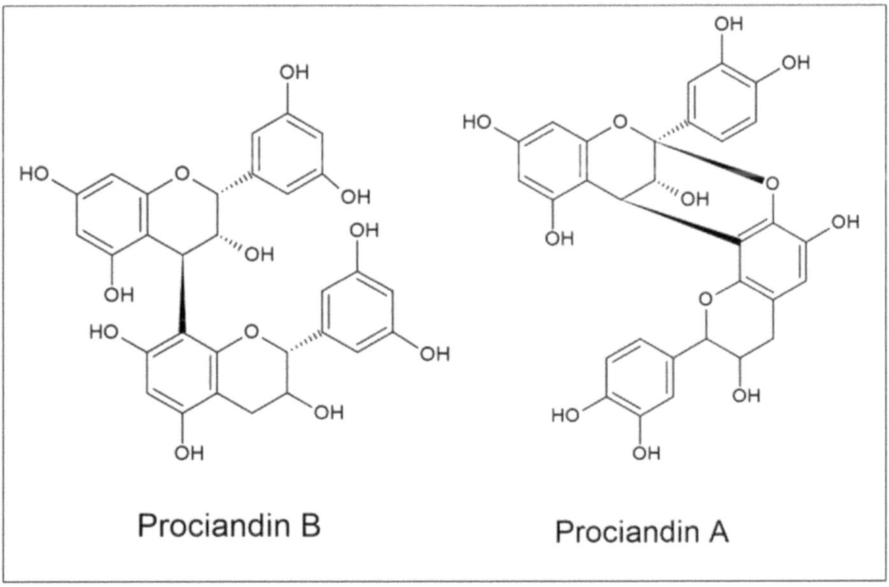

FIGURE 14.2 Chemical structure of procyanidins.

Procyanidins can be classified by their degree of polymerization.[8] Monomers form bonds that lead to oligomers and polymers. Those with 2–7 monomeric units are defined as oligoprocyanidins.[5] Although the degree of polymerization can exceed 30, it is generally the oligomers from 2 to 10 that are of greatest interest.[33] Procyanidins containing combinations of types A and B bonds are called mixed type, the incorporation of type A bonds can add stiffness and stability to the compound.[29]

14.1.3 WALL MATERIALS

The wall materials can isolate and protect the bioactive compound from the external environment factors (e.g., oxygen, moisture, light, pH, heat), and these wall materials can release the active ingredient by various mechanisms such as swelling, dissolution, or degradation. Additionally, the wall material should have no reactivity with the active ingredient, be cheap, show consistent properties during storage, and have food-grade status.[34] The choice of wall material must be compatible with the type of polyphenol to be encapsulated (i.e., physical and chemical characteristics) and with conventional encapsulation methods. For example, polyphenolic compounds are usually microencapsulated with wall materials such as proteins and polysaccharides. It has been found that these macromolecules have properties to protect and preserve nutritional attributes of polyphenols such as antioxidant activity.[35]

Procyanidins are polyphenols that belong to the class of condensed tannins according to Goncalves et al.[36] permits them to be encapsulated by both polysaccharides and proteins. Polysaccharides, that is, pectins, polygalacturonic acid, Arabic gum, Xanthan gum,[36] have demonstrated the best advantages for procyanidin microencapsulation because of their good solubility in water, in addition to their low viscosity at high concentrations, when compared to proteins. On the other hand, advantages such as biocompatibility, emulsifying and foaming capacity, biodegradability, good amphiphilic and functional properties (water solubility) have been reported with the use of proteins extracted from vegetable (cereal proteins, whey proteins, pea proteins), and animal (whey proteins, casein, and gelatin) derived products). However, their use for encapsulating procyanidins has been less studied, probably because of the allergenic effects exhibited overall in animal proteins.[37,38] Nevertheless, the use of vegetable proteins as wall-forming materials in encapsulation reflects the present "green" trend and could respond to the increasing demand for healthy ingredients in the pharmaceutical, cosmetics, and food industries.[39] Recent investigations have shown that polysaccharides mixed with proteins improve the emulsifying and filmogenic properties during microencapsulation[40] and the new material resulted in conjugates covalently cross-linked by the Maillard reaction with interesting functional properties.[39]

14.1.3.1 POLYSACCHARIDES

Polysaccharides are extensively used as wall materials for the microencapsulation of food ingredients due to their capacity to form amorphous glassy solids that provide structural support to delivery systems' wall material. Their ability to bind active ingredients is complemented by their diversity, low cost, and widespread use in food and thus make them the preferred choice for encapsulation.[41] Procyanidins have shown a high affinity for polysaccharides derived from their chemical characteristics.[42,43] These polyphenols are polymers and oligomers of catechin/epicatechin (flavan-3-ol units) and are commonly assembled by β-type bonds, namely, C4–C8 or C4–C6 interflavanic linkages with an average number degree of polymerization varying between 2 and more than 100 monomers.[41] When this degree of polymerization and molar mass increase the level of galloylation of procyanidins, it enhances their affinity for polysaccharides (e.g., pectins) because also increases the number of hydroxyl groups and aromatic rings, which with they could form bonds. In the case of catechin and polymers composed mainly of (+)-catechin units bind more to polysaccharides than -epicatechin and polymer composed mainly of (−)-epicatechin due to the stereochemistry of flavan-3-ols pyran rings.[35] One of the most important reasons why the recent investigations to encapsulate procyanidins has focused on the use of pectins is because it has been found that this wall material avoids astringency, an undesirable mouthfeel of tannins due to specific interactions with salivary proteins. These interactions can result in the formation of (in)soluble aggregates that precipitate, causing a sensation of roughness, dryness, and constriction due to the lack of lubrication.[44]

14.1.3.2 PROTEINS

Among the wall materials, proteins are a suitable selection because they are generally regarded as safe (GRAS) and contain a high nutritional value. Most proteins can be easily digested by the human gastrointestinal tract.[45] Furthermore, proteins are insoluble in acidic conditions as the isoelectric point of most proteins ranges between pH 3 and pH 5, and they are usually dissolved at alkaline pH. Such conditions are ideal for encapsulating compounds of interest if the solution is thereafter acidified to form the microcapsules.[39] Both animal and plant proteins have been used to encapsulate matrices. The most common animal proteins include gelatin, caseins, and whey proteins (mainly β-lactoglobulin). Plant-derived protein matrices include zein, soy protein, etc.

14.1.3.2.1 Animal-Derived Proteins

Caseins are proteins isolated from milk, which naturally form micelles. They can be used to develop gel-like structures by the addition of acids, calcium, or enzymes, or they can be used to build structures with other biopolymers based on their change and hydrophobic characteristics. Caseins have been used to encapsulate various kinds the bioactive agents in molecular complexes, micelles, and hydrogel microcapsules.[46,47] Among wall proteins, gelatin is a water-soluble protein that is relatively cheap and displays interesting functional properties useful for constructing biopolymers microcapsules. Biopolymer microcapsules containing albumin and gelatin have recently been produced by liquid antisolvent precipitation[48–50] or coacervation.[50]

A widely used food ingredient to design biopolymeric microcapsules is whey protein. For example, it has produced microcapsules with β-lactoglobulin[51] and lactoferrin[52] by thermal denaturation under appropriate pH and ionic strength conditions.

14.1.3.2.2 Plant-Derived Proteins

Plant-derived proteins offer some advantages over animal proteins for forming biopolymer microcapsules since the risk of contamination and infection is reduced, they are generally cheaper and can be used in vegetarian or vegan products. Gliadin and zein are two examples of hydrophobic cereal storage proteins utilized to produce biopolymer microcapsules by various methods.[53–55] These microcapsules have also been shown to be suitable for the encapsulation of polyphenols.[54,56]

14.1.4 ENCAPSULATION TECHNIQUES

Encapsulation techniques have been defined as the coating of a bioactive compound either solid, liquid, or a gaseous material inside a miniature capsule composed mainly of a biopolymer,[57,59] which can release its contents in a controlled manner under specific conditions.[60,61] These techniques have been developed for about 60 years[61] and are widely used in the food, pharmaceutical, and cosmetic industries, as well as in agricultural products, veterinary medicine, industrial chemicals, and biotechnology.[62] The main purpose of encapsulation is to protect the core material from degradation by adverse environmental conditions, such as undesirable effects of light, oxygen, and moisture, contributing to improving its stability and shelf life.[59,63,64] In addition, it is used to change liquid solutions to powders for easier handling.[59] It has also been used to improve flavor, odor, nutritional value, and appearance and help fragile materials withstand processing and packaging conditions.[65] In general, for the encapsulation of bioactive agents, there are three steps: (1) the formation of the wall around the material to be encapsulated; (2) ensuring that unwanted leakage does not occur; (3) ensuring that unwanted materials are kept out.[61,66]

The size of microcapsules ranges from 1 μm to several mm.[61,62] Smaller particles, from 1 nm to 1 μm, are referred to as nanoparticles.[62] Many morphologies can occur in encapsulation, but two morphologies are mainly observed in (Figure 14.3).[61] The first is mononuclear capsules, which have a single nucleus enveloped by wall material, while the second is aggregates, which have many nuclei embedded in a matrix.[67,68] The specific shapes and sizes of the capsule depend directly on the selection of the encapsulation method and the core and wall materials from which the capsules are made.[61,65] Several microencapsulation techniques are currently available,[62] in general, they can be divided into 3 groups: chemical methods, physical methods, and physicochemical methods, the main examples of these methods can be seen in Table 14.1.

FIGURE 14.3 Scanning electron images of samples of microcapsules (Own source).

TABLE 14.1 Main Encapsulation Methods of Procyanidis.

Chemical methods	Physical methods	Physicochemical methods
Interfacial polymerization[65,69]	Spray drying[57,62,65,69]	Simple or complex coacervation
Molecular inclusión[57,61,65,69]	Freeze-drying[57,62]	Entrapment in liposomes[57,65,69]
Co-crystallization[57,61]	Extrusion[57,69]	Ionic gelation[57,62,65]
Interfacial polycondensation	Processes using supercritical fluids[62]	Emulsion[61]
In situ polymerization[62]		Methods based on ionic interactions[62]

14.1.4.1 DRYING TECHNIQUES

Drying or dehydration is one of the oldest methods used for food preservation.[70–73] This process is generally understood as water removal from substances that can be solids, liquids, or gases.[71,74] The main objective is to increase product stability.[71,72] Since many years ago, drying techniques have been applied mainly to foods as they help to prolong shelf life. They reduce water to levels that inhibit the growth of microorganisms, chemical, and biochemical reactions.[74,75] However, nowadays, drying has gone from solely for food preservation to being used to develop products, satisfying the demand for convenience and quality.[74]

14.1.4.1.1 Spray Drying

Spray drying is currently one of the most commonly used encapsulation techniques for the food industry.[57,58,61,67] Briefly, the spray drying process consists of converting a wide variety of suspensions into dry particles in a single step.[76,77] The size distribution of these particles varies, but in general, microcapsules between 10 and 100 μm are obtained.[65] These size variations are associated with the variety of droplet sizes in the spray, the geometry of the nozzle, and the viscosity of the solution.[62]

In the operation of spray drying equipment, the operator has a direct influence on several aspects such as inlet temperature, drying air flow rate, liquid stream delivery rate, atomizing air pressure, droplet size atomizer speed or air-liquid ratio, in addition to the use of different wall materials.[58,76] By modifying these different parameters, it is possible to modify the final aspects of the capsule such as appearance, size, bulk density, particle density, porosity, moisture content, flowability, stability, activity retention, aroma, and flavor. There are several advantages of this method over other encapsulation techniques, making it the most widely used in the food industry. Since it is an operation with low processing costs, flexible and continuous, produces good quality particles,[61,62] and is stable, it is an easy process,[57,78] fast, and in addition to its simplicity. It is also suitable for heat-sensitive materials, since the exposure time to elevated temperatures is very short.[65] Furthermore, compared to other methods, spray drying provides the highest efficiency with this method being between 96 and 100%, with higher values than other methods.[57] The disadvantage of this technology is the limited number of available shell materials.[61,78]

14.1.4.1.2 Freeze-Drying

The freeze-drying technique is also known as cryopreservation and is a process used for the dehydration of a large number of materials.[61,79] It is also one of the most commonly used techniques for the protection and preservation of molecules that are thermosensitive[57,62,80] and that are unstable in aqueous solutions.[68]

The chilling or freeze spray technique is a low-temperature dehydration operation that removes water from the product by sublimation.[62] Freeze-drying combines the advantages of freezing and drying to provide a product with stable, dry, active, and readily soluble characteristics.[81] It is an appropriate and fundamental process for materials that require low moisture content (less than 1%) and require a sterile and gentle preservation process.[82]

The description of the freeze-drying process, it is mainly divided into three steps: freezing, primary drying, and secondary drying. However, other sources mention a step prior to these, which is called pretreatment.[82] The use of freeze-dried procyanidin-rich extracts extracted from avocado peel has been reported by Cerda. Also, mention the use of nanoemulsions which are environmentally friendly methods using safe materials.[83]

14.2 CONCLUSIONS AND FINAL REMARKS

The diverse bioactivities of procyanidins are taken into consideration by the scientific community. Also, the use of agricultural byproducts for the extraction of procyanidins is an important topic in order to use and diminish the impact of the wastes on the environment. However, the extraction and recovery of procyanidins have been widely explored. The real challenge is to keep the bioactivities of the molecules by avoiding degradation over time. For that reason, encapsulation technology seems to be a feasible procedure for the protection of phytomolecules from light, air, and any other factors that promote the degradation and the loosing of bioactivities. Once encapsulated, procyanidins can have enormous applications, for example, the development of functional foods, cosmetics, and drugs, among others. Nevertheless, is important to continue exploring new methods for the encapsulation of procyanidins, understand the interaction between procyanidins and the encapsulation matrix, and understand the chemistry process for the oxidation of procyanidins in order to understand how to prevent it.

ACKNOWLEDGMENTS

This project is part of CONACYT No. 301467 titled Modernización de infraestructura científica y tecnológica del Laboratorio de Alimentos del TecNM-Campus Ciudad Valles para la conservación y disposición de ingredientes funcionales alimentarios.

KEYWORDS

- **polyphenols**
- **bioactivities**
- **microencapsulation**

REFERENCES

1. Kumar, V.; Sharma, A.; Kohli, S. K. Bali, S.; Sharma, M.; Kumar, R.; Bhardwaj, R.; Thukral, A. K. Differential Distribution of Polyphenols in Plants using Multivariate Techniques. *Biotechnol. Res. Innov.* **2019,** *3* (1), 1–21.
2. Cory, H.; Passarelli, S.; Szeto, J.; Tamez, M.; Mattei, J. The Role of Polyphenols in Human Health and Food Systems: A Mini-review. *Front. Nutr.* **2018,** *5* (1), 87.
3. Singla, R. K.; Dubey, A. K.; Garg, A.; Sharma, R. K.; Fiorino, M.; Ameen, S. M.; Haddad, M. A.; Al-Hiary, M. Natural Polyphenols: Chemical Classification, Definition of Classes, Subcategories, and Structures. *J AOAC Int.* **2019,** *102* (5), 1397–1400.
4. Tsao, R. Chemistry and Biochemistry of Dietary Polyphenols. *Nutrients* **2010,** *2* (12), 1231–1246.

5. Rue, E. A.; Rush, M. D.; Van Breemen, R. B. Procyanidins: A Comprehensive Review Encompassing Structure Elucidation via Mass Spectrometry. *Phytochem. Rev.* **2018,** *17* (1),1–16.
6. Deprez, S.; Brezillon, C.; Rabot, S.; Philippe, C.; Mila, I.; Lapierre, C.; Scalbert, A. Polymeric Proanthocyanidins Are Catabolized by Human Colonic Microflora into Low-Molecular-Weight Phenolic Acids. *J. Nutr.* **2000,** *130* (11), 2733–2738.
7. Han, X.; Shen, T.; Lou, H. Dietary Polyphenols and Their Biological Significance. *Int. J. Mol. Sci.* **2007,** *8* (4), 950–988.
8. Gu, L.; Kelm, M. A.; Hammerstone, J. F.; Beecher, G.; Holden, J.; Haytowitz, D.; Gebhardt, S.; Prior, R. L. Concentrations of Proanthocyanidins in Common Foods and Estimations of Normal Consumption. *J. Nutr.* **2004,** *134* (3), 613–617.
9. Busse Valverde, N. Proantocianidinas De Uvas y Vinos De Las Variedades Monastrell, Cabernet Sauvignon y Syrah. *All rights Reserv. IJES* **2014,** *281* (4), 1–30.
10. Yue-Wei, G.; Shu, Z.; Kohei, K.; Sheng-Li, W.; Kayo, Y.; Katsuko, K. Molecular Ion Index Assisted Comprehensive Profiling of B-type Oligomeric Proanthocyanidins in Rhubarb by High Performance Liquid Chromatography-tandem Mass Spectrometry. *Anal. Bioanal. Chem.* **2016,** *408* (13), 3555–3570.
11. Wang, Y.; Chung, S. J.; Song, W. O.; Chun, O. K. Estimation of Daily Proanthocyanidin Intake and Major Food Sources in the U.S. Diet. *J. Nutr.* **2011,** *141* (3), 447–452.
12. Wang, H.; Hao, L.; Niu, B.; Jiang, S.; Cheng, J.; Jiang, S. Kinetics and Antioxidant Capacity of Proanthocyanidins Encapsulated in Zein Electrospun Fibers by Cyclic Voltammetry, *J. Agric. Food Chem.* **2016,** *64* (15), 3083–3090.
13. Zhang, L.; Wang, Y.; Li, D.; Ho, C. T.; Li, J.; Wan, X. The Absorption, Distribution, Metabolism and Excretion of Procyanidins. *Food Funct.* **2016,** *7* (3), 1273–1281.
14. Rauf, A.; Imran, M.; Abu-Izneid, T.; Ul-Haq, I.; Patel, S.; Pan, X.; Naz, S.; Sanches-Silva, A.; Saeed, F.; Rasul Suleria, H. A. Proanthocyanidins: A Comprehensive Review. *Biomed. Pharmacother.* **2019,** *116*, 108999.
15. Kamimura, A.; Takahashi, T.; Watanabe, Y. Investigation of Topical Application of Procyanidin B-2 from apple to Identify its Potential Use as a Hair Growing Agent. *Phytomedicine* **2000,** *7* (6), 529–536.
16. Romano, A.; Masi, P.; Pucci, E.; Oliviero, V.; Ferranti, P. Encapsulated proanthocyanidins as novel ingredients. *Chem. Eng. Trans.* **2017,** *57*, 1885–1890.
17. Mitchell, A.; Robertson, D.; Koh, E. Optimizing the Extraction of Procyanidins Oligomers through Decamer. *Nutr. Food Sci. Int. J.* **2017,** *4* (3), 1–7.
18. Fernández-Larrea, J.; Pinent, M.; Bladé, M. C.; Salvadó, M. J.; Blay, M.; Pujadas, G.; Ardévol, A.; Arola, L. Alimentos Ricos en Procianidinas, Alimentatión Funcional Para Prevenir la Aparición de Síndrome Metabólico. *Rev. Esp. Obes.* **2007,** *5* (2), 98–108.
19. Wong-Paz, J. E.; Guyot, S.; Aguilar-Zárate, P.; Muñiz-Márquez, D. B.; Contreras-Esquivel, J. C.; Aguilar, C. N. Separation of Coffee Pulp Bioactive Phenolic Compounds by MPLC Fractionation and Identification by HPLC-ESI-MS. *Innov. Phys. Chem.* **2020,** 217–228.
20. Wong-Paz, J. E.; Guyot, S.; Aguilar-Zárate, P.; Muñiz-Márquez, D. B.; Contreras-Esquivel, J. C.; Aguilar, C. N. Structural Characterization of Native and Oxidized Procyanidins (Condensed Tannins) from Coffee Pulp (Coffea Arabica) using Phloroglucinolysis and Thioglycolysis-HPLC-ESI-MS. *Food Chem.* **2021,** *340*, 127830.
21. Miranda Hernandez, A. M.; Muñiz Marquez, D. B.; Wong Paz, J. E.; Aguilar Zarate, P.; De la Rosa Hernandez, M.; Larios Cruz, R.; Aguilar, C. N. Characterization by

Hplc-Esi-Ms2 of Native and Oxidized Procyanidins from Litchi (Litchi Chinensis) Pericarp. *Food Chem.* **2019,** *291*, 126–131.
22. Khanal, R. C.; Howard, L. R.; Prior, R. L. Effect of Heating on the Stability of Grape and Blueberry Pomace Procyanidins and Total Anthocyanins. *Food Res. Int.* **2010,** *43* (5), 1464–1469.
23. Shi, J.; Yu J.; Pohorly, J. E.; Kakuda, Y. Polyphenolics in Grape Seeds - Biochemistry and Functionality. *J. Med. Food* **2003,** *6* (4), 291–299.
24. Ottaviani, J. I.; Kwik-Uribe, C.; Keen, C. L.; Schroeter, H. Intake of Dietary Procyanidins does not Contribute to the Pool of Circulating Flavanols in Humans. *Am. J. Clin. Nutr.* **2012,** *95* (4), 851–858.
25. Fine, A. M. Oligomeric Proanthocyanidin Complexes: History, Structure, and Phyto-pharmaceutical Applications. *Altern. Med. Rev.* **2000,** *5* (2), 144–15.
26. Bagchi, D.; Sen, C. K.; Ray, S. D.; Das, D. K.; Bagchi, M.; Preuss, H. G.; Vinson, J. A. Molecular Mechanisms of Cardioprotection by a Novel Grape Seed Proanthocyanidin Extract. *Mutat. Res. Fundam. Mol. Mech. Mutagen.* **2003,** *523*, 87–97.
27. Jeong, W.S.; Kong, A.N.T. Biological Properties of Monomeric and Polymeric Catechins: Green Tea Catechins and Procyanidins. *Pharm. Biol.* **2004,** *42*, 84–93.
28. Mendoza-Wilson, A. M.; Castro-Arredondo, S. I.; Balandrán-Quintana, R. R. Computational Study of the Structure-free Radical Scavenging Relationship of Procyanidins. *Food Chem.* **2014,** *161*, 155–161.
29. Rue, E. A.; Glinski, J. A.; Glinski, V. B; van Breemen, R. B. Ion Mobility-mass Spectrometry for the Separation and Analysis of Procyanidins. *J. Mass Spectrom.* **2020,** *55* (2), e4377.
30. Bittner, K.; Rzeppa, S.; Humpf, H. U. Distribution and Quantification of Flavan-3-ols and Procyanidins with Low Degree of Polymerization in Nuts, Cereals, and Legumes. *J. Agric. Food Chem.* **2013,** *61* (38), 9148–9154.
31. Poupard, P.; Sanoner, P.; Baron, A.; Renard, C. M. G.C.; Guyot, S. Characterization of Procyanidin B2 Oxidation Products in an Apple Juice Model Solution and Confirmation of their Presence in Apple Juice by High-Performance Liquid Chromatography Coupled to Electrospray Ion Trap Mass Spectrometry. *J. Mass Spectrom.* **2011,** *46* (11), 1186–1197.
32. Xiao, J. S.; Liu, L.; Wu, H.; Xie, B. J.; Yang, E. N.; Da Sun, Z. Rapid Preparation of Procyanidins B2 and C1 from Granny Smith Apples by using Low Pressure Column Chromatography and Identification of their oligomeric Procyanidins. *J. Agric. Food Chem.* **2008,** *56* (6), 2096–2101.
33. Lea, A. Analysis of Polyphenol Antioxidants in Fortified Foods and Supplements. In *Food Fortification and Supplementation: Technological, Safety and Regulatory Aspects* (Woodhead Publishing Limited, 2008); pp 175–194.
34. Labuschagne, P. Impact of Wall Material Physicochemical Characteristics on the Stability of Encapsulated Phytochemicals: A Review. *Food Res. Int.* **2018,** *107*, 227–247.
35. Bourvellec, C. L.; Watrelot, A. A.; Ginies, C.; Imberty, A.; Renard, C. M. G. C. Impact of Processing on the Noncovalent Interactions between Procyanidin and Apple Cell Wall. *J. Agric. Food Chem.* **2012,** *60*, 9484–9494.
36. Goncalves, R.; Mateus, N.; De Freitas, V. Influence of Carbohydrates on the Interaction of Procyanidin B3 with Trypsin. *J. Agric. Food Chem.* **2011,** *59* (21), 11794–11802.
37. Jenkins, J. A.; Breiteneder, H.; Mills, E. N. C. Evolutionary Distance from Human Homologs Reflects allergenicity of Animal Food Proteins. *J. Allergy Clin. Immunol.* **2007,** *120* (6), 1399–1405.

38. Li, H.; Zhu, K.; Zhou, H.; Peng, W. Effects of High Hydrostatic Pressure Treatment on Allergenicity and Structural Properties of Soybean Protein Isolate for Infant Formula. *Food Chem.* **2012,** *132* (2), 808–814.
39. Nesterenko, A.; Alric, I.; Silvestre, F.; Durrieu, V. Vegetable Proteins in microencapsulation: A Review of Recent Interventions and their Effectiveness. *Ind. Crops Prod.* **2013,** *42* (1), 469–479.
40. Berendsen, R.; Güell, C.; Ferrando, M. A Procyanidin-rich Extract Encapsulated in Water-in-oil-in-water Emulsions Produced by Premix Membrane Emulsification. *Food Hydrocoll.* **2015,** *43*, 636–648.
41. Liu, X.; Renard, C. M. G. C.; Rolland-Sabaté, A. Le Bourvellec, C. Exploring Interactions between Pectins and Procyanidins: Structure-function Relationships. *Food Hydrocoll.* **2021,** 113, 106498.
42. Fernandes, P. A. R.; Le Bourvellec, C.; Renard, C. M. G. C.; Wessel, D. F.; Cardoso, S. M.; Coimbra, M. A. Interactions of Arabinan-rich Pectic Polysaccharides with Polyphenols. *Carbohydr. Polym.* **2019,** *230*, 115–644.
43. Liu, X.; Le Bourvellec, C.; Renard, C.M.G.C. Interactions Between Cell Wall Polysaccharides and Polyphenols: Effect of Molecular Internal Structure. *Compr. Rev. Food Sci. Food Saf.* **2020,** *19* (6), 3574–3617.
44. Watrelot, A. A.; Schulz, D. L.; Kennedy, J. A. Wine Polysaccharides Influence Tannin-protein Interactions. *Food Hydrocoll.* **2017,** *63*, 571–579.
45. Joye, I. J.; McClements, D. J. Biopolymer-based Nanoparticles and Microparticles: Fabrication, Characterization, and Application. *Curr. Opin. Colloid Interface Sci.* **2014,** *19* (5), 417–427.
46. Livney, Y. D. Biopolymeric Amphiphiles and their Assemblies as Functional Food Ingredients and Nutraceutical Delivery Systems. In *Encapsulation Technologies and Delivery Systems for Food Ingredients and Nutraceuticals* (Elsevier, 2012); pp 252–286.
47. Velikov, K. P. Colloidal Emulsions and Particles as Micronutrient and Nutraceutical Delivery Systems. In *Encapsulation Technologies and Delivery Systems for Food Ingredients and Nutraceuticals* (Elsevier, 2012); 319–391.
48. Langer, K.; Balthasar, S.; Vogel, V.; Dinauer, N.; Von Briesen, H.; Schubert, D. Optimization of the Preparation Process for Human Serum Albumin (HSA) Nanoparticles. *Int. J. Pharm.* **2003,** *257* (1–2), 169–180.
49. Huq, T.; Khan, A.; Khan, R. A.; Riedl, B.; Lacroix, M. Encapsulation of Probiotic Bacteria in Biopolymeric System. *Crit. Rev. Food Sci. Nutr.* **2013,** *53* (9), 909–916.
50. Hedayati, R.; Jahanshahi, M; Attar, H. Fabrication and Characterization of Albumin-acacia Nanoparticles based on Complex Coacervation as Potent Nanocarrier. *J. Chem. Technol. Biotechnol.* **2011,** *87* (10), 1401–1408.
51. Jones, O. G.; Decker, E. A.; McClements, D. J. Comparison of Protein-polysaccharide Nanoparticle Fabrication Methods: Impact of Biopolymer Complexation Before or After Particle Formation. *J. Colloid Interface Sci.* **2010,** *344* (1), 21–29.
52. Bengoechea, C.; Jones, O. G.; Guerrero, A.; McClements, D. J. Formation and Characterization of Lactoferrin/Pectin Electrostatic Complexes: Impact of Composition, pH and Thermal Treatment. *Food Hydrocoll.* **2011,** *25* (5), 1227–1232.
53. Regier, M. C.; Taylor, J. D.; Borcyk, T.; Yang, Y.; Pannier, A. K. Fabrication and Characterization of DNA-loaded Zein Nanospheres. *J. Nanobiotechnology* **2012,** *10* (1), 1–13.
54. Patel, A.; Hu, Y.; Tiwari, J. K.; Velikov, K. P. Synthesis and Characterisation of Zein-curcumin Colloidal Particles. *Soft Matter*. **2010,** *6* (24), 6192–6199.

55. Chen, L.; Subirade, M. Elaboration and Characterization of Soy/Zein Protein Microspheres for Controlled Nutraceutical Delivery. *Biomacromolecules* **2009,** *10* (12), 3327–3334.
56. Ezpeleta, I.; Irache, J. M.; Stainmesse, S.; Chabenat, C.; Gueguen, J.; Popineau,Y.; Orecchioni, A. M. Gliadin Nanoparticles for the Controlled Release of All-trans-retinoic Acid. *Int. J. Pharm.* **1996,** *131* (2), 191–200.
57. Parra Huerta, R. A. Revisión: Microencapsulación de Alimentos. *Fac. Nac. Agron.-Medellín.* **2010,** *63* (2), 5669–5684.
58. Rehman, A.; Ahmad, T.; Aadil, R. M.; Spotti, M. J.; Bakry, A. M.; Khan, I. M.; Tong, Q. Pectin Polymers as wall Materials for the Nano-encapsulation of Bioactive Compounds. *Trends Food Sci. Technol.* **2019,** *90*, 35–46.
59. Robert, P.; Gorena, T.; Romero, N.; Sepulveda, E.; Chavez, J.; Saenz, C. Encapsulation of Polyphenols and Anthocyanins from Pomegranate (Punica Granatum) by Spray Drying. *Int. J. Food Sci. Technol.* **2010,** *45* (7), 1386–1394.
60. Bartosz T.; Irene, T. Polyphenols Encapsulation - Application of Innovation Technologies to Improve Stability of Natural Products. *Phys. Sci. Rev.* **2019,** *1* (2), 1–13.
61. Fang, Z.; Bhandari, B. Encapsulation of Polyphenols - A Review. *Trends Food Sci. Technol.* **2010,** *21* (10), 510–523.
62. Munin A.; Edwards-Lévy, F. Encapsulation of Natural Polyphenolic Compounds; A Review. *Pharmaceutics* **2011,** *3* (4), 793–829.
63. Ćujić-Nikolić, N.; Stanisavljević, N.; Šavikin, K.; Kalušević, A.; Nedović, V.; Samardžić, J.; Janković, T. Chokeberry Polyphenols Preservation using Spray Drying: Effect of Encapsulation using Maltodextrin and Skimmed Milk on their Recovery Following in Vitro Digestion. *J. Microencapsul.* **2019,** *36* (8), 693–703.
64. Papoutsis, K.; Golding, J. B.; Vuong, Q.; Pristijono, P.; Stathopoulos, C. E.; Scarlett, C. J.; Bowyer, M. Encapsulation of Citrus Byproduct Extracts by Spray-drying and Freeze-drying using Combinations of Maltodextrin with Soybean Protein and ι-carrageenan. *Foods* **2018,** *7* (7), 1–12.
65. Villena, M. M.; Hernández, M. M.; Lara, V. G.; Martínez, M. R. Técnicas de Microencapsulación: Una Propuesta Para Microencapsular. *ARS Pharm.* **2009,** *50* (1), 43–50.
66. Mozafari, M. R.; Khosravi-Darani, K.; Borazan, G. G.; Cui, J.; Pardakhty, A.; Yurdugul, S. (Encapsulation of Food Ingredients using Nanoliposome Technology. *Int. J. Food Prop.* **2008,** *11* (4), 833–844.
67. Desai, K. G. H.; Park, H. J. Recent Developments in Microencapsulation of Food Ingredients. *Dry. Technol. An Int. J.* **2007,** *0*, 37–41.
68. Madene, A.; Jacquot, M.; Scher, J; Desobry, S. Flavour Encapsulation and Controlled Release - A Review. *Int. J. Food Sci. Technol.* **2006,** *41* (1), 1–21.
69. Yañez Fernandez, J.; Salazar Montoya, L.; Jimenez Hernandez, M.; Marquez Robles, M.; Ramos Ramirez, E. G. Aplicaciones Biotecnológicas de la Microencapsulación. *Avance y perspectiva* **2002,** *21*, 313–319.
70. Dong, W.; Hu, R.; Chu, Z.; Zhao, J.; Tan, L. Effect of Different Drying Techniques on Bioactive Components, Fatty Acid Composition, and Volatile Profile of Robusta Coffee Beans. *Food Chem.* **2017,** *234*, 121–130.
71. Maupoey, P. F.; Andrés Grau, A. M.; Barat Baviera, J. M.; Albors Sorolla, A. M. Introducción al Secado de Alimentos Por Aire Caliente. *Editorial UPV.* 2001.
72. Moses, J. A.; Norton, T.; Alagusundaram, K.; Tiwari, B. K. Novel Drying Techniques for the Food Industry. *Food Eng. Rev.* **2014,** *6* (3), 43–55.

73. Rawson, A.; Tiwari, B. K.; Tuohy, M. G.; O'Donnell, C. P.; Brunton, N. Effect of Ultrasound and Blanching Pretreatments on Polyacetylene and Carotenoid Content of Hot Air and Freeze Dried Carrot Discs. *Ultrason. Sonochem.* **2011,** *18* (5), 1172–1179.
74. Claussen, I. C.; Ustad, T. S.; StrØmmen, I.; Walde, P. M. Atmospheric Freeze Drying—A Review. *Dry. Technol.* **2007,** *25* (6), 947–957.
75. Naidu, M. M.; Vedashree, M.; Satapathy, P.; Khanum, H.; Ramsamy, R.; Hebbar, H. U. Effect of Drying Methods on the Quality Characteristics of Dill (Anethum Graveolens) Greens. *Food Chem.* **2016,** *192*, 849–856.
76. Cal, K.; Sollohub, K. Spray Drying Technique. I: Hardware and Process Parameters. *J. Pharm. Sci.* **2010,** *99* (2), 575–586.
77. Wisniewski, R. In *Spray Drying Technology Review*, 45th International Conference on Environmental Systems, 2015.
78. Gouin, S. Microencapsulation: Industrial Appraisal of Existing Technologies and Trends. *Trends Food Sci. Technol.* **2004,** *15* (7–8), 330–34.
79. Shahidi, F.; Han, X. Q. Encapsulation of Food Ingredients. *Food Sci. Nutr.* **2009,** *33* (6) 501–54.
80. Shukla, S. Freeze Drying Process: A Review. *Int. J. Pharm. Sci. Res.* **2019,** *2* (12), 30561–306.
81. Adams, G. D.; Cook, I.; Ward, K. R. The Principles of Freeze-drying. In *Cryopreservation and Freeze-Drying Protocols*; Springer: New York, NY, 2015, *1257*, pp 121–143.
82. Gaidhani, K. A.;Harwalkar, M.; Bhambere, D.; Nirgude, P. S. Lyophilization/Freeze Drying-A Review. *World J. Pharm. Res.* **2016,** *4* (8), 516–543.
83. Cerda-Opazo, P.; Gotteland, M.; Oyarzun-Ampuero, F. A.; Garcia, L. Design, Development and Evaluation of Nanoemulsion Containing Avocado Peel Extract with Anticancer Potential: A Novel Biological Active Ingredient to Enrich Food. *Food Hydrocolloids* **2021,** *111*, 106370.

CHAPTER 15

Pectin from Byproducts in the Food Industry: Trends and Opportunities

EBER ADDÍ QUINTANA-OBREGÓN[1], MIGUEL ÁNGEL MARTÍNEZ-TÉLLEZ[2], EMMANUEL AISPURO-HERNÁNDEZ[2], and MANUEL ALEJANDRO VARGAS-ORTIZ[3]

[1]*Programa de Investigadoras e Investigadores por México-Centro de Investigación en Alimentación y Desarrollo, A.C., Coordinación de Tecnología de Alimentos de Origen Vegetal. Carretera Gustavo Enrique Astiazarán Rosas # 46 Col. La Victoria, C.P., Hermosillo, Sonora, México*

[2]*Centro de Investigación en Alimentación y Desarrollo, A.C., Coordinación de Tecnología de Alimentos de Origen Vegetal. Carretera Gustavo Enrique Astiazarán Rosas # 46 Col. La Victoria, C.P., Hermosillo, Sonora, México*

[3]*Centro de Investigación en Alimentación y Desarrollo, A.C., Coordinación de Tecnología de Alimentos de Origen Animal. Laboratorio de Calidad, Autenticidad y Trazabilidad de los Alimentos/Química y Biotecnología de Productos Lácteos. Carretera Gustavo Enrique Astiazarán Rosas # 46 Col. La Victoria, C.P., Hermosillo, Sonora, México*

ABSTRACT

Pectin is a polysaccharide with several applications in the pharmaceutical industry, such as stabilizing, encapsulating, gelling, and probiotics. For both, fruits and vegetables, pectin represents between 15–20% of the soluble fiber. These pectins can be obtained from the critics' and apple's pericarp, which are the most known sources. However, recently used byproducts that are coming from the food industry have been studied as pectin alternatives. The pectin obtained from byproducts has shown the capacity to form gels under

Food Byproducts Management and their Utilization. Ricardo Gómez-García, Ana A. Vilas-Boas, Débora A. Campos, Maria Manuela Pintado, & Cristóbal Noé Aguilar (Eds.)

the presence of low sugar concentration and divalent cations; they have been evaluated as edible films to conserve the fruits fresh for long, encapsulant as well. This chapter aims to describe the trends in the byproducts used from food industry waste to obtain pectin and also to open the window of opportunities that pectin may generate for the food chain.

15.1 INTRODUCTION

Pectin is one of the most studied natural polymers due to its wide availability in a variety of plants. However, conventional commercial sources are limited to those obtained from byproducts in the industrial processing of citrus peel, apple pomace, sugar beet, and sunflower seed head oil pulp–juices.[1–5] In chemical terms, pectins are a group of complex heteropolysaccharides rich in galacturonic acid with methanol-esterified carboxyl groups. Pectins are primarily present in the cell wall and middle lamella in plants.[6–9] where they exert diverse functions, including defense, adhesion, and cell hydration. The complexity of pectin structure changes during plant development and varies according to the phenological stages of fruits.[9,10] The release of pectin from the source matrix usually involves exposure to acid and thermal treatments to break the matrix in which it is trapped. However, these processes promote depolymerization, which is favored by physical, chemical, and enzymatic reactions,[11] some of which, depending on the source, are intrinsic to the biochemical changes of the plant material and can be added to those unit operations implemented in the extraction of the polymer. For this reason, after pectin extraction, the characteristics of the variables affected by depolymerization are parameters that must be known.

In commercial terms, pectin is classified according to the number of methyl esters (methoxy) into high methoxy and low methoxy. High methoxy pectin form gels at low pH, while low methoxy pectin does so in the presence of calcium.[1] The esterification degree (low or high methoxy), viscosity, and molecular weight are some of the features that determine the possible pectin applications.[12] Although the main commercial applications of pectin have been used for the food industry, in recent years, its use has expanded to the cosmetic, pharmaceutical, and nutraceutical industries due to its generally recognized as safe (GRAS) classification in the USA and as a food additive in the European Union with E440 code.[13,14]

In 2015 pectin's market value was $1 billion USD,[13] and according to reports from market analysis companies, pectin's world growth value industry from 2020–2025 will be 2.6%,[15] while Fior Market[16] estimates a

6.6% growth for 2027. These growth percentages were reported considering the effects of the Covid-19 pandemic and are a sample of the economic potential in the pectin industry. The constant pectin industry growth has prompted the search for complementary sources of pectin extraction besides processing apples and citrus fruits. Since pectin represents between 15–20% of the soluble fiber in fruits and vegetables,[17] evaluating trends in generating byproducts from the industrialized processing of other plant materials is an added value opportunity. Byproducts rich in pectin are considered waste and can have various applications, in addition to other adjunct components that can be obtained from the pectin extraction process.

15.2 MAIN APPLICATIONS

15.2.1 FOOD

Pectin in food is used mainly as a gelling agent, emulsifier, stabilizer, and substitute for fat and must contain at least 65% units of galacturonic acid to be used as a pectinated ingredient.[10,13,18] Some of the products in which pectin is traditionally used as an ingredient are jams, jellies, fruit conserves, bakery, and confectionery products.[19] On the other hand, pectin's emulsifying capacity has been used in recent years in foods that seek to reduce fat consumption and generate stable emulsions over time, as in mayonnaise.[20,21] although it will not have the same stability as those generated by proteins.[22] Depending on the intrinsic characteristics of pectin, pH, calcium, and other compound concentrations, its use can also be considered to prepare emulsified foods. In addition to its incorporation as an ingredient in food processing, pectin alone, modified or combined with other polymers can be used as packaging or coatings, even of edible nature, to protect against pathogens, ensure product safety, and increase shelf life.[23–26] These applications demonstrate pectin's use growth in the food industry and consequently the increase in demand for its extraction and production.

15.2.2 PHARMACEUTICAL AND NUTRACEUTICAL

Dietary fiber has been extensively studied for its health benefits, particularly related to the digestion processes. Pectin is a soluble fraction of dietary fiber,[27] and due to its great compatibility and non-toxicity, it is considered for health promotion in the pharmaceutical industry.[8] These multiple applications are accomplished through pectin modification by several methods such as chain elongation (cross-linking and grafting) and depolymerization (chemical, physical and enzymatic degradation),[8] which confer properties related to molecular weight, and degree of esterification.[11] Moreover, hydrogel pectin or combined with other polymers can be used for the administration of oral or intravenous medications for colon cancer treatments, prevent and treat diabetes and obesity, reduce cholesterol in the blood, anti-tumor, and treat anemia and bleeding, among other applications.[2,11,28–30] Pectin also has demonstrated prebiotic activity; however, pectin-derived oligosaccharides have shown better prebiotic potential.[31] Its benefits have been seen in probiotics such as *Lactobacillus paracasei* and *Bifidobacterium bifidum*.[32,33] The use of native pectin (or complexes with other polymers) as a vehicle for drugs or bioactive compounds is another opportunity trend. Several studies have demonstrated the advantages of encapsulating with pectin.[34–37]

These trends in the use of pectin for the pharmaceutical industry enhance its value and commercialization and expand the target market for its production.

15.3 PECTIN AND BYPRODUCTS IN THE FOOD INDUSTRY

15.3.1 PECTIN FROM "WASTE"

The production and industrialization of primary food products generate millions of tons of byproducts considered waste in a composition ranging from 30–90% of the processed material and represent an environmental problem.[38–40] For example, in the industrial process of apples, it is estimated that worldwide, 25% of the harvest is discarded as waste[41] and according to Babbar et al.,[42] citrus, apple, sugar beet, olive, potato, and onion byproducts add up to more than 90 million tons of material from which pectin can be extracted. In the wine industry, 50% of the grapes' process is considered a byproduct, corresponding to approximately 32 million tons of byproducts worldwide.[43] In mango, 35–70% of the total weight of the fruit is discarded as waste.[44] Additionally, byproducts generated in the processing of plant materials of the agri-food industry to obtain pulp or other components are either in the process of valorization or have not yet been valued, for instance, pineapple, banana, pomegranate, and tomato,[45] among others that can be seen in Table 15.1.

The byproduct concept motivates the search for opportunities to use generated residues. In general, residues are considered "waste" when the biomass leftover from an agro-industrial process has not been valued. One of the advantages presented by the byproducts obtained from the agri-food industry is their composition in bioactive components with nutritional and health benefits, and sensory characteristics that can be used in the production of high added-value products, in the food, feed, cosmetics, organic fertilizers, and pharmaceuticals, among other industries.[40,46,47] Agro-industrial byproducts found in plant materials, when dehydrated, show an increase in the content of bioactive compounds in the remaining biomass and are rich in dietary fiber.[39] Also, the soluble fraction is rich in pectin.[9]

Several studies have shown pectin yields from byproducts in the agri-food industry. In addition to the traditional sources of apples and citrus fruits, the wide variety of byproduct sources with the potential for pectin extraction motivates its value. Table 15.1 shows the latest studies on the use of byproducts from a wide variety of plant materials.

15.3.2 PECTIN ADJUNCTS AND COMPLEXES

In general, the pectin extraction process is the same regardless of the plant source. During the procedure, residues rich in other compounds can be obtained, which can be considered adjunct byproducts. In oranges, other byproducts from pectin can be obtained, including polyphenols, pigments, and essential oils.[48–50]

In addition to the adjuncts that can be obtained from pectin extraction, there are those matrices or complexes formed by pectin and other molecules. In alcohol, pectin precipitation is sometimes not enough to eliminate other polysaccharides or phenolic compounds that can affect the color when producing a gel.[3] However, the potential phenolic compounds bound to pectin obtained from byproducts increase their functionality in terms of antioxidant capacity, favoring the added value. These complexes can be considered as a final product option without having to purify pectin or adjunct. For instance, the pink guava (*Psidium guajava*) byproducts are rich in colloidal lycopene–pectin complex that can be used as a final product with the advantage of combining the functionality of both compounds.[51]

15.4 PECTIN EXTRACTION

Although the cost of the pectin extraction process can be low, there are other factors that must be considered to evaluate its profitability. Some of the factors to consider according to the type of plant material are the mixture of pectin sources that may be rich in other residues, the variability in the volume of byproducts generated depending on the harvest on supply and demand, competition from other sources with greater profitability in the region (apple and citrus fruits), the consumption trends of fresh or minimally processed fruits and vegetables, and the safety of byproducts handling.[52] In this regard, it is necessary to ensure the sustainability of the byproduct acquisition and prepare unit operations that facilitate pectin extraction under safe conditions. Fortunately, pectin is a widely studied polymer, and a wide variety and combinations of extraction methods have been tested and evaluated.

Pectin's physicochemical properties, in addition to the plant source, depend on the isolation and purification conditions chosen.[53] Therefore, it is essential to select the appropriate extraction method according to the extraction source and the application or pectin's commercial destination. Adetunji et al.[4] classify the procedures for obtaining pectin as conventional and innovative. The conventional ones involve acidification (pH 1.5—3.6) and temperature increase (70–90°C) generally in times of more than 1 h. Acidification, temperature increase, and reaction times increase the yields obtained due to a greater rupture of the matrix that traps pectin in the plant material.[14,18,54] Although these conventional methods contain relatively simple unit operations, the entire process can be complex and consume time and energy.[18,55]

Innovative methods appear and are developed to increase pectin extraction yields, reduce production costs, protect the environment, and achieve sustainability, maintainability, and industrial competitiveness,[56] and can be modified in a conventional method stage or in the substitution stages process. These methods include techniques that involve organic acids, renewable energies, enzymes, and emerging technologies that allow innovating eco-friendly processes with sustainable low consumption of energy and water.[4,14,50,57–59]

In addition to yield and purity, another relevant factor for the methods used, conventional or innovative, is the functionality of pectin as related to its composition and molecular arrangement. It is necessary to monitor its yield and quality during the process to obtain the optimal relationship between performance and characteristics,[14,60] and the application and commercial value will rely on purity and functionality. For instance, the galacturonic acid percentage, degree of methylation, acetylation, and total neutral sugars are factors that can be affected during the extraction process and may influence pectin's functionality.[54,61] Temperature increase and pH decrease can affect the degree of esterification and the final viscosity,[62] which may be a disadvantage for some applications. In this regard, modifications have been evaluated primarily in the type of acids used, the addition of enzymes, and the incorporation of emerging technologies, to obtain better pectin yields and minimize the effect of its functionality on the final product.

15.4.1 ORGANIC ACIDS

Mineral acids are the most frequently used in pectin extraction; a mineral acid is defined online by "A *Dictionary of Science, Oxford*" (2005) as "*a common inorganic acid, such as hydrochloric acid, sulfuric acid, or nitric acid*". Nevertheless, organic acids have been evaluated recently, now defined as low molecular weight compounds that contain one or more carboxylic, sulfonic, alcoholic, or thiol acid groups and that can be covalently linked to groups such as amides, esters, and peptides.[63,64] Both types of acids promote pectic matrix hydrolysis (protopectin) and show the advantage that they are easy to handle, and high extraction yields are obtained.[18] Mineral acids have the disadvantage of generating effluents that require adequate management to avoid polluting the environment.[53] In contrast, organic acids are eco-friendly.[41] Some organic acids used have been acetic, citric, tartaric, lactic malic, and oxalic.[41,65–67] However, some mineral acids generate better yields. Sharma et al.[19] report a pectin extraction yield of 6.3–10.5% and 4.0–8.0% with HCl and oxalate-oxalic acid, respectively. In extractions with nitric acid, better yields and high purity in galacturonic acid have been achieved, but with a decreased methylation degree.[54] On the other hand, although organic acids present lower yields, polymers with high viscosity and molecular weight have been obtained.[41] Although acetic acid has a low yield extraction, it has the advantage of generating pectin with higher degrees of methylation, acetylation, and molecular weight, and containing rhamnogalacturonan I domains.[67] However, citric acid extractions show better extraction yields compared to mineral acids in potato and melon byproduct extractions, but with lower values in methylation and acetylation.[66,67] The most evaluated organic acid is citric acid, which is a safe and inexpensive ingredient with high availability.[66] Other ways to use organic acids are within the matrix in which they are found. Banerjee et al.[68] evaluated the effect of lemon juice on mango byproducts, obtaining higher yields than hydrochloric acid extraction.

A tendency in the extraction of pectin from byproducts is the use of organic acids, being an area of opportunity for the use of acidic from natural resources.

15.4.2 ENZYME-ASSISTED EXTRACTION

The enzymatic extractions present better yields than those obtained by the acid hydrolysis method, with commercial enzymes or enzyme extracts, and reduce energy consumption due to the use of lower temperatures.[69,70] Its use can be to release pectin from the plant matrix from other non-pectin components or obtain pectin fragments.[4] Table 15.1 shows examples of byproducts processed with enzymes. The most used enzymes are cellulases, polygalacturonase, xylanase, α-amylase,[71] and some commercial formulations are Celluclast® Econase®, Neutrase®, Viscoferm®.[72] Enzyme mixtures increase yields than using a single enzyme.[71] In relation to pectin characteristics, Baum et al.[14] report higher percentages of galacturonic acid in the acid extraction (90%) than in the enzymatic extraction (81%). Some of the aspects that must be considered for an enzymatic process are the stirring speed, the optimal enzyme temperature, the permanence time,[70] and the use of bioreactors for enzymatic hydrolysis.

Moving from a laboratory to an industrial level still presents some obstacles due to volume and industrial process conditions; that is why even the use of enzymes for pectin extraction is limited to academic research.[4] In addition, byproducts are generated from the processing of plant materials destined to obtain other products, so obtaining pectin would only be extra to the process. However, in the long term, the costs of handling these byproducts and the waste generated by acid treatments to obtain pectin favor the use of enzymes as an extraction method.

15.4.3 EMERGING TECHNOLOGIES

Looking forward to improving yields, purity, and quality of pectin extraction, as well as decreasing costs and generating eco-friendly processes, the incorporation of equipment as assisted methods has been evaluated, incorporating, or modifying a unit operation in the process or replacing it with an emerging technology. These methods can increase pectin yield and quality, reduce extraction times and temperatures, and reduce the use of toxic or environmentally harmful solvents.[58] An emerging technology involves alternative methods that shorten processing times, save energy, and are generally considered environmentally friendly.[73,74]

15.4.3.1 MICROWAVE-ASSISTED EXTRACTION

Microwaves are in the range of the electromagnetic spectrum of 300 MHz and 300 GHz,[73] extraction times are short and with high yields (like conventional methods), low energy use, a disadvantage is that it requires acid use, although in a smaller volume than conventional methods.[18,74,75] Some byproducts in which good pectin yields have been obtained are orange peel,[76] grapefruit albedo,[77] watermelon[78] papaya peel,[79] among others (Table 15.1) using microwaves for its extraction.

15.4.3.2 ULTRASOUND-ASSISTED EXTRACTION

Ultrasound extraction has the advantage of not requiring acids. However, it works with high operating temperatures, resulting in possible thermal degradation generating pectin with low viscosity, molar mass, and esterification degree.[4,18,80] In addition, it can favor the disintegration of other components that would affect the pectin separation process between the solid and liquid phases.[81] Sonication, although it may not improve the performance of obtaining pectin, it reduces the times of matrix breaking where pectin is.[68] Table 15.1 shows some byproducts where ultrasonic pulses have been used to obtain pectin.

15.4.3.3 OHMIC EXTRACTION

Ohmic heating is an alternative to extract bioactive compounds more efficiently than conventional heating methods.[82] It consists of the application of alternating current through a food matrix using two electrodes,[73] the flow increases the kinetic energy and vibration of molecules generating heat.[83] This method minimally affects the structures of bioactive compounds and is also environmentally friendly.[84] In this method, the use of acid is still necessary, but pectin obtains the following characteristics: High degrees of esterification, galacturonic acid, and emulsifying capacities, such as those obtained by conventional methods.[82,85] Table 15.1 shows some pectin obtained from byproducts using the Ohmic heating method.

15.4.3.4 PULSED ELECTRIC FIELD

A pulsed electric field is a low-temperature method that induces the electroporation of cell membranes in the plant. It consists of applying voltage pulses that create an electric field (1–50 kV/cm) through a liquid medium for a short time.[86,87] This method has been little explored for pectin extraction and depending on the extraction matrix, the applied voltage, and frequency, variable yields have been found; in some cases, good yields have been

TABLE 15.1 Pectin Yields from Byproducts of Various Sources and Extraction Methods (Last Five Years, 2016–2021).

Source	Yield (% w/w)	Most important conditions	Reference
Mango peel	5.40*	**Hot-acid extraction**, pH 2.5 (HCl), 80°C, 2 h, ethanol, drying 45°C	52
Mango Peel	18.4*	**Hot-acid extraction**, **citric acid**[OA] 1.5%, 0.25 h, ethanol, freeze-dried	109
	31.7*		
Apple peel	6.4*	**Hot-acid extraction**, 85°C, 2 h, isopropyl alcohol, lyophilized	41
	6.2*	HCl	
	5.4*	**Tartaric acid**[OA]	
	5.3*	**Malic acid**[OA]	
		Citric acid[OA]	
Potatoes	9.72#	**Hot-acid extraction**, pH 2.04, 90°C, 1 h, ethanol, freeze-dried	67
	8.38#	HCl	
	9.83#	H_2SO_4	
	14.34#	HNO_3	
	4.08#	**Citric acid**[OA]	
		Acetic acid[OA]	
Banana peel	13*	**Hot-acid extraction**, pH 2 (HCl), 90°C, 4 h, ethanol, dried 65°C	109
Carrot pomace	15.6*	**Hot-acid extraction**, pH 1.3 (**citric acid**[OA]), 90°C, 1.33 h, ethanol, drying 40°C	65
Mango peel	5.08*	**Hot-acid extraction**, pH 1.5, 90°C, 2.5 h, isopropyl alcohol, drying 60°C	107
Melon peel	29.48*	**Hot-acid extraction**, pH 1 (**citric acid**[OA]), 95°C, 3.33 h, ethanol, dried under vacuum at 50°C	66
Watermelon rind	19.3+	**Hot-acid extraction**, nitric acid, boiling, 1 h, ethanol, dried under vacuum	110
"Gabsi" pomegranate peel	11.03*	**Hot-acid extraction**, pH 1.7 (nitric acid), 86°C, 1.33 h, ethanol, drying 45°C	54
Tomato peel	15*	**Hot-acid extraction**, ammonium oxalate-**oxalic acid**[OA], 90°C, 24 h, ethanol, drying 40°C	3

TABLE 15.1 *(Continued)*

Source	Yield (% w/w)	Most important conditions	Reference
Lime peel	22*	**Enzyme Extraction**[EET], pH 3.5 (**citric acid**[OA] buffer), 50°C, enzyme	14
	24*	Laminex C2K, 2 h, propan-2-ol, drying 40°C	
	24*	**Hot-acid extraction**, pH 1.7 (nitric acid), 70°C, 4 h, propan-2-ol, drying 40°C	
Cocoa pod husks pericarp	13*	**Enzyme Extraction**[EET], pH 5, 50°C, enzyme Viscozyme L, 1 h, ethanol, drying in the oven	69
Lime peel	15.9*	**Microwave**[EET], 700 W, 3 min, HCl, ethanol, dried 50°C	111
Pomelo peels	2.93*	**Microwave**[EET], 520 W, 5 min, pH 2 (HCl), ethanol, dried 60°C	49
Pistachio green hull	12*	**Ultrasound**[EET], 150 W, pH 1.5, 24 min ethanol, dried 45°C	113
Sisal waste	11.9*	**Ultrasound**[EET], 20 kHz, 450 W, 60 min, ammonium oxalate, ethanol, dried 40°C	112
Grapefruit peel	–	**Ultrasound**[EET]**,** 20 kHz, 800 W, 28 min, ethanol, dried 45°C	114
Tomato peel	10.65*	**Ohmic heating**[EET], 60 V, 81°C, 5 min, HCl, ethanol, dried 55°C	115
Orange juice waste	14.32*	**Ohmic heating**[EET], 50 Hz, 15 V/cm, HCl, 90°C, 30 min, ethanol, dried 55°C	82
Orange peel	≈6*	**Pulsed electric field-assisted acid extraction**[EET], HCl, 60°C, 150 V–100 kHz, 0.5 h, ethanol, drying 45°C	90
	≈5*	pH 2	
	≈2*	pH 3	
		pH 6.8	
Sunflower heads	6.57#	**Subcritical water**[EET], 120°C, 20 min, ethanol, dried 25°C	116
Jackfruit peel	14.96*	**Subcritical water**[EET], 138°C, 9.15 min**,** ethanol, freeze-drying	98
Pomelo peel	19.63*	**Subcritical water**[EET], 90–120°C, dried 65°C	97

[OA]*organic acid;* [EET]*enzyme or emerging technology; *dry matter;* [#]*wet matter;* [+]*fraction insoluble in alcohol.*

[EET]*enzyme or emerging technology; *dry matter;* [#]*wet matter;* [≈]*approximation.*

obtained.[88,89] However, in other cases, these have decreased[90] compared to conventional acid extraction methods and other emerging technologies. A disadvantage is that when increasing voltage intensity, the degree of esterification of pectin obtained decreases, although pectin obtained may present percentages of esterification higher than 65%.[91,92] Another disadvantage is that commercial equipment for pulsed electric field application is still developing, the electrochemical reactions that can induce, and industrial scaling.[86,93]

15.4.3.5 SUBCRITICAL WATER

"Subcritical water" or "pressurized hot water" is defined by Gbashi et al.[94] as "*the expanse of the condensed phase between the boiling point and critical point temperature of water (100–374°C)*". These bioactive compound extracting methods are considered a green technology and can be used in pectin extraction.[95] Although it is a little-studied method, it has been found to achieve comparable or higher pectin yields than conventional ones alone or combined with other methods, but affecting some physicochemical pectin properties such as the degree of esterification and molecular weight, obtaining low methoxy pectin.[96–98]

15.4.3.6 HIGH-PRESSURE PROCESSING

The processes that involve high pressures or high hydrostatic pressures is a method that, when increasing the pressure in short periods, promotes the hydrolysis of plant material for the extraction of compounds of interest. This carries several advantages, such as avoiding the degradation of compounds due to the effect of temperature and shortening hydrolysis times and showing low energy use compared to conventional methods.[99–101] High-pressure processes may be related to the following two different phenomena: The "high hydrostatic pressure" and the "high-pressure homogenization". The difference lies in that high hydrostatic pressure is an adiabatic heating process while high homogenized pressure is a continuous process that works above 400 MPa.[102] These methods have been little studied; however, better pectin yields have been obtained, although in some cases, it has affected other physicochemical characteristics. Potato peel pectin extracted with high pressure at 200 MPa for 5 min increased the galacturonic acid content and decreased the esterification degree and the molecular weight.[103] However, Guo et al.[104] obtained higher molecular weight and viscosity pectin from orange byproducts using high hydrostatic pressures than the conventional method. Nevertheless, when the method has been combined with others, some of the pectin's physicochemical properties have not been affected. Naghshineh et al.[105] combined high hydrostatic pressures with enzymatic extraction, finding that the combined method did not affect pectin's molecular weight and viscosity extracted from lime byproducts. In the case of high homogenized pressures, improvements have been observed in pectin extraction yields in combination with acid extraction.[106]

15.5 CONCLUSION

Currently, pectin demand is growing due to the wide variety of applications, mainly in the food and pharmaceutical industry. For this reason, evaluating alternative sources for pectin extraction has turned the use of byproducts generated from food processing into plant sources. The conventional acid extraction process with mineral acids and high temperatures has been the most widely used but causes a considerable amount of waste and high energy consumption. The modification or substitution of conventional pectin extraction methods has been evaluated to minimize economic and environmental costs by substituting mineral acid for organic acids and using enzymes and emerging technologies. In some cases, although these new methods may affect the pectin's physicochemical properties of pectin-derived oligosaccharides, they can achieve similar or better yields to the conventional ones and can also preserve their functional application properties. However, most of these proposals are currently still in the research stage. Their success on an industrial scale may depend on the byproduct plant source and the physicochemical properties of the obtained pectin.

KEYWORDS

- **fruit and vegetable byproducts**
- **waste valorization**
- **pectin extraction**
- **emerging technologies**

REFERENCES

1. Rolin, C.; De Vries, J. Pectin. In *Food Gels*; Harris P., Ed.; Springer: Dordrecht, 1990.
2. Sriamornsak, P. Chemistry of Pectin and Its Pharmaceutical Uses: A Review. *SUIJ* **2003,** *3*, 206–228.
3. Grassino, A. N.; Halambek, J.; Djaković, S.; Brnčić, S. R.; Dent, M.; Grabarić, Z. Utilization of Tomato Peel Waste from Canning Factory as a Potential Source for Pectin Production and Application as Tin Corrosion Inhibitor. *Food Hydrocoll.* **2016,** *52*, 265–274.
4. Adetunji, L. R.; Adekunle, A.; Orsat, V.; Raghavan, V. Advances in the Pectin Production Process Using Novel Extraction Techniques: A Review. *Food Hydrocoll.* **2017,** *62*, 239–250.
5. Venkatanagaraju, E.; Bharathi, N.; Sindhuja, R. H.; Chowdhury, R. R.; Sreelekha, Y. Extraction and Purification of Pectin from Agro-Industrial Wastes. In *Pectins Extraction, Purification, Characterization and Applications*; Masuelli, M., Ed.; IntechOpen: London, 2019.
6. Thakur, B. R; Singh, R. K.; Handa, A. K. Chemistry and Uses of Pectin - A Review. *Crit. Rev. Food Sci. Nutr.* **1997,** *37*, 47–73.
7. Bonnin, E.; Garnier, C.; Ralet, M. C. Pectin-Modifying Enzymes and Pectin-Derived Materials: Applications and Impacts. *Appl. Microbiol. Biotechnol.* **2014,** *98* (2), 519–532.
8. Chen, J.; Liu, W.; Liu, C.; Li, T.; Liang, R.; Luo, S. Pectin Modifications: A Review. *Crit. Rev. Food Sci. Nutr.* **2015,** *55*, 1684–98.
9. Serna, S. S. O.; Ayala S. F. E. Chemical Composition and Biosynthesis of Dietary Fiber Components. In *Science and Technology of Fibers in Food Systems*; Welti-Chanes, J.; Serna-Saldívar, S. O.; Campanella, O. H.; Tejada-Ortigoza, V., Eds; Springer Nature Switzerland AG: Cham, Switzerland, 2020.
10. Willats, W. G.; McCartney, L.; Mackie, W.; Knox, J. P. Pectin: Cell Biology and Prospects for Functional Analysis. *Plant Mol. Biol.* **2001,** *47* (1), 9–27.
11. Satapathy, S.; Rout, J. R.; Kerry, R. G.; Thatoi, H.; Sahoo, S. L. Biochemical Prospects of Various Microbial Pectinase and Pectin: An Approachable Concept in Pharmaceutical Bioprocessing. *Front. Nutr.* **2020,** *7*, 117.
12. Yapo, B. M.; Robert, C.; Etienne, I.; Wathelet, B.; Paquot, M. Effect of Extraction Conditions on the Yield, Purity and Surface Properties of Sugar Beet Pulp Pectin Extracts. *Food Chem.* **2007,** *100*, 1356–1364.
13. Ciriminna, R; Fidalgo, A.; Dekisi, R.; Ilharco, L.; Plagiaro, M. Pectin Production and Global Market. *Agro. Food Industry Hi. Tech.* **2016,** *27*, 17–20.
14. Baum, A.; Dominiak, M.; Vidal-Melgosa, S.; Willats, W. G. T.; Søndergaard, K. M.; Hansen, P. W.; Mikkelsen, J. D. Prediction of Pectin Yield and Quality by FTIR and Carbohydrate Microarray Analysis. *Food Bioproc. Tech.* **2017,** *10* (1), 143–154.
15. Market Intelligence Data. *Global Pectin Market Growth 2020–2025*. LP Information Inc: Harrisburg, NC, 2020. https://marketintelligencedata.com/reports/156414/global-pectin-market-growth-2020–2025
16. Fior Markets. Pectin market by raw material (Apples, Citrus fruits, Sugar Beet & Others), by type (HM pectin, LM pectin), application (pharmaceutical, food & beverages, personal care products, industry), regions and global forecast 2020–2027. Report 418084. Fior Market Research LLP, Pune, Maharashtra, India https://www.fiormarkets.com/report/pectin-market-by-raw-material-apples-citrus-fruits-418084.html

17. Fuller, S.; Beck, E.; Salman, H.; Tapsell, L. New Horizons for Study of Dietary Fiber and Health: A Review. *Plant Foods Hum. Nutr.* **2016,** *71*, 1–12.
18. Liew, S. Q.; Teoh, W. H.; Yusoff, R.; Ngoh, G. C. Comparisons of Process Intensifying Methods in the Extraction of Pectin from Pomelo Peel. *Chem. Eng. Process. - Process Intensification* **2019,** 107586.
19. Sharma, P. C.; Gupta, A.; Kaushal, P. Optimization of Method for Extraction of Pectin from Apple Pomace. *Indian J. Prod. Resour.* **2014,** *5* (2), 184–189.
20. Chang, C.; Li, J.; Li, X.; Wang, C.; Zhou, B.; Su, Y.; Yang, Y. Effect of Protein Microparticle and Pectin on Properties of Light Mayonnaise. *LWT - Food Sci. Technol.* **2017,** *82*, 8–14.
21. Sun, C.; Liu, R.; Liang, B.; Wu, T.; Sui, W.; Zhang, M. Microparticulated Whey Protein-Pectin Complex: A Texture-Controllable Gel for Low-Fat Mayonnaise. *Food Res. Int.* **2018,** *108*, 151–160.
22. Yang, X.; Nisar, T.; Hou, Y.; Gou, X.; Sun, L.; Guo, Y. Pomegranate Peel Pectin Can Be used as an Effective Emulsifier. *Food Hydrocoll.* **2018,** *85*, 30–38.
23. Tripathi, S.; Mehrotra, G. K.; Dutta, P. K. Preparation and Physicochemical Evaluation of Chitosan/Poly (Vinyl Alcohol)/Pectin Ternary Film for Food-Packaging Applications. *Carbohydr. Polym.* **2010,** *79* (3), 711–716.
24. Espitia, P. J. P.; Du, W. X.; Avena-Bustillos, R. de J.; Soares, N. de F. F.; McHugh, T. H. Edible Films from Pectin: Physical-Mechanical and Antimicrobial Properties - A Review. *Food Hydrocoll.* **2014,** *35*, 287–296.
25. Valdés, A.; Burgos, N.; Jiménez, A.; Garrigós, M. C. Natural Pectin Polysaccharides as Edible Coatings. *Coatings* **2015,** *5* (4), 865–886.
26. Chaiwarit, T.; Masavang, S.; Mahe, J.; Sommano, S.; Ruksiriwanich, W.; Brachais, C. H.; Chambin O.; Jantrawut, P. Mango (cv. Nam Dokmai) Peel as a Source of Pectin and Its Potential use as a Film-Forming Polymer. *Food Hydrocoll.* **2020,** *102*, 105611.
27. Anderson, J. W; Baird, P.; Davis, R. H; Jr. Ferreri, S.; Knudtson, M.; Koraym, A.; Waters, V.; Williams, C. L. Health Benefits of Dietary Fiber. *Nutr. Rev.* **2009,** *67* (4), 188–205.
28. Sriamornsak, P. Application of Pectin in Oral Drug Delivery. *Expert. Opin. Drug Deliv.* **2011,** *8*, 1009–1023.
29. Majee, S. B; Avlani, D.; Ghosh, P.; Biswas, G. R. Therapeutic and Pharmaceutical Benefits of Native and Modified Plant Pectin. *J. Med. Plant Res.* **2018,** *12* (1), 1–6.
30. do Prado, S.; Castro-Alves, V. C.; Ferreira, G. F.; Fabi, J. P. Ingestion of Non-Digestible Carbohydrates from Plant-Source Foods and Decreased Risk of Colorectal Cancer: A Review on the Biological Effects and the Mechanisms of Action. *Front. Nutr.* **2019,** *6*, 72.
31. Gómez, B.; Gullón, B.; Yáñez, R.; Schols, H.; Alonso, J. L. Prebiotic Potential of Pectins and Pectic Oligosaccharides Derived from Lemon Peel Wastes and Sugar Beet Pulp: A Comparative Evaluation. *J. Funct. Foods.* **2016,** *20*, 108–121.
32. Mandalari, G.; Palop, C. N.; Tuohy, K.; Gibson, G. R.; Bennett, R. N.; Waldron, K. W.; Bisignano, G.; Narbad, A.; Faulds, C. B. In Vitro Evaluation of the Prebiotic Activity of a Pectic Oligosaccharide-Rich Extract Enzymatically Derived from Bergamot Peel. *Appl. Microbiol. Biotechnol.* **2007,** *73* (5), 1173–1179.
33. Zhang, S.; Hu, H.; Wang, L.; Liu, F.; Pan, S. Preparation and Prebiotic Potential of Pectin Oligosaccharides Obtained from Citrus Peel Pectin. *Food Chem.* **2018,** *244*, 232–237.
34. Li, W.; Hao, W.; Xiaohua, Z.; Yinchen, H.; Wangwang, L.; Gongming, Y.; Aimin, J. Pectin-Chitosan Complex: Preparation and Application in Colon-Specific Capsule. *Int. J. Agric. Biol. Eng.* **2015,** *8* (4), 151–160.

35. Assadpour, E.; Jafari, S. M.; Maghsoudlou, Y. Evaluation of Folic Acid Release from Spray Dried Powder Particles of Pectin-Whey Protein Nano-Capsules. *Int. J. Biol. Macromol.* **2017,** *95*, 238–247.
36. Rehman, A.; Ahmad, T.; Aadil, R. M.; Spotti, M. J.; Bakry, A. M.; Khan, I. M.; Zhao, L.; Riaz, T.; Tong, Q. Pectin Polymers as Wall Materials for The Nano-Encapsulation of Bioactive Compounds. *Trends Food Sci. Technol.* **2019,** *90*, 35–46.
37. Pieczykolan, E.; Kurek, M. A. Use of Guar Gum, Gum Arabic, Pectin, Beta-Glucan and Inulin for Microencapsulation of Anthocyanins from Chokeberry. *Int. J. Biol. Macromol.* **2019,** *129*, 665–671.
38. Ayala-Zavala, J.; Vega-Vega, V.; Rosas-Domínguez, C.; Palafox-Carlos, H.; Villa-Rodriguez, J. A.; Siddiqui, M. W.; Davila-Aviña, J. E.; González-Aguilar, G. A. Agro-Industrial Potential of Exotic Fruit Byproducts as a Source of Food Additives. *Food Res. Int.* **2011,** *44* (7), 1866–1874.
39. de Moraes, C. T.; Hermes, V. S.; de Oliveira, R. A.; Flôres, S. H. Evaluation of Bioactive Compounds, Chemical and Technological Properties of Fruits Byproducts Powder. *J. Food Sci. Technol.* **2016,** *53* (11), 4067–4075.
40. Szabo, K.; Cătoi, A. F.; Vodnar, D. C. Bioactive Compounds Extracted from Tomato Processing Byproducts as a Source of Valuable Nutrients. *Plant Foods Hum. Nutr.* **2018,** *73*, 268–277.
41. Cho, E. H.; Jung, H. T.; Lee, B. H.; Kim, H. S.; Rhee, J. K.; Yoo, S. H. Green Process Development for Apple-Peel Pectin Production by Organic Acid Extraction. *Carbohydr. Polym.* **2019,** *204*, 97–103.
42. Babbar, N.; Dejonghe, W.; Gatti, M.; Sforza, S.; Elst, K. Pectic Oligosaccharides from Agricultural Byproducts: Production, Characterization and Health Benefits. *Crit. Rev. Biotechnol.* **2016,** *36* (4), 594–606.
43. Iuga, M.; Mironeasa, S. Potential of Grape Byproducts as Functional Ingredients in Baked Goods and Pasta. *Compr. Rev. Food Sci. Food Saf.* **2020,** *19* (5), 2473–2505.
44. O'Shea, N.; Arendt, E. K.; Gallagher, E. Dietary Fiber and Phytochemical Characteristics of Fruit and Vegetable Byproducts and their Recent Applications as Novel Ingredients in Food Products. *Innov. Food Sci. Emerg. Technol.* **2012,** *16*, 1–10.
45. Torres-León, C.; Ramírez-Guzmán, N.; Londoño-Hernandez, L.; Martinez-Medina, G. A.; Díaz-Herrera, R.; Navarro-Macias, V.; Alvarez-Pérez, O. V.; Picazo, B.; Villarreal-Vázquez, M.; Ascacio-Valdes, J.; Aguilar, C. N. Food Waste and Byproducts: An Opportunity to Minimize Malnutrition and Hunger in Developing Countries. *Front. Sustain. Food Syst.* **2018,** *2*, 52.
46. Rizk, E. M.; El-Kady, A. T.; El-Bialy, A. R. Characterization of Carotenoids (Lyco-Red) Extracted from Tomato Peels and Its Uses as Natural Colorants and Antioxidants of Ice Cream. *Ann. Agric. Sci.* **2014,** *59*, 53–6.
47. Faustino, M.; Veiga, M.; Sousa, P.; Costa, E. M.; Silva, S.; Pintado, M. Agro-Food Byproducts as a New Source of Natural Food Additives. *Molecules* **2019,** *24* (6), 1056.
48. Marín, F. R.; Soler-Rivas, C.; Benavente-García, O.; Castillo, J.; Pérez-Alvarez, J. A. Byproducts from Different Citrus Processes as a Source of Customized Functional Fibres. *Food Chem.* **2007,** *100* (2), 736–741.
49. Chen, Q.; Hu, Z.; Yao, F. Y. D.; Liang, H. Study of Two-Stage Microwave Extraction of Essential Oil and Pectin from Pomelo Peels. *LWT - Food Sci. Technol.* **2016,** *66*, 538–545.
50. Hilali, S.; Fabiano-Tixier, A. S.; Ruiz, K.; Hejjaj, A.; Ait Nouh, F.; Idlimam, A.; Bily, A.; Mandi, Laila.; Chemat, F. Green Extraction of Essential Oils, Polyphenols, and Pectins

from Orange Peel Employing Solar Energy: Toward a Zero-Waste Biorefinery. *ACS Sustainable Chem. Eng.* **2019,** *7* (13), 11815–11822.

51. Nagarajan, J.; Krishnamurthy, N. P.; Ramanan Ramakrishnan, N.; Raghunandan, M. E.; Galanakis, C.; Chien, W. O. A Facile Water-Induced Complexation of Lycopene and Pectin from Pink Guava Byproduct: Extraction, Characterization and Kinetic Studies. *Food Chem.* **2019,** *296*, 47–55.
52. San Martín-Hernández, C.; Pérez-Rubio, V.; Muy-Rangel, M. D.; Vargas-Ortiz, M. A.; Quintana-Obregón, E. A. Characterization of Powder and Pectin of Mango Pericarp (*Mangifera Indica* l.) Ripe 'Ataulfo' and Swot Analysis of Its Processing. *TIP Rev. Esp. Cienc. Quim. Biol.* **2020,** *23*, 1–10.
53. Chan, S. Y.; Choo, W. S. Effect of Extraction Conditions on the Yield and Chemical Properties of Pectin from Cocoa Husks. *Food Chem.* **2013,** *141* (4), 3752–3758.
54. Abid, M.; Renard, C. M.; Watrelot, A. A.; Fendri, I.; Attia, H.; Ayadi, M. A. Yield and Composition of Pectin Extracted from Tunisian Pomegranate Peel. *Int. J. Biol. Macromol.* **2016,** *93*, 186–194.
55. Seixas F. L.; Fukuda, D. L.; Turbiani, F. R. B.; García, P. S.; Petkowicz, C. Ld. O.; Jagadevan, M. L.; Gimenes, M. L. Extraction of Pectin from Passion Fruit Peel (*Passiflora Edulis F. Flavicarpa*) by Microwave-Induced Heating. *Food Hidrocoll.* **2014,** *38*, 186–192.
56. Wang, L. Energy Efficiency Technologies for Sustainable Food Processing. *Energy Effic.* **2014,** *7*, 791–810.
57. Knorr, D.; Froehling, A.; Jaeger, H.; Reineke, K.; Schlueter, O.; Schoessler, K. Emerging Technologies in Food Processing. *Annu. Rev. Food Sci. Technol.* **2011,** *2*, 203–235.
58. Marić, M.; Grassino, A. N.; Zhu, Z.; Barba, F. J.; Brnčić, M.; Rimac, B. S. An Overview of the Traditional and Innovative Approaches for Pectin Extraction from Plant Food Wastes and Byproducts: Ultrasound-, Microwaves-, and Enzyme-Assisted Extraction. *Trends Food Sci. Technol.* **2018,** *76*, 28–37.
59. Rudolph B.; Petersen S. A. Methods for steam flash extraction of pectin. U.S. Patent 13/013,709 Pub: US20120190831 A1, 2011.
60. Sirisakulwat, S.; Nagel, A.; Sruamsiri, P.; Carle, R.; Neidhart, S. Yield and Quality of Pectins Extractable from the Peels of Thai Mango Cultivars Depending on Fruit Ripeness. *J. Agric. Food Chem.* **2008,** *56* (22), 10727–10738.
61. Voragen, A. G. J.; Pilnik, W.; Thibault, J. F.; Axelos, M. A. V.; Renard, C. M. G. C. Pectins. In *Food Polysaccharides and Their Applications*; Sephen, A. M., Ed.; Marcel Dekker Inc.: New York, 1995.
62. Andersen, N. M.; Cognet, T.; Santacoloma, P. A.; Larsen, J.; Armagan, I.; Larsen, F. H.; Gernaey, K. V.; Abildskov, J.; Huusom, J. K. Dynamic Modelling of Pectin Extraction Describing Yield and Functional Characteristics. *J. Food Eng.* **2017,** *192*, 61–71.
63. Papagianni, M. Organic Acids. In *Comprehensive Biotechnology*; 2011, 109–120.
64. Lv, X.; Liu, J.; Yin, X.; Gu, L.; Sun, L.; Du, G.; Chen, J.; Liu, L. Microbial Production of Functional Organic Acids. In *Systems and Synthetic Biotechnology for Production of Nutraceuticals*; Liu L.; Chen, J. Eds.; Springer: Singapore, 2019.
65. Jafari, F.; Khodaiyan, F.; Kiani, H.; Hosseini, S. S. Pectin from Carrot Pomace: Optimization of Extraction and Physicochemical Properties. *Carbohydr. Polym.* **2017,** *157*, 1315–1322.
66. Raji, Z.; Khodaiyan, F.; Rezaei, K.; Kiani, H.; Hosseini, S. S. Extraction Optimization and Physicochemical Properties of Pectin from Melon Peel. *Int. J. Biol. Macromol.* **2017,** *98*, 709–716.

67. Yang, J. S.; Mu, T. H.; Ma, M. M. Extraction, Structure, and Emulsifying Properties of Pectin From Potato Pulp. *Food Chem.* **2018,** *244*, 197–205.
68. Banerjee, J.; Vijayaraghavan, R.; Arora, A.; MacFarlane, D. R.; Patti, A. F. Lemon Juice based Extraction of Pectin from Mango Peels: Waste to Wealth by Sustainable Approaches. *ACS Sustain. Chem. Eng.* **2016,** *4* (11), 5915–5920.
69. Mendoza-Vargas, L.; Jiménez-Forero, J.; Ramírez-Niño, M. Evaluación De La Pectina Extraída Enzimáticamente a Partir De Las Cáscaras Del Fruto De Cacao (*Theobroma Cacao* L.). *Rev. U.D.C.A Act. & Div. Cient.* **2017,** *20* (1), 131–138.
70. Vasco-Correa, J.; Zapata Z. A. D. Enzymatic Extraction of Pectin from Passion Fruit Peel (*Passiflora Edulis F. Flavicarpa*) at Laboratory and Bench Scale. *LWT* **2017,** *80*, 280–285.
71. Abou-Elseoud, W. S.; Hassan, E. A.; Hassan, M. L. Extraction of Pectin from Sugar Beet Pulp by Enzymatic and Ultrasound-Assisted Treatments. *Carbohydr. Polym. Technol. Appl.* **2021,** *2*, 100042.
72. Morris, G. A.; Binhamad, H. A. S. Isolation and Characterization of Pectin. In *Pectin Technological and Physiological Properties*; Kontogiorgos, V., Ed.; Springer: Switzerland, AG, 2020.
73. Lee, S. H.; Choi, W.; Jun, S. Conventional and Emerging Combination Technologies for Food Processing. *Food Eng. Rev.* **2016,** *8* (4), 414–434.
74. Freitas, C. M. P.; Sousa, R. C. S.; Dias, M. M. S.; Coimbra, J. S. R. Extraction of Pectin from Passion Fruit Peel. *Food Eng. Rev.* **2020,** *12* (4), 460–472.
75. Urango-Anaya, K. J.; Ortega-Quintana, F. A.; Vélez-Hernández, G.; Pérez-Sierra, Ó. A. Extracción Rápida De Pectina a Partir De Cáscara De Maracuyá (*Passiflora Edulis Flavicarpa*) Empleando Microondas. *Inf. Tecnol.* **2018,** *29* (1), 129–136.
76. Prakash, M. J.; Sivakumar, V.; Thirugnanasambandham, K.; Sridhar, R. Optimization of Microwave Assisted Extraction of Pectin from Orange Peel. *Carbohydr. Polym.* **2013,** *97* (2), 703–709.
77. Bagherian, H.; Zokaee, A. F.; Fouladitajar, A.; Mohtashamy, M. Comparisons between Conventional, Microwave- and Ultrasound-Assisted Methods for Extraction of Pectin from Grapefruit. *Chem. Eng. Process.* **2011,** *50* (11–12), 1237–1243.
78. Prakash, M. J.; Sivakumar, V.; Thirugnanasambandham, K.; Sridhar, R. Microwave Assisted Extraction of Pectin from Waste *Citrullus Lanatus* Fruit Rinds. *Carbohydr. Polym.* **2014,** *101*, 786–791.
79. Maran, J. P.; Prakash, K. A. Process Variables Influence on Microwave Assisted Extraction of Pectin from Waste Carcia Papaya L. Peel. *Int. J. Biol. Macromol.* **2015,** *73*, 202–206.
80. Yan, J. K.; Wang, C.; Qiu, W. Y.; Chen, T. T.; Yang, Y.; Wang, W. H.; Zhang, H. N. Ultrasonic Treatment at Different pH Values Affects the Macromolecular, Structural, and Rheological Characteristics of Citrus Pectin. *Food Chem.* **2020,** 128216.
81. Panchev, I.; Kirchev, N.; Kratchanov, C. Improving Pectin Technology. *Int. J. Food Sci. Technol.* **2007,** *23* (4), 337–341.
82. Saberian, H.; Hamidi-Esfahani, Z.; Ahmadi, G. H.; Barzegar, M. Optimization of Pectin Extraction from Orange Juice Waste Assisted by Ohmic Heating. *Chem. Eng. Process.* **2017,** *117*, 154–161.
83. Rodrigues, R. M.; Genisheva, Z.; Rocha, C. M. R.; Teixeira, J. A.; Vicente, A. A.; Pereira, R. N. Ohmic Heating for Preservation, Transformation, and Extraction. In *Green Food Processing Techniques*. Chemat, F. Ed.; Academic Press: Avignon, France, 2019.
84. Varghese K. S.; Pandey M.; Radhakrishna K.; Bawa, A. Technology, Applications and Modelling of Ohmic Heating: A Review. *J. Food Sci. Technol.* **2014,** *51*, 2304–2317.

85. Saberian, H.; Hamidi-Esfahani, Z.; Ahmadi, G. H., Banakar, A.; Barzegar, M. The Potential of Ohmic Heating for Pectin Extraction from Orange Waste. *J. Food Process. Pres.* **2017,** *42* (2), e13458.
86. Kempkes, M. A. Industrial Pulsed Electric Field Systems. In *Handbook of Electroporation*; Miklavcic, Ed.; Springer: Bedford, MA, 2017.
87. Syed, Q. A.; Ishaq, A; Rahman, U. U.; Aslam, S.; Shukat, R. Pulsed Electric Field Technology in Food Preservation: A Review. *J. Nutr. Health Food Eng.* **2017,** *6* (6), 00219.
88. Yin, Y. G.; Fan, X. D.; Liu, F. X.; Yu, Q. Y.; He, G. D. Fast Extraction of Pectin from Apple Pomace by High Intensity Pulsed Electric Field. *J. Jilin Univ.* **2009,** *5*, 021–029.
89. Qingyu, Y.; Yongguang, Y.; Zemin, X.; Xiangdong, F. Influence on Extraction of Pectin by High Intensity Pulsed Electric Fields [J]. *J. Agric. Mech. Res.* **2009,** 10.
90. Yang, N.; Jin, Y.; Tian, Y.; Jin, Z.; Xu, X. An Experimental System for Extraction of Pectin from Orange Peel Waste Based on the O-Core Transformer Structure. *Biosyst. Eng.* **2016,** *148*, 48–54.
91. Ma, S.; Wang, Z. Pulsed Electric Field-Assisted Modification of Pectin from Sugar Beet Pulp. *Carbohydr. Polym.* **2013,** *92* (2), 1700–1704.
92. de Oliveira, C. F.; Giordani, D.; Gurak, P. D.; Cladera-Olivera, F.; Marczak, L. D. F. Extraction of Pectin from Passion Fruit Peel using Moderate Electric Field and Conventional Heating Extraction Methods. *Innov. Food Sci. Emerg. Technol.* **2015,** *29*, 201–208.
93. Xi, J.; Li, Z.; Fan, Y. Recent Advances in Continuous Extraction of Bioactive Ingredients from Food-Processing Wastes by Pulsed Electric Fields. *Crit. Rev. Food Sci. Nutr.* **2021,** *61* (10), 1738–1750
94. Gbashi, S.; Adebo, O. A.; Piater, L.; Madala, N. E.; Njobeh, P. B. Subcritical Water Extraction of Biological Materials. *Sep. Purif. Rev.* **2017,** *46* (1), 21–34.
95. Zhang, J.; Wen, C.; Zhang, H.; Duan, Y.; Ma, H. Recent Advances in the Extraction of Bioactive Compounds with Subcritical Water: A Review. *Trends Food Sci. Technol.* **2020,** *95*, 183–195.
96. Chen, H.; Fu, X. and Luo, Z. Properties and Extraction of Pectin-Enriched Materials from Sugar Beet Pulp by Ultrasonic-Assisted Treatment Combined with Subcritical Water. *Food Chem.* **2015b,** *168*, 302–310.
97. Liew, S. Q.; Teoh, W. H.; Tan, C. K.; Yusoff, R.; Ngoh, G. C. Subcritical Water Extraction of Low Methoxyl Pectin from Pomelo (*Citrus Grandis* L. Osbeck) Peels. *Int. J. Biol. Macromol.* **2018,** *116*, 128–135.
98. Li, W.; Fan, Z.; Wu, Y.; Jiang, Z.; Shi, R. Eco-Friendly Extraction and Physicochemical Properties of Pectin from Jackfruit Peel Waste by Subcritical Water. *J. Sci. Food Agric.* **2019,** *99* (12), 5283–5292.
99. Farkas, D. F.; Hoover, D. G. High Pressure Processing. *J. Food Sci.* **2000,** *65*, 47–64.
100. Huang, H. W.; Hsu, C. P.; Yang, B. B.; Wang, C. Y. Advances in the Extraction of Natural Ingredients by High Pressure Extraction Technology. *Trends Food Sci. Technol.* **2013,** *33* (1), 54–62.
101. Xie, F.; Li, M.; Lan, X.; Zhang, W.; Gong, S.; Wu, J.; Wang, Z. Modification of Dietary Fibers from Purple-Fleshed Potatoes (*Heimeiren*) with High Hydrostatic Pressure and High Pressure Homogenization Processing: A Comparative Study. *Innov. Food Sci. Emerg. Technol.* **2017,** *42*, 157–164.
102. Santos, L. M.; Oliveira, F. A.; Ferreira, E. H.; Rosenthal, A. Application and Possible Benefits of High Hydrostatic Pressure or High-Pressure Homogenization on Beer Processing: A Review. *Food Sci. Technol. Int.* **2017,** *23* (7), 561–581.

103. Xie, F.; Zhang, W.; Lan, X.; Gong, S.; Wu, J.; Wang, Z. Effects of High Hydrostatic Pressure and High Pressure Homogenization Processing on Characteristics of Potato Peel Waste Pectin. *Carbohydr. Polym.* **2018,** *196*, 474–482.
104. Guo, X.; Han, D.; Xi, H.; Rao, L.; Liao, X.; Hu, X.; Wu, J. Extraction of Pectin from Navel Orange Peel Assisted by Ultra-High Pressure, Microwave or Traditional Heating: A Comparison. *Carbohydr. Polym.* **2012,** *88* (2), 441–448.
105. Naghshineh, M.; Olsen, K.; Georgiou, C. A. Sustainable Production of Pectin from Lime Peel by High Hydrostatic Pressure Treatment. *Food Chem.* **2013,** *136* (2), 472–478.
106. Van Audenhove, J.; Bernaerts, T.; De Smet, V.; Delbaere, S.; Van Loey, A. M.; Hendrickx, M. E. The Structure and Composition of Extracted Pectin and Residual Cell Wall Material from Processing Tomato: The Role of A Stepwise Approach Versus High-Pressure Homogenization-Facilitated Acid Extraction. *Foods* **2021,** *10*, 1064.
107. Nagel, A.; Winkler, C.; Carle, R.; Endress, H. U.; Rentschler, C.; Neidhart, S. Processes Involving Selective Precipitation for the Recovery of Purified Pectins from Mango Peel. *Carbohydr. Polym.* **2017,** *174*, 1144–1155.
108. Nguyen, H. D.; Nguyen, H. V.; Savage, G. P. Properties of Pectin Extracted from Vietnamese Mango Peels. *Foods.* **2019,** *8* (12), 629.
109. Kamble, P. B.; Gawande, S.; Patil, T. S. Extraction of Pectin from Unripe Banana Peel. *Int. J. Res. Eng. Technol.* **2017,** *4* (7), 2259–2264.
110. Petkowicz, C. L. O.; Vriesmann, L. C.; Williams, P. A. Pectins from Food Waste: Extraction, Characterization and Properties of Watermelon Rind Pectin. *Food Hydrocoll.* **2017,** *65*, 57–67.
111. Rodsamran, P.; Sothornvit, R. Microwave Heating Extraction of Pectin from Lime Peel: Characterization and Properties Compared with the Conventional Heating Method. *Food Chem.* **2019,** *278*, 364–372.
112. Yang, Y.; Wang, Z.; Hu, D.; Xiao, K.; Wu, J. Y. Efficient Extraction of Pectin from Sisal Waste by Combined Enzymatic and Ultrasonic Process. *Food Hydrocoll.* **2018,** *79*, 189–196.
113. Kazemi, M.; Khodaiyan, F.; Labbafi, M.; Hosseini, S. S. Ultrasonic and Heating Extraction of Pistachio Byproduct Pectin: Physicochemical, Structural Characterization and Functional Measurement. *Food Meas.* **2020,** *14*, 679–693.
114. 114. Wang, W.; Ma, X.; Jiang, P.; Hu, L.; Zhi, Z.; Chen, J.; Ding, J.; Ye, X.; Liu, D. Characterization of Pectin from Grapefruit Peel: A Comparison of Ultrasound-Assisted and Conventional Heating Extractions. *Food Hydrocoll.* **2016,** *61*, 730–739.
115. Sengar, A. S.; Rawson, A.; Muthiah, M.; Kalakandan, S. K. Comparison of Different Ultrasound Assisted Extraction Techniques for Pectin from Tomato Processing Waste. *Ultrason. Sonochem.* **2020,** *61*, 104812.
116. Ma, X.; Jing, J.; Wang, J.; Xu, J.; Hu, Z. Extraction of Low Methoxyl Pectin from Fresh Sunflower Heads by Subcritical Water Extraction. *ACS Omega* **2020,** *5* (25), 15095–15104.

CHAPTER 16

Agave Byproducts: As Sources of Phytochemicals with Functional Activities and Their Management for Industrial Applications

L. ALFONSO JIMÉNEZ-ORTEGA, OCTAVIO VALDEZ-BARO, MANUEL J. BERNAL-MILLÁN, M. MAGDALENA RIVERA-SALAS, and J. BASILIO HEREDIA

Centro de Investigación en Alimentación y Desarrollo, AC. Functional Foods and Nutraceuticals Department, Carretera a Eldorado Km 5.5, Campo el Diez, Culiacán, Sinaloa, Mexico

ABSTRACT

One of the food industry challenges is the use and management of waste and byproducts. The *Agave* is a plant native to America, with significant economic and cultural value. It is the main raw material for making tequila and other distilled and undistilled beverages. However, its transformation generates byproducts such as leaves, vinasses, and bagasse, which are maybe relevant for the food and pharmaceutical industry since they are sources of nutraceutical compounds. The leaves represent 50% of the total *Agave* weight and are composed of polyphenols, saponins, and terpenes that have shown in vitro and in vivo pharmacological effects. Fructans are extracted from the bagasse fibers and the leaves, which are carbohydrates that can exert prebiotic effects due to their bioactivities for example as antioxidants and anti-inflammatory. *Agave* syrup has shown sweetening capacity, providing a low caloric content, in addition to the presence of prebiotics.

Food Byproducts Management and their Utilization. Ricardo Gómez-García, Ana A. Vilas-Boas, Débora A. Campos, Maria Manuela Pintado, & Cristóbal Noé Aguilar (Eds.)

These biomasses can be transformed into other products potentially applicable to the food, agriculture, animal feed, biofuels, pharmaceutical, textile industries, making it a field of opportunity to manage the *Agave* byproducts (AbP), helping to minimize their environmental impact. The growing demand for functional products encourages science and industrial groups to find new sources to develop food ingredients. Besides, the correct handling of byproducts increases the local economy of the *Agave*-producing regions.

16.1 INTRODUCTION

The *Agave* is one of the most characteristic and representative crops of Mexico. It represents the most extensive genus of the Agavaceae family with 210 species. More than 75% of these are in Mexico, of which approximately 135 species are endemic to the country, being the center of domestication and diversification. Traditionally from pre-Hispanic times to the present, it has been used as food (leaves, stems, flowers), drink (mead: *Agave* nectar; pulque: fermented nectar; tequila and mezcal: distilled must), medicinal (leaves and nectar), and for manufacturing of textile fibers (sheets and bagasse). The species that stand out for their diversity of uses and economic potential are *Agave tequilana* Weber var. Azul, *A. salmiana*, *A. americana, A sisalana, A. angustifolia*, *A. potatorum*, *A. esperrimi*, *A. lechuguilla,* and *A. mapisaga.* The yield and composition of the biomass of each species can vary.[1,2] Figure 16.1 shows the traditional Jima of *Agave* in Tequila, Jalisco, Mexico.

Tequila can only be produced in specific states and regions of Mexico (181 municipalities), because of its protected designation of origin. The tequila industry is one of the most profitable, producing, and economically important. In 2020, 374 million liters were produced, for which 1.47 million tons of *Agave* were needed. The major producing states are Jalisco, Guanajuato, Michoacán, Nayarit, Zacatecas, Morelos, Puebla, Guerrero, Oaxaca, and Tamaulipas. In 2019, 103,000 ha of *Agave* were planted, representing a production volume of 1.513 million tons, with a value of 29,058 million Mexican pesos. The global demand for tequila is rising. In 2020, 286.7 million liters were destined to export, representing 76% of the total national production. During 2019 it was the fifth-largest agro-food product exported, leaving an economic spill of 1874 million dollars. The major importing countries are the United States, Japan, United Kingdom, Germany, Spain, and Canada.[3,4]

FIGURE 16.1 Traditional Jima of Agave in Tequila, Jalisco, Mexico (CC license). (a) Common agave field with young pineapple plants; (b) cross-sections of agave leaves or spines; (c) cutting leaves from Agave "pineapple" ("Jimado"); (d) Agave leaf-cutting person ("Jimador").

Undoubtedly, the tequila industry, fermented and distilled beverages from *Agave*, such as bacanora, sotol, mezcal, raicilla, mead, and pulque, will continue to grow. In 2019, there were 162 tequila-exporting companies. Projections to 2030 estimate 347.32 million liters of tequila production, for which 2.12 million tons of *Agave* will be needed.[5,6] However, 1735 million kg of bagasse and 4164 million liters of vinasse will also be generated, derived from the production process. Hence, it is necessary to revolutionize programs for the use of AbP and strict regulations on their treatments, as well as an integral and decisive sanitary and environmental surveillance by the corresponding authorities.

16.2 KEY CONSIDERATIONS OF MANAGEMENT OF *AGAVE* BY PRODUCTS (ABP)

Global climate change, coupled with the growing demand for food, directly impacts agricultural production. The *Agave* is a xerophytic plant and an excellent CO_2 sequestrant, it develops in limited water conditions; however, high levels of CO_2 can change its anatomy and morphology. It has been observed that higher concentrations of CO_2 increase the accumulation of biomass to be potentially useful for producing bioethanol or other applications of these AbP. Therefore, developing efficient alternatives for the use of AbP is an opportunity for industry and research, especially in arid regions where there are marginal lands, taking into account future climatological projections that indicate an increase in temperature and water scarcity.[2]

16.2.1 AGAVE BAGASSE

One of the main AbP generated from the distilled alcoholic beverages industry is bagasse. It represents more than 40% of the total weight (wet basis) of the caulescent stem. For each liter of mezcal, 15–20 kg (wet base) are generated, and for each liter of tequila, 5 kg (wet base) are generated. Therefore, based on the annual tequila volume production, approximately 1870 million tons of bagasse are generated. Bagasse is made up of lignocellulose (cellulose, hemicellulose, and lignin) and is usually used in fertilizers, construction materials, textile fibers, paper, handicrafts, bio-energy production, and livestock feed; the technical viability of its use as a solid fuel in arid regions has been supported.[7,8]

The valorization and administration of AbP must be efficient, considering the cost and CO_2 emissions of their transfer, from the generating industries to the revalorization plants. These factors must be considered in a circular economy; hence, the strategic planning of the location of both must be considered. Case studies have been observed in Mexico, applying mathematical models to strategically plan the AbP management of the tequila industry for the production of bioethanol, considering the location, fuel demand, sales, transportation, and production costs; concluding that the implementation of biofuel production plants from this biomass is feasible and simple, satisfying the national demand for bioethanol by 40%.[9]

16.2.2 AGAVE LEAVES

The leaves are an AbP derived from the Jima (*Agave* stem harvesting process) (Figure 16.1), representing 66% in dry weight of the biomass in 3-year-old plants, where the leaves with the stem ratio are 4:1; however, over the years it increases. They are rich in secondary metabolites such as saponins and phenolics, which can contribute to the treatment of chronic degenerative diseases (see in Section 16.4). It stands out for its high moisture content and volatile compounds (85%), complex carbohydrates such as xylan, heteroxylan, mannan, arabinoxylan, soluble such as glucose, fructose, fructans, and sucrose, insoluble such as cellulose and lignin. Therefore, the yield in ethanol production would be compared with the corn biomass estimating 401 L/t, with a productivity of 5.7–19 t/ha/year (assuming that 56.7% of the dry weight of the total plant are leaves) and 2300–7900 L/ha/year, this with leaves of *A. tequilana* which is used specifically for the production of tequila. However, the same model could be used for *A. americana* leaves.[10]

16.2.3 VINASSES

Vinasse is a highly toxic AbP to the ecosystem, which is released from the distillation of tequila. Its composition contains high concentrations of suspended organic materials, heavy metals, higher alcohols, and phenols. It has an acid pH and is usually discharged into the environment, bodies of water, or soils at high temperatures >85°C, so its contact is highly damaging, reducing productivity and present life. Around 10–12 L of vinasses are produced per liter of tequila; therefore, based on the 2020 production, 4488 million liters was estimated.[8,11]

From its treatment, biosolids are released, which can be used for the agricultural sector since they are rich in phytonutrients such as nitrogen, phosphorus, and potassium. Even bagasse and biosolids from stillage could replace commercial substrates due to their content of chemical parameters similar to or higher than those currently marketed. Other treatment alternatives are chemical and biotechnological; however, they are not efficient for large production volumes.[8,11] Therefore, it represents an opportunity for the scientific and industrial sector to provide efficient solutions, such as the generation of biofuels, extraction of bioactive compounds and vitamins, and applications in the metalworking industry, among others.

16.2.4 PRODUCTION OPTIMIZATION AND ENVIRONMENTAL IMPACT OF ABP

The environmental impact begins from the sowing of the *Agave* (CO_2 and fuels), where during the growth, it releases CO_2 and N_2O. In the Jima, the leaves are discarded, and during the transport to the industries, CO_2 is released. Once the production process has started, which involves heating the stem and subsequent tearing, CO_2 and bagasse are produced, while in the fermentation and distillation, CO_2 and vinasse are generated. Later in the formulation, maturation, bottling, packing, transportation, and distribution can also cause environmental impacts using fuels and electricity.[12]

The composition of AbP is wide (see Section 16.3), providing versatility for industrial uses. These bio-products must be made considering the environmental, sustainable, and economic factors. The European market for bioproducts is on the rise. In 2015, it was estimated that the monetary return for these industries was 695 billion euros, highlighting the paper, wood, and furniture sector, followed by biomass processing, chemical industry, standing out the latter being the cosmetic and surfactant sectors, followed by the production of bioactive compounds.[13]

Due to the environmental impact of the high production of tequila in Mexico, clean production methodologies could be adopted, in particular, the one proposed by the United Nations Environmental Program (UNEP) and United Nations Industrial Development Organization (UNIDO), which have the objective of minimizing inputs and outputs, to reduce environmental impacts at the end of the process. Performance, water and energy consumption, wastewater generation, and waste per kilogram of the final product can be considered indicators. In the processing food industry, a useful strategy to optimize resources is the implementation of a 4.0 production system, which also contributes to the development of the fourth industrial revolution. This system proposes to interconnect all the links of the production chain, from the field to the table, passing through the supply chains, labor, logistics, systems, processes, and consumers, to integrate all the byproducts in different industries. Additionally, a link between the consumer and the producer should be created to raise awareness of all the links and resources used to produce the inputs.[14]

The concept of agro-extractivism can potentially be applied to the production of *Agave;* in recent years, the export market for agave products has grown strongly and therefore the contamination of AbP.[15] The production of *Agave* and tequila is a phenomenon that encompasses cultural, traditional, economic, political, social aspects, and multinational interests, for which the support of the government sector must be reinforced in applying rigorous regulations regarding environmental emissions, labor rights, and support to the Mexican countryside and industry. Models and sustainability analyses have been proposed in environmental and economic terms, monetizing the indicators of AbP treatment and use for the generation of bioethanol and bioenergy through biorefineries, which can represent an optimal solution to the environmental problems of tequila production in Mexico.[16]

16.3 SECONDARY METABOLITES AND NUTRACEUTICALS FROM AbP

AbP contains bioactive compounds such as some carbohydrates with antioxidant and prebiotic activity, as well as phytochemical compounds such as flavonoids, saponins, and terpenes,[17] which, depending on their extraction method, may have biological potential and can be used as a medical alternative as an auxiliary treatment against some diseases. Figure 16.2 shows the relevant phytochemicals and nutraceutical compounds isolated and identified in *Agave* spp.

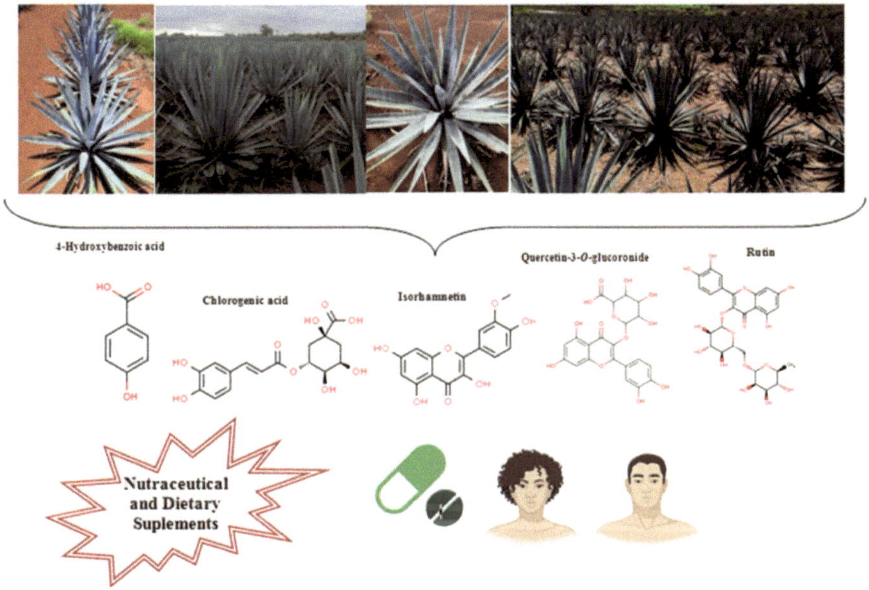

FIGURE 16.2 Main phytochemicals isolated and identified in *Agave* spp. (From BioRender and PhenolExplorer).

16.3.1 PREBIOTICS (DIETARY FIBERS AND FRUCTOOLIGOSACCHARIDES)

Prebiotics are a nondigestible food ingredient that beneficially reverberates the host by selectively stimulating the growth and/or activity of one or a limited number of bacteria in the colon, and thus improving host health.[18] Fructans, one of the isolated compounds from *Agave, are* considered prebiotic due to their structure and type of bonds (heterogeneous mixture of fructose polymers linked by fructose–fructose glycosidic bonds). If the fructan is constituted by a polymer of 2–10 fructose molecules, it is known as fructooligosaccharide (FOS), but if it is greater than 10, it is called inulin.[19,20] Furthermore, according to the major structural units, the classification of fructans into five main groups is inulin, levan, mixed levan, inulin neo series, and levan neo series.[21]

Fructans are not digestible through the intestinal tract enzymes; therefore, the entire molecules go into the colon. These polymers are fermented through the beneficial intestinal microflora, also known as probiotics or lactic acid bacteria (LAB), which favors the growth of these microorganisms.[22] Most of these fructans are present in the genus *Agave,* mainly inulin, and some FOS are counted as low molecular weight dietary fiber, a group of compounds quantified as total carbohydrates.[23]

A variety of fructan compounds of the *Agave* genus have been identified and isolated. Escobedo-Garcia et al.[19] determined the carbohydrate profiles of *agave* bagasse; thin-layer chromatography (TLC) analysis showed that agave bagasse samples have mainly FOS and simple sugars. They also performed the carbohydrate profile of *Agave* bagasse by High-Performance Anion-Exchange Chromatography (HPAEC), corroborating that glucose, fructose, and sucrose were the most abundant carbohydrates, in addition to indicating the occurrence of 1-N, 1-K, and 1-F (nystose, kestose, and fructo-furanosylnystose) in these AbP.

Avila-Fernandez et al.[24] carried out a process to produce functional oligosaccharides from the isolation of fructans of *A. tequilana* Weber var. Azul. Also, Michel-Cuello et al.[25] isolated fructans from *A. salmiana* by enzymatic hydrolysis. In another study, Montañez-Soto et al.[26] performed the extraction, characterization, and quantification of fructans in *A. tequilana* Weber var. Azul heads and leaves. The content of total and direct reducing sugars, inulin, FOS, glucose, and fructose was determined, finding high percentages of inulin (43.24%) and FOS (24.96%). *Agave,* especially the species *A. tequilana* Weber var. Azul, is recognized for its high content of fructans (inulin and FOS), making it a very important plant for the industry due to its extensive uses in foods and medical treatments with claims of health benefits.

16.3.2 FLAVONOIDS

Flavonoids are a type of secondary metabolites produced by plants; they stand out for the wide spectrum of bioactivities such as immunomodulatory, antitumor, antioxidant, among others. They are classified into anthocyanidins, catechins, flavanones, flavonols, flavones, chalcones, and isoflavones.

They are made up of three rings, two of the benzenoid type and a third of the heterocyclic pyrane ring.[27,28]

These compounds differ in the degree of hydroxylation or methoxylation of the condensed rings, and they naturally occur as aglycones and methylated and glycoside derivatives.[29] Different studies have extracted and identified flavonoids from *Agave*. The studies from Rizwan et al.[30] reported total flavonoid content extracted with different solvents from the leaves of *A. attenuate*. These values ranged from 304.8 to 43.35 catechin equivalents (CE) (mg/100 g of dry plant matter). The extraction power of solvents on total flavonoid content values was found in the following order: methanol (304.8%), ethyl acetate (197.2%), chloroform (79.8%), n-butanol (71.5%), and n-hexane (43.35%).

Morreeuw et al.[31] identified the flavonoid profile in the residual biomass of *A. lechuguilla*. Specific HPLC-UV-MS analyses confirmed the concentration of diverse polyphenols mainly flavonoids such as flavanones (hesperidin), anthocyanidins (delphinidin, cyanidin), flavonols (quercetin, isorhamnetin, and kaempferol), and flavones (apigenin). Almaraz-Abarca et al.[32] detected a great number of flavonoid compounds in the leaves of *Agave victoriae reginae*, *A. lechuguilla*, and *A. striata*, showing flavanols (26 compounds), flavones (6 compounds), and dihydro flavonoids (19 compounds). Same compounds were identified in *A. duranguensis* flowers.[33] Santos et al.,[17] isolated chemical components of *A. sisalanea* biomass, among which the flavonoid kaempferol stood out. Flavonoids are generally found in *Agave* leaves; this was demonstrated in the research by Ahumada-Santos et al.,[34] where a phytochemical screening showed flavonoids in the *A. impressa*, *A. ornithobroma*, *A. rzedowskiana*, *A. schidigera*, *A. tequilana*, and *A. angustifolia* species.

16.3.3 SAPONINS

Saponins are part of a large family of compounds containing a steroid aglycone (sapogenin) linking to glucan or more oligosaccharides. The chemical structure of sapogenin defines the classification of saponins as triterpenoid saponins. They can be present bioactivities such as anticancer and antihypercholesterolemic.[35]

Different studies have shown that *Agave* is an important source of steroidal spirostanol-type sapogenins, while agavegenin D is the only cholestane-type sapogenin reported in *Agave*. Spirostanols have been isolated from leaves, flowers, leaf juice, rhizomes, and callus cultures of *Agave* plants[36] Leal-Diaz et al.[37] detected in *A. salmiana* concentrated mead, saponins derived from kammogenin, mannogenin, gentrogenin, and hecogenin, which represented 74, 11, 8, and 7% of total saponins, respectively. In this same species, *A. salmiana,* Puente-Garza et al.[38] identified five different saponins in the samples: chlorogenin glycoside 1 (CG1), chlorogenin glycoside 2 (CG2), tigogenin glycoside (TG), hecogenin glycoside 1 (HG1), and hecogenin glycoside 2 (HG2).

Peréz[39] isolated five steroidal saponins (Magueyoside A-E) from *A. offoyana* flowers. Besides, 11 steroidal saponins were obtained from the leaves of *A. offoyana* after multiple separation procedures of the active fractions; six of these compounds were previously reported as agabrittonoside E, magueyosides A, and B, agabrittonosides D and A, and cantalasaponin-1. According to studies by Nasri et al.,[40] saponins are the main type of bioactive compounds present in *A. americana* extract, with a content of 80 g of diosgenin equivalent per kg of dry weight (d.w.). Leal-Diaz et al.[41] analyzed mead samples of the species *A. salmiana and A. americana* and identified the presence of saponins derived from kammogenin, mannogenin, gentrogenin, and hecogenin. The saponin content in the mead of immature *A. salmiana* was two times higher (478.3 equivalents of protodioscin (PE) μg/g mead (DM) compared to *A. americana* (179.0 PE μg/g mead d.w.).

16.3.4 TERPENES

Terpenes are the most abundant group of phytochemicals in nature, due to the great variety of substrates that terpene synthase can use to biosynthesize them. These are classified depending on the carbon atoms that they present in their molecular structures. For example, hemiterpenoids (C5), monoterpenes (C10), homoterpenes (C11,16), sesquiterpenes (C15), diterpenes (C20), sesterpenes (C25), triterpenes (C30), tetraterpenes (C40), and those containing more than 40 atoms are called polyterpenoids. Their main function in plants is defense, thanks to their volatile nature; they release metabolites capable of repelling certain predators. For human beings, they are of great relevance since the pharmaceutical (cosmetics and additives), food (additives and essences), and agricultural (biopesticides) industries are used to mention a few. In addition, medicinal activities have been demonstrated as antioxidants.[42] Villanueva-Rodríguez et al.[43] mentioned that *Agave* species used in mezcal production provide mostly terpene compounds, which can be spiced (terpinene and terpineol) or herbal terpenes. Ade-Ajayi et al.[44] also detected the presence of terpenoid compounds in *Agave sisalana* Perrine juice (waste) through phytochemical screening.

In other studies, Rios-Deras et al.[45] identified limonene and α-terpineol compounds in *A. duranguensis* wort. Álvarez-Ainza et al.[46] found the alpha-terpineol terpene compound in mezcal produced with *A. angustifolia*. López-Romero et al.[29] described one study where they analyzed different *Agave* extracts and reported various types of terpenes. They identified nine terpenes in *A. salmiana*, eight in *A. angustifolia,* and 32 in *A. tequilana*, observing that linalool is the main terpene in the three *Agave* plants. In an *A. angustifolia* extract, the terpenes identified were *p*-cymene, limonene, β-trans-ocimene, linalool, α-terpineol, nerol, geraniol, and trans-nerolidol. Terpenes found in *A. tequilana* Weber var. Azul included α-linalool, α-terpinene, p-cymene, limonene, β-trans-ocimene, β-cis-ocimene, sabinene, linalool, 2,4,6-octatriene, 4-terpineol, α-terpineol, nerol, bornyl formate, geraniol, α-cubebene, among others. *A. salmiana* contained α-linalool, α-terpinene, p-cymene, limonene, β-trans-ocimene, linalool, 4-terpineol, geraniol, and trans-nerolidol.

16.3.5 COUMARINS

Coumarin and its derivatives are of great importance in synthetic organic and natural product chemistry. Coumarins have been a very important platform as a structural element of several compounds with biological activities, such as anticancer, antiproliferative, antioxidant, antifungal, anticoagulant, anti-inflammatory, antiviral, photosensitizing, antibacterial, antineurodegenerative compounds.[47] The presence of coumarins has been detected in *A. sisalana*, *A. fourcroydes*, and *A. americana.*[48–51]

16.4 FUNCTIONAL ACTIVITIES OF ABP

The bioactive potential of AbP is mainly due to the phytochemical compounds that are available. These compounds have been identified in multiple studies as potential health benefit effects. Besides, various biological activities have been determined in AbP, such as antioxidant, antimicrobial, anticancer, anti-inflammatory, antiparasitic, and ulcer protective.

16.4.1 ANTIOXIDANT ACTIVITY (AOX)

The AOX activity of the different *Agave* species varies depending on the number of bioactive compounds they present and the method for determining the antioxidant activity, such as ORAC, ABTS, DPPH, and β-Carotene bleaching, among others.[34]

The bioactive compounds to which this antioxidant activity is attributed in agave extracts are phenolic compounds and terpenes, specifically tannins, phenolic acids, flavonoids, saponins. These compounds can be isolated and tested separately or tested in synergy. More positive results are shown concerning the antioxidant activity when there is a mixture of phenolic compounds and saponins than when only the potential of saponins or phenolic compounds is tested.[52] Other compounds, such as polysaccharides from *Agave sisalana,* presented a strong antioxidant activity. The determining factors in this activity can be the monosaccharide composition, molecular weight, branching.[53] When the different parts of the *Agave* are analyzed separately, the antioxidant activity may be different because, in some parts, the tissue is greener than in others. Such is the case of the *Agave golden*, where the green leaves parts have a higher antioxidant activity than the albino part; this can be justified because phytochemicals are more present in tissues that carry out more photosynthesis; therefore, there is a greater antioxidant activity.[54]

In the case of *Agave attenuata*, methanol extract had the highest antioxidant activity above chloroform, ethyl acetate, butanol, and hexane. This can be explained since methanol is a polar solvent and can dissolve phenolic compounds of polar nature, such as phenolic acids and flavonoids, responsible for great antioxidant activity.[30]

When subjected to stress conditions, the *Agave* tends to synthesize a greater amount of phytochemicals in response to this stimulus, which causes a greater antioxidant activity. *Agave salmiana*, when subjected to drought stress, more saponins are synthesized, which are correlated with antioxidant activity.[55] Usually, the largest amount of AbP are the leaves, but there are also flowers. The flowers have a different phytochemical composition than the leaves, being the phenolic compounds the ones with the highest antioxidant activity. Phytochemicals and antioxidant activity in agave flowers vary depending on the stage of development and the section where it is analyzed.[56] The temperature influences antioxidant activity; for example, in *Agave americana,* when the temperature and extraction time are increased, there is a greater release of phenolic compounds such as flavonoids responsible for this antioxidant activity.[57]

16.4.2 ANTIBACTERIAL ACTIVITY

Extracts with apolar solvents such as hexane or dichloromethane have an affinity to phytochemicals of the terpenoid type, such as saponins and phytosterols. *Agave cupreata* presents these compounds that can inhibit bacteria such as *S. aureus* and *P. aeruginosa.*[58]

Regarding the more polar solvent extracts such as methanol, the antibacterial activity may vary depending on the *Agave* species since they all have a different phytochemical composition. In the case of the Mexican species *Agave impress*, *Agave ornithobrom*, *Agave rzedowskian*, *Agave tequilana*, *Agave schidigera*, and *Agave angustifolia*, they show differences regarding the minimum inhibitory concentration (MIC) and what type of bacteria they can inhibit. It is observed that the most active species are *Agave tequilana* and *Agave schidigera*, managing to inhibit *Streptococcus* group A-4, *Salmonella enterica typhi*, *Shigella dysenteriae*, *Escherichia coli* 25922, and *Pseudomonas aeruginosa* 27853. Continuing with less active species such as *Agave impress*, *Agave ornithobrom*, and *Agave angustifolia*, they only managed to inhibit *Streptococcus* group A-4 and *Shigella dysenteriae*. The least active species was *Agave rzedowskian*, managing to inhibit only *Shigella dysenteriae.*[34]

When two polar solvents, such as water and methanol, are compared, a similarity is observed in the phytochemical composition. Although they can inhibit the same bacteria, the concentration used is lower in methanol since they present a different phytochemical composition. In the case of *Agave sisalana,* both extracts can inhibit *Staphylococcus aureus, Salmonella typhi, Escherichia coli,* and *Streptococcus pyogenes* bacteria.[59] It could be said that the antibacterial activity depends on multiple factors such as the species of *Agave*, type of solvent, and type of bacteria.

16.4.3 ANTIFUNGAL ACTIVITY

Extracts using ethanol and hexane in *Agave lechuguilla* inhibit *R. stolonifer*, *C. gloeosporioides,* and *P. digitatum*. The hexane extract presented a greater inhibition in *R. stolonifer* and *P. digitatum*; however, the ethanol extract was effective only in *C. gloeosporioides*. This means that *R. stolonifer* and *P. digitatum* are inhibited in the presence of compounds of a nonpolar nature such as terpenoids and that *C. gloeosporioides* is more vulnerable to compounds of a polar nature such as phenolic compounds.[60] Regarding *Agave attenuata*, extracts with hexane, chloroform, and butanol inhibit *A. niger*, *A. flavus*, *A. alternata*, and *R. solani*. Methanol does not inhibit *A. niger*, but it does inhibit the rest, while ethyl acetate only inhibits *A. niger* and not the rest.[30]

16.4.4 INSECTICIDAL ACTIVITY

Normally organic solvents such as ethanol, methanol, acetone, among other chlorinated ones, have been used for the extraction of AbP metabolites; this is due to the polarity of the metabolites of interest, and however, these may present traces in the final products, damaging the environment and health. Some investigations have evaluated aqueous extracts of juice from the leaves of *A. sisalana* against *Aedes aegypti*, observing an ovicidal and cytotoxic effect; aqueous extracts of *A. americana* leaves have shown an insecticidal effect against *Brassica olearace.*[61,62]

Methanol extracts of *Agave americana* show lethality in the insects *Sitophilus oryzae* adults.[63] The contribution of this insecticidal activity is mainly by saponins and phenolic compounds, where saponins increase the permeability of the membranes, which leads to phenolic compounds with insecticidal characteristics entering and exerting this function.[63] Hexane extracts of *Agave tequilana* are effective insecticides against *B. tobacco*, where if the concentration of the extract increases, the activity increases.[64]

16.4.5 ANTIPARASITIC ACTIVITY

The interaction of the cell membrane of the parasites with lipophilic compounds such as saponins and phytosterols causes the solubilization or rupture of the membrane, which implies a greater passage of phytochemicals that are toxic to the parasites, causing their death and decrease.[65] Hexane or ethyl acetate is suitable for a higher content of saponins and other phytochemicals with antiparasitic action. The effectiveness of ethyl acetate extracts of *Agave tequilana* shows a linear behavior with the concentration administered, where the higher the concentration, the higher the antiparasitic activity. Saponins destabilize membranes, and phenolic compounds such as flavonoids exert toxicity in parasites such as nematodes *P. redivivus.*[64] Aqueous extract of *Agave sisalana* shows greater antiparasitic activity when the exposure time is longer for the parasites *Haemonchus contortus*, *Oesophagostomum columbianum*, and *Trichostrongylus colubriformis*. The antiparasitic activity depends on the number, age, and type of nematodes.[65]

16.4.6 ANTICANCER ACTIVITY

The inhibition of cancer cells MCF-7 of the methanol extract of *Agave americana* shows a linear behavior with the applied dose. If the dose of extract increases, the inhibition of cancer cells increases.[66] Saponins are one of the main phytochemicals that are credited with anticancer activity. Saponins isolated from *Agave salmiana* showed an apoptosis effect in Colon cancer HT-29.[67] Fructans from the stems of *Agave angustifolia* have prebiotic potential, which improves the colon's health, causing a positive function on the intestinal wall, which prevents the formation of colon cancer.[68]

16.4.7 ANTI-INFLAMMATORY ACTIVITY

Agave cupreata extracts with acetone and dichloromethane show inhibition of mouse ear edema, whereas the extract with dichloromethane shows greater inhibition. In the case of the extract with acetone, saponins may be the cause of the anti-inflammatory activity. On the other hand, the extract with dichloromethane presents alkaloids, to which the said activity is attributed.[58]

The species *A. americana*, *A. tequilana*, and *A. angustifolia* show inhibition of ear edema with 81, 68, and 51%, respectively. This activity may be higher in *A. americana* due to the presence of steroidal triterpenes and saponins.[69] The same case applies to the species *A. angustifolia* and *A. pygmaea*, but this time in inhibiting pro-inflammatory cytokines (TNF-α and IL-6) and increasing anti-inflammatory cytokine IL-10. Both species showed a decrease in TNF-α, IL-6, and an increase in IL-10; therefore, both species have an anti-inflammatory effect. However, due to its variation in phytochemical compounds, *A. pygmaea* had a higher anti-inflammatory activity, where it inhibited TNF-α, IL-6, and increased more IL-1.[70]

Agave americana ethanol extract shows increasing edema inhibition when both the concentration and the exposure time are increased. The increased anti-inflammatory activity is because the extract contains flavonoids, which, being in higher concentrations and longer exposure time, will be more effective in inhibiting edema.[50]

16.4.8 ULCER PROTECTIVE ACTIVITY

The mechanism of ulcer protective activity occurs when lipophilic phytochemicals such as phytosterols, phospholipids, and sterol esters protect the gastrointestinal mucosa from acids, ethanol, nonsteroidal anti-inflammatory drugs (NSAIDs), and other substances that cause ulcers. These lipid compounds increase the viscosity of the mucosa, preventing acidification of the cells.[71] The presence of saponins and homoisoflavonoids promotes ulceroproter activity.

Methanol extract of *A. angustifolia* and *A. pygmaea* presents both homoisoflavonoids and saponins. These species show a high percentage of protection against the number and severity of ulcers, *A. angustifolia* being superior with 6% more protection in both number and severity of ulcers.[70]

16.5 MANAGEMENT OF INDUSTRIAL APPLICATIONS OF ABP

Byproducts and residues refer to applying adequate procedures that allow obtaining a benefit due to the management and transformation of certain residues. In the use of AbP, in addition to the environmental benefit that it represents, there are other benefits that are related to the industrial applications that can be given to AbP; such as its use as animal feed, fertilizer, and biomass for the production of enzymes and biofuels or as raw material to obtain high added-products, for potential applications in nutraceuticals, medical, and pharmaceutical industries.

16.5.1 PHARMACEUTICAL AND NUTRACEUTICAL POTENTIAL USES OF ABP

Carbohydrates, flavonoids, and saponins are among the important AbP compounds extracted from *A. sisalana* residues after sisal fiber extraction. A polysaccharide-enriched fraction obtained from the crude extract of the leaves of *A. sisalana* has been used in the development of a moisturizing cosmetic nanoemulsion.[72,73] Sapogenins (tigogenin and hecogenin) obtained from *A. sisalana* residues have been used as precursors to synthesize steroidal hormones.[74] Sapogenins have great potential in different pharmaceutical applications, which have made them an object of study, mainly for their anti-inflammatory, anticancer, and antimicrobial properties.[74,75] Flavonoids in *Agave* species have a significant efficiency to be applied in nutraceutical and pharmacology because agave extracts have demonstrated various biological activities[34] (see Section 16.4).

16.5.2 AGRICULTURE APPLICATION OF ABP

Another potential use of bagasse is as a substrate in agriculture. *Agave* bagasse (AB) composting has been studied using ammonium nitrate as a nitrogen source to adjust the AB C: N ratio of 25:1, the recommended value

TABLE 16.1 Main Biological Activities of AbP.

Agave specie and part	Extract	Compound	Biological activity	Method	Results	Ref.
Agave americana leaves	MeOH	Phenolic, flavonoids	AOX	DPPH	3.71–23.30 EC_{50}	[57]
	H_2O	Phenolic, flavonoids	AOX	DPPH, β-carotene-linoleic acid Bleaching test	8.993 mg/mL IC_{50} (DPPH), 1.229 mg/mL IC_{50} β–carotene	[82]
			Antibacterial	MIC	*S. typhymurium* ATCC 14028 (0.078 mg/mL), *S. aureus* ATCC 6538 (0.156 mg/mL), *P. aeruginosa* ATCC 49189 (0.156 mg/mL), *E. coli* ATCC 8739 (0.156 mg/mL), *L. monocytogenes* ATCC 19117 (0.312 mg/mL)	
	MeOH	Alkaloids	Anticancer	MTT assay cell, SRB assay cell	Breast cancer MCF-7 (MTT IC_{50}: 826.1 µg/mL, SRB IC_{50}: 545.9 µg/mL)	[66]
	Acetone	Saponins	Anti-inflammatory	Edema inhibition	81% edema inhibition	[69]
	H_2O and organic solvents	Saponins	Antibacterial	MBC	*E. coli* (MBC: 28–202 mg/mL), *S. aureus* (MBC: 20–100 mg/mL)	[83]
	MeOH	Saponins	Insecticidal	Lethal concentration (LC_{50})	*S. oryzae* adults (LC_{50}: 8.55–12.35 µg/cm^2)	[63]
	EtOH	Saponins, flavonoids	Anti-inflammatory	Edema inhibition	55.12–76.92% edema inhibition	[50]
	MeOH	Saponins	Antibacterial	MIC	*S. aureus* (MIC: 2.5 mg/mL), *E. coli* (MIC: 10 mg/mL), *P. aeruginosa* (MIC: 2.5 mg/mL), *S. typhi* (MIC: 2.5 mg/mL)	[84]

TABLE 16.1 *(Continued)*

Agave specie and part	Extract	Compound	Biological activity	Method	Results	Ref.
Agave salmiana leaves	MeOH	Phenolic, flavonoids	AOX	ORAC	146 μmol TE/g d.w.	[55]
	H_2O and organic solvent extracts of flowers	Phenolic compounds	AOX	DPPH	Extract: EtOH (40.9–74.78%), MeOH: (23.92–66.10%), H_2O (29.44–78.55%), Acetone: (15.05–50.97%), Hex: (10.22–49.27%)	[56]
			Antibacterial, antifungal	MIC	Extract: EtOH (*E. coli*: 13.47–17.85 mg/mL, *S. sonnei*: 6.83 mg/mL), H_2O (*E. coli*: 6.98–27.78 mg/mL, *S. sonnei*: 13.22–20.56 mg/mL, *S. typhimurium*: 18.33–18.89 mg/mL), Acetone (*E. coli*: 17.41–30.26 mg/mL), Hex (*E. coli*: 15.08 mg/mL, *S. sonnei*: 13.69–21.42 mg/mL, *S. typhimurium*: 9.39–9.70 mg/mL)	[56]
	MeOH	Saponins	AOX	ORAC	976 μmol TE/g d.w.	[55]
	Acetone	Saponins	Anticancer	Colon cancer HT-29	IC_{50}: 82.7 μg/mL	[20]
Agave attenuata leaves	MeOH chloroform, EtOAc, n-but, n-hex	Phenolics, flavonoids	AOX	DPPH and Inhibition in a linoleic acid system	61.41–73.97% inhibition DPPH and 50.12–70.35% inhibition of linoleic acid peroxidation	[30]
			Antibacterial, antifungal	MIC	*B. subtilis* (158–250 mg/mL), *P. multocida* (27.4–250 mg/mL), *S. aureus* (89.3–252 mg/mL), *E. coli*	

TABLE 16.1 *(Continued)*

Agave specie and part	Extract	Compound	Biological activity	Method	Results	Ref.
					(15.2–152 mg/mL), *A. niger* (115–244 mg/mL), *A. flavus* (18.4–110 mg/mL), *A. alternata* (69.4–189 mg/mL), *R. solani* (20.4–140 mg/mL).	
Agave golden leaves	EtOH/ Acetone 1:1	Glutathione/ ascorbate	AOX	FRAP	4.3 mmol/g	[54]
Agave sisalana leaves	EtOH H_2O	Hydrophilic polysaccharide	AOX	DPPH	3.2 mg/mL EC_{50}	[53]
			Immunological	In vitro lymphocyte proliferation (A570)	0.292–0.502 absorbance lymphocyte T cell	
	MeOH, Butanol	Phenolic, saponins	Antibacterial, antifungal	MIC	*P. aeruginosa* ATCC 9027, *E. coli* ATCC 8739, *S. aureus* ATCC 6538, *B. subtilis* ATCC 6633, *S. choleraesuis* ATCC 10708, *C. albicans* ATCC 10231, *A. niger* ATCC 16404 (MIC: >12.5 mg/mL).	[52]
			AOX	DPPH	IC_{50}: 1.452 µg/mL	
	Acid extraction of waste juice	Saponins	Anti-inflammatory	Edema inhibition	88.7% inhibition of abdominal writhing, 67.3% inhibition of hind paw edema, 48% inhibition of ear edema	[85]
	MeOH, H_2O	Alkaloids, flavonoids, tannins, saponins	Antibacterial	MIC, MBC	*S. dysenteriae* (MBC: 20 mg/mL, MIC: 10 mg/mL), *B. atrophaeus* (MBC: 40 mg/mL, MIC: 20 mg/mL), *E. faecalis* (MBC: 40 mg/mL, MIC: 20 mg/mL), *P. aeruginosa* (MBC: 20 mg/mL, MIC: 10 mg/mL)	[24]

TABLE 16.1 *(Continued)*

Agave specie and part	Extract	Compound	Biological activity	Method	Results	Ref.
	MeOH, H_2O	Tannins, saponins, flavonoids	Antibacterial	MIC, MBC	*S. aureus* (MBC: 20 mg/mL, MIC: 10 mg/mL), *S. typhi* (MBC: 40 mg/mL, MIC: 20 mg/mL), *E. coli* (MBC: 20 mg/mL, MIC: 10 mg/mL), *S. pyogenes* (MBC: 20 mg/mL, MIC: 10 mg/mL)	[59]
	Waste	Saponins	Antiparasitic	Coproculture	*Haemonchus* spp. (82.6% de reduction *Oesophagostomum* spp (79.6%) reduction), *Trichostrongylus* spp. (53% reduction)	[65]
Agave pygmaea leaves	MeOH	Saponins	Anti- inflammatory	Edema inhibition	70.2% edema inhibition	[70]
			Immuno-modulatory	Inhibition of pro-inflammatory cytokines	41% inhibition of TNF-α and 43% inhibition of IL-6	
			Ulcer protective	Protection against number and severity of ulcers	81% protection from number of ulcers and 85% protection from ulcer severity	
Agave angustifolia leaves	MeOH	Saponins, homoiso-flavonoids	Anti- inflammatory	Edema inhibition	55% edema inhibition	[70]
			Immuno-modulatory	Inhibition of pro-inflammatory cytokines	59% inhibition of TNF-α and 57% inhibition of IL-6	
			Ulcer protective	Protection against number and severity of ulcers	84% protection from number of ulcers and 91% protection from ulcer severity	

TABLE 16.1 *(Continued)*

***Agave* specie and part**	**Extract**	**Compound**	**Biological activity**	**Method**	**Results**	**Ref.**
	MeOH	Triterpenes, flavonoids, tannins	Antibacterial	MIC	*Streptococcus group A-4* (MIC: 15 mg/mL), *S. dysenteriae* (MIC: 15 mg/mL)	[34]
			AOX	DPPH	6.46 μM TE/g d.w.	
	Acetone	Saponins	Anti- inflammatory	Edema inhibition	51% edema inhibition	[69]
	H_2O extract of stems	Fructans	Anticancer	TER assay	Colon cancer (CaCo-2) 12.7% Monolayer resistance and 19.3% paracellular permeability	[68]
Agave impressa leaves	MeOH	Triterpenes, flavonoids, tannins	Antibacterial	MIC	*Streptococcus group A-4* (MIC: 15 mg/mL), *S. dysenteriae* (MIC: 15 mg/mL)	[34]
			AOX	DPPH	8 μM TE/g d.w.	
Agave ornithobroma leaves	MeOH	Triterpenes, flavonoids, tannins	Antibacterial	MIC	*Streptococcus group A-4* (MIC: 15 mg/mL), *S. dysenteriae* (MIC: 15 mg/mL)	[34]
			AOX	DPPH	19.93 μM TE/g d.w.	
Agave rzedowskiana leaves	MeOH	Triterpenes, flavonoids, tannins	Antibacterial	MIC	*S. dysenteriae* (MIC: 15 mg/mL)	[34]
			AOX	DPPH	27.41 μM TE/g d.w.	
Agave tequilana leaves	MeOH	Triterpenes, flavonoids, tannins	Antibacterial	MIC, MBC	*Streptococcus group A-4* (MIC: 5 mg/mL), *S. dysenteriae* (MIC: 5 mg/mL), *S. enterica typhi* (MBC: 10 mg/mL, MIC: 5 mg/mL), *E. coli* 25922 (MIC:	[34]

TABLE 16.1 *(Continued)*

Agave specie and part	Extract	Compound	Biological activity	Method	Results	Ref.
					10 mg/mL), *P. aeruginosa* (MBC: 15 mg/mL, MIC: 5 mg/mL)	
			AOX	DPPH	9.86 µM TE/g d.w.	
	Acetone	Saponins	Anti-inflammatory	Edema inhibition	68% edema inhibition	[69]
	Hx and ethyl acetate	Alkaloids, flavonoids, saponins	Insecticidal	Mortality (%)	69.6–94% mortality in *B. tabaci*	[64]
			Antiparasitic	Mortality (%)	62–100% mortality in *P. redivivus*	
Agave schidigera leaves	MeOH	Triterpenes, flavonoids, tannins	Antibacterial	MIC, MBC	*Streptococcus group A-4* (MIC: 5 mg/mL), *S. dysenteriae* (MIC: 5 mg/mL), *S. enterica typhi* (MBC: 10 mg/mL, MIC: 10 mg/mL), *P. aeruginosa* (MBC: 15 mg/mL, MIC: 10 mg/mL)	[34]
			AOX	DPPH	15.79 µM TE/g d.w.	
Agave lechuguilla leaves	EtOH /Hx	Phenolic, essential oils, sapogenins	Antifungal	Inhibition of sporulation (%) and mycelial inhibition (%)	Inhibition of sporulation: *R. stolonifer* (EtOH: 77.1%, Hx: 100%), *C. gloesporoides* (EtOH: 100%, Hx: 99.7%), *P. digitatum* (EtOH: 52.6%, Hx: 90.5%) Mycelial inhibition: *R. stolonifer* (EtOH: 36%, Hx: 100%), *C. gloesporoides* (EtOH: 100%, Hx: 88%), *P. digitatum* (EtOH: 82%, Hx: 82%)	[60]

TABLE 16.1 *(Continued)*

***Agave* specie and part**	**Extract**	**Compound**	**Biological activity**	**Method**	**Results**	**Ref.**
Agave cupreata	DCM	Saponins, alkaloids	Antibacterial	MIC	*S. aureus* (MIC: 16 mg/mL), *P. aeruginosa* (16 mg/mL)	[58]
			Anti-inflammatory	Edema inhibition (%)	Dichloromethane (31.25–64.29%), Acetone (34.19–48.82%)	

AOX, antioxidant; ATCC, American Type Culture Collection; DCM, Dichloromethane; DPPH= 1, 1-diphenyl-2-picrylhydrazyl; d.w., dry weight; FRAP, ferric antioxidant power; MBC, minimum bactericidal concentration; MIC, minimum inhibitory concentration; ORAC, oxygen radical absorbance capacity.

for excellent biodegradability. Composting is a destructive process of organic matter to convert it into CO_2 and water. However, in AB composting, the complete mineralization of the bagasse was avoided to obtain a substitute for commercial substrates. No statistically significant differences were found in comparing commercial substrates with AB compost in tomato production under greenhouse conditions. These results show the potential use of AbP as a substrate for agriculture cultivations.[76]

16.5.3 ANIMAL FEED APPLICATION OF ABP

The AB is rich in cellulose that is the primary source of carbon and energy in ruminants, for this could be considered a good food source for ruminants. Alkaline pretreatments solubilize lignin and hemicelluloses, thus increasing the digestibility of lignocellulosic residues.[77] Digestibility value (36%) of bagasse (*Agave tequilana*) without any treatment is located between the values obtained for sunflower husks (26%) and wheat straw (39%).

Improved digestibility of dry AB can be achieved by the treatment with $Ca(OH)_2$ at a concentration of 10% at 50°C with 80% moisture and a contact time of 24 h. The digestibility values of AB increase from 36% to 54% after the treatment with $Ca(OH)_2$. After the alkaline treatment, AB digestibility reached values equivalent to maize byproducts (45% in stalks, leaves, and cobs) and lower values to alfalfa (65%) or soybean meal (84%). In addition to the increase in the digestibility of the AB, the alkaline treatment with $Ca(OH)_2$ represents a source of calcium, which is beneficial in animal feed.[78]

16.5.4 PAPER-DERIVED PRODUCTS OF ABP

Agave leaves present high concentrations of cellulose that allow obtaining the pulp with a higher yield and present a lower content of lignin, which is a desirable characteristic since that would mean a decrease in the use of chemicals during the pulping process. The paper produced from *Agave tequilana* leaves through a chemical pulping process presented a tensile strength comparable to the paper made with pine and eucalyptus fibers.[77]

16.5.5 BIOFUELS PRODUCTION

Second-generation biofuels do not need additional land cultivation because they use agricultural or forestry residues as biomass sources. *Agave* leaves are a potential source of monosaccharides. However, the presence of lignin prevents the hydrolysis of cellulose and hemicellulose.[79]

Bioconversion processes generally have four main components: (1) a pretreatment process in which biomass is partially delignified; (2) an enzymatic process in which enzymes (i.e., cellulases, hemicellulases, pectinases, and polysaccharide oxygenases) hydrolyze polysaccharides to sugars; (3) a fermentation process in which sugars are converted into fuel, usually ethanol; and (4) a separation process, usually distillation, in which fuel is separated from water and residual solids.[80]

Pretreatment is necessary to maximize the extraction of sugars and improve the biodigestibility of the lignocellulosic material for bioethanol production. In the pretreatment process, the biomass is ground and treated with hot water or steam that may also contain a dilute acid or base.[80,81]

Agave atrovirens bagasse obtained from the pulque industry was hydrolyzed with dilute base and steam treatment and subsequently subjected to enzymatic hydrolysis presenting a yield of 56% in reducing sugars. This yield was similar to the yield obtained by converting other lignocellulosic materials (sugarcane bagasse) into sugars. Although the yield of reducing sugars was similar between sugarcane bagasse and *Agave atroviens* bagasse, there were significant differences in alcohol yield by fermentation, which was higher in cane bagasse (32.6%); both lignocellulosic materials continue to be important sources of reducing sugars for ethanol production.[81]

16.5.6 PRODUCTION OF ENZYMES

Enzymes have applications in many fields, including chemistry, fermentation, agricultural, pharmaceutical, and food production. AbP, particularly *Agave tequilana* leaves, have been exploited as an inexpensive carbon and energy source for ligninolytic and lignocellulolytic enzyme production such as ligninases (lignin peroxidase, manganase peroxidase, versatile peroxidase, phenol oxidoreductases, laccases, cellulases, xylanases, pectinases, and inulinases).[77] Different pretreatments have been studied for the depolymerization and deslignification of lignocellulosic materials used as substrates in enzyme production. The use of ultrasound on the *A. durangensis* leaves increased the production of cellulases and inulinases, while the use of heat pretreatment on the same material increased the output of β-fructofuranosides.[73]

16.6 CONCLUSIONS AND FUTURE PERSPECTIVES

The *Agave* is a resource of great importance for the Mexican economy and is extensively used for Tequila production. Tequila, the main alcoholic beverage made in Mexico, provides a great economic return to the country, occupying first place in exports. However, the production of tequila and *Agave* contains a socioeconomic, cultural, environmental, and multinational interests context since the most important companies are foreign. Future production projections are ambitious, with mainly high export volumes and subsequently high accumulation of AbP. In this regard, a set of innovative green solvents, such as last generation deep eutectic solvents for the extraction of nutraceuticals with high added value, could conduct to a circular industrial economy by exploiting the AbP from the tequila industry to develop functional compounds that could be added to distillates in the future as functional additives. It must also be designed and planned from the beginning of byproducts production to mitigate the environmental impact, regulate the decontamination processes of AbP, and generate support and training on the implementation of bioenergy production plants, nutraceuticals, food additives, food supplements, functional feed for livestock and pets. AbP, as reflected throughout the chapter, contains a great variety of phytochemicals that contribute to health; however, among those that stand out for their novelty (i.e., an opportunity for research, industrial, and medicinal uses), are saponins, whose presence is limited to certain families and species of plants; one of them abundant in the composition is Agave. All these are in harmony with sustainable production, taking into account the principles of the circular economy.

KEYWORDS

- **agave biomass**
- **antioxidants**
- **biofunctionality**
- **phytochemicals**
- **waste recovery**

REFERENCES

1. Álvarez-Ríos, G. D.; Pacheco-Torres, F.; Figueredo-Urbina, C. G.; Casas, A. Management, Morphological and Genetic Diversity of Domesticated Agaves in Michoacán, México. *J. Ethnobiol. Ethnomed.* **2020,** *16* (1), 3.
2. Bader, A. B. D. Exploiting the Potential of Agave for Bioenergy in Marginal Lands. Doctor of Philosophy Doctoral, Newcastle University, Newcastle United Kingdom, 2015.
3. SIAP. Panorama Agroalimentario 2020 [Online] 2020. https://nube.siap.gob.mx/gobmx_publicaciones_siap/pag/2020/Atlas-Agroalimentario-2020
4. CRT. *Informaciónestadística* [Online] 2021. https://www.crt.org.mx/EstadisticasCRTweb/
5. IIEG. *Exportaciones de tequila: un estudio de caso para la obtención de pondedadores de comercio exterior* [Online] 2020. https://iieg.gob.mx/ns/wp-content/uploads/2020/05/Estudio-de-exportaciones-de-tequila-mayo-2020.pdf
6. SAGARPA. *Agave tequilero y mezcalero mexicano* [Online] 2017. https://www.gob.mx/cms/uploads/attachment/file/257066/Potencial-Agave_Tequilero_y_Mezcalero.pdf
7. Leonardo, G. C. Uso de bagazo de la industria mezcalera como materia prima para generar energía. *Ingenierías* **2010,** *8* (47), 8–16.
8. CONACYT. *Industria del tequila y generación de residuos* [Online] 2019. https://www.cyd.conacyt.gob.mx/?p=articulo&id=287
9. Ponce-Ortega, J. M.; Santibañez-Aguilar, J. E. Chapter 5; Optimization of the Supply Chain Associated to the Production of Bioethanol From Residues of Agave From the Tequila Process in Mexico. In *Strategic Planning for the Sustainable Production of Biofuels*; Ponce-Ortega, J. M., Santibañez-Aguilar, J. E., Eds.; Elsevier, 2019; pp 115–146.
10. Corbin, K. R.; Byrt, C. S.; Bauer, S.; DeBolt, S.; Chambers, D.; Holtum, J. A. M.; Karem, G.; Henderson, M.; Lahnstein, J.; Beahan, C. T.; Bacic, A.; Fincher, G. B.; Betts, N. S.; Burton, R. A. Prospecting for Energy-Rich Renewable Raw Materials: Agave Leaf Case Study. *PLoS One* **2015,** *10* (8), e0135382.
11. Rafael, C. G. M.; Raymundo, G. E. D.; Ramón, R. M.; Ariel, R. C. J.; Noé, D. P. Caracterización química y física del bagazo de agave tequilero compostado con biosólidos de vinaza como componente de sustratos para cultivos en contenedor. *Rev. Int. de Contam. Ambient.* **2018,** *34* (3), 373–282.
12. CRT. *Estrategia de Sustentabilidad de la Cadena Agave-Tequila.* [Online] 2016. https://www.crt.org.mx/images/Documentos/Estrategia_Sustentabilidad.pdf
13. Moreno, A. D.; Susmozas, A.; Oliva, J. M.; Negro, M. J. Chapter 1; Overview of Bio-Based Industries. In *Biobased Products and Industries*; Galanakis, C. M., Ed.; Elsevier, 2020; pp 1–40.
14. Ramos, S.; Etxebarria, S.; Cidad, M.; Gutierrez, M.; San Martín, D.; Iñarra, B.; Olabarrieta, I.; Melado-Herreros, A.; Zufia J. 1 - Cleaner Production Strategies for the Food Industry. In *The Interaction of Food Industry and Environment*; Galanakis, C., Ed.; Academic Press, 2020; pp 1–34.
15. Tetreault, D.; McCulligh, C.; Lucio, C. Distilling Agro-extractivism: Agave and Tequila Production in Mexico. *J. Agrar. Change.* **2021,** *21* (2), 219–241.
16. Sanchez, A.; Sanchez, S.; Dueñas, P.; Hernandez-Sanchez, P.; Guadalajara, Y. The Role of Sustainability Analysis in the Revalorization of Tequila Residues and Wastes Using Biorefineries. *Waste Biomass Valorization* **2020,** *11* (2), 701–713.

17. Santos, J. D. G.; Vieira, I. J. C.; Braz-Filho, R.; Branco, A. Chemicals from *Agave sisalana* Biomass: Isolation and Identification. *Int. J. Mol. Sci.* **2015,** *16* (4), 8761–8771.
18. Gibson, G. R.; Hutkins, R.; Sanders, M. E.; Prescott, S. L.; Reimer, R. A.; Salmien, S. J.; Scott, K.; Stanton, C.; Swanson, K. S.; Cani, P. D.; Verbeke, K.; Reid, G. Expert Consensus Document: The International Scientific Association for Probiotics and Prebiotics (ISAPP) Consensus Statement on the Definition and Scope of Prebiotics. *Nat. Rev. Gastroenterol. Hepatol.* **2017,** *14* (8) 491–502.
19. Escobedo-García, S.; Salas-Tovar, J. A.; Flores-Gallegos, A. C.; Contreras-Esquivel, J. C.; González-Montemayor, A. M.; López, M. G.; Rodríguez-Herrera, R. Functionality of *Agave* Bagasse as Supplement for the Development of Prebiotics-Enriched Foods. *Plant Foods Hum. Nutr.* **2020,** *75* (1), 96–102.
20. Santos-Zea, L.; Leal-Diaz, A. M.; Cortes-Ceballos, E.; Gutierrez-Uribe, J. A.; Agave (*Agave spp.*) and its Traditional Products as a Source of Bioactive Compounds. *Curr. Bioact. Compd.* **2012,** *8* (3), 218–231.
21. Nava-Cruz, N. Y.; Medina-Morales, M. A.; Martinez, J. L.; Rodriguez, R.; Aguilar, C. N. *Agave* Biotechnology: An Overview. *Crit. Rev. Biotechnol.* **2015,** *35* (4), 546–559.
22. Moreno-Vilet, L.; García-Hernández, M. H.; Delgado-Portales, R. E.; Corral-Fernandez, N. E.; Cortez-Espinoza, N.; Ruiz-Cabrera, M. A.; Portales-Pérez, D. P. *In vitro* Assessment of Agave Fructans (*Agave salmiana*) as Prebiotics and Immune System Activators. *Int. J. Biol. Macromol.* **2014,** *63*, 181–187.
23. Jasso-Padilla, I.; Juarez-Flores, B.; Alvarez-Fuentes, G.; de la Cruz-Martínez, A.; González-Ramírez, J.; Moscosa-Santillan, M.; González-Chávez, M.; Oros-Ovalle, C.; Prell, F.; Czermak, P.; Martínez-Gutiérrez, F. Effect of Prebiotics of *Agave salmiana* Fed to Healthy Wistar Rats. *J. Sci. Food Agric.* **2017,** *97* (2), 556–563.
24. Ávila-Fernández, A.; Galicia-Lagunas, N.; Rodríguez-Alegría, M. E.; Olvera, C.; López-Munguía, A. Production of Functional Oligosaccharides Through Limited Acid Hydrolysis of Agave Fructans. *Food Chem.* **2011,** *129* (2), 380–386.
25. Michel-Cuello, C.; Ortiz-Cerda, I.; Moreno-Vilet, L.; Grajales-Lagunes, A.; Moscosa-Santillan, M.; Bonnin, J.; González-Chávez, M. M.; Ruiz-Cabrera, M. Study of Enzymatic Hydrolysis of Fructans from *Agave salmiana* Characterization and Kinetic Assessment. *Sci. World J.* **2012.**
26. Montañez-Soto, J.; Venegas-González, J.; Vivar-Vera, M.; Ramos-Ramírez, E. Extracción, caracterización y cuantificación de los fructanos contenidos en la cabeza y en las hojas del *Agave tequilana* Weber azul. *Bioagro* **2011,** *23* (3), 199–206.
27. Yang, B.; Liu, H.; Yang, J.; Gupta, V. K.; Jiang, Y. New Insights on Bioactivities and Biosynthesis of Flavonoid Glycosides. *Trends Food Sci. Technol.* **2018,** *79*, 116–124.
28. Imran, M.; et al. Luteolin, a Flavonoid, as an Anticancer Agent: A Review. *Biomed. Pharmacother.* **2019,** *112*, 108612.
29. López-Romero, J. C. Ayala-Zavala, J. F.; González-Aguilar, G. A.; Peña-Ramos, E. A.; González-Ríos, H. Biological Activities of *Agave* byproducts and Their Possible Applications in Food and Pharmaceuticals. *J. Sci. Food Agric.* **2018,** *98* (7), 2461–2474.
30. Rizwan, K.; Zubair, M.; Rasool, N.; Riaz, M.; Zia-Ul-Haq, M.; De Feo, V. Phytochemical and Biological Studies of *Agave attenuata*. *Int. J. Mol. Sci.* **2012,** *13* (5), 6440–6451.
31. Morreeuw, Z. P.; Escobedo-Fregoso, C.; Ríos-González, L. J.; Castillo-Quiroz, D.; Reyes, A. G. Transcriptome-Based Metabolic Profiling of Flavonoids in *Agave lechuguilla* Waste Biomass. *Plant Sci.* **2021,** *305*, 110748.

32. Almaraz-Abarca, N.; del Socorro M.; González-Elizondo, M.; da Graça Campos, Z. E. Ávila-Sevilla.; Delgado-Alvarado, E. A.; Ávila-Reyes J. A. Variability of the Foliar Phenol Profiles of the *Agave victoriae-reginae* Complex (Agavaceae). *Bot. Sci.* **2013,** *91* (3), 295–306.
33. Barriada-Bernal, L. G.; Almaraz-Abarca, N.; Delgado-Alvarado, E. A.; Gallardo-Velazquez, T.; Avila-Reyes, J. A.; Torres-Morán, M. I.; González-Elizondo, M. S.; Herrerra-Arrieta, Y. Flavonoid Composition and Antioxidant Capacity of the Edible Flowers of *Agave durangensis* (Agavaceae). *CYTA J. Food* **2014,** *12* (2), 105–114.
34. Ahumada-Santos, Y. P.; Montes-Avila, J.; Uribe-Beltrán, M. J.; Diaz-Camacho, S. P.; López-Angulo, G.; Vega-Aviña, R.; López-Valenzuela, J. A.; Heredia, J. B.; Delgado-vargas, F. Chemical Characterization, Antioxidant and Antibacterial Activities of Six *Agave* Species from Sinaloa, Mexico. *Ind Crops Prod.* **2013,** *49*,143–149.
35. Faizal, A.; Geelen, D. Saponins and Their Role in Biological Processes in Plants. *Phytochem Rev.* **2013,** *12* (4), 877–893.
36. Sidana, J.; Singh, B.; Sharma, O. P. Saponins of *Agave*: Chemistry and Bioactivity. *Phytochemistry* **2016,** *130*, 22–46.
37. Leal-Díaz, A. M.; Noriega, L. G.; Torres-Villalvazo, I.; Torres, N.; Aleman-Escondrillas, G.; López-Romero, P.; Sánchez-Tapia, M.; Aguilar-López, M.; Furuzawa-Carballeda, J.; Velázquez-Villegas, L. A.; Avila-Nava, A.; Ordaz, G.; Gutiérrez-Uribe, J. A.; Serna-Saldivar, S. O.; Tovar, A. R. Aguamiel Concentrate from *Agave salmiana* and its Extracted Saponins Attenuated Obesity and Hepatic Steatosis and Increased Akkermansia Muciniphila in C57BL6 Mice. *Sci. Rep.* **2016,** *6* (1), 1–15.
38. Puente-Garza, C. A.; Espinosa-Leal, C. A.; García-Lara, S. Steroidal Saponin and Flavonol Content and Antioxidant Activity During Sporophyte Development of Maguey (*Agave salmiana*). *Plant Foods Hum Nutr.* **2018,** *73* (4), 287–294.
39. Pérez, A. J.; Calle, J. M.; Simonet, A. M.; Guerra, J. O.; Stochmal, A.; Macías, F. A. Bioactive Steroidal Saponins from *Agave offoyana* Flowers. *Phytochemistry* **2013,** *95*, 298–307.
40. Nasri, S.; Salem, H. B. Effect of Oral Administration of *Agave americana* or *Quillaja saponaria* Extracts on Digestion and Growth of Barbarine Female Lamb. *Livest. Sci.* **2012,** *147* (1–3), 59–65.
41. Leal-Díaz, A. M.; Santos-Zea, L.; Martínez-Escobedo, H. C.; Guajardo-Flores, D.; Gutiérrez-Uribe, J. A.; Serna-Saldivar, S. O. N. Effect of *Agave americana* and *Agave salmiana* Ripeness on Saponin Content from Aguamiel (agave sap). *J. Agric. Food Chem.* **2015,** *63* (15), 3924–3930.
42. Boncan, D. A. T.; Tsang, S. S. K.; Li, C.; Lee, I. H. T.; Lam, H. M.; Chan, T. F.; Hui, J. H. L. Terpenes and Terpenoids in Plants: Interactions with Environment and Insects. *Int. J. Mol. Sci.* **2020,** *21* (19).
43. Villanueva-Rodríguez, S.; Escalona-Buendia, H. Tequila and Mezcal: Sensory Attributes and Sensory Evaluation. In *Alcoholic Beverages*; Elsevier, 2012; pp 359–378.
44. Ade-Ajayi, A. F.; Hammuel, C.; Ezeayanaso, C.; Ogabiela, E. E.; Udiba, U. U.; Angym, B.; Olabanji, O. Preliminary Phytochemical and Antimicrobial Screening of *Agave sisalana* Perrine Juice (waste). *J. Environ. Chem. Ecotoxicol.* **2011,** *3* (7), 180–183.
45. Los Rios-Deras, D.; Rutiaga-Quiñones, O.; López-Miranda, J.; Páez-Lerma, J.; López, M.; Soto-Cruz, N. Improving *Agave duranguensis* Must for Enhanced Fermentation: C/N Ratio Effects on Mezcal Composition and Sensory Properties. *Rev. Mex. Ing. Quim.* **2015,** *14* (2), 363–371.

46. Alvarez Ainza, L. M.; García Galaz, A.; González Ríos, H.; Prado Jaramillo, N.; Acedo-Félix, E. Sensory Analysis and Minor Volatil Compounds of Distilled from *Agave angustifolia* Haw (Bacanora). *Biotecnia.* **2015,** *17* (3), 22–29.
47. Cao, D.; Liu, Z.; Verwilst, P.; Koo, S.; Jangjili, P.; Kim, J. S.; Lin, W. Coumarin-Based Small-Molecule Fluorescent Chemosensors. *Chem. Rev.* **2019,** *119* (18), 10403–10519.
48. de Andrade, T. G. C. S.; da Silva, Y. E. A.; Espinoza, D. G.; de Lima, L. M.; de O Cezar, A. K.; de Andrade, I. G. C. S.; de Oliva Neto, P.; Palacios, J. L.; dos Santos, L. Preliminary Phytochemical Analysis and the Effect of *Agave sisalana* on Body Weight and Defensive Behaviours in Ovariectomized Rats. *J. Med. Plant Res.* **2017,** *11* (34), 538–548.
49. Iser, M.; Valdivié, M.; Figueredo, L.; Nuñez, E.; Más, D.; Martínez, Y. Secondary Metabolites, Quality Indicators and Organoleptic Characteristics of Stems Meal from *Agave fourcroydes* (Henequen). *Cuba. J. Agric Sci.* **2020,** *54* (1).
50. Misra, A. K.; Varma, S. K.; Kumar, R. Anti-inflammatory Effect of an Extract of *Agave americana* on Experimental Animals. *Pharmacogn Res.* **2018,** *10* (1), 104.
51. Muthangya, M., Chigodi, M. O.; Samoei, D. K. Phytochemical Screening of *Agave sisalana* Perrine Leaves (Waste), South Estern Kenya University, 2013.
52. Ribeiro, B. D.; Alviano, D. S.; Barreto, D. W.; Zarur Coelho, M. A. Functional Properties of Saponins from Sisal (*Agave sisalana*) and Jua (*Ziziphus joazeiro*): Critical Micellar Concentration, Antioxidant and Antimicrobial Activities. *Colloids Surf. A Physicochem Eng. Asp.* **2013,** *436*, 736–743.
53. Zhang, X. Liu, L.; Lin, C. Structural Features, Antioxidant and Immunological Activity of a New Polysaccharide (SP1) from Sisal Residue. *Int J. Biol. Macromol.* **2013,** *59*, 184–191.
54. Deng, B. Antioxidative Response of Golden Agave Leaves with Different Degrees of Variegation Under High Light Exposure. *Acta Physiol. Plant.* **2012,** *34* (5), 1925–1933.
55. Puente-Garza, C. A.; Meza-Miranda, C.; Ochoa-Martínez, D.; García-Lara, S. Effect of *in Vitro* Drought Stress on Phenolic Acids, Flavonols, Saponins, and Antioxidant Activity in *Agave salmiana. Plant Physiol. Biochem.* **2017,** *115*, 400–407.
56. Medina-Galvan, M. I.; Bernardino-Nicanor, A.; Castro-Rosas, J.; Negrete-Rodriguez, M. D. X.; Conde-Barajas, E.; Gonzalez-Cruz, L. Antimicrobial and Antioxidant Activity of Flower Scape Extracts of *Agave salmiana*: Effect of the Extraction Solvent and Development Stage. *Res. J. Biotechnol.* **2018,** *13* (12), 1–9.
57. Ben Hamissa, A. M. Seffen, M. Aliakbarian, B. Casazza, A. A. Perego, P. Converti, A. Phenolics Extraction from *Agave americana* (L.) Leaves Using High-Temperature, High-Pressure Reactor. *Food Bioprod. Process.* **2012,** *90* (1), 17–21.
58. Salazar-Pineda, D. T.; Castro-Alarcon, N.; Moreno-Godinez, M. E.; Nicasio-Torres, M. D.; Perez-Hernandez, J.; Alvarez-Fitz, P. Antibacterial and Anti-inflammatory Activity of Extracts and Fractions from *Agave cupreata. Int. J. Pharmacol.* **2017,** *13* (8),1063–1070.
59. Hammuel, C.; Yebpella, G. G.; Shallangwa, G. A.; Magomya, A. M.; Agbajp, A. S. Phytochemical and Antimicrobial Screening of Methanol and Aqueous Extracts of *Agave sisalana. Acta Pol. Pharm.* **2011,** *68* (4), 535–539.
60. de Rodríguez, D. J.; Rodriguez Garcia, R.; Hernandez Castillo, F. D.; Aguilar Gonzalez, C. N.; Saenz Galindo, A.; Villareal Quintanilla J. A. Moreno *In Vitro* Antifungal Activity of Extracts of Mexican Chihuahuan Desert Plants Against Postharvest Fruit Fungi. *Ind. Crops Prod.* **2011,** *34* (1), 960–966.
61. Bermúdez-Bazán, M.; Castillo-Herrera, G. A.; Urias-Silvas, J. E.; Escobedo-Reyes, A.; Estarrón-Espinosa, M. Hunting Bioactive Molecules from the Agave Genus: An Update on Extraction and Biological Potential. *Molecules* **2021,** *26* (22).

62. Pereira, A. J.; Cardoso, I. M.; Araújo, H. D.; Santana, F. C.; Carneiro, A. P. S.; Coelho, S. P.; Pereira, F. J. Control of *Brevicoryne brassicae* (Hemiptera: Aphididae) with Extracts of Agave americana var. Marginata Trel. in Brassica Oleracea Crops. *Ann. Appl. Biol.* **2019,** *174* (1), 14–19.
63. Maazoun, A. M.; Hamdi, S. H.; Belhadj, F.; Ben Jemaa, J. M.; Messaoud, C.; Marzouki, M. N. Phytochemical Profile and Insecticidal Activity of *Agave americana* Leaf Extract Towards *Sitophilus oryzae* (L.) (Coleoptera: Curculionidae). *Environ. Sci. Pollut. Res.* **2019,** *26* (19), 19468–19480.
64. Herbert-Doctor, L. A.; Saavedra-Aguilar, M.; Villarreal, M. L.; Cardoso-Taketa, A.; Vite-Vallejo, O. Insecticidal and Nematicidal Effects of *Agave tequilana* Juice Against *Bemisia tabaci* and *Panagrellus redivivus*. *Southwest Entomol.* **2016,** *41* (1), 27–40.
65. Botura, M. B.; Silva, G. D.; Lima, H. G.; Oliveira, J. V. A.; Souza, T. S.; Santos, J. D. G.; Branco, A.; Moreira, E. L. T.; Almeida, M. A. O.; Batatinha, M. J. M. *In Vivo* Anthelmintic Activity of an Aqueous Extract from Sisal Waste (*Agave sisalana* Perr.) Against Gastrointestinal Nematodes in Goats. *Vet. Parasitol.* **2011,** *177* (1), 104–110.
66. Anajwala, C. C.; Patel, R. M.; Dakhara, S. L.; Jariwala, J. K. *In Vitro* Cytotoxicity Study of *Agave americana*, *Strychnos nuxvomica* and *Areca catechu* Extracts Using MCF-7 Cell Line. *J. Adv. Pharm. Technol.* **2010,** *1* (2), 245–252.
67. Santos-Zea, L.; Fajardo-Ramirez, O. R.; Romo-Lopez, I.; Gutierrez-Uribe, J. A.; Fast Centrifugal Partition Chromatography Fractionation of Concentrated *Agave (Agave salmiana)* Sap to Obtain Saponins with Apoptotic Effect on Colon Cancer Cells. *Plant Foods Hum. Nutr.* **2016,** *71* (1), 57–63.
68. Allsopp, P.; Possemiers, S.; Campbell, D.; Oyarzábal, I. S.; Gill, C.; Rowland, I. An Exploratory Study into the Putative Prebiotic Activity of Fructans Isolated from *Agave angustifolia* and the Associated Anticancer Activity. *Anaerobe* **2013,** *22*, 38–44.
69. Monterrosas-Brisson, N.; Arenas Ocampo, M. L.; Jiménez-Ferrer, E.; Jiménez-Aparicio, A. R.; Zamilpa, A.; González-Cortazar, M.; Tortoriello, J.; Herrera-Ruiz, M. Anti-Inflammatory Activity of Different Agave Plants and the Compound Cantalasaponin-1. *Molecules* **2013,** *18* (7), 8136–8146.
70. El-Hawary, S. S.; El-Kammar, H. A.; Farag, M. A.; Saleh, D. O.; El Dine, R. S. Metabolomic Profiling of Five *Agave* Leaf Taxa via UHPLC/PDA/ESI-MS Inrelation to Their Anti-Inflammatory, Immunomodulatory and Ulceroprotective Activities. *Steroids* 2020, *160*, 108648.
71. Tovey, F. Chapter 8; Duodenal Ulcer and Diet in Sub-Saharan Africa. In *Digestive Diseases in Sub-Saharan Africa*; Segal, I., Ed.; Academic Press, 2019; pp 53–65.
72. Gomes, B. S. M. A.; Saatman M, M.; Moreira, B. A.; Becerra, S. A. H. R.; Leite-Silva, V. R.; Darrocha-Filho, P. A.; Fernandez, N. M. M.; de Oliveira, R. H. A.; Arantes, O. E.; Santos, L. P.; de Farias, S. V. S.; Brandt, G. R.; Ferrari, M. Evaluation of *In Vitro* and *In Vivo* Safety of the byproduct of *Agave sisalana* as a New Cosmetic Raw Material: Development and Clinical Evaluation of a Nanoemulsion to Improve Skin Moisturizing. *Ind. Crops Prod.* **2017,** *108*, 470–479.
73. Contreras-Hernández, M. G.; Ochoa-Martínez, L. A:, Rutiaga-Quiñones, J. G.; Rocha-Guzmán, N. E.; Lara-Ceniceros, T. E.; Contreras-Esquivel, J. C.; Prado, B. L. A.; Rutiaga-Quiñones, O. M. Effect of Ultrasound Pre-Treatment on the Physicochemical Composition of *Agave durangensis* Leaves and Potential Enzyme Production. *Bioresour. Technol.* **2018,** *249*, 439–446.
74. Cushman, J. C.; Davis, S. C.; Yang, X.; Borland, A. M. Development and use of Bioenergy Feedstocks for Semi-Arid and Arid Lands. *J. Exp. Bot.* **2015,** *66* (14), 4177–4193.

75. Santos, J.; Branco, A. GC-MS Characterisation of Sapogenins from Sisal Waste and a Method to Isolate Pure Hecogenin. *BioResources* **2014,** *9* (1), 1325–1333.
76. Iñiguez, G.; Martínez, G.; Flores, P.; Virgen, G. Utilización de subproductos de la industria tequilera. Parte 9. Monitoreo de la evolución del compostaje de dos fuentes distintas de bagazo de *Agave* para la obtención de un substrato para jitomate. *Rev. Int. de Cont. Ambien.* **2011,** *27* (1), 47–59.
77. Alemán-Nava, G. S. Gatti, I. A.; Parra-Saldivar, R.; Dallemand, J. F.; Rittmann, B. E. Iqbal, H. M. N. Biotechnological Revalorization of Tequila Waste and byproduct Streams for Cleaner Production–A Review from Bio-Refinery Perspective. *J. Clean. Prod.* **2018,** *172*, 3713–3720.
78. Ramirez-Cortina, C.; Maria, A. G.; Luc, R. Valorización de residuos agroindustriales del tequila para alimentación de rumiantes. *Rev. Chapingo, Ser. Cienc. For. y del Ambient.* **2012,** 449–557.
79. Avila-Gaxiola, J. Treatments to Improve Obtention of Reducing Sugars from *Agave* Leaves Powder. *Ind. Crops Prod.* **2018,** *112*, 577–583.
80. Sorek, N.; Yeats, T. H.; Szemenyei, H. Youngs, H.; Somerville, C. R. The Implications of Lignocellulosic Biomass Chemical Composition for the Production of Advanced Biofuels. *BioScience* **2014,** *64* (3),192–201.
81. Hernández-Salas, J. M.; Villa-Ramírez, M. S.; Veloz-Rendón, J. S.; Rivera-Hernández, K. N.; González-César, R. A.; Plascencia-Espinosa, M. A.; Trejo-Estrada, S. R. Comparative Hydrolysis and Fermentation of Sugarcane and Agave Bagasse. *Bioresour. Technol.* **2009,** *100* (3), 1238–1245.
82. Bouaziz, M. A.; Bchir, B.; Chalbi, H.; Sebii, H.; Karra, S.; Smaoui, S.; Attia, H.; Besbes, S. Techno-Functional Characterization and Biological Potential of *Agave americana* Leaves: Impact on Yoghurt Qualities. *J. Food Meas. Charact.* **2021,** *15* (1), 309–326.
83. Babu, M. P.; Dang, R. Das, K. Phytochemical Investigations and Characterization of Antimicrobial Activity of Bioguided Fractionated Leaves of *Agave americana* L. *AP.* **2015,** *4* (1), 61–67.
84. Shegute, T.; Wasihun, Y. Antibacterial Activity and Phytochemical Components of Leaf Extracts of *Agave americana. J. Exp. Pharmacol.* **2020,** *12*, 447–454.
85. Dunder, R. J.; Quaglio, A. E. V.; Maciel, R. P.; Luiz-Ferreira, A.; Almeida, A. C. A.; Takayama, C.; de Faria, F. M.; Souza-Brito, A. R. M. Anti-inflammatory and Analgesic Potential of Hydrolyzed Extract of *Agave sisalana* Perrine ex Engelm., Asparagaceae. *Rev Bras Farmacog.* **2010,** *20* (3), 376–381.

CHAPTER 17

Food Byproducts as Potential Functional Ingredients for Animal Feeding

OCTAVIO VALDEZ-BARO[1], L. ALFONSO JIMÉNEZ-ORTEGA[1], L. ÁNGEL CABANILLAS-BOJÓRQUEZ[1], ERICK P. GUTIÉRREZ-GRIJALVA[2], and J. BASILIO HEREDIA[1]

[1]*Centro de Investigación en Alimentación y Desarrollo, A.C., Carretera a Eldorado Km. 5.5, Col. Campo El Diez, Culiacán, Sinaloa, Mexico*

[2]*Catedras CONACYT-Centro de Investigación en Alimentación y Desarrollo, A.C., Carretera a Eldorado Km. 5.5, Col. Campo El Diez, Culiacán, Sinaloa, Mexico*

ABSTRACT

Food production generates a large amount of waste from primary production and industrial transformation until it reaches the consumers. About 30% of agricultural and food production results in waste or byproducts that can be reprocessed for other industrial interest applications for example the livestock industry, which is one of the most significant worldwide due to its high growing demand and the new challenges to ensure food security. Production systems are diversified, seeking economic and sustainable food options for their feeding. The byproducts of the agricultural industry, mainly fruits and vegetables, stand out because they could be used as foods because their provide a diversity of functional compounds such as flavonoids, flavonols, anthocyanins, tannins, triterpenoids, carotenoids, and fiber. These compounds can exert different biological activities like antimicrobial, antioxidant, prebiotic, and anti-inflammatory, which could improve the animal intestinal microbiota and immunology. Moreover, this search for functional

Food Byproducts Management and their Utilization. Ricardo Gómez-García, Ana A. Vilas-Boas, Débora A. Campos, Maria Manuela Pintado, & Cristóbal Noé Aguilar (Eds.)

ingredients provides better management of byproducts and wastes from the food industry toward a circular economy and zero residues. This chapter aims to show the byproducts of some agricultural interest crops, their main bioactive compounds, and their beneficial effects on animal health, including the related regulations, the global market for livestock products, and the impact on the environment and food safety.

17.1 INTRODUCTION

The great demand for food in the world is constantly increasing because it seeks to ensure that the world's entire population can access complete and safe food. These objectives are forecast toward 2050, so to achieve them, food production will be increased, generating a greater amount of byproducts FAO.[1] Regarding food production, the crops with the highest demand, such as corn, rice, wheat, soybeans, sugarcane, oil crops, beans, banana, citrus, tomato, and onion, are those that generate a large amount of byproducts.[2,3] In addition to these crops, processing industries like fruit and vegetables, grain, brewery, and winery contribute significant amounts of byproducts. These byproducts can be seeds, stems, leaves, peel, bran, pomace, trimmings, pulp waste, husks, spent grain, and wastewater, among others.[4] The large quantity of these byproducts presents a great pollution problem since they are not totally used or treated. Instead, byproducts are managed in a nonsustainable manner, either burying them, throwing them into water bodies or burning them, which lead to the emission of greenhouse gases due to their decomposition.[5] Various investigations have shown that food byproducts have great potential for health benefits due to their bioactive compounds, such as phenolic compounds, terpenes, alkaloids, oligosaccharides, and peptides.[6] Based on the presence of these compounds, various studies have shown that incorporating these byproducts in the diets of animals has beneficial potential in their health, showing effects such as antioxidant activity, antimicrobial activity, anti-inflammatory activity, improvement of biochemical parameters, and improved absorption of essential nutrients, among others. In addition, the growth performance is not changed.[7,8]

Once the beneficial effect on animal health has been demonstrated, it is necessary to find the appropriate way of processing guarantees the quality and safety of fortified foods with byproducts. In addition, nutritional, antinutritional, and nutraceutical factors present in the diet must be considered to prepare the ideal formulation for animal feed. These processes must comply with the regulations of the competent authorities since it must be based on scientific studies that byproducts do not cause adverse effects on the health of the animals that consume them.[9]

17.2 GLOBAL MARKET FOR ANIMAL FEED

The global animal feed market size was USD 37.83 billion in 2019, and this number is expected to grow 3.8% from 2020 to 2027. This market faces nutritional demands as the consumption of animal-based products continues to grow. The animal feed market categorizes its needs based on type, species, and form.[10]

- Type: Probiotics, Enzyme, Antioxidants, Antibiotics, Amino acids, Vitamins, Minerals
- Species: Poultry, Ruminants, Fish, Swine
- Form: Pellets, Crumbles, Mash

Moreover, the United Nations Food and Agriculture Organization (FAO)[11] estimates that the demand for food will grow around 60% by 2050, and also, animal protein demand will grow by 1.7% per year (The International Feed Industry Federation).[12]

The global needs for animal protein are not equally distributed. For instance, around 26 developing countries consume less than 10 kg per capita, while others range between 10 and 20 kg *per capita*. On the other hand, developed countries consume nearly 90 kg of meat *per capita*.[13] Based on these global scenarios, food byproducts could respond to the needs by formulating novel food ingredients such as vitamins, antioxidants, amino acids, and growth promoter alternatives. The latter is pivotal in public health given the rise in antibiotic resistance for the excessive use of antibiotics as growth promoters.[14–17]

17.3 ANIMAL FEED REGULATION

The regulation for animal feed is still under development. Institutions such as the European Commission[18] and the Food and Drug Administration (FDA)[19] have sections describing the characteristics that the feed used for animals must contain but are not entirely clear. Likewise, despite the advances in research and technology for the potential use of food byproducts in animal feed and the growing environmental problem due to the accumulation of wastes, the possible inclusion of food byproducts in the formulation of animal feed (as long as it has been shown in vitro and in vivo studies that the intake of these does not cause any damage) has led to starting regulatory programs for animal nutrition and protection such as the byproducts sterilization for animal feed and the prohibition of mammalian proteins for ruminants feed.[9]

17.4 QUALITY OF ANIMAL FEED

It has been established that the quality of feed nutrition depends on factors like feed presentation, hygiene, antinutritional factors, digestibility, and palatability. Also, feed additives can influence animal growth and metabolism. Additionally, in 2015, the International Agency for Research on Cancer classified meat as a carcinogen (group 1) and red meat as a probable carcinogen (group 2A). Some studies showed that meat consumption might be related to colorectal cancer.[13] Thus, new strategies are being developed to minimize the risks and improve meat's nutritional and health properties. On this subject, studies have associated meat with health issues mainly due to additives during processing, like added salt, heme iron, and the oxidation reactions of these molecules. The oxidation produces potentially harmful molecules like the carbonyl-based cytotoxic and genotoxic 4-hydroxynonenal and malondialdehyde.[13] Additionally, there are some mutagenic amines formed during cooking like 2-amino-1-methyl-6-phenylimidazo[4,5-b]pyridine and 2-amino-3,8-dimethylimidazo[4,5-f]quinoxaline (8-MelQx).[20,21]

For the above, recent studies have evaluated strategies to improve the meat quality. One of the most promising alternatives is the integration of antioxidant molecules from byproducts, in order to reduce the environmental damage of agriculture waste and incorporate antioxidant molecules as feed additives of natural origin, which is in contrast to synthetic antioxidants that are usually under careful use for some potential genotoxic effects.

In this sense, using plant-based antioxidant molecules as feed additives is necessary to assess their safe use in livestock, poultry, and humans. Thus, one of the main aspects to evaluate is the present inconsistency of the effect of these molecules on the livestock. Therefore, their mode of action is not fully known, and one important step before using an antioxidant of plant origin is to standardize some factors of the process. These factors are related to their content in plants, which might vary as these antioxidant molecules are the plant's secondary metabolites, produced mainly as defense metabolites against biotic and abiotic factors. Hence, their content in plants will depend on environmental conditions, physiological variations, geographic variation, genetic factors, storage conditions, the amount of plant material, and the extraction techniques of the antioxidants.[22,23] To overcome the factor of the variations of antioxidant content in plants or plant byproducts, metabolomic tools to isolate and identify these constituents are a key aspect to incorporate before their inclusion as a part of the animal feeding market.

17.5 AGRICULTURAL BYPRODUCTS

Agricultural products play a fundamental role in human nutrition due to their multiple health benefits, easy consumption, and accessibility. However, with the arrival of SARS-Cov-2, the supply chain was seriously affected due to the mobility restrictions implemented worldwide, and modifications were made to the import and export policies of grains and seeds to preserve food safety.[2] The FAO decreed 2021 as the international year of fruits and vegetables, emphasizing their nutritional qualities and the relevance of their inclusion in the diet. However, it highlighted the aspect of creating awareness among consumers about the value chain and the importance of reducing the byproducts and waste that they give off throughout the production chain (from farm to table), as well as the importance of maintaining standards of quality that guarantee the safety and with it the accessibility of the products to the final consumers.[3]

17.5.1 PRODUCTION AND LOSSES

In 2018, 868 million tons of fruits and 1080 million tons of vegetables were produced, predominantly banana, citrus, watermelon, melon, apple, pear, grape, mango, guava, tomato, onion, garlic, leek, pumpkin, *brassicas* sp, eggplant, beans, legumes, carrot, among other crops. The major producing countries belong to East and South Asia, being as well the regions with the highest percentage of postharvest losses (>20%), followed by North America and Europe (>15%), being roots, tubers, oil crops, fruits, and vegetables, the products with the highest percentage of losses.[2,3]

17.5.2 SOURCES OF AGRO BYPRODUCTS

Agricultural byproducts can derive from two concepts. One is food loss, which refers to the decrease in the quantity and/or quality of available food due to the decision-making and actions taken by the suppliers of the supply chains. These losses are usually made in postharvest, processing, and distribution stages. The other concept is food waste, which refers to the decrease in the quantity and/or quality of food due to the decisions of retailers, food chains, and end consumers. In the main producing countries, losses and waste occur in storage, followed by processing and packaging.[3] Some other causes for which losses or waste occurs in the primary sector are the lack of modern machinery, untrained personnel, and low prices at harvest. In terms of the end customer, waste occurs by the discrimination of the products by their appearance, texture, consistency, or excess in their purchase, which ends up in the trash. Therefore, the correct management and use of these byproducts are of vital importance since it contributes to the economy of the producers, improves food and nutritional security, and creates a sustainable production chain with the environment.[24]

17.5.3 MANAGEMENT AND ENVIRONMENTAL FACTORS OF AGRO BYPRODUCTS

Usually, the agricultural byproducts destined for animal feed do not comply with the quality standards for human consumption, mainly due to sensory issues or because they may be waste from the agro-industrial transformation. However, they still could be useful as a source of nutrients. For instance, the wheat husk is derived by obtaining flours or derivatives from fruits and vegetables such as seeds, pulps, discarded from the manufacture of juices, porridges, and even processed products such as bakery and confectionery which do not meet quality criteria. In all cases, they must preserve their safety.[25] The European Union (EU)[26] recommends mainly for primary production, channeling its byproducts and losses toward animal feed, encouraging public authorities to provide the resources to efficiently carry out this transfer, and justifying the action and effect due to their multiple benefits on animal health. It also makes the same recommendation for food manufacturing, because it is estimated that 19% of the total waste generated in the EU derives from agro-industrial transformation.

Currently, consumers demand food produced under ecological systems, circular economies, and innovative eco-technologies, promoting zero waste, in addition to a natural, clean, and sustainable production. Therefore, using agricultural byproducts to feed livestock is an option with advantages and benefits for farmers. Approximately 30% of fruit and vegetable production is turned into waste, which can potentially be used by other industries as raw materials, subjecting them to transformation processes for their valorization through functional ingredients development.[27,28] These kinds of wastes are a source of phytochemicals that exert biological activities, which could act beneficially on animal health and nutrition, increasing their quality, and at the same time reducing pollution. Some of the food industries that emit the greatest amount of waste and byproducts are oil companies (70%), sugar production (85%), juices, and wines (50%), where peels, pulps, fibers, and seeds predominate. From these, antioxidant, antibacterial, and anti-inflammatory compounds have been identified, such as polyphenols, flavonoids, carotenoids, vitamins, prebiotics, and phytosterols, to name a few.[28–30]

An example is the use of banana inflorescence that contains a great diversity of antioxidant and functional compounds such as phenolic acids, flavonoids, triterpenes, phytosterols, and arabinogalactans, which can be used for different industrial sectors like nutraceuticals, animal and human feed additives, and dietary supplements, due to the large worldwide production (>125 million tons per year).[31] Agricultural byproducts and derivatives of livestock production have been used in the production of bio-energies, giving positive results from a technical, environmental, and economic point of view, in addition to the fact that new sources of material susceptible to anaerobic degradation are being analyzed, to take advantage of different agricultural residues, not just maize silage.[15]

Pectin and cellulose from fruit and vegetable byproducts can be used to produce edible coatings for other agricultural products to extend their shelf life, preserving their sensory and microbiological qualities. Positive results have been observed, in reducing both emissions to the environment and the physicochemical and nutritional characteristics of the treated products.[32]

The environmental footprint derived from food losses and waste is high, where the highest share is from fruits and vegetables, followed by tubers and roots. Emissions to the environment are blue-water consumption and carbon footprint.[24] Eliminating food losses and waste could significantly reduce (8%) greenhouse gas emissions, in addition to saving a quarter of the water currently used for agriculture. Mathematical models have been proposed that link food waste with economic, environmental, social, and marketing factors, suggesting that the problem should be approached from different points of view, involving all the links in the production and marketing chain.[5]

Greenhouse gases are produced throughout the agricultural chain, as they derive from the transport of seeds, sowing, maintenance, harvesting, transportation of the final product, distribution to sales, and consumption chains. This generated carbon footprint is commonly expressed in the equivalent amount of carbon dioxide (CO_2). According to FAO,[24] global CO_2 emissions from food losses and waste are approximately 3.3 gigatons, which is equivalent to 7% of total greenhouse gas emissions. The production of seeds and grains contributes more than 60% of CO_2 emissions, while roots, tubers, and oilseeds contribute 10%; fruits and vegetables only 5%. The region with the largest participation in this production is Central and South Asia, followed by North America and Europe. In the case of corn, its consumption generates the largest carbon footprint with more than 3.50 tons of CO_2 per ton of corn, while its field production only generates just over 0.50 tons.[24]

17.6 BIOACTIVE AND FUNCTIONAL COMPOUNDS FROM FRUIT AND VEGETABLE BYPRODUCTS

Food byproducts are a rich source of bioactive compounds such as carotenoids, vitamins, peptides, and polyphenols related to beneficial effects on human health. For instance, carotenoids such as β-carotene and β-cryptoxanthin play an important role in the photoprotection of the human eye by their ability to transform into a provitamin A. Also, ascorbic acid, tocopherol, thiamine, niacin, and riboflavin improve the physiological performance as they act as direct or indirect catalysts in various metabolic processes. Likewise, peptides and polyphenols are compounds with antihypertensive, antioxidant, hypocholesterolemic, anticarcinogenic, and immunomodulatory potential.[33–35]

Likewise, these bioactive compounds from food byproducts have been studied as potential ingredients in different animal feed formulations.[33] In this sense, carotenoids and phenolic compounds have been used as food additive to improve animal health and nutrition.[33–35] It has been reported that astaxanthin, β-carotene, lycopene, and lutein increase the skin and tissue pigmentation of fish; also, they improve the metabolic and physiological functions such as positive influence on growth and reproduction, as well as the sensory properties in food products, and enhance the health of animal by the antioxidant activity.[34] Moreover, overripe fruits such as avocado (skin and seed), cantaloupe melons (peels and seeds), mango (paste), pomegranate (peel), orange (waste), and pomelo (peel) are sources of bioactive compounds.[34] Also, vegetables and other food byproducts have bioactive compounds such as carotenoids and polyphenols; therefore fruits, vegetables, and other food byproducts have been studied for food additive.[34] Table 17.1 summarizes the bioactive compounds found in food byproducts that can be used as food for animals.

TABLE 17.1 Bioactive Compounds from Food Waste with Potential for the Development of Animal Feed Ingredients.

Food waste	Quantities of food waste used for animal feed	Phytochemicals found in food waste by spectrometric analysis	Reference
Agave lechuguilla byproduct	0–0.6% of by product addition	Diosgenin, smilagenin, hecogenin, manogenin, tigogenin hexose, yucagenin, chlorogenin, diosgenin diglucoside, and quercetin	[51]
Avocado skin and seed	–	Tocopherols, unsaturated fatty acids, flavonoids, and phenolic acids	[42]
Broccoli stems and leaves	0, 30, 60, and 90 g/kg of dried broccoli stems and leaves meal	Carotenoids, vitamin E, vitamin C	[47]
Cantaloupe melons peels and seeds	20 mg of cantaloupe melons pells and seeds	Gallic, chlorogenic, caffeic, syringic, ferulic and ellagic acids, rutin, quercetin, kaempferol, isorhamnetin, and tannins.	[38]
Carrot leaves	0.1 g of dried carrot leaf	Lutein, zeaxanthin, β-cryptoxanthin, β-carotene, and α-carotene	[46]
Carrot pomace	30–40% of carrot pomace	Carotenoids	[45]
Carrot rejects and wastes	0.5 g of carrot rejects and wastes	α-caroteno y β-caroteno	[44]
Gingko biloba leaves	10 g of *Gingko biloba* leaves	Quercetin, kaempferol, and isorhamnetin	[52]
Grape pomace	15 g of Grape pomace	Linolenic, linoleic, and oleic acid, as well as resveratrol, gallic, vanillic, and ferulic acids	[41]
Mango paste	100 g of mango paste	Phenolic acids, flavonoids, carotenoids, and tocopherols	[39]
Mesquite leaves, bark, pods, and roots	20–60% of mesquite leaves, bark, pods, and roots	Phenolic compounds, alkaloids, proteins, sugar, and carotenoids	[50]
Moringa leaves, flowers, roots, and seeds	5–20% of moringa leaves, flowers, roots, and seeds	Kaempferol, quercetin, rhamnose, isothiocyanate, glucosinolates, isoquercitrin, rhamnetin, ferulic acid, gallic acid, epicatechin, catechin, vanillin, protocatechuic acid, caffeic acid, cinnamic acid	[54]
Pomegranate peel	5 g of pomegranate peels	Gallic acid and ellagic acids, as well as punicalagin α and β	[43]
Pomelo peel	–	Carotenoids, flavonoids, phenolic compounds, coumarins	[36]

TABLE 17.1 *(Continued)*

Food waste	Quantities of food waste used for animal feed	Phytochemicals found in food waste by spectrometric analysis	Reference
Potato peels	–	Caffeic acid, p-coumaric acid, ferulic acid, gallic acid, protocatechuic acid, gentisic acid, p-hydroxybenzoic acid, chlorogenic acid, and vanillic acid	[49]
Onion peel	0.1 g of onion peel	Oxalic acid, linolenic acid, malonic acid, palmitic, acid, and oleic acid	[40]
Orange waste	240 g of orange waste	Naringenin, hesperitin, chrysoeriol, sinensetin, nobiletin, isosakuranetin	[37]
Overripe fruit and vegetables	–	β-carotene, zeaxanthin, lutein, gallic acid, quercetin, and chlorogenic acid.	[35]
Tomato skin and seeds	10 g of tomato skin and seeds	Lycopene	[48]

17.6.1 FRUIT BYPRODUCTS

Fruit byproducts have been used as animal feed in several studies due to their phytochemical composition. In this sense, the pomelo peel was reported that it could be used for animal feed production because it is an important source of phytochemicals such as carotenoids, flavonoids, phenolic compounds, and coumarins, which are important in animal health.[36] Likewise, the phytochemical composition of orange waste was studied, finding that orange waste is an important source of flavonoids.[37]

Vella et al.[38] studied the bioactive compounds of cantaloupe melons byproducts, and they found that seeds and peel are a source of polyphenolics, tannins, and flavonoids. Similarly, Sumaya-Martinez et al.[39] reported that mango paste (a byproduct of industrial processing of mango) has phenolic acids, carotenoids, tocopherols, and flavonoids.

On the other hand, watermelon rind, honeydew peel, banana peel, apple pomace, strawberry pomace, plum pomace, grape pomace, and pineapple skin have shown to be important sources of phenolic and flavonoid compounds.[40] Also, grape pomaces have high phytochemical compounds such as phenolic compounds and polyunsaturated fatty acids.[41] Likewise, Achilonu et al.[42] reported that avocado byproducts (skin and seeds) have bioactive compounds such as unsaturated fatty acids, tocopherols, and polyphenolic compounds; it has been proved that the avocado waste addition in pig feed does not affect its health negatively. Recently, Wu et al.[43] reported that pomegranate peel has a rich source of phenolic compounds. Therefore, fruit byproducts have high phytochemical compounds related to higher antioxidant capacity and could be used as ingredients in animal feed.

17.6.2 VEGETABLE BYPRODUCTS

The phytochemical composition of vegetable byproducts such as carrot pomace and pulp, onion peel, radish peel, garlic peel, broccoli stems, and tomato waste has been studied due to their use in animal feed. In this sense, carrot rejects, and wastes have been reported as rich sources of carotenoids such as α-carotene and β-carotene.[44] Likewise, Tiwari et al.[45] reported that carrot pomace is an important source of carotenoids. Similarly, carrot leaves were found to have carotenoids that are important in poultry intake to improve the nutritional composition of egg yolks.[46] Also, Hu et al.[47] reported that broccoli stems and leaves are a source of carotenoids and vitamins, finding that the addition of broccoli byproducts on roman brown-shell laying hens feed increased the xanthophyll levels of egg yolks, improved egg quality, and reduced cholesterol levels.

On the other hand, Strati and Oreopoulou[48] studied the bioactive composition of tomato byproducts, showing that tomato skin and seeds are sources of carotenoids; their composition could be varied by the extraction method. Besides, Torres and Domínguez[49] showed that peels of potato, onion, radish, and garlic are an important source of phenolic and flavonoid compounds.[40] Therefore, vegetable byproducts could be used as animal feed.

17.6.3 OTHER FOOD BYPRODUCTS

Food byproducts that are not considered fruits or vegetables have been studied for their potential use in animal feed due to their phytochemical composition. In this sense, Ruiz-Nieto et al.[50] summarize that different parts of the mesquite tree have been used for animal feed due to its phytochemical composition. They reported that secondary metabolites such as phenolic compounds, alkaloids, proteins, sugar, and carotenoids are the principal compounds found in the different parts of the mesquite tree. Pena-Rodriguez et al.[51] demonstrated that *Agave lechuguilla* byproducts (namely guishe) incorporated into animal feed enhanced the growth and improved the health and nutritional status of shrimp, which could be due to the saponins and flavonols composition of guishe. Likewise, Wang et al.[52] reported that *Gingko biloba* leaves have bioactive compounds such as flavonoids, proteins, and sugar; they found that solid-state fermentation may be a promising strategy to enhance the bioactive compounds of these byproducts, which could be used as an ingredient to animal feed.

On the other hand, corn husks are an important byproduct that has been used as animal feed due to the high volumes generated by the corn industry. Zuorro et al.[53] reported that corn husks are a source of flavonoids such as anthocyanins which could be enhanced by enzyme-assisted extraction. Similarly, moringa is considered an underestimated tree with high phytochemical content related to its health potential. In this sense, Abd El-Hack et al.[54] reported that all moringa tree parts (leaves, flowers, roots, and seeds) have phytochemical compounds such as alkaloids, tannins, carotenoids, and polyphenolic compounds, which have been used as food in animals and showed that it improves the health of birds and poultry.

17.7 BENEFITS OF FOOD BYPRODUCTS ON ANIMAL HEALTH

The consumption of food byproducts by animals such as poultry, cattle, and aquatic animals can have certain health benefits due to the high content of bioactive compounds such as phenolic compounds, terpenes, alkaloids, peptides, and oligosaccharides. These compounds are responsible for providing benefits to animal health, such as improving biochemical parameters, antioxidant activity, anti-inflammatory activity, antimicrobial activity, microbiota, and nutrient absorption. Table 17.2 summarizes the beneficial effects on the health of animals by the ingestion of diets with food byproducts.

17.7.1 FRUIT BYPRODUCTS

An animal diet rich in fruit byproducts can improve important aspects regarding the health conditions of livestock. The fruit byproducts come from the processing of fruits like apple, grape, pomegranate, mango, and tamarind, where these fruits are some of the most produced and processed.

17.7.1.1 APPLE BYPRODUCTS

Apple byproducts, such as apple pulp waste, have broad potential in reducing blood biochemical parameters such as uric acid and cholesterol in laying hens. Although the concentration administered must be taken care of, since exceeding 10% in the diet results in adverse effects on the health of laying hens and the production of eggs due to the high fiber content, which interrupts the absorption of nutrients essential.[55]

Regarding apple pomace, it has an anti-inflammatory and proliferative effect in piglets, in which the beneficial effect depends on the organ of the animal. In positive factors such as anti-inflammatory and proliferative, the expression of interleukin 10 (IL-10) had an increase in kidneys and liver; in the expression of cyclin D1, there was an increase in the liver. In the case of the factors with negative effects, there was a decrease in the stomach in the expression of nuclear factor kappa-light-chain-enhancer of activated B cells (NF-kB) and tumor necrosis factor alpha (TNF-α). This combination of increasing IL-10 and cyclin D1 values and decreasing values of TNF-α and NF-kB shows beneficial effects such as anti-inflammatory and proliferative, which indicates that piglet feeding supplemented with apple pomace has a positive effect on health.[56]

TABLE 17.2 Benefits to the Health of Animals Through the Consumption of Food Byproducts.

Food byproducts	Animal	Best dose	Health benefit	Reference
Fruit byproducts				
Apple pulp waste	Laying hens	10%(w/w) in the diet	Reduction of uric acid and cholesterol	[55]
Apple and red wine pomace	Piglets	3.5% (w/w) in the diet	• Anti-inflammatory activity Proliferative activity	[56]
Avocado pulp waste	Pigs	20% (w/w) in the diet	• Antioxidant effect of lipids and proteins Increase in polyunsaturated fats	[67]
Banana stem and leaves	Lambs	7 kg/animal/day	• Antiparasitic activity	[64]
Citrus pulp waste, potato wastewater, pea pods, and tomato pulp waste	Calves	20% (w/w) citrus pulp waste, 20% (w/w) potato wastewater, 30% (w/w) pea pods, and 30% (w/w) tomato pulp waste	• Immunological activity • Improvement of biochemical parameters	[103]
Cocoa husks	Pigs	10% (w/w) in the diet	• Immunoglobulin increased • HDL increased	[69]
Condensed tannins isolated from grape seed	Seabass	400 mg/kg	• Antioxidant activity Anti-inflammatory activity	[60]
Ensiled pineapple waste	Cattle	100%	Digestibility improved	[65]
Grape pomace concentrate	Chickens	60 g/kg	• Antioxidant activity Inhibition of lipid peroxidation	[57]
Grape pomace and grape seed extract	Broilers	60 g/kg grape pomace and 7.2 g/kg grape seed extract	• Antimicrobial activity Microbiota growth	[59]
Grape seed extract	Broilers	450 mg/kg	• Antibacterial activity Antioxidant activity Improvement of the microbiota	[58]
Olive mill	Crayfish	5 g/kg	• Antioxidant activity Improved immune response	[62]

TABLE 17.2 *(Continued)*

Food byproducts	Animal	Best dose	Health benefit	Reference
Olive mill	Seabream	5% (w/w) in the diet	• Inhibition of lipid peroxidation	[63]
Papaya leaves	Free-range Chickens	15% (w/w) in the diet	• Antiparasitic activity	[68]
Pomegranate extract	Calves	10 g/d	Produced more lymphocyte-derived cytokines Increased titers against antigen vaccination	[104]
Tamarind seed	Broilers	50% (w/w) in the diet	• Microbiota growth • Improvement of hematological parameters	[66]
Wine grape pomace	Lambs	10% (w/w) in the diet	• Antioxidant activity • Inhibition of lipid peroxidation	[61]
Vegetable byproducts				
Broccoli stems and leaves	Laying hens	90 g/kg	• Increased carotenoids • Cholesterol-lowering	[47]
Corn stalks fermented	Sheep	100%	Reduction of cholesterol, triglycerides, urea, and creatinine	[70]
Pumpkin seed	Poultry	33–66 g/kg	• Weight gain • Fertility increased • Improvement of biochemical parameters	[71]
Plant byproducts				
Camelina oil byproducts	Pigs	12% (w/w) in the diet	• Glucose decrease • Antioxidant activity • Anti-inflammatory activity	[80]

TABLE 17.2 *(Continued)*

Food byproducts	Animal	Best dose	Health benefit	Reference
Cottonseed	Tilapia	15% (w/w) in the diet	Improved hematological indexes	[72]
Cottonseed meal protein hydrolysate	Mitten crab	0.3 and 0.6% (w/w) in diet	• Antioxidant activity Improved immune response	[73]
Cottonseed meal protein hydrolysate	Crucian carp	50 g/kg	Improved absorption of crude protein, Zn, and Fe	[74]
Fermented cottonseed	Broilers	6% (w/w) in the diet	• Decrease in lipid biochemical parameters • Increased metabolic and hormonal activity	[105]
Fermented cottonseed	Broilers	8.9% (w/w) in the diet	• Antioxidant activity • Microbiota growth • Increased immunoglobulin level	[77]
Fermented cottonseed	Yellow-feathered broilers	8% (w/w) in the diet	• Growth increased • Increased immune responses	[76]
Green tea byproducts	Goats	2% (w/w) in the diet	• Weight gain and intake • Improvement of biochemical parameters • Spleen cells proliferation	[78]
Sunflower seed residue	Broilers	20% (w/w) in the diet	Small intestine improvement	[79]

17.7.1.2 GRAPE BYPRODUCTS

Grape byproducts are very varied since they involve fresh fruit residues like seeds and pomace and residues from the wine industry, such as wine pomace.

In the case of grape pomace, it has an antioxidant effect on chickens in the ileum, serum, excreta, and muscle tissue, both by ABTS and by FRAP methods, in addition to an inhibitory effect on lipid peroxidation, preventing the formation of malondialdehyde (MDA). This antioxidant activity is attributed to the high content of polyphenols present in grape pomace, where the adequate concentration of 60 g/kg does not affect the growth and yield of chickens and provides greater protection against oxidative effects.[57]

Grape seed extracts added to the diet of broilers cause beneficial effects such as antibacterial, microbiota improvement, and antioxidant activity. Regarding the latter, adding 450 mg/kg of grape extract to the diet presents the best antioxidant activity, even surpassing vitamin C by the FRAP and ABTS methods. In addition to showing antioxidant effects, there is also the best antibacterial activity against *E. coli* and a decrease in coliforms and *E. coli* in the ileal microbiota, which causes a better performance of the microbiota without these pathogens.[58]

The use of grape pomace and grape seed extract incorporated in broiler chicks' feed shows beneficial effects, improving the microbiota and providing an antimicrobial effect against pathogenic bacteria. The inclusion of 60 g/kg of grape pomace and 7.2 g/kg of grape seed extract increases the growth of *Lactobacillus* sp and a decrease in *Clostridium* sp, demonstrating the elimination of pathogens and promotion of beneficial bacteria.[59]

Supplementation of condensed tannins extracted from grape seeds in the seabass diet shows positive effects on antioxidant and anti-inflammatory activity. Adding 400 mg/kg shows the best beneficial effects. At this concentration, the antioxidant activity is manifested in an increase in the production of antioxidant enzymes such as catalase (CAT), glutathione peroxidase (GSH-Px) and superoxide dismutase (SOD), and inhibition of lipid peroxidation, greatly reducing the production of MDA. Anti-inflammatory activity occurs through a decrease in the pro-inflammatory cytokines TNF-α and interleukin 6 (IL-6). As it was seen, condensed tannins can avoid the oxidative and pro-inflammatory effects caused by hypoxic stress that is very common in fish.[60]

The use of wine grape pomace in the diet of ram lambs has an antioxidant effect, reducing free radicals and inhibiting lipid peroxidation. The 10% of wine grape pomace incorporated into the diet shows increased body weight and antioxidant enzymes glutathione peroxidase (GPX) and SOD; a decrease in reactive oxygen species (ROS) and MDA is also shown. This combination of an increase in antioxidant enzymes and a decrease in substances produced by oxidative stress causes an antioxidant activity that improves the quality of meat, health, and growth of ram lambs.[61]

17.7.1.3 OLIVE BYPRODUCTS

Olive mil is the main byproduct derived from the industry responsible for the extraction of olive oil. Olive mill wastewater incorporated into the crayfish diet in both low and high concentrations has antioxidant effects, manifesting itself in an increase in the antioxidant enzymes glutathione peroxidase, glutathione reductase, and catalase. Also, this fortified diet improves immunological parameters. These positive effects are due to polyphenols, which can prevent diseases in the production of marine animals.[62]

Olive mill vegetation water incorporated into the seabream feed does not show an alteration in fish growth. This diet increased antioxidant activity, albeit slightly, and increased lipid peroxidation as storage time increased. This combination of antioxidant effects causes fish fillets to have a higher proportion of omegas by inhibiting their oxidation, therefore a better nutritional value.[63]

17.7.1.4 BANANA BYPRODUCTS

In the production of bananas, multiple byproducts are generated, where two of the most wasted in the harvest are the stem and the leaves. Incorporating banana foliage (based on leaves and stem) in the feeding of lambs infected with the parasite *H. contortus* shows an antiparasitic effect. This effect can be attributed to the bioactive compounds that these byproducts present as flavonoids and terpenoids. As the nutritional value of this foliage is low, only the leaves can be used since they are more responsible for the antiparasitic effect, reducing the negative nutritional effects and having a beneficial effect on lambs' health.[64]

17.7.1.5 PINEAPPLE BYPRODUCTS

Ensiled pineapple waste is one of the bulkier byproducts of the pineapple harvest. The incorporation of 100% ensiled pineapple waste in the cattle feed shows an improvement in digestion compared with the feed commonly used based on hay, without adverse effects on other parameters such as nitrogen content, excretion of feces and urine, absorption, retention, pH, blood metabolites, ammonia, and temperature. Moreover, there was an increase in bacteria population in the rumen in the species *F. succinogenes* and *R. flavefaciens*, which are bacteria that have a high degradation of cellulose, achieving greater use of the feed, obtaining better results in the digestibility and growth of cattle.[65]

17.7.1.6 TAMARIND BYPRODUCTS

Once the tamarind pulp is used, the seed remains as a residue, which is one of the largest byproducts of the tamarind processing industry. Tannins are highly present in tamarind seeds, and they must be removed before this byproduct can be used. The best results are obtained when the tannins are removed by fermentation with *B. subtilis*. The incorporation of 50% of fermented tamarind seed into the diet of broiler chickens does not affect or minimally affect growth performance. However, this diet improves biochemical parameters, decreasing the albumin content and increasing the globulin content in serum. Furthermore, there is an increase in the number of beneficial bacteria *Lactobacillus* sp present in the large intestine and a decrease in pathogenic bacteria *E. coli* and *Bacteroidetes* sp. in the small intestine. Combining all these effects indicates that incorporating fermented tamarind seed into the conventional diet of broiler chickens improves their health without negative effects.[66]

17.7.1.7 OTHER FRUIT BYPRODUCTS

Avocado pulp waste is a byproduct derived from the processing and packaging of the fruit. This byproduct can be incorporated into the diet of finishing pigs. Adding 20% avocado pulp waste to pigs' diet inhibits lipid peroxidation, preventing MDA production, which gives greater stability to pig lipids. This inhibition may be due to the high presence of vitamin E in avocado, which has antioxidant effects. These beneficial effects are also reflected in the meat quality, where there is a greater presence of polyunsaturated fats and with high stability.[67]

The incorporation of 15% of papaya leaves in the diet of free-range chickens has an antiparasitic effect avoiding the disease coccidiosis, which is one of the most common diseases in chickens. This beneficial effect can be explained by the presence of proteases from papaya leaves such as papain, chymopapain, and lysozyme, where they digest protozoa; the presence of vitamin A improves the protection of intestinal cells repairing the damage caused by coccidiosis.[68]

The byproducts of cocoa husks included in a pig diet in a proportion of 10% triggered an increase in high-density lipoprotein (HDL), immunoglobulins A, G, and M. These effects can be caused by the polyphenols present in cocoa husks, which can help the animals endure stress conditions, in addition to reducing pathogen infections so that pigs will need fewer antibiotic treatments.[69]

17.7.2 VEGETABLE BYPRODUCTS

Like fruit byproducts, vegetable byproducts contain many bioactive compounds that can help the health of animals that consume diets enriched with these byproducts. Most vegetable byproducts come from leaves, stems, seeds, or vegetables that are not accepted for human consumption.

17.7.2.1 CORN BYPRODUCTS

Corn stalks are byproducts of the corn harvest. Sheep feed based on 100% corn stalks fermented by *T. reesei* shows no adverse effects. It also shows improvement in blood biochemical parameters such as decreased cholesterol, triglycerides, urea, and creatinine. There is also a weight gain, increased digestibility. The combination of improvements in blood biochemical parameters and increase in weight and digestibility cause a higher yield since greater gains are obtained by rapid growth, and a benefit to the health of sheep is guaranteed.[70]

17.7.2.2 BROCCOLI BYPRODUCTS

Usually, the byproducts of broccoli are the stems and leaves. Incorporating broccoli stems and leaves in an amount of 90 g/kg to laying hens' diet improves the quality of the egg without affecting its performance, improving the carotene content in the yolk, and reducing cholesterol in the egg. In laying hens, cecal fermentation is improved, manifesting a greater production of short-chain acids such as acetic, propionic, and butyric acid, causing greater activity of the microbiota.[47]

17.7.2.3 PUMPKIN BYPRODUCTS

One of the most abundant byproducts from both industrial processing and consumption is pumpkin seeds. Including pumpkin seeds in the diet of poultry provides health benefits without adverse effects. The consumption of diets with pumpkin seeds in broilers improves yields since they eat more food and, therefore, their weight is greater. Broilers also improve biochemical parameters, where triglycerides, glucose, very low-density lipoprotein (VLDL), and low-density lipoprotein (LDL) cholesterol are decreased. In Turkey, pumpkin seed diets improve fertility rate and hatchability. The improvement in health, fertility, and weight gain can be attributed to the phytochemicals present in the pumpkin seed, such as carotenes, phenolic compounds, saponins, vitamin C, and dietary fiber.[71]

17.7.3 PLANT BYPRODUCTS

Byproducts of plants are all those that are not consumed as a fruit or vegetable. These byproducts present powerful beneficial health effects when incorporated into the diet of animals. The processing or use of plants such as green tea, sunflower, cotton, and generated byproducts has beneficial properties in animal health, such as improving biochemical parameters, antioxidant capacity, better absorption of nutrients, and immunological properties, among others.

17.7.3.1 COTTON BYPRODUCTS

After the use of cotton fibers, cottonseed is the main byproduct of the cotton industry. When a percentage of a common food is replaced by cottonseed, an improvement in hematological values is observed. Such is the case of a diet containing 15% cottonseed in tilapia, where hematological values such as red blood cell, white blood cell, and hemoglobin are increased. However, cottonseed concentrations must be considered since values greater than 45% result in a decrease in hematological values, so it is only recommended to add a maximum of 15–30%, where said percentages do not show adverse effects.[72]

Cottonseed meal protein hydrolysate is used to fortify diets in animals. Low concentrations of a protein hydrolysate, around 0.3–0.6% in the diet of mitten crab, present important antioxidant effects, increase in the concentration of the antioxidant enzymes SOD and CAT, and the inhibition of lipid peroxidation, reducing the production of MDA. The immune response also benefits since there is a greater expression of genes related to the immune system. The mechanism of these beneficial effects is not clear, but it is speculated that the bioactive peptides of cottonseed meal protein hydrolysate have an influence on the gene modulation of the enzymes and proteins of the immune system.[73]

Adding cottonseed meal protein hydrolysate is consistent with an increase in protein absorption. This effect occurs when 50 g/kg of cottonseed meal protein hydrolysate is added to the crucian carp diet. In addition to better absorption of proteins, greater absorption of minerals such as Zn, Fe, and Ca occurs. This effect is due to the peptides of cottonseed, which, before being hydrolyzed and absorbed, act as modulators of the absorption of certain nutrients in particular.[74]

The fermentation of cottonseed can result in greater improvements than the unfermented product. The inclusion in the diet of broiler chickens with 6% cottonseed fermented by *C. tropicalis* causes effects of improvement of biochemical parameters with a decrease of glucose, triglycerides, and LDL. The hormonal activity of hormones such as epinephrine, growth hormone, and sensitive lipase hormone, and increased metabolic enzymes such as male enzymes and lipoprotein lipase also increase. These combined effects result in a decrease in abdominal fat since there was greater oxidation of fatty acids without adverse effects on growth performance.[75]

Including 8% of cottonseed fermented by *B. subtilis* to yellow-feathered broilers' diet has weight gain effects. There was an increase in immunoglobulins M and G and an increase in complement C3 and C4. These results indicate a greater immunological activity, which can protect broilers from diseases and even increase yields.[76]

Adding to the broiler diet, 8.9% of cottonseed fermented by *B. subtilis* and *Saccharomycetes* N5 shows an increase in immunoglobulins M and G. Regarding the antioxidant activity, there is an increase in the antioxidant enzymes GSH-Px, SOD and a decrease in MDA, achieving inhibition of lipid peroxidation. There was a growth effect of *Lactobacilli*. The combination of these effects indicates that the health of broilers is improved since there is antioxidant activity, increased beneficial bacteria, which can produce beneficial metabolites such as immunoglobulins. Nevertheless, the percentage of inclusion should be carefully considered since more than 8.9% shows negative effects on the growth of broilers.[77]

17.7.3.2 GREEN TEA BYPRODUCTS

When preparing green tea from *C. sinensis* leaves, the infusion is consumed, leaving the leaves as a byproduct. The incorporation of 2% of a byproduct of green tea leaves in the diet of goats results in weight gain, in addition to a better quality of meat, since the content of polyunsaturated fatty acids is increased; these same acids are maintained to a greater degree with this incorporation, reducing lipid peroxidation, reducing the amount of MDA produced. Regarding the biochemical parameters, there was a decrease in glucose and cholesterol. Moreover, the proliferation of spleen cells increases. These beneficial effects can be attributed to the presence of catechins since it has antioxidant and inhibitory properties of some enzymes that reduce the concentration of glucose and cholesterol.[78]

17.7.3.3 SUNFLOWER BYPRODUCTS

The byproduct of obtaining edible oil from sunflower seeds is the main waste of sunflower. In broiler chickens, the addition of 20% of sunflower seed byproduct presents an improvement in their microbiota, reducing the number of pathogenic bacteria such as *Clostridium* spp and increasing the number of beneficial bacteria such as *Lactobacillus* spp. The increase in *Lactobacillus* spp causes an increase in lactic, butyric, and isobutyric acid. An increase in protein digestibility is also seen. Due to this good digestibility, the amount of *Clostridium* spp decreases since this microorganism feeds on proteins. This combination of effects demonstrates intestinal health in broiler chickens.[79]

17.7.3.4 OTHER PLANT BYPRODUCTS

Camelina oil byproducts can be used to improve health in fattening pigs. A diet with 12% camelina oil byproducts shows a decrease in glucose. Regarding the antioxidant activity, there is an increase in the spleen of antioxidant enzymes SOD, CAT, and GPX. The expression of pro-inflammatory cytokines interleukin-8 (IL-8), TNF-α, IL-6, and interleukin-1 beta (IL-1β)

decreases in the spleen, increasing the expression of the anti-inflammatory cytokine IL-4. In addition to the regulation of cytosines, there was also inhibition of the cyclooxygenase-2 (COX-2) enzyme that causes pro-inflammatory effects and an increase in nitrogen oxide, demonstrating the immune response. These beneficial effects do not influence the adverse effects against pig performance. These benefits are due to the presence of polyunsaturated fatty acids and other bioactive compounds such as glucosinolates and erucic acid from camelina oil byproducts.[80]

17.8 FUNCTIONAL FOODS FOR PETS

17.8.1 TRENDS IN THE PRODUCTION AND MARKET OF PETS AND ENVIRONMENTAL FACTORS

The trend of buying and adopting pets around the world is increasing considerably, coupled with the factor of new forms of treatment and humanization of animals; in particular, those of companions that include dogs, cats, small mammals, horses, mini pigs, birds, among others, where their owners try to get the best food and treat for their pets. This encourages an increase in the demand for food made with select meat products, which have a large environmental footprint due to greenhouse gas emissions, fuel use, soil erosion, and water consumption, leading to an imbalance in environmental sustainability. More than 60% of the American population has pets, and it is projected that it will continue to increase. In 2015, there were an estimated 163.4 million dogs and cats in the United States alone, so it is necessary to take advantage of the byproducts of the agricultural industry to mitigate food demand, considering the environmental impact and health benefits to pets.[81]

In the United States, dogs and cats consume approximately 21% of the amount of dietary energy produced by humans and approximately 40% of the energy derived from animals, along with the amount of stool produced (about 43% by mass as in the US population). The environmental impact produced by their diet makes up 30% of those generated by animal production, mainly affecting the use of land, fossil fuels, water, biocides, and phosphates. An outstanding fact is that food production for dogs and cats produces around 64 million tons of CO_2 (equivalents of nitric oxide and methane). Looking for sustainable production alternatives and circular economies in the sector is vital due to the strong economic spillovers of dry food for dogs and cats. In 2012, this market was valued at around 12.3 billion dollars, being the most active and dominant in the sector, in addition to being highly diversifiable foods susceptible to fortification and/or enrichment in their formulas. So the addition of bioactive compounds from plant sources can be an alternative to minimize the demanding impact of this sector.[81,82]

The behavior and buying habits of pet owners are relevant in the design of new functional products. Reference [83] mentioned that owners of pets under 20 years of age have a higher percentage of purchasing vegan and vegetarian food, followed by the group of 20–29 years. In the same way, the eating habits of the owners influence decision-making in the food of theirs pet, being vegans the ones who stated in the highest percentage that they could feed their dog and/or cat a vegan diet, followed by vegetarians for dogs and pescetarians for cats. As for the reasons why other pet owners (omnivores) have not opted for plant-based foods, it is because of the lack of scientific evidence to justify their change that research in this area is important due to the large economic spillovers from the sector.

Current US regulations (FDA) allow the incorporation of nutraceutical ingredients, such as vitamins (C, E), amino acids (methionine, thiamine), flavorings from plant sources (essential oils). Current trends continue to encourage pet owners to purchase plant-based foods, improve the quality of life of their pets, and increase their years of life.[82,84] The trend in the improvement of food and health for pets is so great that novel technologies have even been applied from the chemical, medical, and pharmaceutical sectors commonly used for food and human health. Among them are the Near-Infrared Reflectance Spectroscopy, nanotechnology, gastrointestinal simulation models to determine the bioavailability and digestibility of nutrients, nutrigenomics, studies of the gastrointestinal microbiome, metabolomics, which together include molecular, spectroscopic, chromatographic techniques that are often complex and expensive to perform.[82]

17.8.2 HEALTH BENEFITS FOR PET OWNERS

There is a risk of contracting zoonotic diseases if sick animals are acquired, so promoting correct animal nutrition based on foods of plant origin is of vital importance since the latter contain beneficial phytochemicals for their health, in particular antioxidants and immunostimulators. These can prevent the animal from contracting diseases, which can later transmit them to the owner, because most of the time they are part of the family, having close contact between both. Besides, this situation is worrisome for public health; therefore, the authorities are encouraged to implement safety and quality systems in the places of sale of pets and exotic animals to control the origin, transmission, and spread of this class of pathologies.[85,86]

17.8.3 HEALTH BENEFITS FOR PETS FOCUS ON DOGS AND CATS

It has been observed in diets for Beagle breed dogs, based on vegetables, that the latter is a good source of protein and energy and contains a uniform composition in both micronutrients and macronutrients, improving the digestibility of the food. Regarding the sensory acceptance of diets based on meat and vegetables, no significant differences have been observed in the parameters of ranges of consumption, distraction, hesitation, level of anticipation, and interest after consumption by dogs.[87]

Some phytobiotics and natural products regulate or stimulate the immunomodulatory system of companions and livestock animals, highlighting the use of vitamins with antioxidant activity such as E, which is capable of capturing reactive oxygen and nitrogen species, as well as modulating its signaling. Likewise, this vitamin is of vital importance for protection against oxidative damage. On the other hand, vitamin D is involved in developing the bone system, proliferation of monocytes, and phagocytic activity of macrophages. Zinc is essential for growth, development, metabolism, and the adaptive immune system, among other important functionalities in the body. Polyunsaturated fatty acids such as Omega-3 are closely related to increased immune system function, particularly alpha-linolenic acid, eicosapentaenoic acid, and docosahexaenoic acid. On the other hand, it has also been observed that the use of essential oils and phytobiotics such as cinnamon (*Cinnamomum verum*), thyme (*Thymus* spp), turmeric (*Curcuma longa*), and oregano (*Origanum vulgare*) has improved the immune response of livestock animals. Epigallocatechin-3-gallate isolated from green tea (*Camellia sinensis*) has shown a potent immunomodulatory effect as well as probiotics, which promote the increase of short-chain fatty acids in the large intestine, such as butyrate, which increases the immunoglobulin A, diets supplemented or rich in these compounds can help improve animal health, reducing the use of veterinary antibiotics.[88]

In the case of probiotics, the improvement of intestinal health and the prevention of inflammatory diseases have been shown. However, some factors can limit their use and commercialization, which are linked to the gaps in characterizing the microbiota of dogs and cats. The commercialized products are commonly not clear in the declaration of the ingredients and probiotic bacteria. Likewise, the use of strains is not fully studied and their mode of action is uncertain since it depends on the microbiota of the different species.[89]

Recent research has shown the efficacy of chia seed supplementation (*Salvia hispanica* L.) as a possible adjuvant to treat feline idiopathic hypercalcemia. It improves symptoms and normalizes serum ionized calcium levels in just four weeks of supplementing 2 g of chia seeds per day.[90] The effects of turmeric (*Curcuma longa*), glasswort (*S. herbacea*), and *Ganghwa mugwort* (*A. princeps* Pampanini sc. Sajabal), subjected to fermentation by *Enterococcus faecium*, as additives for Beagles dog food were studied. The antioxidant capacity (ABTS) increased by fermentation, and there was a reduction in the levels of superoxide in MAC-T cells; the total phenolic compounds and total flavonoids increased in the mixture of the three plants. *Ganghwa mugwort* showed the highest total polyphenols content, total flavonoids, and antioxidant activity (ABTS); its inclusion in the food for Beagles was efficient for its high phytochemicals content and antioxidant activity. Also, its sensory acceptance was good, without presenting statistical difference with the control diet so the addition of medicinal plants could be a good option as functional additives in dog food.[91]

The supplementation with faba beans has been studied in food for Beagles, observing that the dogs remained at the initial weight in all the experimental and control groups; however, the feces were softer with the treatment. Besides, a linear increase was observed between the frequency of defecation and the increase of faba beans. No side effects or health complications were observed during the study, so supplementation with this natural ingredient can improve bowel frequency in Beagles.[92] Similar effects have been observed with the supplementation of sorghum mill feed, observing a better digestibility with the supplementation of sorghum flour. In contrast, with the supplementation of whole sorghum, there was an increase in antioxidant activity (ORAC) because the diversity of sorghum byproducts can increase the antioxidant activity as well as the digestibility of dog food.[93] Reference [94] mentions using food supplements to treat idiopathic epilepsy in dogs, where those of plant origin predominate like coconut oils, fish, milk thistle, and cannabis. The use of medicinal plants such as herbs of origin was also observed in Chinese, garlic, parsley, to name a few,[95] a wide variety of medicinal plants and vegetables (*Allium sativum, Aloe vera, Calendula officinalis, Ulmus fulva, Plantago major, Usnea longissima, Urtica dioica, Borago officinalis, Curcubita pepo, Zingiber officinale, Cedronella canariensis, Pinus pondedorsa, Humulus lupulus, Avena sativa, Valeriana officinalis, Passiflora incarnata, Ganoderma lucidum, Rumex acetosella, Rheum palmatum*, and *Plantago psyllium*, among others) are used in British Columbia in Canada to treat different diseases and disorders such as anal gland, rheumatoid arthritis, joint pain, asthma, heart problems, epilepsy, cancer, tumors in the prostate gland, in pets in particular dogs and cats. Most of these plants contain bioactive secondary metabolites, in particular flavonoids and triterpenes, responsible for the physiological activities, which could suggest the addition of some of these compounds to dry foods, to exert a chronic antioxidant effect and/or develop new pharmacological treatments, with fewer side effects than the current ones. Table 17.3 shows some advantages and disadvantages of the use of ingredients derived from agricultural byproducts for pet food compared with conventional foods.

TABLE 17.3 Advantages and Disadvantages of Using Functional Ingredients Derived from Agricultural Byproducts for Pet Food Compared to Conventional Food.[96,98]

Advantages	Disadvantages
• Bioactive ingredients (antioxidants) provide beneficial effects to the health of the animal	• Palatability may be decreased
• The phytochemicals contained in the ingredients can reduce the risk of diseases	• There is a lack of scientific research that can be scaled up to a pilot or commercial level
• The health of pets can contribute to the health of the caregiver and household members	• There is misinformation about its functionalities
• There is a recovery of biomass and byproducts without value in the market	• Presence of xenobiotic residues
• Its use helps to minimize the environmental impact derived from the final dispositions of agricultural byproducts	

Source: Adapted from ref. [96, 98]

17.8.4 TRENDS IN PET FOOD RESEARCH

Di Cerbo et al.[96] reflected in a review (some of the research from 1941 to 2017) aspects of functional food for dogs and cats, such as the addition and/or supplementation of phytochemicals like flavonoids, carotenoids, fatty acids, dietary fiber (fructooligosaccharides), extracts of medicinal plants, fruits, and vegetables. They highlighted bioactive activities like antioxidant, anticancer, antitumor, immunostimulators, and neurotransmitter regulators. Viana et al.[97] mentioned the increase in the research trend about pet food, linked with terms such as natural, health, care, organic, giving 2484 scientific publications indexed in SCOPUS and Web of Science by 2019, observing an increase in the trend since 1997, highlighting 2018 with at least 70 publications. The United States led the number of registered patents (1826) until 2019. Most investigations addressed aspects of health, antioxidant activity, plant extracts, prebiotics, and probiotics, which are the strongest trends in animal nutrition and the inclusion of exotic ingredients, to improve health and increase the longevity of pets. The foregoing suggests the effective use and administration of agricultural and agroindustrial byproducts since they are a source of phytochemicals that can improve animal nutrition, particularly that of pets.

Current trends in the pet food market are the inclusion of human superfoods into animal nutrition, such as turmeric, coconut, honey, beta glucans, vitamins, minerals, antioxidants, probiotics, protein derived from lamb meat, salmon, and others. Natural ingredients such as sweet potato, cannabidiol (CBD), products made from vegetables and fruits (vegetarian) and with insect proteins are currently available on the market. The market for wet food for cats has increased considerably; however, dry food remains the market leader, for both cats and dogs. On the other hand, fresh food is a new segment that continues to gain strength in the market during 2021, it grew by 44%. The correct additives such as the inclusion of gums and stabilizers can generate better palatability and emulsion of the ingredients.[98–102]

17.9 CONCLUSION

Incorporating a certain amount of food byproducts in the diet of animals presents noticeable and promising improvements to their health. Although the maximum allowed amount must be taken into consideration since exceeding this amount may present antinutritional factors, where there are adverse effects on animal health, in addition to a decrease in growth. The way they are incorporated will have differences in their effectiveness, where factors such as the bioavailability of nutrients and bioactive compounds must be fully studied and understood. Food byproducts contain multiple bioactive compounds of different natures; due to this, it is necessary to study the isolation of certain groups of compounds to observe which of all is the one that presents the health benefits. This can lead to an improvement in the incorporation into diets since we can incorporate the fractions that exercise beneficial functions, avoid antinutritional compounds, guarantee beneficial effects on health, and increase or maintain yields, avoiding the large number of antibiotics supplied to all kinds of animals.

The formulation of diets with food byproducts must be technologically viable since the challenge is to find the optimal way to produce these diets, guaranteeing quality, safety, as well as being economically viable.

KEYWORDS

- **animal health**
- **bio-wastes**
- **feed materials**
- **natural-derived**
- **value-added**

REFERENCES

1. FAO. The Future of Food and Agriculture, 2018.
2. FAO. Food Outlook Biannual Report on Global Food Markets [Online] 2021. http://www.fao.org/3/ca9509en/CA9509EN.pdf
3. FAO. Fruit and Vegetables–Your Dietary EDssentials. The International Year of Fruits and Vegetables [Online] 2021. http://www.fao.org/3/cb2395en/cb2395en.pdf
4. Mateos-Aparicio, I.; Matias, A. Chapter 9; Food Industry Processing byproducts in Foods. In *The Role of Alternative and Innovative Food Ingredients and Products in Consumer Wellness*; Galanakis, C. M., Ed.; Academic Press, 2019; pp 239–281.
5. Cattaneo, A.; Federighi, G.; Vaz, S. The Environmental Impact of Reducing Food Loss and Waste: A Critical Assessment. *Food Policy* **2021,** *98*, 101890.
6. Georganas, A.; Giamouri, E.; Pappas, A. C.; Papadomichelakis, G.; Galliou, F.; Manios, T.; Tsiplakou, E.; Fegeros, K.; Zervas, G. Bioactive Compounds in Food Waste: A Review on the Transformation of Food Waste to Animal Feed. *Foods* **2020,** *9*, 291.
7. Guil-Guerrero, J. L.; Ramos, L.; Moreno, C.; Zúñiga-Paredes, J. C.; Carlosama-Yépez, M.; Ruales, P. Plant-Food byproducts to Improve Farm-Animal Health. *Anim. Feed Sci. Technol.* **2016,** *220*, 121–135.
8. Čolović, D.; Rakita, S.; Banjac, V.; Đuragić, O.; Čabarkapa, I. Plant Food byproducts as Feed: Characteristics, Possibilities, Environmental Benefits, and Negative Sides. *Food Rev. Int.* **2019,** *35*, 363–389.
9. Waste, U. S. F. Federal Animal Feed Policy [Online] 2021. https://policyfinder.refed.com/federal-policy/federal-animal-feed-policy#:~:text=Under%20the%20federal%20Food%20Drug,or%20held%20in%20unsanitary%20conditions
10. DBMR. Global Animal Feed Market–Industry Trends and Forecast to 2027 [Online] 2021. https://www.databridgemarketresearch.com/reports/global-animal-feed-market

11. Food and Agriculture Organization of the United Nations. Faostat Statistical Database [Online] 2021. https://www.fao.org/faostat/en/#data/QCL
12. IFIF. Global feed statistics [Online]. https://ifif.org/global-feed/statistics/
13. Jiang, J.; Xiong, Y. L. Natural Antioxidants as Food and Feed Additives to Promote Health Benefits and Quality of Meat Products: A Review. *Meat Sci.* **2016**, *120*, 107–117.
14. Data Bridge Market Research. Global Animal Feed Market–Industry Trends and Forecast to 2027 [Online]. https://www.databridgemarketresearch.com/reports/global-animal-feed-market
15. Venanzi, S.; Pezzolla, D.; Cecchini, L.; Pauselli, M.; Ricci, A.; Sordi, A.; Torquati, B.; Gigliotti, G. Use of Agricultural byproducts in the Development of an Agro-Energy Chain: A Case Study from the Umbria Region. *Sci. Total Environ.* **2018,** *627*, 494–505.
16. Wilkinson, J. M.; Lee, M. R. F. Review: Use of Human-Edible Animal Feeds by Ruminant Livestock. *Animal* **2018,** *12*, 1735–1743.
17. Zeng, Z.; Zhang, S.; Wang, H.; Piao, X. Essential Oil and Aromatic Plants as Feed Additives in Non-Ruminant Nutrition: A Review. *J. Anim. Sci. Biotechnol.* **2015,** *6*.
18. EC. Regulation (ec) no 882/2004 of the European Parliament and of the Council on 29 April 2004 on Official Controls Performed to Ensure the Verification on Compliance with Feed and Food Law, Animal Health and Animal Welfare Rules [Online] 2021. https://eur-lex.europa.eu/legal-content/EN/TXT/PDF/?uri=CELEX:02004R0882-20140630&from=EN (accessed April 9, 2021)
19. FDA. Food and Drug Administrators Regulation on Pet Food [Online] 2021. https://www.fda.gov/animal-veterinary/animal-health-literacy/fdas-regulation-pet-food
20. Kanner, J. Dietary Advanced Lipid Oxidation Endproducts are Risk Factors to Human Health. *Mol. Nutr. Food Res.* **2007,** *51*, 1094–1101.
21. Negre-Salvayre, A.; Coatrieux, C.; Ingueneau, C.; Salvayre, R. Advanced Lipid Peroxidation End Products in Oxidative Damage to Proteins. Potential Role in Diseases and Therapeutic Prospects for the Inhibitors. *British J. Pharmacol.* **2008,** *153*, 6–20.
22. Stevanović, Z. D.; Bošnjak-Neumüller, J.; Pajić-Lijaković, I.; Raj, J.; Vasiljević, M. Essential Oils as Feed Additives—Future Perspectives. *Molecules* **2018,** *23*.
23. Figueiredo, A. C.; Barroso, J. G.; Pedro, L. G.; Scheffer, J. J. C. Factors Affecting Secondary Metabolite Production in Plants: Volatile Components and Essential Oils. *Flavour Fragr. J.* **2008,** *23*, 213–226.
24. FAO. The State of Food and Agriculture 2019 Moving Forward on Food Loss and Waste Reduction [Online] 2021. http://www.fao.org/3/ca6030en/ca6030en.pdf
25. FDA. Guidance for industry Human Food byproducts for Use as Animal Food [Online] 2021. http://www.fda.gov/AnimalVeterinary/default.htm
26. EU. Recommendations for Action in Food Waste Prevention [Online] 2021. https://ec.europa.eu/food/sites/food/files/safety/docs/fs_eu-actions_action_implementation_platform_key_recommendations.pdf
27. Kosseva, M. R. Chapter 3; Sources, Characteristics and Treatment of Plant-Based Food Waste. In *Food Industry Wastes*, 2nd ed.; Kosseva, M. R., Webb, C., Eds.; Academic Press, 2020; pp 37–66.
28. Kasapidou, E.; Sossidou, E.; Mitlianga, P. Fruit and Vegetable Co-Products as Functional Feed Ingredients in Farm Animal Nutrition for Improved Product Quality. *Agriculture* **2015,** *5*, 1020–1034.
29. Prathamesh, B. H.; Sahoo, A.; Patil, N. Review: Food Industry Byproducts Used as a Functional Food Ingredients. *Int. J. Waste Resour.* **2016,** *6*, 1000248.

30. Lai, W. T.; Khong, N. M. H.; Lim, S. S.; Hee, Y. Y.; Sim, B. I.; Lau, K. Y.; Lai, O. M. A Review: Modified Agricultural Byproducts for the Development and Fortification of Food Products and Nutraceuticals. *Trends Food Sci. Technol.* **2017,** *59*, 148–160.
31. Lau, B. F.; Kong, K. W.; Leong, K. H.; Sun, J.; He, X.; Wang, Z.; Mustafa, M. R.; Ling, T. C.; Ismail, A. Banana Inflorescence: Its Bio-Prospects as an Ingredient for Functional Foods. *Trends Food Sci. Technol.* **2020,** *97*, 14–28.
32. Hernalsteens, S. Chapter 24; Edible Films and Coatings Made up of Fruits and Vegetables. In *Biopolymer Membranes and Films*; de Moraes, M. A., da Silva, C. F., Vieira, R. S., Eds.; Elsevier, 2020; pp 575–588.
33. Georganas, A.; Giamouri, E.; Pappas, A. C.; Papadomichelakis, G.; Galliou, F.; Manios, T.; Tsiplakou, E.; Fegeros, K.; Zervas, G. Bioactive Compounds in Food Waste: A Review on the Transformation of Food Waste to Animal Feed. *Foods* **2020,** *9*, 18.
34. Pasarin, D.; Rovinaru, C. Sources of Carotenoids and Their Uses as Animal Feed Additives - A Review. *Sci. Pap. Ser. D-Anim. Sci.* **2018,** *61*, 74–85.
35. Utczas, M.; Cacciola, F.; Giuffrida, D.; Russo, M.; Bonaccorsi, I.; Dugo, P.; Mondello, L. Bioactives Screening in Overripe Fruits and Vegetables by Liquid Chromatography Coupled to Photodiode Array and Mass Spectrometry Detection. *Food Anal. Meth.* **2018,** *11*, 3053–3070.
36. Tocmo, R.; Pena-Fronteras, J.; Calumba, K. F.; Mendoza, M.; Johnson, J. J. Valorization of Pomelo (*citrus grandis osbeck*) Peel: A Review of Current Utilization, Phytochemistry, Bioactivities, and Mechanisms of Action. *Compr. Rev. Food. Sci. Food Saf.* **2020,** *19*, 1969–2012.
37. Bellete, B. S.; Ramin, L. Z.; Porto, D.; Ribeiro, A. I.; Forim, M. R.; Zuin, V. G.; Fernandes, J. B.; Silva, M. An Environmentally Friendly Procedure to Obtain Flavonoids from Brazilian Citrus Waste. *J. Braz. Chem. Soc.* **2018,** *29*, 1123–1129.
38. Vella, F. M.; Cautela, D.; Laratta, B. Characterization of Polyphenolic Compounds in Cantaloupe Melon Byproducts. *Foods* **2019,** *8*, 10.
39. Sumaya-Martinez, M. T.; Medina-Carrillo, R. E.; Gonzalez-Ocegueda, E.; Jimenez-Ruiz, E. I.; Balois-Morales, R.; Sanchez-Herrera, L. M.; Lopez-Nahuatt, G. Mango (*mangifera indica l.*) Pulping Byproducts: Antioxidant Activity and Bioactive Compounds of Three Mango Cultivars. *Rev. Bio Cienc.* **2019,** *6*, 20.
40. Kuppusamy, S.; Venkateswarlu, K.; Megharaj, M. Examining the Polyphenol Content, Antioxidant Activity and Fatty Acid Composition of Twenty-One Different Wastes of Fruits, Vegetables, Oilseeds and Beverages. *SN Appl. Sci.* **2020,** *2*, 13.
41. Iora, S. R. F.; Maciel, G. M.; Zielinski, A. A. F.; da Silva, M. V.; Pontes, P. V. D.; Haminiuk, C. W. I.; Granato, D. Evaluation of the Bioactive Compounds and the Antioxidant Capacity of Grape Pomace. *Int. J. Food Sci. Technol.* **2015,** *50*, 62–69.
42. Achilonu, M.; Shale, K.; Arthur, G.; Naidoo, K.; Mbatha, M. Phytochemical Benefits of Agroresidues as Alternative Nutritive Dietary Resource for Pig and Poultry Farming. *J. Chem.* **2018,** *2018*, 15.
43. Wu, X.; Venkitasamy, C.; McHugh, T.; Pan, Z. Process Development for Antioxidant Extraction from Wet Pomegranate Peel. *Trans. ASABE* **2021,** *64*, 191–202.
44. Kaur, G. J.; Kumar, D.; Orsat, V.; Singh, A. Assessment of Carrot Rejects and Wastes for Food Product Development and as a Biofuel. *Biomass Convers. Biorefinery* **2020,** 12.
45. Tiwari, S.; Upadhyay, N.; Malhotra, R. Three Way Anova for Emulsion of Carotenoids Extracted in Flaxseed Oil from Carrot Bio-Waste. *Waste Manage.* **2021,** *121*, 67–76.
46. Titcomb, T. J.; Kaeppler, M. S.; Cook, M. E.; Simon, P. W.; Tanumihardjo, S. A. Carrot Leaves Improve Color and Xanthophyll Content of Egg Yolk in Laying Hens but are

not as Effective as Commercially Available Marigold Fortificant1. *Poult. Sci.* **2019,** *98*, 5208–5213.
47. Hu, C. H.; Zuo, A. Y.; Wang, D. G.; Pan, H. Y.; Zheng, W. B.; Qian, Z. C.; Zou, X. T. Effects of Broccoli Stems and Leaves Meal on Production Performance and Egg Quality of Laying Hens. *Anim. Feed Sci. Technol.* **2011,** *170*, 117–121.
48. Strati, I. F.; Oreopoulou, V. Recovery and Isomerization of Carotenoids from Tomato Processing byproducts. *Waste Biomass Valorization* **2016,** *7*, 843–850.
49. Torres, M. D.; Domínguez, H. Valorisation of Potato Wastes. *Int. J. Food Sci. Technol.* **2020,** *55*, 2296–2304.
50. Ruiz-Nieto, J. E.; Hernandez-Ruiz, J.; Hernandez-Marin, J.; Mendoza-Carrillo, J.; Abraham-Juarez, M.; Isiordia-Lachica, P. M.; Mireles-Arriaga, A. I. Mesquite (*prosopis spp.*) Tree as a Feed Resource for Animal Growth. *Agrofor. Syst.* **2020,** *94*, 1139–1149.
51. Pena-Rodriguez, A.; Pelletier-Morreeuw, Z.; Garcia-Lujan, J.; Rodriguez-Jaramillo, M. D.; Guzman-Villanueva, L.; Escobedo-Fregoso, C.; Tovar-Ramirez, D.; Reyes, A. G. Evaluation of *agave lechuguilla* Byproduct Crude Extract as a Feed Additive for Juvenile Shrimp Litopenaeus Vannamei. *Aquac. Res.* **2020,** *51*, 1336–1345.
52. Wang, J. H.; Cao, F. L.; Su, E. Z.; Zhao, L. G.; Qin, W. S. Improvement of Animal Feed Additives of Ginkgo Leaves Through Solid-State Fermentation Using *aspergillus niger*. *Int. J. Biol. Sci.* **2018,** *14*, 736–747.
53. Zuorro, A.; Lavecchia, R.; Gonzalez-Delgado, A. D.; Garcia-Martinez, J. B.; L'Abbate, P. Optimization of Enzyme-Assisted Extraction of Flavonoids from Corn Husks. *Processes* **2019,** *7*, 14.
54. Abd El-Hack, M. E.; Alagawany, M.; Elrys, A. S.; Desoky, E. S. M.; Tolba, H. M. N.; Elnahal, A. S. M.; Elnesr, S. S.; Swelum, A. A. Effect of Forage *moringa oleifera l.* (moringa) on Animal Health and Nutrition and its Beneficial Applications in Soil, Plants and Water Purification. *Agriculture* **2018,** *8*, 22.
55. Hadi, G.; Ali, N.; Sarayen, R. The Effect of Apple Pulp and Multi Enzyme on Performance and Blood Parameters in Native Laying Hens, 2014.
56. Sehm, J.; Lindermayer, H.; Meyer, H. H. D.; Pfaffl, M. W. The Influence of Apple- and Red-Wine Pomace Rich Diet on Mrna Expression of Inflammatory and Apoptotic Markers in Different Piglet Organs. *Anim. Sci.* **2006,** *82*, 877–887.
57. Brenes, A.; Viveros, A.; Goñi, I.; Centeno, C.; Sáyago-Ayerdy, S. G.; Arija, I.; Saura-Calixto, F. Effect of Grape Pomace Concentrate and Vitamin e on Digestibility of Polyphenols and Antioxidant Activity in Chickens. *Poult. Sci.* **2008,** *87*, 307–316.
58. Hajati, H.; Hassanabadi, A.; Golian, A.; Nassiri-Moghaddam, H.; Nassiri, M. The Effect of Grape Seed Extract and Vitamin c Feed Supplements on Carcass Characteristics, Gut Morphology and Ileal Microflora in Broiler Chickens Exposed to Chronic Heat Stress. *Iran J. Appl. Anim. Sci.* **2015,** *5*, 155–165.
59. Viveros, A.; Chamorro, S.; Pizarro, M.; Arija, I.; Centeno, C.; Brenes, A. Effects of Dietary Polyphenol-Rich Grape Products on Intestinal Microflora and Gut Morphology in Broiler Chicks. *Poult. Sci.* **2011,** *90*, 566–578.
60. Peng, K.; Wang, G.; Wang, Y.; Chen, B.; Sun, Y.; Mo, W.; Li, G.; Huang, Y. Condensed Tannins Enhanced Antioxidant Capacity and Hypoxic Stress Survivability but not Growth Performance and Fatty Acid Profile of Juvenile Japanese Seabass *(lateolabrax japonicus)*. *Anim. Feed Sci. Technol.* **2020,** *269*, 114671.

61. Zhao, J. X.; Li, Q.; Zhang, R. X.; Liu, W. Z.; Ren, Y. S.; Zhang, C. X.; Zhang, J. X. Effect of Dietary Grape Pomace on Growth Performance, Meat Quality and Antioxidant Activity in Ram Lambs. *Anim. Feed Sci. Technol.* **2018,** *236*, 76–85.
62. Parrillo, L.; Coccia, E.; Volpe, M. G.; Siano, F.; Pagliarulo, C.; Scioscia, E.; Varricchio, E.; Safari, O.; Eroldogan, T.; Paolucci, M. Olive Mill Wastewater-Enriched Diet Positively Affects Growth, Oxidative and Immune Status and Intestinal Microbiota in the Crayfish, *astacus leptodactylus*. *Aquaculture* **2017,** *473*, 161–168.
63. Sicuro, B.; Daprà, F.; Gai, F.; Palmegiano, G.; Schiavone, R.; Zilli, L.; Vilella, S. Olive Oil byproduct as a Natural Antioxidant in Gilthead Sea Bream (*sparusaurata*) Nutrition. *Aquacult. Int.* **2010,** *18*, 511–522.
64. Marie-Magdeleine, C.; Boval, M.; Philibert, L.; Borde, A.; Archimède, H. Effect of Banana Foliage (*musa x paradisiaca*) on Nutrition, Parasite Infection and Growth of Lambs. *Livest. Sci.* **2010,** *131*, 234–239.
65. Suksathit, S.; Wachirapakorn, C.; Opatpatanakit, Y. Effects of Levels of Ensiled Pineapple Waste and Pangola Hay Fed as Roughage Sources on Feed Intake, Nutrient Digestibility and Ruminal Fermentation of Southern Thai Native Cattle. *Songklanakarin J. Sci. Technol.* **2011,** *33*, 281–289.
66. Jana, A.; Adak, A.; Halder, S. K.; Das, A.; Paul, T.; Mondal, K.; Das Mohapatra, P. A New Strategy for Improvement of Tamarind Seed Based Chicken Diet After Microbial Detannification and Assessment of its Safety Aspects. *Acta Biologica Szegediensis* **2015,** *59*, 1–9.
67. Hernández-López, S. H.; Rodríguez-Carpena, J. G.; Lemus-Flores, C.; Grageola-Nuñez, F.; Estévez, M. Avocado Waste for Finishing Pigs: Impact on Muscle Composition and Oxidative Stability During Chilled Storage. *Meat Sci.* **2016,** *116*, 186–192.
68. Al-Fifi, Z. Effect of Leaves Extract of *carica papaya*, *vernonia amigdalina* and *azadiratcha indica* on the Coccidiosis in Free-Range Chickens. *Asian J. Anim. Sci.* **2007,** *1*, 26–32.
69. Damiano, M.; Malagutti, L.; Rosi, F. Cocoa Husks Feeding Improves Some Indices of Health and Welfare of Pigs. *Indian J. Res.* **2014,** *3*, 210–213.
70. Omer, H.; Ali, F.; Gad, S. Replacement of Clover Hay by Biologically Treated Corn Stalks in Growing Sheep Rations. *J. Agric. Sci.* **2012,** *4*.
71. Achilonu, M. C.; Nwafor, I. C.; Umesiobi, D. O.; Sedibe, M. M. Biochemical Proximates of Pumpkin (*cucurbitaeae* spp.) and Their Beneficial Effects on the General Well-Being of Poultry Species. *J. Anim. Physiol. Anim. Nutr.* **2018,** *102*, 5–16.
72. Yue, Y.-R.; Zhou, Q.-C. Effect of Replacing Soybean Meal with Cottonseed Meal on Growth, Feed Utilization, and Hematological Indexes for Juvenile Hybrid Tilapia, *oreochromis niloticus* × *o. Aureus*. *Aquaculture* **2008,** *284*, 185–189.
73. Cheng, H.; Liu, W.; Yuan, X.; Jia, E.; Zhang, D.; Jiang, G. Effects of Dietary Cottonseed Meal Protein Hydrolysate on Growth, Antioxidants and Immunity of Chinese Mitten Crab *eriocheir sinensis*. *J. Oceanol. Limnol.* **2020,** *38*, 869–882.
74. Gui, D.; Liu, W.; Shao, X.; Xu, W. Effects of Different Dietary Levels of Cottonseed Meal Protein Hydrolysate on Growth, Digestibility, Body Composition and Serum Biochemical Indices in Crucian Carp (*carassius auratus gibelio*). *Anim. Feed Sci. Technol.* **2010,** *156*, 112–120.
75. Niu, J.-L.; Zhang, J.; Wei, L.-Q.; Zhang, W.-J.; Nie, C.-X. Effect of Fermented Cottonseed Meal on the Lipid-Related Indices and Serum Metabolic Profiles in Broiler Chickens. *Animals* **2019,** *9*, 930.
76. Tang, J. W.; Sun, H.; Yao, X. H.; Wu, Y. F.; Wang, X.; Feng, J. Effects of Replacement of Soybean Meal by Fermented Cottonseed Meal on Growth Performance, Serum

Biochemical Parameters and Immune Function of Yellow-Feathered Broilers. *Asian-Australas J. Anim. Sci.* **2012,** *25*, 393–400.
77. Wang, Y.; Deng, Q.; Song, D.; Wang, W.; Zhou, H.; Wang, L.; Li, A. Effects of Fermented Cottonseed Meal on Growth Performance, Serum Biochemical Parameters, Immune Functions, Antioxidative Abilities, and Cecal Microflora in Broilers. *Food Agric. Immunol.* **2017,** *28*, 725–738.
78. Ahmed, S. T.; Lee, J.-W.; Mun, H.-S.; Yang, C.-J. Effects of Supplementation with Green Tea Byproducts on Growth Performance, Meat Quality, Blood Metabolites and Immune Cell Proliferation in Goats. *J. Anim. Physiol. Anim. Nutr.* **2015,** *99*, 1127–1137.
79. Kalmendal, R.; Elwinger, K.; Holm, L.; Tauson, R. High-Fibre Sunflower Cake Affects Small Intestinal Digestion and Health in Broiler Chickens. *Br. Poult. Sci.* **2011,** *52*, 86–96.
80. Taranu, I.; Gras, M.; Pistol, G. C.; Motiu, M.; Marin, D. E.; Lefter, N.; Ropota, M.; Habeanu, M. Ω-3 Pufa Rich Camelina Oil Byproducts Improve the Systemic Metabolism and Spleen Cell Functions in Fattening Pigs. *PLoS One* **2014,** *9*, e110186.
81. Okin, G. S. Environmental Impacts of Food Consumption by Dogs and Cats. *PLoS One* **2017,** *12*, e0181301.
82. Godoy, M. R. C. d.; Hervera, M.; Swanson, K. S.; Jr., G. C. F. Innovations in Canine and Feline Nutrition: Technologies for Food and Nutrition Assessment. *Annu. Rev. Anim. Biosci.* **2016,** *4*, 311–333.
83. Dodd, S. A. S.; Cave, N. J.; Adolphe, J. L.; Shoveller, A. K.; Verbrugghe, A. Plant-Based (Vegan) Diets for Pets: A Survey of Pet Owner Attitudes and Feeding Practices. *PLoS One* **2019,** *14*, e0210806.
84. Radosevich, J.; McGee, N.; Rawson, N. E. Chapter 38; Functional Ingredients in the Pet Food Industry: Regulatory Considerations. In *Nutraceutical and Functional Food Regulations in the United States and Around the World*; 3rd ed.; Bagchi, D., Ed.; Academic Press, 2019; pp 611–616.
85. Halsby, K. D.; Walsh, A. L.; Campbell, C.; Hewitt, K.; Morgan, D. Healthy Animals, Healthy People: Zoonosis Risk from Animal Contact in Pet Shops, A Systematic Review of the Literature. *PLoS One* **2014,** *9*, e89309.
86. McGee, N.; Radosevich, J.; Rawson, N. E. Chapter 27; Functional Ingredients in the Pet Food Industry: Regulatory Considerations. In *Nutraceutical and Functional Food Regulations in the United States and Around the World*; 2nd ed.; Bagchi, D., Ed.; Academic Press: San Diego, 2014; pp 497–502.
87. Callon, M. C.; Cargo-Froom, C.; DeVries, T. J.; Shoveller, A. K. Canine Food Preference Assessment of Animal and Vegetable Ingredient-Based Diets Using Single-Pan Tests and Behavioral Observation. *Front. Vet. Sci.* **2017,** *4*.
88. Bobeck, E. A. Nutrition and Health: Companion Animal Applications: Functional Nutrition in Livestock and Companion Animals to Modulate the Immune Response. *J. Anim. Sci.* **2020,** *98*.
89. Baffoni, L. Probiotics and Prebiotics for the Health of Companion Animals. In *Probiotics and Prebiotics in Animal Health and Food Safety*; Di Gioia, D., Biavati, B., Eds.; Springer International Publishing: Cham, 2018; pp 175–195.
90. Fantinati, M.; Priymenko, N. Managing Feline Idiopathic Hypercalcemia with Chia Seeds (*salvia hispanica l.*): A Case Series. *Front. Vet. Sci.* **2020,** *7*.
91. Park, D. H.; Kothari, D.; Niu, K.-M.; Han, S. G.; Yoon, J. E.; Lee, H.-G.; Kim, S.-K. Effect of Fermented Medicinal Plants as Dietary Additives on Food Preference and Fecal Microbial Quality in Dogs. *Animals* **2019,** *9*, 690.

92. Corsato Alvarenga, I.; Holt, D.; Aldrich, C. G. Evaluation of Faba Beans as an Ingredient in Dog Diets: Apparent Total Tract Digestibility of Extruded Diets with Graded Levels of Dehulled Faba Beans (*vicia faba l.*) by Dogs. *J. Anim. Sci.* **2020,** *98.*
93. Corsato Alvarenga, I.; Aldrich, C. G. The Effect of Sorghum Fractions on Apparent Total Tract Digestibility and Antioxidant Capacity by Dogs. *PLoS One* **2018,** *13*, e0206090.
94. Berk, B. A.; Packer, R. M. A.; Law, T. H.; Volk, H. A. Investigating Owner use of Dietary Supplements in Dogs with Idiopathic Epilepsy. *Res. Vet. Sci.* **2018,** *119*, 276–284.
95. Lans, C. Do Recent Research Studies Validate the Medicinal Plants used in British Columbia, Canada for Pet Diseases and Wild Animals Taken into Temporary Care? *J. Ethnopharmacol.* **2019,** *236*, 366–392.
96. Di Cerbo, A.; Morales-Medina, J. C.; Palmieri, B.; Pezzuto, F.; Cocco, R.; Flores, G.; Iannitti, T. Functional Foods in Pet Nutrition: Focus on Dogs and Cats. *Res. Vet. Sci.* **2017,** *112*, 161–166.
97. Viana, L. M.; Mothé, C. G.; Mothé, M. G. Natural Food for Domestic Animals: A National and International Technological Review. *Res. Vet. Sci.* **2020,** *130*, 11–18.
98. PFI. 7 Pet Food Solutions for Companion Animal Health [Online]. https://www.petfood-industry.com/articles/6442-functional-foods-effects-on-dog-cat-health-summarized
99. PFI. Adventures in Pet Food [Online]. https://www.petfoodindustry.com/blogs/7-adven-tures-in-pet-food/post/10859-wet-pet-food-continues-ascent-especially-for-cats
100. PFI. State of the us Pet Food and Treat Industry, 2021 [Online]. https://www.petfoodpro-cessing.net/articles/15310-state-of-the-us-pet-food-and-treat-industry-2021
101. PFP. Understanding Texture and How to Formulate Pet Food, Treats for Success [Online]. https://www.petfoodprocessing.net/articles/14938-understanding-texture-and-how-to-formulate-pet-food-treats-for-success
102. PFI. Functional Foods and Advances in Pet Nutrition [Online]. https://pfiaa.com.au/functional-foods-and-advances-in-pet-nutrition/
103. Alnaimy, A. Using of Citrus Byproducts in Farm Animals Feeding. *Open Access J. Sci.* **2017,** *1.*
104. Oliveira, R. A.; Narciso, C. D.; Bisinotto, R. S.; Perdomo, M. C.; Ballou, M. A.; Dreher, M.; Santos, J. E. Effects of Feeding Polyphenols from Pomegranate Extract on Health, Growth, Nutrient Digestion, and Immunocompetence of Calves. *J. Dairy Sci.* **2010,** *93*, 4280–4291.
105. Niu, J. L.; Zhang, J.; Wei, L. Q.; Zhang, W. J.; Nie, C. X. Effect of Fermented Cottonseed Meal on the Lipid-Related Indices and Serum Metabolic Profiles in Broiler Chickens. *Animals (Basel)* **2019,** *9.*

CHAPTER 18

Bioactive Properties and Potential Utility of Extractable and Non-Extractable Phytochemical Fractions from Selected Fruits and Horticultural Wastes

YASUNORI HAMAUZU

Faculty of Agriculture, Department of Agricultural and Life Sciences, Shinshu University, Nagano, Japan

ABSTRACT

Fruits and vegetables are rich in phytochemicals, including polyphenols, carotenoids, and pectic polysaccharides, which are often present at high concentrations in rather inedible materials, such as medicinal fruits and agro-industrial byproducts and wastes. In horticultural or agro-industrial byproducts, over-matured organs of vegetables, immature fruits from thinning practice or peel, seeds, and pomace obtained after food processing tend to contain tough cell wall fiber components and secondary metabolites such as polyphenols because of plant protection strategies. These plant materials have values as phytochemical sources and as functional materials with potential biological activities. When considering the extraction of phytochemicals, the differences among maturation stages and/or plant parts as well as the changes that occurred during postharvest and processing conditions, should also be weighed. Moreover, the biological activity of a fraction cannot always be explained by existing phytochemicals. Still, it should be related to the chemical composition that is affected by the handling of materials or extraction procedures. In this

Food Byproducts Management and their Utilization. Ricardo Gómez-García, Ana A. Vilas-Boas, Débora A. Campos, Maria Manuela Pintado, & Cristóbal Noé Aguilar (Eds.)

chapter, phytochemical profiles, mainly focusing on polyphenols, and the biological activities of extractable and non-extractable fractions from fruits and agro-industrial byproducts and wastes are described. For extractable fractions, studies on the biological activities of extracts from various agro-industrial wastes and selected plant materials (immature apple, quince fruit and seeds, Chinese quince fruit, grape bunch stem, and persimmon peel) that are closely related to our regional industry and research are mainly described. For non-extractable fractions, the expected health benefits and potential importance of non-extractable polyphenols/fiber complexes are explained.

18.1 INTRODUCTION

Fruits and vegetables contain various bioactive compounds (phytochemicals), such as polyphenols, carotenoids, ascorbic acid, organosulfur compounds, and pectic polysaccharides that are, at least partly, thought to be related to the health effects of fruit and vegetable intake.[1,2] Phytochemical content and composition vary with different commodities or cultivars, but even within the same commodities or cultivars, it varies among tissue parts and growth, maturation, and ripening stages. Phytochemicals are often found at high concentrations in inedible medicinal plants or unutilized parts/stages of fruits and vegetables usually discarded. For instance, epidermal tissues such as skin and seed coats contain abundant polyphenols, carotenoids, and cell wall fiber components.[3] These unutilized or low-utilized parts/stages of fruit and vegetables result from regular production processes followed by various agro-industries, including fruit and vegetable production and food-processing industries. As rich sources of phytochemicals, fruit byproducts, agro-industrial wastes, and some medicinal fruits can be valuable materials for health-related industries. It is also known that phytochemicals in fruits and vegetables and in their tissue parts are changed by postharvest handling conditions, including storage and processing. In addition to phytochemical features, various bioactivities of extracts prepared from byproducts and wastes have been reported.[4]

This chapter provides information about phytochemical characteristics and bioactive properties of extractable and non-extractable fractions from some selected fruits and agro-industrial byproducts.

18.2 BIOLOGICAL ACTIVITIES AND POTENTIAL UTILITY OF EXTRACTABLE FRACTIONS OF FRUIT AND AGRO-INDUSTRIAL BYPRODUCTS AND WASTES

During the management of fruit production and related agro-industries, unutilized, or under-utilized parts/stages of fruits, such as thinned immature fruits, skins, cores, seeds, and pomace from ripe fruits, are generated as byproducts and wastes. In the case of vegetables, thinned seedlings and unutilized plant parts, including over-matured leaves and stalks, also occur as horticultural wastes. These byproducts and wastes are often rich in various phytochemicals and can be used as sources of such compounds.[5–7]

Although extracts from byproducts and wastes have some functional properties, such as anti-oxidant, anti-inflammatory, anti-cancer, anti-atherogenic, and anti-diabetic activities,[4] a direct association between these properties and the phytochemicals in the extract has sometimes been challenging to achieve. Moreover, the biological functions of such extracts might be explained by the combined action of one or more known phytochemicals and unknown factors. Therefore, besides being valuable as phytochemical sources, fruit byproducts, and horticultural wastes have functional properties that should also be considered important values.

In this section, the screening studies that compare various fruit byproducts and horticultural wastes will be first introduced, followed by the phytochemical characteristics and functional properties of extracts from selected plant materials (immature apple, apple pomace, quince, Chinese quince fruit and seed, grape bunch stem, and persimmon peel) that are particularly related to our research and regional industries. The phytochemical characteristics and bioactive properties of the extractable fractions from selected fruits and byproducts are summarized in Tables 18.1 and 18.2, respectively.

TABLE 18.1 Phytochemical Characteristics from Selected Fruits and Byproducts.

Fruit/ Byproduct	Phytochemical characteristic	Reference
Immature apple (*Malus pumila*)	Catechin, Epicatechin, Procyanidins (B1, B2, C1),	[12,14,16]
	Chlorogenic acid (5-Caffeoylquinic acid),	
	p-Coumaroylquinic acid	
	Dihydrochalcones (Phloridzin and Phloretin-xyloglucoside)	
	Quercetin glycosides, Cyanidin 3-galactoside	
	Total EPPs by HPLC 3.9–6.0% of DW	
	[Ratio of dihydrochalcones tends to be higher than ripe fruit]	

TABLE 18.1 *(Continued)*

Fruit/ Byproduct	Phytochemical characteristic	Reference
Apple pomace (*M. pumila*)	Similar polyphenols exist but ratio of flavonols is higher than whole fruit.	[22,24,110]
	Total EPPs 0.17–0.99% of DM	
	Pectin 3.5–14.3% w/w of DM	
	Total Dietary Fiber approx. 60–90% of DM	
Quince fruit (*Cydonia oblonga*)	3-Caffeoylquinic acid, 5-Caffeoylquinic acid,	[36,39,42]
	Procyanidin polymers,	
	Flavonol glycosides (peel)	
	Total EPPs 0.3% of FW, NEPAs 0.19% of FW	
	Pectin 0.79% of FW	
Quince seed (*C. oblonga*)	3-Caffeoylquinic acid, 5-Caffeoylquinic acid,	[36,39,40]
	Flavone derivatives (Vicenin-2, Stellarin-2, Schaftoside, Isoschaftoside, Lucenin-2, Chrysoeriol di-C-glycosides)	
	Mucilage (mainly Cellulose microfibrils + Glucuronoxylan)	
Chinese quince (*Pseudocydonia sinensis*)	3-Caffeoylquinic acid,	[42,77]
	Procyanidins (various degree of polymerization)	
	[Mainly highly polymerized procyanidins]	
	Total EPPs 1.3% of FW, NEPAs 0.19% of FW	
	Pectin 0.84% of FW, Cellulose + Lignin 4% of FW	
Grape bunch stem (*Vitis vinifera*)	Catechin, Procyanidin B1, Proanthocyanidin polymers	[59,61,62]
	Caftaric acid, Flavonols (mainly quercetin 3-glycosides)	
	Total EPPs by HPLC 2.2% of DW	
	Fiber components 60–90% of DM	
Grape seed (*Vitis spp.*)	Catechin, Epicatechin, Epicatechin gallate, Procyanidins	[4,111,112]
	[Monomers ≈ 50% of total polyphenols]	
	Gallic acid	
	Total EPP 7.9–15.5% of DW	
Persimmon peel (*Diospyros kaki*)	Tannin 2.8% of DM [as NEPAs in dried peel]	[66,67,70, 72]
	Epicatechin, Epigallocatechin,	
	Gallic acid,	
	Flavonol glycosides	
	β-Cryptoxanthin, β-Carotene, Lycopene	
	Dietary fiber 17.2% of DM	
	Chitinase	

DM, dry matter; DW, dry weight; EPPs, extractable polyphenols; FW, fresh weight; NEPAs, non-extractable proanthocyanidins.

TABLE 18.2 Bioactive Properties of Extractable Fraction from Selected Fruits and Agro-Industrial byproducts/wastes.

Fruit/Byproduct	Type of extract	Bioactivity	Experiment	Reference
Immature apple	Hot-water extract	Cytoprotection	AAPH-induced cytotoxicity in MA104 cells	[8]
		Gastroprotection	Oral administration to rats before induction of gastric injury by HCl-EtOH	[9]
	Crude polyphenols	Anti-allergic effect	Inhibitory effect on degranulation of RBL-2H3 and peritoneal mast cells	[18]
Apple pomace	Hot-water extract	Gastroprotection	Oral administration to rats before induction of gastric injury by HCl-EtOH	[9]
	Polyphenol fraction	Anti-cancer property	Cytotoxic effects against human cancer cells	[25]
Quince fruit	Hot-water extract	Cardioprotective and anti-diabetic effect	Daily oral administration of extract to rats induced diabetes by streptozotocin	[43]
		Gastroprotection	Oral administration to rats before induction of gastric injury by HCl-EtOH	[45]
		Anti-allergic effect	Atopic dermatitis model mice given diet containing the extract	[44]
	70% EtOH extract	Aphrodisiac effect	Daily oral administration to male rats and several sexual behaviors were monitored	[113]
Quince seed	Mucilage (Hot-water extract)	Wound healing	Cream containing mucilage was applied to full-thickness excision wound in rabbits	[46]
	Ethanol extract	Anti-bacterial effect	Extracts were intraperitoneally injected to mice 3 days after injection of bacterial suspension	[49]
		Anti-allergic effect	Topical application to the skin of atopic dermatitis model mice	[50]
Chinese quince fruit	Hot-water extract	Gastroprotection	Oral administration to rats before induction of gastric injury by HCl-EtOH	[45]

TABLE 18.2 *(Continued)*

Fruit/Byproduct	Type of extract	Bioactivity	Experiment	Reference
	Crude polyphenols	Anti-influenza viral activity	Inhibitory effect on hemagglutination of chicken erythrocytes	[42]
Chinese quince seed	Hot-water (40°C) extract	Anti-fungal activity	Inhibitory effect on mycelial growth of *Colletotrichum gloeosporioides*	[55]
Grape bunch stem	Hot-water extract	Cytoprotection	AAPH-induced cytotoxicity in MA104 cells	[8]
	Ethanol extract	Cytoprotection	AAPH-induced cytotoxicity in MA104 cells	[8]
	proanthocyanidin octamer	Anti-cancer activity	Inhibitory effect on cell proliferation and metastasis in cancer cell lines	[64]
Grape seed	Hot-water extract	Cytoprotection	AAPH-induced cytotoxicity in MA104 cells	[8]
Persimmon peel	Ethyl acetate fraction from 40% ethanol extract	Antioxidant and neuroprotective effect	Neuronal cells (PC-12 and SH-SY5Y) exposed to oxidative stress	[69]
	2-methoxy-4-vinylphenol from methanol extract	Tyrosinase inhibiting activity	Inhibitory effect on enzymatic oxidation of L-tyrosine and L-DOPA	[66]

AAPH, 2,2'-azobis-(2-amidinopropane) dihydrochloride; L-DOPA, L-3,4-dihydroxyphenylalanine.

18.2.1 FRUIT BYPRODUCTS AND HORTICULTURAL WASTES-SCREENING STUDIES

Although there are many studies concerning the bioactive potential of fruit and vegetable byproducts and wastes, reports on the bioactivity of extracts under the same conditions seem to be limited. Kabir et al.[8] studied the total polyphenol content, antioxidant activity, and cytoprotective effect of extracts from 26 fruit and vegetable wastes and byproducts. They found, the EC_{50} value of the grape seed (*Vitis labruscana*: *V. labrusca* × *V. vinifera*) hot-water extracts (HW), measured by the DPPH system for determining antioxidant activity, was the highest (10.5 ± 0.4 μg/mL) among those of all samples, followed by those of the immature prune (*Prunus domestica*) ethanol extracts (Et) (15.2 ± 0.5 g/mL) and the buckwheat hull (*Fagopyrum esculentum*) Et (37.6 ± 6.7 g/mL). Using an AAPH-induced cytotoxicity test in MA104 cells, the grape seed HW, which had high polyphenol content and antioxidant activity, showed high cytoprotective activity (89.9 ± 1.6%) (Figure 18.1). However, tests done with broccoli leaf (*Brassica oleracea* Italica group) HW also showed high-cell viability, comparable to that with the grape seed HW, although the total polyphenol content and antioxidant activity in the broccoli leaf HW were not as high. The authors also made an interesting observation, that the black adzuki bean hull (*Vigna angularis*) HW stimulated cell viability (210.9 ± 11.5%). This result suggests that some compounds in black adzuki bean hull can induce cell proliferation and that such compounds can be extracted effectively using hot water.

Our research group also studied the gastroprotective properties of selected extracts from the same set of samples described in the work of Kabir et al.,[8] and observed that several extracts had a significant protective effect, including Chinese quince (*Pseudocydonia sinensis*) HW (100% prevention of injury), quince (*Cydonia oblonga*) HW (79.3%), immature apple (*Malus pumila*) HW (68.0%), and apple pomace HW (67.9%) when 200 mg extract was orally administered to each experimental rat 30 min before the induction of gastric injury with acidified ethanol.[9] Additionally, grape skin Et, apple pomace Et, perilla pomace (*Perilla frutescens*) HW, broccoli leaf HW, broccoli basal stem HW, cabbage outer leaf (*B. oleracea* Capitata group) HW and Et, and buckwheat hull HW and Et also showed a significant effect on the prevention of ethanol-induced gastric lesions (unpublished results).

Widsten et al.[10] studied the antibacterial and antioxidant activities of extracts (extracted by acetone–water, 80:20, v/v) from byproducts of the fruit processing industry to determine their usability as materials for antimicrobial packaging or antioxidant dietary supplements. In this study, 19 skin and seed extracts (including apple, avocado, banana, feijoa, grape, kiwi fruit, lemon, mango, mandarin, mangosteen, orange, passionfruit, pineapple, pawpaw, blackberry, and blackcurrant byproducts) and five fruit juicing residue extracts (from boysenberry, blueberry, raspberry, strawberry, and tomato) were investigated and compared with five tree tannins. The authors concluded that the most promising natural substances for antibacterial packaging or antioxidant dietary supplements were mango seed extract and tannic acid, which are rich in polygalloyl glucose type phenolics. This study also observed moderate antibacterial activity of extracts from feijoa skin powder or heat-modified tannic acid (200°C, 260 min).

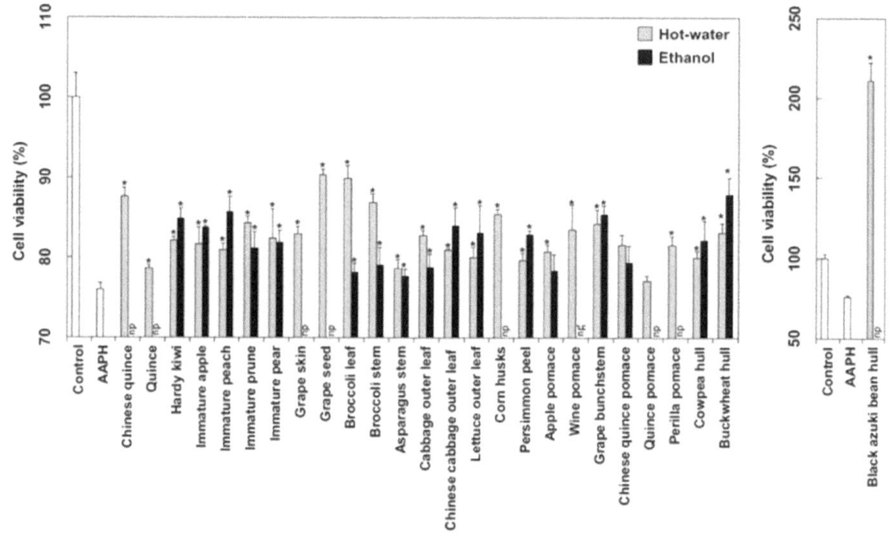

FIGURE 18.1 Effects of extracts from fruits and vegetables wastes and byproducts on MA104 cell viability. MA104 cells were treated with sample extracts for 24 h. Cell viability was determined by MTT assays and expressed as percentages of viable cells compared to the untreated control. Data are presented as mean ± SD ($n = 6$). Asterisks (*) indicate significant differences ($P < 0.05$) between sample extract-treated cells and cells treated with 2,2'-azobis-(2-amidinopropane) dihydrochloride (AAPH) only.

Source: Reprinted with permission from Ref. [8] © 2015 Elsevier

18.2.2 APPLE FRUIT AND BYPRODUCTS

Apples are produced worldwide (about 86 million tons in 2020).[11] Consequently, large amounts of fruit byproducts and wastes are also globally produced. During apple production, fruit thinning is conducted to control the leaf-to-fruit ratio, which affects the final fruit quality; therefore, abundant immature thinned fruits are generated in apple orchards as a fruit waste. Thinned immature apples contain high amounts of polyphenols which decreases over the fruit maturation period.[12,13] The polyphenol content of immature apples (approximately 30 days after pollination) was approximately 10 times higher than that of the ripe fruit. Therefore, a higher antioxidant activity was reported in immature apples than in ripe fruit.[14] HW from immature apples exhibited cytoprotective activity in an AAPH-induced cytotoxicity cell system[8] and gastroprotective effects against ethanol-induced gastric injury in rats.[9]

The main classes of polyphenols in apples are flavan-3-ols (catechin, epicatechin, and proanthocyanidin), hydroxycinnamic acids (chlorogenic acid and *p*-coumaroylquinic acid), dihydrochalcones (phloridzin and phloretin xyloglucoside), and flavonols (quercetin glycosides).[15–17] Kanda et al.[18] studied the anti-allergic activities of polyphenol fractions extracted from immature apples and reported that the crude apple polyphenol fraction strongly inhibited histamine release from rat basophilic leukemia (RBL-2H3) cells and rat peritoneal mast cells. The study showed that the most effective components of the crude polyphenol fraction were proanthocyanidins (condensed tannins).

Dihydrochalcones (phloretin 2′-*O*-glucoside = phloridzin, and phloretin 2′-*O*-xyloglucoside) are among the characteristic compounds of apples. Phloridzin has been shown to inhibit sodium/glucose cotransporters in the intestine (SGLT1) and kidney (SGLT2), and is expected to be an antihyperglycemic and antihyperlipidemic agent in diabetes.[19] Phloridzin content is higher in immature fruits than in ripe fruits, but it seems to be distributed more largely in seeds[14] and leaves.[16] In dry leaves (cv. Ozark Gold), the phloridzin content was reported to be very high, representing more than 11% of the dry weight.

Apple pomace, a byproduct of apple juice production, comprising skin, core, seed, and pulp fibers, is a noteworthy material containing abundant phytochemicals, including polyphenol/pectin mixtures.[20,21] Although some polyphenols are present in the juice fraction, a large proportion of polyphenols, such as flavonols or procyanidin polymers, remain in the pomace.[22,23] Polyphenols and pectin contents have been estimated to be 170–990 mg/100 g DW and 3.5–14.32 g/100 g DW, respectively.[24] Rana et al.[25] extracted polyphenols from industrial apple pomace and showed that dihydrochalcones and flavonols were predominant. The extracted polyphenol fractions showed significant anticancer and anti-inflammatory activities.

Besides polyphenols, pectin is also regarded as an important phytochemical in apple fruit. Apple polyphenols and pectin have been shown to have functions related to cardiovascular health. In a rat experiment, Aprikian et al.[26] observed that apple pectin and the polyphenol-rich fraction of an apple extract were more beneficial for blood lipid metabolism when fed together than when fed separately. A human crossover study has shown that the LDL-cholesterol lowering effect was observed after cloudy apple juice intake, but an inverse effect (increased LDL-cholesterol) was found after clear apple juice intake.[27] Therefore, a mixture of polyphenols and fiber components, such as pectin, is an important ingredient of apple fruit to reduce cardiovascular disease. Because apple pomace contains a high amount of polyphenol/pectin mixture, it can be expected to be a health-promoting food material.

18.2.3 QUINCE FRUIT AND BYPRODUCTS

Quince fruit is a medicinal fruit that has been used in folk medicine in various parts of the world. Quince fruits are used as traditional remedies to treat diabetes (either raw or cooked fruits),[28] cough (decoction),[29] fever (hot-water extract),[30] inflammatory bowel disease,[31] cystitis (cooked and applied externally),[32] and constipation (raw or as jelly, direct or as tisane)[33]; they have also been used as skin emollients (decoction).[34] In Islamic literature (Ahadith), the consumption of quince fruit is recommended with claims of beneficial effects on heart ailments.[35]

In addition to flesh parts, leaves, seeds, and flowering stems have been used as traditional folk medicines.[29,36,37] Leaf extracts (infusions) are used for cold, flu, cough, and as bronchial calmative, and the decoction of leaves is used as diuretic and to treat kidney stones, cough, asthma, and headache.[29] Seed decoction has been used to treat diarrhea,[29] cough, bronchitis, and constipation,[36] and as a constituent of medicines for conjunctivitis.[38]

Although the chemical components involved in the health-promoting functions in the traditional usage of quince are not specified, polyphenols have been listed as important phytochemicals in the utilized parts of this plant.[36,39] One characteristic of the polyphenolic profile in quince is the existence of 3-*O*-caffeoylquinic and 5-*O*-caffeoylquinic acids as the two main phenolic acids. These two caffeoylquinic derivatives are common to the pulp, peel, seed, and leaf, although other polyphenol components are different among these parts. The peels and leaves contain kaempferol glycosides and quercetin glycosides, including kaempferol 3-*O*-rutinoside and quercetin 3-*O*-rutinoside (rutin) as characteristic flavonoids. The seeds contain characteristic flavone derivatives such as vicenin-2 (apigenin 6,8-di-C-glucoside), stellarin-2 (chrysoeriol 6,8-di-C-glucoside), schaftoside (apigenin 6-C-glucoside-8-C-arabinoside), isoschaftoside (apigenin 6-C-arabinoside-8-C-glucoside), lucenin-2 (luteolin 6,8-di-C-glucoside), 6-C-pentosyl-8-C-glucosyl chrysoeriol, and 6-C-glucosyl-8-C-pentosyl chrysoeriol.[36,40] Concerning polyphenols in the pulp, 3-*O*-caffeoylquinic and 5-*O*-caffeoylquinic acids have been reported to account for most of total polyphenols (in methanolic extracts).[41] However, a large amount of procyanidins (mean degree of polymerization 35) could be extracted in our study (cultivar "Smyrna") when 60–70% (v/v) acetone was used as the extractant.[42] No mention to polymeric procyanidins could be found in several reports, this might be due to a series of factors, for example, (1) only low-molecular-weight compounds are focused, (2) determination using HPLC without thioacidolysis is especially difficult for polymeric procyanidins, and (3) procyanidin polymers may bind to cell wall polysaccharides during freezing or drying and change to a non-extractable form. In addition to polyphenols, other compounds such as steroids, terpenoids, and polysaccharides, including a component of seed mucilage, are also important phytochemicals in quince.[39]

18.2.3.1 QUINCE FRUIT (WHOLE FRUIT OR FLESH)

Hot-water extracts of quince fruits, including decoction, have often been used as folk medicines for treating diabetes, cough, and fever as well as for heart disease, as described above.

Mirmohammadlu et al.[43] reported the cardioprotective and anti-diabetic effects of an aqueous extract (decoction) of quince fruit. Briefly, fresh quince fruits were dried in the shade and decocted in water for 30 min. The extract was filtered and concentrated until a honey-like viscosity was achieved, and then stored at –20°C until use. The extract (containing 24% water) was diluted in distilled water and once-daily doses of 80, 160, and 240 mg/kg body weight were given for 6 weeks to male Sprague-Dawley rats in which diabetes was induced by a single intraperitoneal dose of streptozotocin (60 mg/kg) dissolved in citrate buffer (1 mL, pH 4.5). The results clearly showed that quince fruit extract successfully reduced serum triglycerides, total cholesterol, LDL-cholesterol, liver damage markers (ALT, AST, and ALP), and renal dysfunction markers (urea and creatinine), all of which were elevated in the diabetic rats. In addition, HDL-cholesterol, which was reduced in diabetic rats, increased in the rats administered the quince extract.

These observations strongly support the use of quince hot-water extracts as a folk medicine to treat heart disease or diabetes.

In our study, hot-water extracts of quince fruit showed anti-allergic[44] and gastroprotective effects.[45] Shinomiya et al.[44] reported the inhibitory effect of quince hot-water extracts on atopic dermatitis (AD)-like skin lesion symptoms in NC/Nga mice. In this study, experimental diets were prepared by adding 2.5 and 5.0% of hot-water extracts from quince fruit (that were obtained by boiling peel, pulp, and seeds for 1 h). AD model mice were fed such diets for 63 days. The results clearly demonstrated that the mice fed the experimental diets containing quince hot-water extracts, especially at 5.0%, had a significant reduction of AD-like skin lesions as well as in serum IgE level. Moreover, quince hot-water extracts showed an inhibitory effect on the release of β-hexosaminidase from the rat basophilic leukemia (RBL-2H3) cells. The study also showed that the small molecular fraction (<3 kDa) from quince hot-water extracts reduced the mRNA expression of the high-affinity IgE receptor (FcεRI) γ subunit. These results suggest that hot-water extracts from quince fruit have an inhibitory effect on type I allergy by suppressing IgE production and IgE-mediated degranulation. Concerning the gastroprotective effect, oral administration (a dose of 200 mg per rat) of quince hot-water extracts from the pulp effectively reduced (80% inhibition) ethanol-induced gastric injuries in rats.[45] Regarding this extract, the yield was 9.1% of quince pulp after 1 h of boiling. Total polyphenol and polyuronide contents were 22.3 mg (epicatechin equivalent) and 41.0 mg (galacturonic acid equivalent) per 1 g of extract, respectively.

18.2.3.2 QUINCE SEEDS

18.2.3.2.1 Quince Seed Mucilage

Quince seeds extrude mucilage when soaked in water. Seed mucilage has been traditionally used in the treatment of cough, sore throat[35], and respiratory diseases, especially in children, and also applied externally to minor burns.[46] In the scientific literature, quince seed mucilage has been shown to heal wounds more rapidly than a commercial wound healing cream (1% phenytoin).[47] Quince seed mucilage acts by increasing the wound fluid levels of four growth factors, the epidermal growth factor (EGF), transforming growth factor (TGF)-β1, vascular endothelial growth factor (VEGF), and platelet-derived growth factor (PDGF), which play a crucial role in the wound healing process.[46]

Quince seed mucilage is mainly composed of cellulose microfibrils associated with glucuronoxylan, with a high proportion of glucuronic acid residues.[39,48] Because of this, the mucilage has a high water-retention capacity in deionized water or buffer solution around physiological pH 6.8–7.4.[48] The application of mucilage to a wound provides a moist environment, which is the best environment for healing. In addition to this suitable environment, polyphenolic antioxidants also act as accelerators for wound healing because free radicals in wound sites are cytotoxic and delay wound healing.[46] Quince seed polyphenols can be, at least partly, extracted by hot water heated around 50–60°C; these conditions are often used for mucilage preparation.

18.2.3.2.2 Quince Seed Ethanol Extract

When soaked in ethanol, quince seeds cannot extrude the mucilage, and the phytochemical composition of the ethanol extracts may differ significantly from that of the hot-water extracts. Quince seed ethanol extracts have anti-bacterial effects on Enterobacteriaceae (*Escherichia coli*, *Klebsiella pneumoniae*, and *Enterobacter aerogenes*), and the effect is stronger than those of acetone or aqueous extracts.[49] In this study, the anti-bacterial effect was also confirmed by *in vivo* experiments using albino mice that were inoculated intraperitoneally with a bacterial suspension.

In our study, ethanol extracts of quince seeds showed a preventive effect on atopic dermatitis (AD)-like symptoms in NC/Nga mice by topical application to the skin.[50] In this study, quince seed ethanol extracts were topically administered to the upper dorsal skin at concentrations of 0.1 or 1.0% (in ethanol) twice a week for 4 weeks. AD-like skin lesions were induced by topical application of Biostir-AD, a hydrophilic petrolatum-based ointment-containing extract of house dust mite (*Dermatophagoides farinae*). The dermatitis induction started 1 week after the beginning of the administration of the quince seed extract. For the following 3 weeks, the Biostir-AD was applied to the skin area twice a week immediately after the administration of the quince seed ethanol extracts. The results clearly showed that the administration of quince seed ethanol extracts to mice significantly reduced total clinical severity scores, especially in the 1.0% group (score = 0.5), while mice administered only vehicle (ethanol) resulted in severe AD-like lesions (score = 8.2). Interestingly, the expression of thymus- and activation-regulated chemokine (TARC) in the dorsal skin was downregulated. The direct effect of quince seed ethanol extracts on keratinocytes was also observed in an *in vitro* evaluation using the human keratinocyte cell line HaCaT. In this experiment, the expression and production of TARC in HaCaT cells were suppressed by the extract. Although the active compounds have not been fully elucidated, it is known that the quince seed ethanol extracts contained 10.7 mg (epicatechin equivalent) of total polyphenols per 1 g of extract and exerted 88.8 μmol TE (Trolox equivalent) per 1 g extract in free radical scavenging activity (unpublished data).

18.2.4 CHINESE QUINCE–THE UNDERUTILIZED FRUIT

Chinese quince (*Pseudocydonia sinensis*) is believed to be native to China (Zhejiang province), and is now widely planted in Japan, China, and Korea. Chinese quince fruit has been used in traditional medicine in Asian countries. Because of its hard flesh, strong astringency, and high acidity, the fruit is not edible in its raw form. These characteristics are similar to those of quince fruit (*Cydonia oblonga*), but Chinese quince has numerous stone cells as well as a high content of insoluble fibers (cellulose and lignin) that are larger than those in the quince fruit, making the flesh more unpleasant to eat. Therefore, Chinese quince fruit is usually consumed as processed food products such as concentrated juice extracts, liquor, jam, jelly, glutinous starch syrup, crystallized fruit, and throat lozenges. Dried fruits have also been used in traditional medicine in the form of hot-water extracts because of their antitussive and/or expectorant properties. Thus, homemade medicines from Chinese quince fruit (including fruit liquor, decoction, syrup, and paste) have been reported to have antitussive, expectorant, antispasmodic, and antidiuretic actions, and have been used to combat respiratory infections, intestinal dysregulation, and diuresis as well as for treating people with a weak constitution.

Despite the existence of utilities as folk medicines or processed foods with some health claims, this fruit that requires too much effort for processing or cooking is sometimes discarded on a street corner without any use. Therefore, we have been studying new aspects on the utility of this underutilized fruit, supported by scientific research.

18.2.4.1 CHINESE QUINCE FLESH

Chinese quince fruit (flesh part) contains high amounts of polyphenols (>1280 mg/100 g FW); a concentration that is approximately four times higher than that in quince fruit (cv. Smyrna; 303 mg/100 g FW), and 20 times higher than that in apple fruit (cv. Fuji; 61 mg/100 g FW).[42] Polyphenol composition of Chinese quince fruit is characterized by a procyanidin mixture with various degrees of polymerization. We studied the effect of polyphenolic fraction (at the same dose) extracted with 70% (v/v) acetone from Chinese quince, quince, and apple flesh on *in vitro* anti-flu viral activity, which was assessed by hemagglutination inhibition test using chicken erythrocytes. The highest efficacy was observed for Chinese quince polyphenols. In another study, heat-treated Chinese quince polyphenols (4 and 12 h boiling with 5% citric acid) showed a decreased anti-influenza viral activity, but moderate activity was retained.[51] Interestingly, the *in vitro* antioxidant activity of the polyphenols increased after an extended boiling time. This may be due to an increase in small molecular compounds, including dimeric and oligomeric procyanidins, formed by the degradation of higher polymers.[52] When heat-treated Chinese quince polyphenols were administered to rats, the *in vivo* antioxidant activity (assessed by the ferric reducing antioxidant power or FRAP method) was elevated, and plasma vanillic acid, a phenolic metabolite of procyanidins produced by intestinal bacterial degradation and enzymes in the body, also increased significantly.[52] Thus, heat treatment of Chinese quince polyphenols seems to increase health benefits, supporting the effectiveness of traditionally used Chinese quince fruit hot-water extract or decoction.

When Chinese quince fruit was boiled for an extended period of time, not only polyphenols but also other components, such as sugars, organic acids, and pectic polysaccharides (pectin), were also extracted. Especially for pectin extraction, long-time boiling is effective because the pectin of Chinese quince fruit mostly exists in an insoluble form.[53] Besides polyphenols, pectin also has various pharmacological activities, including gastroprotective effects. We investigated the chemical characteristics and preventive efficacy of boiling-water extracts from Chinese quince fruits on HCl/ethanol-induced gastric lesions in rats. The administration of the fruit extract obtained by boiling for 2 h significantly prevented gastric mucosal lesions; however, protection was not achieved with extracts obtained after boiling for 1 h only. Because boiling for extended periods of time has the advantage of breaking cell wall polysaccharides and extracting chemical components from the fruit tissue, the extract obtained by boiling for 2 h had more phytochemicals such as antioxidants than those obtained by boiling for 1 h. In fact, the amount of polyphenols extracted from 100 g of fruit tissue after 1 and 2 h of boiling were 791 mg (62.3%) and 985 mg (77.6%), respectively. Likewise, the amount of pectin extracted from 100 g of tissue after boiling for 1 and 2 h was 291 mg (34.6%) and 365 mg (43.5%), respectively. Therefore, extended periods of boiling time in Chinese quince fruit processing may be beneficial from the viewpoint of obtaining phytochemicals with strong gastroprotective activity.[9]

18.2.4.2 CHINESE QUINCE SEEDS

The seeds of Chinese quince also extrude mucilage when they are soaked in water, similar to that of quince seeds. Although there are few studies on Chinese quince seeds, the number of seeds per fruit (around 200) seems to be higher than that in quince fruit (around 25–40 seeds in cultivar "Smyrna"); therefore, the collecting efficiency of Chinese quince seeds might be higher than that of quince seeds. Polyphenols of Chinese quince seeds have been shown to be, at least partly, similar to those of quince seeds. Vicenin-2, schaftoside, isoschaftoside, and lucenin-2, the characteristic flavone derivatives, have been detected in hot-water extracts (40°C, 24 h) of Chinese quince seeds.[54] This seed extract has potential utility as an edible coating material that may extend fresh produce shelf life. Angraeni et al.[55] reported that the dipping treatment of mango fruit (cv. Nam Dok Mai Si Thong) in Chinese quince seed extract reduced necrotic lesion expansion (anthracnose disease) caused by *Colletotrichum gloeosporioides*. Moreover, cherry tomatoes treated with coatings containing 3% seed mucilage showed higher flesh firmness and lesser weight loss than uncoated fruits.[56]

18.2.5 GRAPE BUNCH STEMS

Grapes are one of the most produced fruit globally, and approximately 50% of grapes are used to make wine.[57] Viticulture produces more than 70 million tons of berries per year in a worldwide cultivated area of 7.5 million ha[58]; therefore, a significant volume of grapevine woody derivatives, such as bunch stems and canes, are produced as horticultural- or agro-industrial wastes.[59] Bunch stems accumulate during the winemaking process and constitute 5% of the processed grapes at harvest time,[60] accounting for up to 25% of the total byproducts (residues) generated by the wine industry.[61] Grape bunch stems contain dietary fiber (60–90% of total dry matter) as a crucial component that is composed mainly of cellulose (40–49% of the total cell wall polymer) and lignin (approximately 22.5% of total dietary fiber).[61] Polyphenols are regarded as important phytochemicals, reaching up to 5.8% of their dry weight.[61] The main constituents of bunch stem polyphenols have been reported to be flavanols (mainly (+)-catechin and procyanidin B1), hydroxycinnamic acids (mainly caftaric acid), and flavonols (mainly quercetin 3-*O*-glycosides)[59,61] as contrasted with the polyphenol composition of grape canes that mainly comprises monomeric catechins and ε-viniferin (dehydrodimers of resveratrol).[59] However, according to a recent study, polymeric procyanidins seem to be the most important polyphenols in grape bunch stems.[62] Nieto et al.[62] studied the effects of different extraction conditions on grape (cv. Merlot) bunch stem polyphenols using an environmentally friendly solvent (aqueous ethanol). In this study, the optimal extraction condition for obtaining an extract with the highest polyphenol content (187 mg GAE/g extract) was 30% (v/v) ethanol at 120°C for 10 min, and the procyanidin content quantified for this extract was 80 mg/g (mDP = 12). Moreover, the mean degree of polymerization of these procyanidins was 12, indicating that polymers (and not monomers or dimers) were the predominant flavanols in grape bunch stems.

Concerning the biological activities of extracts from grape bunch stems, HW and Et from "Chardonnay" bunch stems showed a strong protective effect on AAPH-induced cytotoxicity in a cell system.[8] Both extracts contained moderate levels of proanthocyanidins and showed strong anti-allergic activity (inhibitory effect on the β-hexosaminidase release from RBL-2H3 cells) in our additional experiments.[63] In another study on grape bunch stem hot-water extracts, Kawahara et al.[64] isolated and characterized a novel proanthocyanidin octamer, which comprised one epigallocatechin and seven epicatechins with one gallate ester, that showed significant anti-cancer activity. In this study, the proanthocyanidin octamer suppressed cell growth of PC-3 prostate cancer cells, the expression of the metastasis-promoting gene *FABP5*, and the invasion of cancer cells. This effect was stronger than that of epigallocatechin gallate, a well-known cancer-preventing agent. Since the proanthocyanidin octamer inhibited cell proliferation not only in a prostate cancer cell line (PC-3) but also in a colorectal cancer cell line (Lovo) and a breast cancer cell line (MDA-MB-231), this compound could be a potential anti-cancer agent for various cancers.

18.2.6 PERSIMMON PEELS

Persimmon (*Diospyros kaki*) is a deciduous plant native to China, Korea, and Japan, and fruit production is high in Asian countries, Spain, and Brazil.[65] Persimmon fruit is used as a folk medicine and is believed to have pharmacological effects, such as blood pressure-lowering and diuretic effects, and may be used as a hangover remedy and for prophylaxis against the cold.[66] Persimmon is rich in many nutritious and bioactive components, such as proteins, dietary fibers, sugars, minerals, vitamins (including ascorbic acid and tocopherols), carotenoids, and polyphenols (including flavonoids and tannins).[65–67] Roughly, there are two types of persimmon fruits: astringent and non-astringent. Astringent persimmon fruits remain astringent even when they are ripe. Therefore, the fruits should be subjected to an astringent removal treatment before eating.

Drying is a de-astringent treatment traditionally conducted in the southern area of Nagano, Japan. When astringent persimmons are processed into dried fruits, large quantities of peel are generated as byproducts. Persimmon peel has been known to have a higher polyphenol and dietary fiber content as well as stronger antioxidant and chitinase activities than the pulp.[67–70]

Et and HW of persimmon peel showed antioxidant and cytoprotective activities.[8] Jeong et al.[69] also reported that the ethyl acetate fraction from 40% (v/v) ethanol extracts showed antioxidant and neuroprotective effects on neuronal cell lines (PC-12 and SH-SY5Y) exposed to oxidative stress with H_2O_2. Additionally, Fukai et al.[66] reported that 2-methoxy-4-vinylphenol isolated from methanol extracts from persimmon peel showed higher tyrosinase inhibitory activity than arbutin, suggesting that persimmon peel can be used as a material with a whitening effect.

However, in my opinion, the value of persimmon peel might be higher in the non-extractable fraction than in the extracted fraction because of the existence of a large amount of insoluble tannins associated with dietary fiber. Preventive effects on atherosclerosis in low-density lipoprotein receptor-deficient mice fed a high-fat diet supplemented with persimmon peel powder (used in whole) has been reported[71] although the functional constituents have not been identified.

Persimmon peel is rich in chitinases. According to Takii et al.,[68] persimmon peel (cv. Fuji-gaki) showed the highest chitinase activity among 46 plant samples from 24 families investigated (it was 1.08 U/100 g live skin weight and 68.2 U/mg protein in the skin). Sangkaew and Koh[72] studied the utilization of persimmon peel as a beneficial ingredient in poultry diets to improve the digestibility of chitin-rich shrimp meals (meals supplemented with shrimp heads and shells). They reported that in an *in vitro* digestion experiment, addition of persimmon peel (up to 6%) effectively improved the digestibility of dry matter, crude protein, and chitin in shrimp meals. This study is interesting because the usability of a byproduct (shrimp waste) was improved by using another byproduct (persimmon peel). Moreover, chitinases from persimmon fruit pulp have been confirmed to have antifungal activity against the plant pathogenic fungus *Trichoderma viride.*[73] Because of the higher activity of chitinases in persimmon peel than in pulp,[68] persimmon peel can also be used as an environmentally friendly pest repellent.

18.3 BIOACTIVE POTENTIAL OF NON-EXTRACTABLE FRACTIONS

18.3.1 PHYTOCHEMICALS OF NON-EXTRACTABLE FRACTIONS

Here, "non-extractable fraction" refers to the fraction comprised of phytochemicals that cannot be extracted by mild extraction conditions (i.e., without special degradative chemical treatments). This fraction mainly includes cell wall components, such as cellulose, hemicellulose, pectic polysaccharides (pectin), and lignin as dietary fibers, and non-extractable polyphenols (NEPPs) as the most important antioxidants.

NEPPs are defined as polyphenols that are not significantly released from the plant tissue or food matrix by mastication, the acidic pH of the stomach, or the action of digestive enzymes.[74] NEPPs include polyphenols that bind strongly to cell wall materials (such as high-molecular-weight tannins) and those linked to cell walls (such as insoluble-bound phenolic acids). Although many studies on polyphenols as typical bioactive compounds have been conducted, most studies have focused only on extractable polyphenols. In fact, the NEPPs of fruits and vegetables had been ignored until relatively recently.[75] However, it is known that a certain amount of NEPPs exist in fruits and vegetables and that they are quite abundant in some cases. For example, the content of non-extractable proanthocyanidins (NEPAs) and their ratios to total proanthocyanidins in some berries have been shown to be as follows: blackcurrants (*Ribes nigrum*; 202 mg/100 g FW, 50.2%), sea buckthorn (*Hippophae rhamnoides*; 199 mg/100 g FW, 81.5%), cranberries (*Vaccinium oxycoccos*; 181 mg/100 g FW, 44.6%), and redcurrants (*Ribes rubrum*; 86.8 mg/100 g FW, 70.8%).[76] In bananas (*Musa sapientum*), most proanthocyanidins exist as NEPAs (64.3 mg/100 g FW, 100%). In our study, we found that quince fruit contained 190 mg/100 g FW (43.4% of total proanthocyanidins) and Chinese quince fruit contained 267 mg/100 g FW (16.1%).[77] Pérez-Jiménez and Saura-Calixto[78] reported the importance of peels as rich sources of NEPPs (including hydrolyzable polyphenols and NEPAs). However, it is known that their content seems to be affected by pretreatment for sample storage, such as freezing and/or drying. Drying causes an increase in NEPA content in some fruits or waste.[79–81] Thus, the non-extractable fractions of fruits and agro-industrial byproducts contain certain amounts of bioactive compounds, that are also expected to exert certain bioactive functions.

18.3.2 BIOLOGICAL FUNCTION OF NON-EXTRACTABLE POLYPHENOL/DIETARY FIBER COMPLEXES

As already mentioned, the non-extractable fractions of fruits and agro-industrial byproducts are composed mainly of dietary fiber components and associated polyphenols (i.e., NEPP/fiber complexes). There have been many studies on the biological activities of NEPPs "after extraction" from associated macromolecular components.[82–84] Ding et al.[85] reviewed various extraction methods, characteristics, and antioxidant properties of NEPPs from food byproducts. To extract NEPPs, acid or base hydrolysis have often been used to decompose the macromolecular components associated with them; base hydrolysis (such as 2–4 M NaOH, for 1–4 h at room temperature) seems to be more effective than acid hydrolysis.[85,86]

However, studies on the bioactive potential of NEPP/fiber complexes in their unchanged forms are still scarce. Moreover, from the viewpoint of utilizing all available resources, utilization of the entire NEPP/fiber complex may be more important than using NEPPs only. Here, I introduced some potential functions of the NEPP/fiber complex. The biological functions of the NEPP/fiber complexes from selected fruit and agro-industrial byproducts and wastes are summarized in Table 18.3.

18.3.2.1 ANTIOXIDANT ACTIVITY

The NEPPs have antioxidant activity even when they are associated with the cell wall fibers and, therefore, NEPP/fiber complexes are also called "anti-oxidant dietary fibers"[87] or "macromolecular antioxidants"[75] although their effects are not limited to antioxidant activity. Concerning the antioxidant capacity of the NEPP/fiber complex, the antioxidant capacity or reducing ability of those polyphenols associated with the fibers may be weaker than that of free polyphenols.[79] It has been shown that the antioxidant activity of polyphenols associated with fiber or bovine serum albumin is lower than those of their equivalent amounts in free form.[88,89]

TABLE 18.3 Biological Function of Non-Extractable Polyphenols/Dietary Fiber Complexes from Selected Fruits and Agro-Industrial byproducts/Wastes.

Fruit/Byproduct	Type of fraction	Bioactivity/Function	Experiment	Reference
Quince fruit	70% acetone insoluble solids (AIS)	• Free radical scavenging • Bile acid binding (*in vitro*)	• DPPH assay AIS was incubated with bile acid mixture after simulated digestion	[77]
Chinese quince fruit	70% acetone insoluble solids (AIS)	• Free radical scavenging • Bile acid binding (*in vitro*)	• DPPH assay AIS was incubated with bile acid mixture after simulated digestion	[77]
Grape berry	0.1% TFA-50% methanol insoluble	• Prebiotic effect • Reduction of detrimental effect of high-fat intake	Mice were given high-fat diet supplemented non-extractable fraction	[105]
Grape seed	Grape antioxidant dietary fiber (GADF)	• Prebiotic effect • (increase of *Lactobacillus*)	Rats were given diet containing GADF for 4 week	[114]
	defatted, acidified 70% acetone insoluble fraction of GADF	• Provide flavanol metabolites bioavailable at least 24 h after ingestion	Oral gavage of a suspension of NEPA-rich fraction to rats	[92]
Dried persimmon	Ethanol insoluble	• Antioxidant effect (*in vivo*)	Rats were given diet supplemented non-extracted persimmon fraction	[91]
	70% acetone insoluble solids (AIS)	• Bile acid binding (*in vitro*)	AIS was incubated with bile acid mixture after simulated digestion	[79]
Wheat bran	Whole bran	• Antioxidant effect • Source of ferulic acid and the metabolites circulating continuously	Rats were given diet supplemented 26% wheat bran	[93]

Although its antioxidant capacity is lower than that of the free form, the NEPP/fiber complex may exert antioxidant activity in the stomach and small intestine, especially when NEPPs are associated with dietary fiber. However, the antioxidant capacity derived from NEPPs is mostly released in the colon through the action of the microbiota, which can either break the linkage between polyphenols and the fibers or transform them into other metabolites.[90] Because the released metabolites may also exhibit antioxidant activity and can be absorbed, the metabolites may have a systemic effect, improving the antioxidant status of the body. Matsumura et al.[91] showed that the plasma oxygen radical absorbance capacity (ORAC) values in rats fed a diet supplemented with 5% of the non-extractable fraction (ethanol insoluble fraction) of dried persimmon fruit was 1.5 times higher than that of control rats after 1 week.

18.3.2.2 AVAILABILITY OF FUNCTIONAL METABOLITES

Although phenolic metabolites originating from NEPPs may be the same as those from EPPs, and the metabolic fates of NEPPs have some specific advantages that may affect their health-related properties.[74] Because of their association with other macromolecules, such as pectic polysaccharides, NEPPs have two specific features that distinguish them from EPPs during colonic transformation. The first is the time and duration of metabolite availability; the fermentation of the macromolecules associated with NEPPs takes time, and thus the release, metabolism, and absorption of NEPPs take more time than those of EPPs. Therefore, the metabolites derived from NEPPs circulate in the body for longer periods than those derived from EPPs.[92,93] In fact, ferulic acid and its metabolites circulated longer in rats fed wheat bran (containing insoluble-bound ferulic acid) than those in rats fed free ferulic acid.[93] In another study, flavanol metabolites (epicatechin monomers and microbial metabolites) in rats orally administered NEPA-rich fractions from grape seeds were reported to be bioavailable at least 24 h after ingestion because of progressive depolymerization of the NEPA-rich fraction during their transit along the intestinal tract.[92] González-Sarrías et al.[94] reported that the production of these metabolites from NEPPs can persist for up to 3–4 days in the bloodstream, where they can exert mainly anti-inflammatory and antioxidant systemic effects. Thus, longer colonic transformation times and promotion of sustained circulation of beneficial metabolites in the human body are important advantages of the NEPP/fiber complex.

The second feature is the extent of fermentation. Because NEPPs are bound to other fermentable dietary components, such as polysaccharides and proteins, the activity of the colonic microbiota can be affected, which may result in higher levels of fermentation. Saura-Calixto et al.[95] reported that the *in vitro* fermentability (assessed by the production of short-chain fatty acids; SCFA) of polymeric NEPAs isolated from carob pods was 23%, whereas that of NEPAs from grape seeds and peel associated with dietary fiber was 50%. Therefore, plant materials containing NEPPs with fermentable fibers, such as peels and pomaces as well as medicinal fruits seem to have an advantage in the bioavailability of functional metabolites.

18.3.2.3 BILE ACID BINDING CAPACITY

The binding of bile acids and increasing their excretion are thought to be important functions of dietary fiber that contribute to reductions in cholesterol in the body. This is because accelerated excretion of bile acids reduces the enterohepatic bile acid pool, which may stimulate the conversion of cholesterol to additional bile acids in the liver.[96–98] Thus, bile acid-binding and increased fecal excretion have been hypothesized as possible mechanisms by which dietary fiber lowers cholesterol levels.

Cell wall fibers, such as pectin and lignin, are well-known bile acid-binding agents[99–101] but the associated polyphenols, including NEPPs, seem to contribute significantly to the bile acid-binding capacity of dietary fiber. We prepared NEPP/fiber complex (70% acetone-insoluble solids, AISs) from eight fruit samples (apple, unripe pear, ripe pear, quince, Chinese quince, hawthorn, unripe blueberry, and ripe blueberry) by removing the extractable procyanidins (EPCs) with 70% (v/v) acetone and measured an *in vitro* bile acid-binding activity.[77,102] The results showed that the bile acid-binding activity of fruit AISs had a strong positive relationship ($R^2 = 0.7084$) with an index of cell wall components (polyuronides × lignin × NEPC × mDP^{-1} × $cellulose^{-1}$).[102] This indicates that the AISs (including cell wall materials) composed of large amounts of pectin (polyuronides), lignin, and NEPCs with low mDP values and small amounts of cellulose tended to have large bile acid-binding capacities. The existence of NEPCs among the AISs had an impact on bile acid-binding activity because their relationship would be relatively weak if only the pectin, lignin, and cellulose contents were considered ($R^2 = 0.49612$). In another study, the effect of NEPAs on the bile acid-binding capacity of AISs was shown more clearly by using dried persimmon AISs, which have an extremely high content of NEPAs (524 mg/g

AIS).[79] The bile acid-binding capacity of dried persimmon AISs was 40.2 μmol/g AIS, which corresponded to 29% of the activity of cholestyramine, a bile acid sequestrant. Although this was an *in vitro* experiment, we are currently investigating the *in vivo* effect of AISs of dried persimmon on the acceleration of bile acid excretion.

18.3.2.4 STIMULATION OF BENEFICIAL BACTERIA

The dietary fiber fraction has a prebiotic effect and can modulate beneficial gut microbiota because of the existence of fermentable polysaccharides such as pectin.[103] Iwata et al.[104] investigated the effects of fibers from inedible parts of vegetables on the growth of bifidobacteria. In this study, plant materials were lyophilized and milled (<0.8 mm), and dietary fibers were extracted using the Prosky method. As a result, fibers from the peel of Chinese yam, pea pod, and broad bean pod significantly stimulated *Bifidobacterium longum* growth.

The NEPP/fiber fraction from grapes has been shown to modulate probiotic bacteria *in vivo*.[105] *Lachnospiraceae* (genus *Coprococcus*) and *Mollicutes* have been shown to increase in mice fed a high-fat diet supplemented with NEPP/fiber fractions of table grapes (0.1% TFA-50% methanol insoluble fraction), while *Peptococcaceae* decreased compared with those in mice fed a high-fat diet only.

Polyphenols released from the NEPP/fiber complex in the small intestine have also been reported to stimulate intestinal bacteria. *In vitro* studies have shown that two *Bifidobacterium* strains (*B. adolescentis* and *B. bifidum*) were stimulated by rutin, sinapic acid, caffeic acid, and vanillic acid,[106] while *Clostridium coccoides* (*Blautia coccoides*) were stimulated by flavanol monomers[107]; some of the compounds may be released from NEPPs. The level of bioaccessibility of associated polyphenols varies depending on the type of NEPPs used. It has been shown that a high proportion of hydrolyzable polyphenols (58%) is released by treatment with digestive enzymes, while NEPAs are not.[108] Therefore, most NEPAs reach the colon and undergo colonic fermentation.[74]

18.4 CONCLUSIONS

Fruits and agro-industrial byproducts are rich sources of phytochemicals, especially fibers, EPPs, and NEPPs, that have various biological or pharmaceutical activities, such as antioxidant, anti-inflammatory, anti-cancer, anti-diabetic, anti-allergic, and gastroprotective activities. Most of the hot-water extracts or ethanol extracts from fruit and agro-industrial byproducts and wastes show cytoprotective antioxidant activity. The selected plant materials have similar properties in terms of being rich in fiber and flavanol-type polyphenols, although they also have different characteristics. Immature apples tend to have polyphenols with a higher ratio of dihydrochalcones than ripe fruits. Apple pomace contains higher pectin and polyphenols with a higher ratio of flavonoids than the whole fruit. Extracts from quince fruit or seeds, traditionally used in folk medicine, are useful materials with preventive effects against heart disease, for gastroprotection and wound healing, and with anti-allergic effects. Chinese quince fruit, which has a higher content of procyanidins and insoluble fibers than quince fruit, shows strong anti-flu and gastroprotective activities. Mucilage from Chinese quince seed is potentially usable as an edible coating material to extend the shelf life of fresh produce. Grape bunch stems contain proanthocyanidin octamers with anti-cancer activity. Persimmon peel can be used as a chitin-degrading material in shrimp meal for poultry, with the expectation of enhancing the nutritional value of the meals by benefiting from the function of the phytochemicals in the peel.

The NEPP/fiber complex exerts beneficial effects in the digestive tract, such as antioxidant activity, bile acid-binding activity, and stimulation of beneficial bacterial growth as well as beneficial systemic effects due to longer circulation of functional metabolites. Some plant parts, such as fruit peels, are a good source of NEPP/fiber complexes. Although many extraction methods for NEPPs have been developed, utilizing them together with all other solids (including fibers) may be recommended for achieving optimal benefits. Even though agro-industrial byproducts and horticultural wastes may be valuable materials, they are not always regarded or treated as food materials. Therefore, to maintain safety and security of food standards in such byproducts and wastes, a strict control of pesticide residues is also needed. Most extraction methods used to prepare the extracts introduced in this chapter are not extremely complicated or sophisticated, but are rather simple and cottage industrial methods. When thinking about using immature fruits, fruit peel, matured leaves of vegetables, or any other byproduct or horticultural waste, environmentally friendly agriculture practices, such as organic horticulture or natural farming, seem to be effective not only for safety and security, but also for enhancing the secondary metabolite content of plant materials. This is because secondary metabolites of fruits and vegetables, such as polyphenols, are stimulated and enhanced under appropriate levels of biotic and/or abiotic stresses.[109]

KEYWORDS

- **biological activities**
- **fruits byproducts**
- **horticultural wastes**
- **phytochemicals**
- **proanthocyanidins**

REFERENCES

1. Cui, J.; Lian, Y.; Zhao, C.; Du, H.; Han, Y.; Gao, W.; Xiao, H.; Zheng, J. Dietary Fibers from Fruits and Vegetables and Their Health Benefits Via Modulation of Gut Microbiota. *Compr. Rev. Food Sci. Food Saf.* **2019,** *18* (5), 1514–1532.
2. Liu, R. H. Health-Promoting Components of Fruits and Vegetables in the Diet. *Adv. Nutr.* **2013,** *4* (3), 384S–392S. https://doi.org/10.3945/an.112.003517
3. Fierascu, R. C.; Sieniawska, E.; Ortan, A.; Fierascu, I.; Xiao, J. Fruits byproducts - A Source of Valuable Active Principles. A Short Review. *Front. Bioeng. Biotechnol.* **2020,** *8*, Article 319. https://doi.org/10.3389/fbioe.2020.00319
4. Coman, V.; Teleky, B. E.; Mitrea, L.; Martău, G. A.; Szabo, K.; Călinoiu, L. F.; Vodnar, D. C. Bioactive Potential of Fruit and Vegetable Wastes. *Adv. Food Nutr. Res.* **2020,** *91*, 157–225.
5. Ben-Othman, S.; Jõudu, I.; Bhat, R. Bioactives from Agri-Food Wastes: Present Insights and Future Challenges. *Molecules* **2020,** *25* (3), Article 510. https://doi.org/10.3390/molecules25030510
6. Campos, D. A.; Gómez-García, R.; Vilas-Boas, A. A.; Madureira, A. R.; Pintado, M. M. Management of Fruit Industrial byproducts–A Case Study on Circular Economy Approach. *Molecules* **2020,** *25* (2), Article 320. https://doi.org/10.3390/molecules25020320
7. Sagar, N. A.; Pareek, S.; Sharma, S.; Yahia, E. M.; Lobo, M. G. Fruit and Vegetable Waste: Bioactive Compounds, Their Extraction, and Possible Utilization. *Compr. Rev. Food Sci. Food Saf.* **2018,** *17*, 512–531.
8. Kabir, F.; Tow, W. W.; Hamauzu, Y.; Katayama, S.; Tanaka, S.; Nakamura, S. Antioxidant and Cytoprotective Activities of Extracts Prepared from Fruit and Vegetable Wastes and byproducts. *Food Chem.* **2015,** *167*, 358–362.
9. Hamauzu, Y. Anti-Ulcerative Potential of Some Fruits and the Extracts. In *Peptic Ulcer Disease*; Chai, J., Ed.; Intech Open: Croatia, 2011; pp 401–418.
10. Widsten, P.; Cruz, C. D.; Fletcher, G. C.; Pajak, M. A.; McGhie, T. K. Tannins and Extracts of Fruit byproducts: Antibacterial Activity Against Foodborne Bacteria and Antioxidant Capacity. *J. Agric. Food Chem.* **2014,** *62* (46), 11146–11156.
11. FAOSTAT. Food and Agriculture Organization of the United Nations (FAO). FAOSTAT Database [Online] 2022. https://www.fao.org/faostat/en/#data/QCL.

12. Renard, C. M.; Dupont, N.; Guillermin, P. Concentrations and Characteristics of Procyanidins and Other Phenolics in Apples During Fruit Growth. *Phytochemistry* **2007,** *68* (8), 1128–1138.
13. Hamauzu, Y.; Iijima, E.; Banno, K. Changes in Catechin and Procyanidin Contents During Fruit Development of Two Apple Cultivars (in Japanese, with an English Summary). *J. Jpn. Soc. Hortic. Sci.* **1999,** *68*, 1184–1193.
14. Duda-Chodak, A.; Tarko, T.; Tuszyński, T. Antioxidant Activity of Apples--An Impact of Maturity Stage and Fruit Part. *Acta Sci. Pol. Technol. Aliment.* **2011,** *10* (4), 443–454.
15. Patocka, J.; Bhardwaj, K.; Klimova, B.; Nepovimova, E.; Wu, Q.; Landi, M.; Kuca, K.; Valis, M.; Wu, W. *Malus domestica*: A review on Nutritional Features, Chemical Composition, Traditional and Medicinal Value. *Plants* **2020,** *9* (11), Article 1408. https://doi.org/10.3390/plants9111408
16. Wojdyło, A.; Oszmiański, J. Antioxidant Activity Modulated by Polyphenol Contents in Apple and Leaves During Fruit Development and Ripening. *Antioxidants* **2020,** *9* (7), 567. https://doi.org/10.3390/antiox9070567
17. Zheng, H. Z.; Hwang, I. W.; Chung, S. K. Enhancing Polyphenol Extraction from Unripe Apples by Carbohydrate-Hydrolyzing Enzymes. *J. Zhejiang Univ. Sci. B* **2009,** *10* (12), 912–919.
18. Kanda, T.; Akiyama, H.; Yanagida, A.; Tanabe, M.; Goda, Y.; Toyoda, M.; Teshima, R.; Saito, Y. Inhibitory Effects of Apple Polyphenol on Induced Histamine Release from RBL-2H3 Cells and Rat Mast Cells. *Biosci. Biotechnol. Biochem.* **1998,** *62* (7), 1284–1289.
19. Najafian, M.; Jahromi, M. Z.; Nowroznejhad, M. J.; Khajeaian, P.; Kargar, M. M.; Sadeghi, M.; Arasteh, A. Phloridzin Reduces Blood Glucose Levels and Improves Lipids Metabolism in Streptozotocin-Induced Diabetic Rats. *Mol. Biol. Rep.* **2012,** *39* (5), 5299–5306.
20. Maragò, E.; Iacopini, P.; Camangi, F.; Scattino, C.; Ranieri, A.; Stefani, A.; Sebastiani, L. Phenolic Profile and Antioxidant Activity in Apple Juice and Pomace : Effects of Different Storage Conditions. *Fruits* **2015,** *70*, 213–223.
21. Rydzak, L.; Kobus, Z.; Nadulski, R.; Wilczyński, K.; Pecyna, A.; Santoro, F.; Sagan, A.; Starek-Wójcicka, A.; Krzywicka, M. Analysis of Selected Physicochemical Properties of Commercial Apple Juices. *Processes* **2020,** *8*, 1457. https://doi.org/10.3390/pr8111457
22. Persic, M.; Mikulic-Petkovsek, M.; Slatnar, A.; Veberic, R. Chemical Composition of Apple Fruit, Juice and Pomace and the Correlation Between Phenolic Content, Enzymatic Activity and Browning. *LWT* **2017,** *82*, 23–31.
23. Renard, C. M. G. C.; Watrelot, A. A.; Le Bourvellec, C. Interactions Between Polyphenols and Polysaccharides: Mechanisms and Consequences in Food Processing and Digestion. *Trends Food Sci. Technol.* **2017,** *60*, 43–51.
24. Antonic, B.; Jancikova, S.; Dordevic, D.; Tremlova, B. Apple Pomace as Food Fortification Ingredient: A Systematic Review and Meta-Analysis. *J. Food Sci.* **2020,** *85* (10), 2977–2985.
25. Rana, S.; Kumar, S.; Rana, A.; Padwad, Y.; Bhushan, S. Biological Activity of Phenolics Enriched Extracts from Industrial Apple Pomace. *Ind. Crops Prod.* **2021,** *160*, 113158. 10.1016/j.indcrop.2020.113158
26. Aprikian, O.; Duclos, V.; Guyot, S.; Besson, C.; Manach, C.; Bernalier, A.; Morand, C.; Rémésy, C.; Demigné, C. Apple Pectin and a Polyphenol-Rich Apple Concentrate are More Effective Together than Separately on Cecal Fermentations and Plasma Lipids in Rats. *J. Nutr.* **2003,** *133* (6), 1860–1865.

27. Ravn-Haren, G.; Dragsted, L. O.; Buch-Andersen, T.; Jensen, E. N.; Jensen, R. I.; Németh-Balogh, M.; Paulovicsová, B.; Bergström, A.; Wilcks, A.; Licht, T. R.; Markowski, J.; Bügel, S. Intake of Whole Apples or Clear Apple Juice has Contrasting Effects on Plasma Lipids in Healthy Volunteers. *Eur. J. Nutr.* **2013,** *52* (8), 1875–1889.
28. Tahraoui, A.; El-Hilaly, J.; Israili, Z. H.; Lyoussi, B. Ethnopharmacological Survey of Plants Used in the Traditional Treatment of Hypertension and Diabetes in South-Eastern Morocco (Errachidia province). *J. Ethnopharmacol.* **2007,** *110* (1), 105–117.
29. Kültür, S. Medicinal Plants Used in Kirklareli Province (Turkey). *J. Ethnopharmacol.* **2007,** *111* (2), 341–364.
30. Hilgert, N. I. Plants used in Home Medicine in the Zenta River Basin, Northwest Argentina. *J. Ethnopharmacol.* **2001,** *76* (1), 11–34.
31. Rahimi, R.; Shams-Ardekani, M. R.; Abdollahi, M. A Review of the Efficacy of Traditional Iranian Medicine for Inflammatory Bowel Disease. *World J. Gastroenterol.* **2010,** *16* (36), 4504–4514.
32. Sezik, E.; Yeşilada, E.; Honda, G.; Takaishi, Y.; Takeda, Y.; Tanaka, T. Traditional Medicine in Turkey X. Folk Medicine in Central Anatolia. *J. Ethnopharmacol.* **2001,** *75* (2–3), 95–115.
33. Agelet, A.; Vallès, J. Studies on Pharmaceutical Ethnobotany in the Region of Pallars (Pyrenees, Catalonia, Iberian Peninsula). Part II. New or Very Rare Uses of Previously Known Medicinal Plants. *J. Ethnopharmacol.* **2003,** *84* (2–3), 211–227.
34. Pieroni, A.; Quave, C. L.; Villanelli, M. L.; Mangino, P.; Sabbatini, G.; Santini, L.; Boccetti, T.; Profili, M.; Ciccioli, T.; Rampa, L. G.; Antonini, G.; Girolamini, C.; Cecchi, M.; Tomasi, M. Ethnopharmacognostic Survey on the Natural Ingredients used in Folk Cosmetics, Cosmeceuticals and Remedies for Healing Skin Diseases in the Inland Marches, Central-Eastern Italy. *J. Ethnopharmacol.* **2004,** *91* (2–3), 331–344.
35. Marwat, S. K.; Khan, M. A.; Khan, M. A.; Ahmad, M.; Zafar, M.; Sultana, S. Fruit Plant Species Mentioned in the Holy Quran and Ahadith and Their Ethnomedicinal Importance. *Am. Eurasian J. Agric. Environ. Sci.* **2009,** *5*, 284–295.
36. Khoubnasabjafari, M.; Jouyban, A. A Review of Phytochemistry and Bioactivity of Quince (*Cydonia oblonga* Mill.). *J. Med. Plant Res.* **2011,** *5* (16), 3577–3594.
37. Sajid, S.; Zubair, M.; Waqas, M.; Nawaz, M.; Ahmad, Z.; Kashmir, A. A Review on Quince (*Cydonia oblonga*): A Useful Medicinal Plant. *Glob. Vet.* **2015,** *14* (4), 517–524.
38. Siddiqui, T. A.; Zafar, S.; Iqbal, N. Comparative Double-Blind Randomized Placebo-Controlled Clinical Trial of a Herbal Eye Drop Formulation (Qatoor Ramad) of Unani Medicine in Conjunctivitis. *J. Ethnopharmacol.* **2002,** *83* (1–2), 13–17.
39. Ashraf, M. U.; Muhammad, G.; Hussain, M. A.; Bukhari, S. N. Cydonia oblonga M., A Medicinal Plant Rich in Phytonutrients for Pharmaceuticals. *Front. Pharmacol.* **2016,** *7*, Article 163. https://doi.org/10.3389/fphar.2016.00163
40. Ferreres, F.; Silva, B. M.; Andrade, P. B.; Seabra, R. M.; Ferreira, M. A. Approach to the Study of C-glycosyl Flavones by Ion Trap HPLC-PAD-ESI/MS/MS: Application to Seeds of Quince (*Cydonia oblonga*). *Phytochem. Anal.* **2003,** *14* (6), 352–359.
41. Carvalho, M.; Silva, B. M.; Silva, R.; Valentão, P.; Andrade, P. B.; Bastos, M. L. First Report on *Cydonia oblonga* Miller Anti-Cancer Potential: Differential Antiproliferative Effect Against Human Kidney and Colon Cancer Cells. *J. Agric. Food Chem.* **2010,** *58* (6), 3366–3370.
42. Hamauzu, Y.; Yasui, H.; Inno, T.; Kume, C.; Omanyuda, M. Phenolic Profile, Antioxidant Property, and Anti-Influenza Viral Activity of Chinese Quince (*Pseudocydonia sinensis*

Schneid.), Quince (*Cydonia oblonga* Mill.), and Apple (*Malus domestica* Mill.) Fruits. *J. Agric. Food Chem.* **2005,** *53* (4), 928–934.
43. Mirmohammadlu, M.; Hosseini, S. H.; Kamalinejad, M.; Esmaeili Gavgani, M.; Noubarani, M.; Eskandari, M. R. Hypolipidemic, Hepatoprotective and Renoprotective Effects of *Cydonia Oblonga* Mill. Fruit in Streptozotocin-Induced Diabetic Rats. *Iran. J. Pharm. Res.* **2015,** *14* (4), 1207–1214.
44. Shinomiya, F.; Hamauzu, Y.; Kawahara, T. Anti-Allergic Effect of a Hot-Water Extract of Quince (Cydonia oblonga). *Biosci. Biotechnol. Biochem.* **2009,** *73* (8), 1773–1778.
45. Hamauzu, Y.; Nakamura, K.; Ooji, Y. Long-Time Heating Improves the Functionality of Chinese Quince and Quince Fruit Products. *Acta Hortic.* **2012,** *939*, 345–351.
46. Tamri, P.; Hemmati, A.; Boroujerdnia, M. G. Wound Healing Properties of Quince Seed Mucilage: In Vivo Evaluation in Rabbit Full-Thickness Wound Model. *Int. J. Surg.* **2014,** *12* (8), 843–847.
47. Hemmati, A. A.; Mohammadian, F. An Investigation into the Effects of Mucilage of Quince Seeds on Wound Healing in Rabbit. *J. Herbs Spices Med. Plants* **2000,** *7* (4), 41–46.
48. Ashraf, M. U.; Hussain, M. A.; Muhammad, G.; Haseeb, M. T.; Bashir, S.; Hussain, S. Z.; Hussain, I. A Superporous and Superabsorbent Glucuronoxylan Hydrogel from Quince (*Cydonia oblanga*): Stimuli Responsive Swelling, on-off Switching and Drug Release. *Int. J. Biol. Macromol.* **2017,** *95*, 138–144.
49. Alizadeh, H.; Rahnema, M.; Semnanai, S. N.; Hajizadeh, N. Detection of Compounds and Antibacterial Effect of Quince (*Cydonia oblonga* Miller) Extracts *in Vitro* and *in Vivo*. *JBAPN* **2013,** *3*, 303–309.
50. Kawahara, T.; Tsutsui, K.; Nakanishi, E.; Inoue, T.; Hamauzu, Y. Effect of the Topical Application of an Ethanol Extract of Quince Seeds on the Development of Atopic Dermatitis-Like Symptoms in NC/Nga Mice. *BMC Complement. Altern. Med.* **2017,** *17* (1), 80. https://doi.org/10.1186/s12906-017-1606-6
51. Hamauzu, Y.; Kume, C.; Yasui, H.; Fujita, T. Reddish Coloration of Chinese Quince (*Pseudocydonia sinensis*) Procyanidins During Heat Treatment and Effect on Antioxidant and Antiinfluenza Viral Activities. *J. Agric. Food Chem.* **2007,** *55*, 1221–1226.
52. Hamauzu, Y.; Takedachi, N.; Miyasaka, R.; Makabe, H. Heat Treatment of Chinese Quince Polyphenols Increases Rat Plasma Levels of Protocatechuic and Vanillic Acids. *Food Chem.* **2010,** *118*, 757–763.
53. Hamauzu, Y.; Tsujitani, T. Changes in Pectic Polysaccharides of Cell-Wall Preparation from Chinese Quince Fruit During Heat Treatment. *J. Food Biochem.* **2013,** *37*, 262–269.
54. Angraeni, L.; Sari, L. K.; Hamauzu, Y.; Tongdeesoontorn, W.; Naradisorn, M. Identification of Phenolic Profile of Chinese Quince Seed (*Pseudocydonia sinensis* Schneid.). *Agricultural Sci. J.* **2016,** *47* (2)(Suppl.), 61–64.
55. Angraeni, L.; Sari, L. K.; Hamauzu, Y.; Tongdeesoontorn, W.; Naradisorn, M. Control of Mango Anthracnose by Using Chinese Quince (*Pseudocydonia sinensis*) Seed Extract. *FABJ* **2019,** *7* (3), 72–89.
56. Inthamat, P.; Hamauzu, Y.; Tongdeesoontorn, W. Effect of Lemon Basil (*Ocimum basilicum*) Seed Mucilage and Chinese Quince (*Pseudocydonia sinensis*) Seed Mucilage Coating to Cherry Tomato (*Solanum lycopersicum* 'Sida') Fruits. *Acta Hortic.* **2020,** *1278*, 7–12.
57. OIV - FAO. *Table and Dried Grapes. FAO-OIV Focus 2016. Non-Alcoholic Products of the Vitivini-Cultural Sector Intended for Human Consumption.* Published by the Food

and Agriculture Organization of the United Nations: Rome, [Online] 2016. http://www.fao.org/3/i7042e/i7042e.pdf
58. OIV. *2017 World Vitiviniculture Situation. OIV Statistical Report on World Vitiviniculture*, International Organization of Vine and Wine (OIV): Paris, 2017. http://www.oiv.int/public/medias/5479/oiv-en-bilan-2017.pdf (Retrieved March 28, 2021)
59. Ferreyra, S.; Torres-Palazzolo, C.; Bottini, R.; Camargo, A.; Fontana, A. Assessment of in-Vitro Bioaccessibility and Antioxidant Capacity of Phenolic Compounds Extracts Recovered from Grapevine Bunch Stem and Cane byproducts. *Food Chem.* **2021,** *348*, Article 129063. https://doi.org/10.1016/j.foodchem.2021.129063
60. Ferreyra, S.; Bottini, R.; Fontana, A. Assessment of Grapevine Stems as Source of Phenolics with Antioxidant Properties. *Rev. Fac. Cienc. Agrar. Univ. Nac. Cuyo* **2019,** *51* (2), 417–426.
61. Barros, A.; Gironés-Vilaplana, A.; Texeira, A.; Baenas, N.; Domínguez-Perles, R. Grape Stems as a Source of Bioactive Compounds: Application Towards Added-Value Commodities and Significance for Human Health. *Phytochem. Rev.* **2015,** *14*, 921–931.
62. Nieto, J. A.; Santoyo, S.; Prodanov, M.; Reglero, G.; Jaime, L. Valorisation of Grape Stems as a Source of Phenolic Antioxidants by Using a Sustainable Extraction Methodology. *Foods* **2020,** *9* (5), 604. https://doi.org/10.3390/foods9050604
63. Hamauzu, Y.; Kawahara, T. Anti-Degranulation and Bile Acid-Binding Activity of Extracts from Fruits and Agro-Industrial byproducts. *Food Biosci.* **2021,** *43*, 101244.
64. Kawahara, S.; Toda, K.; Matsumoto, K.; Ishihara, C.; Nonobe, S.; Matsunaga, C.; Gomi, Y. K.; Senga, S.; Kawaguchi, K.; Yamamoto, A.; Suwannachot, J.; Noda, R.; Kobayashi, S.; Hamauzu, Y.; Makabe, H.; Fujii, H. Isolation and Characterization of a Novel Oligomeric Proanthocyanidin with Significant Anti-Cancer Activities from Grape Stems (*Vitis vinifera*). *Sci. Rep.* **2019,** *9* (1), 12046. https://doi.org/10.1038/s41598-019-48603-5
65. Kashif, M.; Akhtar, N.; Mustafa, R. An Overview of Dermatological and Cosmeceutical Benefits of *Diospyros kaki* and its Phytoconstituents. *Rev. Bras. Farmacogn.* **2017,** *27*, 650–662.
66. Fukai, S.; Tanimoto, S.; Maeda, A.; Fukuda, H.; Okada, Y.; Nomura, M. Pharmacological Activity of Compounds Extracted from Persimmon Peel (*Diospyros kaki* THUNB.). *J. Oleo Sci.* **2009,** *58* (4), 213–219.
67. Gorinstein, S.; Zachwieja, Z.; Folta, M.; Barton, H.; Piotrowicz, J.; Zemser, M.; Weisz, M.; Trakhtenberg, S.; Màrtín-Belloso, O. Comparative Contents of Dietary Fiber, Total Phenolics, and Minerals in Persimmons and Apples. *J. Agric. Food Chem.* **2001,** *49* (2), 952–957.
68. Takii, Y.; Yano, M.; Tsujimoto, S.; Sakaide, Y.; Nagaike, A.; Nishimura, S. Differences in Chitinase Activity in the Anatomies of Plant Species. *Jpn J. Food Chem.* **2010,** *17* (3), 227–230.
69. Jeong, D. W.; Cho, C. H.; Lee, J. S.; Lee, S. H.; Kim, T.; Kim, D. O. Deastringent Peel Extracts of Persimmon (*Diospyros kaki* Thunb. cv. Cheongdo-Bansi) Protect Neuronal PC-12 and SH-SY5Y Cells Against Oxidative Stress. *J. Microbiol. Biotechnol.* **2018,** *28* (7), 1094–1104.
70. Yaqub, S.; Farooq, U.; Shafi, A.; Akram, K.; Murtaza, M. A.; Kausar, T.; Siddique, F. Chemistry and Functionality of Bioactive Compounds Present in Persimmon. *J. Chem.* **2016,** *2016*, Article 3424025. http://dx.doi.org/10.1155/2016/3424025
71. Quan, N.; Kobayashi, K.; Matsunami, Y.; Ide, M.; Makarova, M.; Shen, L.; Ohno, S.; Zheng, Y.; Kobayashi, H.; Lopez, L. R.; Matsuura, E. Persimmon (*Diospyros kaki*

Thunb 'Saijo') Peel Improved Dyslipidemia and its Related Production of Atherogenic Autoantigen Complexes in Low-Density Lipoprotein Receptor-Deficient Mice. *Open Nutr. J.* **2012,** *6* (1), 12–20.
72. Sangkaew, M.; Koh, K. Improvement in the *in Vitro* Digestibility of Shrimp Meal by the Addition of Persimmon Peel. *J. Poult. Sci.* **2021,** *58* (1), 51–57.
73. Zhang, J.; Kopparapu, N. K.; Yan, Q.; Yang, S.; Jiang, Z. Purification and Zcharacterization of a Novel Chitinase from Persimmon (*Diospyros kaki*) with Antifungal Activity. *Food Chem.* **2013,** *138* (2–3), 1225–1232.
74. Pérez-Jiménez, J.; Díaz-Rubio, M. E.; Saura-Calixto, F. Non-Extractable Polyphenols, A Major Dietary Antioxidant: Occurrence, Metabolic Fate and Health Effects. *Nutr. Res. Rev.* **2013,** *26* (2), 118–129.
75. Saura-Calixto, F. The Story of the Introduction of Non-Extractable Polyphenols into Polyphenol Research: Origin, Development and Perspectives. In *Non-Extractable Polyphenols and Carotenoids: Importance in Human Nutrition and Health*; Saura-Calixto, F.; Pérez-Jiménez, J., Eds.; The Royal Society of Chemistry: Cambridge, UK, 2018; pp 1–16.
76. Hellström, J. K.; Törrönen, A. R.; Mattila, P. H. Proanthocyanidins in Common Food Products of Plant Origin. *J. Agric. Food Chem.* **2009,** *57* (17), 7899–7906.
77. Hamauzu, Y.; Mizuno, Y. Non-Extractable Procyanidins and Lignin are Important Factors in the Bile Acid Binding and Radical Scavenging Properties of Cell Wall Material in Some Fruits. *Plant Foods Hum. Nutr.* **2011,** *66* (1), 70–77.
78. Pérez-Jiménez, J.; Saura-Calixto, F. Fruit Peels as Sources of Non-Extractable Polyphenols or Macromolecular Antioxidants: Analysis and Nutritional Implications. *Food Res. Int.* **2018,** *111*, 148–152.
79. Hamauzu, Y.; Suwannachot, J. Non-Extractable Polyphenols and in Vitro Bile Acid-Binding Capacity of Dried Persimmon (*Diospyros kaki*) Fruit. *Food Chem.* **2019,** *293*, 127–133.
80. Ferreira, D.; Guyot, S.; Marnet, N.; Delgadillo, I.; Renard, C. M.; Coimbra, M. A. Composition of Phenolic Compounds in a Portuguese Pear (*Pyrus communis* L. var. S. Bartolomeu) and Changes After Sun-Drying. *J. Agric. Food Chem.* **2002,** *50* (16), 4537–4544.
81. Esparza-Martínez, F. J.; Miranda-López, R.; Mata-Sánchez, S. M.; Guzmán-Maldonado, S. H. Extractable and Non-Extractable Phenolics and Antioxidant Capacity of Mandarin Waste Dried at Different Temperatures. *Plant Foods Hum. Nutr.* **2016,** *71* (3), 294–300.
82. Gulsunoglu, Z.; Karbancioglu-Guler, F.; Raes, K.; Kilic-Akyilmaz, M. Soluble and Insoluble-Bound Phenolics and Antioxidant Activity of Various Industrial Plant Wastes. *Int. J. Food Prop.* **2019,** *22* (1), 1501–1510.
83. Luo, J.; Zhang, P.; Li, S.; Shah, N. P. Antioxidant, Antibacterial, and Antiproliferative Activities of Free and Bound Phenolics from Peel and Flesh of Fuji Apple. *J. Food Sci.* **2016,** *81* (7), M1735–M1742.
84. Wang, Z.; Li, S.; Ge, S.; Lin, S. Review of Distribution, Extraction Methods, and Health Benefits of Bound Phenolics in Food Plants. *J. Agric. Food Chem.* **2020,** *68* (11), 3330–3343.
85. Ding, Y.; Morozova, K.; Scampicchio, M.; Ferrentino, G. Non-Extractable Polyphenols from Food byproducts: Current Knowledge on Recovery, Characterization, and Potential Applications. *Processes* **2020,** *8* (8), Article 925. https://doi.org/10.3390/pr8080925

86. Li, W.; Yang, R.; Ying, D.; Yu, J.; Sanguansri, L.; Augustin, M. A. Analysis of Polyphenols in Apple Pomace: A Comparative Study of Different Extraction and Hydrolysis Procedures. *Ind. Crops Prod.* **2020,** *147*, Article 112250. https://doi.org/10.1016/j.indcrop.2020.112250
87. Das, A. K.; Nanda, P. K.; Madane, P.; Biswas, S.; Das, A.; Zhang, W.; Lorenzo, J. M. A Comprehensive Review on Antioxidant Dietary Fibre Enriched Meat-Based Functional Foods. *Trends Food Sci. Technol.* **2020,** *99*, 323–336.
88. Rohn, S.; Rawel, H. M.; Kroll, J. Antioxidant Activity of Protein-Bound Quercetin. *J. Agric. Food Chem.* **2004,** *52* (15), 4725–4729.
89. Wu, Z.; Ming, J.; Gao, R.; Wang, Y.; Liang, Q.; Yu, H.; Zhao, G. Characterization and Antioxidant Activity of the Complex of Tea Polyphenols and Oat β-glucan. *J. Agric. Food Chem.* **2011,** *59*, 10737–10746.
90. Pérez-Jiménez, J. Antioxidant Capacity of Non-Extractable Polyphenols as a Key Dietary Parameter. In *Non-Extractable Polyphenols and Carotenoids: Importance in Human Nutrition and Health*; Saura-Calixto, F.; Pérez-Jiménez, J., Eds.; The Royal Society of Chemistry: Cambridge, UK, 2018; pp 263–283.
91. Matsumura, Y.; Ito, T.; Yano, H.; Kita, E.; Mikasa, K.; Okada, M.; Furutani, A.; Murono, Y.; Shibata, M.; Nishii, Y.; Kayano, S. Antioxidant Potential in Non-Extractable Fractions of Dried Persimmon (*Diospyros kaki* Thunb.). *Food Chem.* **2016,** *202*, 99–103.
92. Mateos-Martín, M. L.; Pérez-Jiménez, J.; Fuguet, E.; Torres, J. L. Non-Extractable Proanthocyanidins from Grapes are a Source of Bioavailable (epi)catechin and Derived Metabolites in Rats. *Br. J. Nutr.* **2012,** *108* (2), 290–297.
93. Rondini, L.; Peyrat-Maillard, M. N.; Marsset-Baglieri, A.; Fromentin, G.; Durand, P.; Tomé, D.; Prost, M.; Berset, C. Bound Ferulic Acid from Bran is more Bioavailable than the Free Compound in Rat. *J. Agric. Food Chem.* **2004,** *52* (13), 4338–4343.
94. González-Sarrías, A.; Espín, J. C.; Tomás-Barberán, F. A. Non-Extractable Polyphenols Produce Gut Microbiota Metabolites that Persist in Circulation and Show Anti-Inflammatory and Free Radical-Scavenging Effects. *Trends Food Sci. Technol.* **2017,** *69*, 281–288.
95. Saura-Calixto, F.; Pérez-Jiménez, J.; Touriño, S.; Serrano, J.; Fuguet, E.; Torres, J. L.; Goñi, I. Proanthocyanidin Metabolites Associated with Dietary Fibre from in Vitro Colonic Fermentation and Proanthocyanidin Metabolites in Human Plasma. *Mol. Nutr. Food Res.* **2010,** *54* (7), 939–946.
96. Brouns, F.; Theuwissen, E.; Adam, A.; Bell, M.; Berger, A.; Mensink, R. P. Cholesterol-Lowering Properties of Different Pectin Types in Mildly Hyper-Cholesterolemic Men and Women. *Eur. J. Clin. Nutr.* **2012,** *66* (5), 591–599.
97. Naqash, F.; Masoodi, F. A.; Rather, S. A.; Wani, S. M.; Gani, A. Emerging Concepts in the Nutraceutical and Functional Properties of Pectin-A Review. *Carbohydr. Polym.* **2017,** *168*, 227–239.
98. Nishida, S.; Horinouchi, A.; Higashimura, Y.; Akahori, R.; Matsumoto, K. Cholestyramine, A Bile Acid Sequestrant, Increases Cecal Short Chain Fatty Acids and Intestinal Immunoglobulin A in Mice. *Biol. Pharm. Bull.* **2020,** *43* (3), 565–568.
99. Garcia-Diez, F.; Garcia-Mediavilla, V.; Bayon, J. E.; Gonzalez-Gallego, J. Pectin Feeding Influences Fecal Bile Acid Excretion, Hepatic Bile Acid and Cholesterol Synthesis and Serum Cholesterol in Rats. *J. Nutr.* **1996,** *126* (7), 1766–1771.
100. Liu, C.; Lin, X. L.; Wan, Z.; Zou, Y.; Cheng, F. F.; Yang, X. Q. The Physicochemical Properties, *in Vitro* Binding Capacities and *in Vivo* Hypocholesterolemic Activity of Soluble Dietary Fiber Extracted from Soy Hulls. *Food Funct.* **2016,** *7* (12), 4830–4840.

101. Sayar, S.; Jannink, J. L.; White, P. J. *In Vitro* Bile Acid Binding Activity Within Flour Fractions from Oat Lines with Typical and High Beta-Glucan Amounts. *J. Agric. Food Chem.* **2006,** *54* (14), 5142–5148.
102. Hamauzu, Y. Non-Extractable Polyphenols in Fruit: Distribution, Changes, and Potential Health Effects. In *Non-Extractable Polyphenols and Carotenoids: Importance in Human Nutrition and Health*; Saura-Calixto, F., Pérez-Jiménez, J., Eds.; The Royal Society of Chemistry: Cambridge, UK, 2018; pp 284–306.
103. Larsen, L. R.; Buerschaper, J.; Schieber, A.; Weber, F. Interactions of Anthocyanins with Pectin and Pectin Fragments in Model Solutions. *J. Agric. Food Chem.* **2019,** *67* (33), 9344–9353.
104. Iwata, E.; Hotta, H.; Goto, M. The Screening Method of a Bifidogenic Dietary Fiber Extracted from Inedible Parts of Vegetables. *J. Nutr. Sci. Vitaminol.* **2009,** *55* (4), 385–388.
105. Collins, B.; Hoffman, J.; Martinez, K.; Grace, M.; Lila, M. A.; Cockrell, C.; Nadimpalli, A.; Chang, E.; Chuang, C. C.; Zhong, W.; Mackert, J.; Shen, W.; Cooney, P.; Hopkins, R.; McIntosh, M. A Polyphenol-Rich Fraction Obtained from Table Grapes Decreases Adiposity, Insulin Resistance and Markers of Inflammation and Impacts Gut Microbiota in High-Fat-Fed Mice. *J. Nutr. Biochem.* **2016,** *31*, 150–165.
106. Gwiazdowska, D.; Juś, K.; Jasnowska-Małecka, J.; Kluczyńska, K. The Impact of Polyphenols on Bifidobacterium Growth. *Acta Biochim. Pol.* **2015,** *62* (4), 895–901.
107. Tzounis, X.; Vulevic, J.; Kuhnle, G. G.; George, T.; Leonczak, J.; Gibson, G. R.; Kwik-Uribe, C.; Spencer, J. P. Flavanol Monomer-Induced Changes to the Human Faecal Microflora. *Br. J. Nutr.* **2008,** *99* (4), 782–792.
108. Saura-Calixto, F.; Serrano, J.; Gonñi, I. Intake and Bioaccessibility of Total Polyphenols in a Whole Diet. *Food Chem.* **2007,** *101*, 492–501.
109. Barański, M.; Srednicka-Tober, D.; Volakakis, N.; Seal, C.; Sanderson, R.; Stewart, G. B.; Benbrook, C.; Biavati, B.; Markellou, E.; Giotis, C.; Gromadzka-Ostrowska, J.; Rembiałkowska, E.; Skwarło-Sońta, K.; Tahvonen, R.; Janovská, D.; Niggli, U.; Nicot, P.; Leifert, C. Higher Antioxidant and Lower Cadmium Concentrations and Lower Incidence of Pesticide Residues in Organically Grown Crops: A Systematic Literature Review and Meta-Analyses. *Br. J. Nutr.* **2014,** *112* (5), 794–811.
110. Garcia-Amezquita, L. E.; Tejada-Ortigoza, V.; Serna-Saldivar, S. O.; Welti-Chanes, J. Dietary Fiber Concentrates from Fruit and Vegetable Byproducts: Processing, Modification, and Application as Functional Ingredients. *Food Bioproc. Tech.* **2018,** *11*, 1439–1463.
111. Bozan, B.; Tosun, G.; Özcan, D. Study of Polyphenol Content in the Seeds of Red Grape (*Vitis vinifera* L.) Varieties Cultivated in Turkey and Their Antiradical Activity. *Food Chem.* **2008,** *109* (2), 426–430.
112. Cádiz-Gurrea, M. L.; Borrás-Linares, I.; Lozano-Sánchez, J.; Joven, J.; Fernández-Arroyo, S.; Segura-Carretero, A. Cocoa and Grape Seed Byproducts as a Source of Antioxidant and Anti-Inflammatory Proanthocyanidins. *Int. J. Mol. Sci.* **2017,** *18* (2), Article 376. https://doi.org/10.3390/ijms18020376.
113. Aslam, M.; Sial, A. A. Effect of Hydroalcoholic Extract of *Cydonia oblonga* Miller (quince) on Sexual Behaviour of Wistar Rats. *Adv. Pharmacol. Sci.* **2014,** Article 282698. https://doi.org/10.1155/2014/282698
114. Pozuelo, M. J.; Agis-Torres, A.; Hervert-Hernández, D.; Elvira López-Oliva, M.; Muñoz-Martínez, E.; Rotger, R.; Goñi, I. Grape Antioxidant Dietary Fiber Stimulates Lactobacillus Growth in Rat Cecum. *J. Food Sci.* **2012,** *77* (2), H59–H62.

Index

W

Y